Advancing Variable Star Astronomy

Founded in 1911, the AAVSO boasts over 1200 members and observers, and is the world's largest nonprofit organization dedicated to variable star observation. This timely book marks the AAVSO's centennial year, presenting an authoritative and accurate history of this important association. Writing in an engaging and accessible style, the authors move chronologically through five eras of the AAVSO, discussing the evolution of its structure and purpose. Throughout the text, the main focus is on the thousands of individuals whose contributions have made the AAVSO's progress possible. Describing a century of interaction between amateur and professional astronomers, the authors celebrate the collaborative relationships that have existed over the years. As the definitive history of the first 100 years of the AAVSO, this text has broad appeal and will be of interest to amateur and professional astronomers, as well as to historians and sociologists of science in general.

Tom Williams is the AAVSO Historian, and has served in AAVSO leadership positions for over 20 years, including twice as President. He earned a Ph.D. in History from Rice University in 2000, after spending over 30 years as an industrial chemist and manufacturing manager. Since then, Dr. Williams has served as a senior editor for the *Biographical Encyclopedia of Astronomers* (Springer, 2007) and wrote 43 entries for that work. Over the past 25 years, he has published book chapters, encyclopedia entries, and more than 30 articles and book reviews.

Michael Saladyga is Technical Assistant and Archivist at the AAVSO, where he has written articles on variable star astronomy history for *JAAVSO*, and contributed to a number of AAVSO technical monographs and articles on variable stars. He earned a Ph.D. in English and American Literature from Brandeis University in 1995. As AAVSO Archivist, he has arranged and cataloged the AAVSO's historical papers – the foundation for much of this book.

Annual Meeting of the AAVSO, November 10, 1917. Left to right: Mary H. Vann, Forrest H. Spinney, William J. Delmhorst, Dorothy W. Block, Mrs. E. T. Brewster, Leon Campbell, Anne S. Young, Susan Raymond, Leah Allen, John J. Crane, Dorothy W. Reed, Annie J. Cannon, Edward C. Pickering, E. S. McColl, Henrietta S. Leavitt, Edwin T. Brewster, Solon I. Bailey, William T. Olcott, Ida E. Woods, Rev. T. C. H. Bouton, David H. Wilson, Henry. R. Schulmaier, Michael J. Jordan, Alan B. Burbeck, I. F. Conant, Francis L. Ducharme, W. H. Reardon.

Advancing Variable Star Astronomy

The Centennial History of the American Association of Variable Star Observers

Thomas R. Williams
American Association of Variable Star Observers

Michael Saladyga
American Association of Variable Star Observers

Shaftesbury Road, Cambridge CB2 8EA, United Kingdom

One Liberty Plaza, 20th Floor, New York, NY 10006, USA

477 Williamstown Road, Port Melbourne, VIC 3207, Australia

314–321, 3rd Floor, Plot 3, Splendor Forum, Jasola District Centre, New Delhi – 110025, India

103 Penang Road, #05–06/07, Visioncrest Commercial, Singapore 238467

Cambridge University Press is part of Cambridge University Press & Assessment, a department of the University of Cambridge.

We share the University's mission to contribute to society through the pursuit of education, learning and research at the highest international levels of excellence.

www.cambridge.org
Information on this title: www.cambridge.org/9780521519120

First published 2011

A catalogue record for this publication is available from the British Library

Library of Congress Cataloging-in-Publication data
Williams, Thomas R. (Thomas Roger)
Advancing Variable Star Astronomy : The Centennial History of the American Association of Variable Star Observers / Thomas R. Williams, Michael Saladyga.
p. cm
Includes bibliographical references and index.
ISBN 978-0-521-51912-0
1. American Association of Variable Star Observers. 2. Astronomical observations – United States – History 3. Variable stars – Observations – History. I. Saladyga, Michael, 1950– II. Title.
QB4.9.U6W55 2011
523.8′44 – dc22 2010046599

ISBN 978-0-521-51912-0 Hardback

The whole field of variable star astronomy could not exist without the dedicated effort of volunteer observers who contribute their time, talent, and energy to creating and maintaining the database, without which the field would be paralyzed for lack of data. The effort required to create these observations is the solitary work of individuals in backyard observatories or with portable instruments who have labored with remarkably little recognition over many years. We hope this book has solidified an appreciation especially of the value of their efforts.

We hope, too, that anyone who reads this book also gains an appreciation of the important, truly outsized role that women have played in variable star astronomy. More than any other sub-discipline within astronomy, variable star astronomy has benefited from the special insights and, especially, the dedication of women to this field. Women's characteristic patience and attention to detail have amplified the value of their labors in the tabulating, plotting, and analyzing of data gathered from individual observers.

Four women, in particular, deserve special recognition in the context of the history of the AAVSO: Margaret Walton Mayall and Janet Akyüz Mattei, who played especially important roles in saving the association from oblivion in the middle of its first century; and Ellen Dorrit Hoffleit and Martha Locke Hazen, whose efforts in counseling especially Janet Mattei were of incalculable benefit at critical times to the organization and its leadership.

To these, we must add our wives, Anna Fay Williams and Ann Maureen Saladyga, whose patience and forbearance have made our labors more tolerable.

To all of them, the variable star observers, the women who have labored over their data, and our wives who facilitated our efforts, we dedicate this book.

Contents

Foreword

A century ago, in the spring of 1911, the amateur astronomer and popular science writer William Tyler Olcott called on his fellow enthusiasts to begin to monitor the magnitudes of about a hundred stars whose light varied more or less unpredictably. Such observations of variable stars would aid professional astronomers, and particularly the Harvard College Observatory. By that fall, Olcott reported enlisting 13 members (including three women) to an organization he had formed and named The American Association of Variable Star Observers. The opening third of this volume presents a revealing glimpse into the heroic early days of the AAVSO. Today the association has become the preeminent amateur astronomy society, with more than 1200 members around the world and an archive of more than 19 million individual variable star observations.

My own membership spans more than half of the life of this venerable organization. I joined up in 1947 as a teenager on the Kansas prairies, determined to become one of the core group of observers who racked up more than a thousand observations per year. Alas! I never achieved that goal, but nevertheless the association and I have been mutually helpful ever since. Indeed, the AAVSO has played a very significant role in my career trajectory, in a way I could not have imagined when I signed up as a life member.

In 1948 I hitchhiked to Milwaukee for the annual meeting of the Astronomical League, the newly founded umbrella organization of amateur astronomers. We took a field trip to Yerkes Observatory to see the 40-inch telescope, the world's largest refractor, and there an amateur astronomer, hired as a summer employee, demonstrated the awesome instrument to us. The idea that an amateur could actually get a summer job at an observatory opened a whole new vista for me. How exciting, I thought, if I could get a summer job at Harvard College Observatory, at that time the home of the AAVSO! Emboldened, I sent an inquiry to Leon Campbell, then the Recorder and, essentially, the CEO of the organization. Amazingly, I got an invitation for the following summer from Harlow Shapley, the director of the observatory, godfather of the AAVSO, and the most famous astronomer in America.

In Cambridge I promptly visited the headquarters of the AAVSO. To me, as a teenager, the diminutive Leon Campbell seemed as old and welcoming as Santa Claus. He was ensconced in a crowded nineteenth-century drawing room in the far southern corner of the old frame building that constituted the west wing of the original observatory. He shared the space with his secretary, Helen Stephansky, and his scientific assistant, Margaret Mayall, a kind, motherly sort who would eventually become his successor. Only later did I realize that she could be intensely determined and stubborn, precisely the qualities needed to hold the organization together in perilous times.

The most memorable character I met that summer was Harlow Shapley himself. He was a ball of energy, a gifted conversationalist, and a man of very wide interests, from ants and fungi to baseball and galaxies. When he came to Cambridge in 1921, he was probably the world's leading observer of variable stars. As a graduate student at Princeton, he had made nearly 10 000 observations of eclipsing binary stars, and he eventually derived orbits for 90 binaries, compared with scarcely 10 analyzed previously. Then at Mt. Wilson Observatory, he had used the 60-inch reflector to study variable stars in globular clusters and also Cepheid variables in the Milky Way in order to calibrate Henrietta Leavitt's period-luminosity relation. At Harvard he took an almost proprietary interest in the AAVSO.

My own desk that summer was directly across the hall from Shapley's office. Among many miscellaneous assignments as his summer assistant, I learned how to find and fetch glass plates from the plate stacks, and one project in particular involved locating the photographs of what we called "Luyten's flare star," which is now more formally the variable star UV Ceti. The route from Shapley's office to the AAVSO Headquarters led across a covered wooden passage that bridged the small valley separating the newer fireproof buildings from the original frame structures. It passed through the rotunda below the historic "Great Refractor," for nearly two decades (1847–1866) the largest telescope in America, and circling around the refractor's pier was a memorable exhibit from the AAVSO: the light curve of the dwarf nova SS Cygni, long a favorite of the Association's observers. The light curve recorded all the irregularly spaced outbursts since 1911, though one during the World War II slump in observations was almost missed. According to a caption accompanying the serpentine display, after the war a lone observation arrived, capturing the otherwise undocumented outburst.

The early 1950s brought great changes to Harvard College Observatory and the AAVSO. In 1949, shortly after I had met Leon Campbell, he crossed Harvard's typical retirement age and then, in the spring of 1951, he died, having been succeeded by Margaret Mayall. The following year Harlow Shapley reached 67 and retirement as the Observatory's director. The University administration was not in a hurry to recruit Shapley's replacement, as they wanted to assess the future directions of astronomical education and research, but they made the senior professor, Donald Menzel, acting director. Menzel quickly discovered that the observatory was greatly overextended compared with its financial resources, that many of the staff members were scandalously underpaid, and that the physical plant required substantial upgrading. One of his first proposals was to replace the old frame building that, among other things, housed the AAVSO; moreover, he planned to evict the AAVSO and capture the substantial endowment raised earlier by the Observatory, which had been used to support the AAVSO. These were indeed fraught times for the AAVSO, and the middle third of this volume covers this traumatic event extensively with compelling and detailed archival documentation.

By that time, I had become a graduate student in astronomy. There were about 20 of us, under-informed, but aware that massive changes were going on around us. Bart J. Bok was our favorite professor because he was always congenial and alert to the welfare of the students. Rumor had it that Bok had been offered an important directorship elsewhere, but that Shapley had urged him to stay as heir-apparent at Harvard. Menzel, who tended to be more aloof, had only a few months more seniority than Bok. The overwhelming majority of us felt that both Bok and the AAVSO were getting a bum deal.

The story told in Chapters 10 through 14 in this book presents these happenings from the point of view of the AAVSO. It is, without intended deceit, a highly biased account. Shapley, to whom the AAVSO owed an enormous amount, comes in for a certain amount of disparagement. But Shapley, who had piloted the Observatory through the depths of the Great Depression when fundraising was difficult, nevertheless had established a premier graduate school in astronomy and a highly seminal summer school. He had coped during the war years, and his internationalism had been battered in the McCarthy era. As an ex-director, he made a strong point of keeping out of the way and letting others take command, and thus his defense of the AAVSO was very low key. It is Donald Menzel who emerges as the bogeyman in this retelling. Menzel was the astrophysicist who introduced the new quantum mechanics into astronomy and the coronagraph into America. He had long been interested in popular science writing, and from 1937 had edited *The Telescope* before it merged with *Sky* to form *Sky and Telescope*. He had for many years been an AAVSO member and at the time was even a vice president. Menzel was clearly caught in a financial crunch when he was appointed acting director of Harvard College Observatory, and he struggled to find a way through the tunnel. His options were limited and with unpleasant consequences.

I know both authors of this comprehensive centennial history, and I know that neither harbors ill feelings toward either Shapley or Menzel. They wish to caution readers not to extrapolate the events presented in this book to a more general view of either astronomer in the larger history of astronomy. As Nietzsche famously said, whatever does not kill strengthens, and eventually the AAVSO emerged as a far stronger independent organization.

In the years following the strains of the eviction, I gradually became better acquainted with Menzel. He seemed less and less remote as we found more interests in common, and he was personally very helpful to me. The reconciliation of the AAVSO and the observatory came slowly as others took over and hard feelings ebbed. The AAVSO was welcomed back to Harvard for its annual meetings, and with my role as mid-wife, the Smithsonian Observatory freely offered many thousands of dollars worth of computer and staff time to facilitate the organization's entry into the cyber world.

The final third of this volume describes the challenges and growing pains as the AAVSO expanded its scientific mission to accommodate many more kinds of variable stars besides the long period ones that had been the central mission for many decades. Particularly exciting are such collaborations between amateur and professional astronomers such as monitoring gamma ray bursts, which became expanded into the AAVSO High-Energy Network. The Citizen Sky Internet project has introduced more than 3000 registrants to amateur astronomy. Perhaps no better illustration of the growth of the organization is the fact that thousands of charts can now be downloaded from the AAVSO's Variable Star Plotter. Members can glow with pride over the hard-won success and stability of the organization as it contributes to serious cosmic science.

Advancing Variable Star Astronomy gives a rich, thoroughly documented account of a century of progress, through times of exhilaration and a period of traumatic stress. It is both fascinating and instructive. *Ad astra per aspera!*

Owen Gingerich
Cambridge, Massachusetts
September 2010

Preface

The American Association of Variable Star Observers (AAVSO) completes 100 years of significant contributions to science in 2011. This centennial history makes available for the first time an accurate and thorough institutional history for scholars while celebrating the achievement for AAVSO members.

The AAVSO is a private, nonprofit international scientific research and educational organization of amateur and professional astronomers who are interested in stars that change in brightness – variable stars. Founded in Connecticut in 1911 to coordinate variable star observations made largely by amateur astronomers, by 1919 the AAVSO had become affiliated with Harvard College Observatory in Cambridge, Massachusetts. In 1954, the AAVSO once again became independent of Harvard and opened its headquarters in Cambridge. Today, with more than 1200 members in 56 countries and a database of more than 19 million observations, the AAVSO is the world's largest association of variable star observers.

Over the course of this century, the scientific agenda of the AAVSO changed, slowly at first, adding a few nonvariable star–related scientific observing projects to its basic program of long-period variable star observation. Over time, a few new categories of variable stars were added, gradually broadening the numbers and types of variable stars observed. In the past four decades, the changes have come more rapidly as more sophisticated instruments became available and especially as the needs of the professional astronomical community broadened. Today, the AAVSO Headquarters and its observers play key roles in the real-time coordination of orbiting observatories managed by NASA and other agencies. Most recently, the availability of many long-time series of variable star observations, sometimes with more than a hundred years of high-quality data in a digital format, is proving extraordinarily valuable to theoretical studies of stellar evolution and other processes.

The primary organizing principle for this book is the five directorships in AAVSO history. The first part covers the history of variable star astronomy before the founding of the AAVSO. The second part is devoted to the founding of the organization by William Tyler Olcott, who was essentially its first director. Four subsequent sections are devoted to the directorships of Leon Campbell, Margaret Mayall, Janet Mattei, and Arne Henden. Biographical sketches in each part cover the life of each of these key individuals before their AAVSO involvement and include historical context for the period covered.

Underlying themes woven into the history presented here include the following five ideas: (1) AAVSO activities focus on the observational science of variable star astronomy; (2) the AAVSO is a member-led organization in which there is a strong focus on the people involved, especially the observers; (3) the strong involvement and cooperation of professional astronomers with amateurs in the AAVSO organizational leadership as well as in the science; (4) the AAVSO as an international organization from the beginning because of its focus on a specific scientific topic as well as its proximity to and relationship with a major center of astronomical research in that topic; and (5) how organizational flexibility and evolution over time have served the AAVSO well.

A centennial is a time for celebration, and the AAVSO has much to celebrate – not least its splendid survival and growth after a crisis in which most volunteer organizations would likely have collapsed. Thus, this book is intended, at a minimum, as a celebration of a century of contributions to science by AAVSO members, its volunteer observers, and dedicated staff workers.

We hope this history will become a point of pride as well as an inspiration to AAVSO members and observers. But we need also to point out that previous brief histories of the AAVSO are not comprehensive, not easily accessed, and are in need of updating. That updating is needed not only in a chronological sense, but also for the correction of known fundamental errors in those earlier historical sketches. We uncovered these other problematic points in previous history sketches mainly through our use of the enormous resource now available for the first time in the AAVSO archives.

Our view as authors of what we could achieve in this book has changed since we first proposed an outline to Cambridge University Press. Our first intent was to cover the history of the American Association of Variable Star Observers at a fairly uniform level of detail throughout its first nine decades and thus to remain true to its intended purpose as a history of the first century of the AAVSO's existence. Two problems have since interjected themselves in that project.

First, we became increasingly aware, as more archival resources became available, that even the understanding we had of that history from previous writing was more than a little at odds with the facts that came to light from various archives and how those facts meshed with what we already had at hand for resources. A concerted effort has been made in the past decade to organize, as a formal archive, both all the available records of the association since its beginning and the archival resources that have, over time, been donated to the association by various key figures from our history. Thus, the history presented here is even more at odds with the extant literature than we originally anticipated. We present, as an appendix, a discussion of the historiography of the AAVSO and what we believe to be the motivations that led each of the previous writers to present that history as they saw it. We hope that by doing so, we make clear in the process how this history differs from all previous attempts, and why.

Second, as each of us thought more about writing the same level of detail into the period of the last 25 years that we each agreed was essential for historical understanding of the first 75 years, we both developed reservations about doing so. Since we both have been intimately involved with the AAVSO for more than 25 years, we came to realize that we really could not present a completely unbiased view of the period from an historical perspective. Thus readers who see a measurable difference in detail presented in the last 25 years of this history will be making a valid observation. Without apology, we present an understanding of the events of the period, without the same level of detail and historical documentation that is included for the first 75 years, and for the most part eliminating our personal involvement in that history to the extent possible. We make no claim with respect to any absence of bias, but have attempted to the best of our ability to avoid stressing personal involvement and resulting personal biases. Errata to the book will be made available at www.cambridge.org/9780521519120.

Cambridge, Massachusetts
September 2010

Acknowledgments

The authors shared equally in this book's planning, research, and writing but were helped greatly nearly every step of the way by others. We were especially fortunate in that those who were willing to help shared in some measure our intense interest in the history of the AAVSO and the history of variable star astronomy.

AAVSO Past President David B. Williams deserves a special note of thanks. We not only needed his astute editing skills when we least expected it, but we also benefited from his insightful comments and questions on every chapter – thanks to his years of experience as a skilled observer and his years as an AAVSO Council member.

Anna Faye Williams helped substantially during the final editing stages. Her painstaking work helped improve the readability and conciseness of this book when it was needed most.

Elizabeth O. Waagen, AAVSO Senior Technical Assistant, was the one person we sought out for help whenever the archives were mute. With more than 30 years at AAVSO Headquarters, she has seen it all, heard it all, and done it all. Elizabeth is an incredible font of knowledge about all things AAVSO.

We salute the professional archivists and librarians who have helped us not only with documents and books, but also through their kindness and patience as well: the Archives Staff, Nathan Pusey Library, Harvard University, Cambridge, Massachusetts; Peter J. Knapp, Special Collections Librarian and College Archivist, Watkinson Library, Trinity College, Hartford, Connecticut; Michael McCarthy, Reference Librarian, New York Law School; Mount Holyoke College Archives and Special Collections; Diane Norman, Genealogy/Local History Specialist, Otis Library, Norwich, Connecticut; Dale Plummer, Historian, Norwich, Connecticut; University of California Santa Cruz, Mary Lea Shane Archives of the Lick Observatory; Cheryl Turkington, Archivist, Morristown Library, New Jersey; and Greg Shelton, Assistant Librarian and Archivist, USNO Library, Washington, DC.

Except where noted, the photographs reproduced in this book were drawn from collections in the AAVSO Archives. We thank the AAVSO for allowing us to use whatever we needed from its archives.

We are also grateful to these historians and researchers: Jeanne Larzalaere Bloom, Freelance Researcher and Genealogist, Chicago, Illinois; Charles Sullivan, Director, and his staff, Cambridge Historical Commission, Cambridge, Massachusetts; Peter Broughton, RASC Historian, Toronto, Canada; John Toone, BAA-VSS historian; and Michael Dixon, Historical Society, Cecil County, Maryland.

We thank warmly the following individuals who were willing to share their memories, thoughts, and insights: Yussef Akyüz, Sudbury, Massachusetts; Michael Mattei, Littleton, Massachusetts; Barbara Howell, Cambridge, Maryland; Charles Whitney; Charles Scovil; John Bortle; Robert Evans and his colleague, Thomas Richards, RASNZ; and Richard Berry.

Our thanks to those who responded to our call for their comments on particular chapters: Diane Norman; Dale Plummer; John Toone; John Isles; David B. Williams; John Percy; Elizabeth O. Waagen; Charles Whitney; John Bortle; Charles Scovil; Arne Henden; Aaron Price; Rebecca Turner; Matthew Templeton; and Mike Simonsen. We also thank the production staff of Cambridge University Press.

Part I

PIONEERS IN VARIABLE STAR ASTRONOMY PRIOR TO 1909

1 • The emergence of variable star astronomy – a need for observations

We can not expect to accomplish such hopes in a moment, for we have just begun our investigation.
– Friedrich W. A. Argelander, 1844[1]

With Tycho Brahe's discovery of a "new star" in 1572, a powerful realization occurred. No longer could one assume the unchanging character of the skies, the belief since the time of Aristotle and reinforced by Christian dogma. As historian Helen Lewis Thomas remarked, "for the first time it was realized that a continual watch must be kept on the heavens and that observation must be continuous and systematic."[2]

This idea is the essence of variable star astronomy; it is important not only to observe what is new or changing, but also to keep a careful record of those events. The early history of variable star astronomy is a story of catalogs and charts. The publication and availability of these tools brought attention to the phenomenon of stellar variability, or changes in brightness, and led to the discovery of many more variables.

The awareness of a star's variability came with the discovery of a previously uncharted third-magnitude star in the constellation Cetus on August 13, 1596, by David Fabricius (1564–1617), a Protestant minister in Osteel, Ostfriesland. It became fainter over time and then disappeared after October. Later, the star was named o Ceti in 1603 by Johann Bayer as he prepared his celestial atlas *Uranometria*. The star had also been called Mira, a name later applied to the class of long-period variable stars called Miras.

Early discoveries of variable stars were accidental, not the result of any deliberate program but a recognition that stars differed in brightness from previous observations. Variability was suspected for several stars: β Persei in 1669 by Geminiano Montanari (1633–1687) and Giacomo Filippo Maraldi (1665–1729); R Hydrae in 1670 by Montanari; and χ Cygni in 1686 by Gottfried Kirche (1639–1710), who determined its precise period.[3]

Evidence of a star's periodic variability was not recognized until 1639, when differences in o Ceti's brightness were observed in December 1638 and later in November 1639. Johannes Phocylides Holwarda, also cited as Jan Fokkens (1618–1651), estimated its period as about 11 months. In 1667, Ismaël Boulliau (1605–1694), a Catholic priest of Loudun, France, more accurately calculated its period as 333 days.[4] Others – Boulliau, and later, Jacques Cassini (1677–1756) – also noted o Ceti's variations in amplitude.[5]

Variable star astronomy improved with the systematic work of Edward Pigott (1753–1825) and John Goodricke (1764–1786). Between 1782 and 1784, they confirmed the variability of β Persei, determined its period, and deduced that it was likely an eclipsing system in which a planet occults the star.[6] In 1784, Pigott discovered the variability of η Aquilae, and Goodricke identified δ Cephei as a variable star. They also determined more precise periods for the other known variables.[7] Pigott published what might be called the first catalog of variable stars in 1786, confirming 12 variables and 39 suspected variables.[8]

A contemporary of Pigott and Goodricke, William Herschel (1738–1822), observed differences in the brightness of the star 44 ι Bootis in 1781 and 1787, and α Herculis in 1795.[9] In 1796, he published an article titled "On the method of observing the changes that happen to the fixed stars."[10] This work includes his first "Catalogue of the comparative brightness, for ascertaining the permanency of the lustre of stars." Herschel describes a method for noting the comparative brightness of stars that became the basis for Friedrich W. A. Argelander's step method of estimating brightness in 1844. Herschel recognized the need for determining the precise magnitudes that previous atlases (Bayer, 1603; Flamsteed, about 1689) failed to provide, although they were significant for the accuracies of their stellar coordinates. Herschel's method differed from these earlier compilations by describing a star's brightness relative to that of other stars, rather than assigning magnitudes to stars on a fixed scale. Herschel prepared four catalogs through 1799, all published in the *Philosophical Transactions of the Royal Society, London*.

The interest in publishing stellar position catalogs led to the accidental discovery of the first asteroids and prompted a more intensive search for additional asteroids. In turn, the asteroid searches produced the next cycle of variable star discoveries. Carl Ludwig Harding (1765–1834) discovered five variable stars as he cataloged star positions along the ecliptic from 1809 to 1831. By 1831, his discoveries increased the total number of known variable stars by more than 25 percent to 24 stars.[11]

William Herschel's son, John Frederick William Herschel (1792–1871), recognized the value in having amateur astronomers study variable stars. He made this appeal in his *A Treatise on Astronomy* of 1833:

> This is a branch of practical astronomy which has been too little followed up, and it is precisely that in which amateurs of science, provided only with good eyes, or moderate instruments, might employ their time to excellent advantage. It holds out a sure promise of rich discovery, and is one in which astronomers in established observatories are almost of necessity precluded from taking a part by the nature of the observations required.[12]

Figure 1.1. Friedrich W. A. Argelander (courtesy, Special Collections, University Library, University of California Santa Cruz, Mary Lea Shane Archives of the Lick Observatory).

ARGELANDER'S APPEALS AND CRITERIA

A more systematic approach to variable star astronomy began in 1843, when Friedrich Wilhelm August Argelander (1799–1875) published his *Uranometria Nova* atlas and catalog, followed in 1844 by his discussion of methods and theory. As director of Bonn Observatory, he began observing variable stars in 1838. His observations from 1838 to 1867 were published in *Astronomische Nachrichten* (*AN*), Bonn observatory publications, and elsewhere.[13]

Uranometria Nova was important because it filled a need for practical charts with reliable magnitudes for studying the variability of suspected stars.[14] The charts were also convenient for viewing stars visible to the unaided eye. *Uranometria Nova* shows 3256 stars visible from central Europe to declination –32° in 19 degrees of brightness down to sixth magnitude. There are 18 variable stars in his compilation, which included a description of the charts and a catalog of the stars.[15]

Argelander followed his atlas with an "Appeal to the friends of astronomy" in *Schumacher's Jahrbuch* of 1844, devoting his last chapter to "The variable stars."[16] He summarized the brief history of discoveries and variable star work, listing his 18 known variables along with position, period, and magnitude range; possible causes of stellar variability, in general; and the peculiarities of specific variable stars.

In the same article, Argelander stressed the importance of studying eight relationships to understand why stars vary in brightness and what laws govern "these apparitions." These include (1) a precise period; (2) variations between times of maxima and minima; (3) the degree of irregularity of period; (4) variations of amplitude; (5) relationships between variations and regularities; (6) a light curve from the observations; (7) regularities and deviations in the light curve with respect to what is known about the star's period and amplitude; and (8) the relationship of the star's change in magnitude to any change in its color.

Finally, Argelander concluded his article with clear instructions on how to observe variable stars. He emphasized the importance of using a suitable range of comparison stars and how to make magnitude estimates using the step method and stressed the value of publishing observations. These criteria remain primary considerations in the field of variable star astronomy. With publication of his *Uranometria Nova* and his "Appeal to the friends of astronomy," Argelander intended to provide astronomers with the tools and knowledge to make accurate and consistent variable star observations that would be useful in scientific analysis.

Argelander's most important work was his compilation of the *Bonner Durchmusterung* (*BD*), a catalog of 324 189 stars located between declination –2° and +90° and as faint as ninth magnitude. The first sections appeared in 1859, and the last were published in 1862.[17] Based on the catalog, 185 known variables were marked on charts published in 1863. The 1859 *BD* was the start of an all-sky survey that would not be completed until 1914.

The *BD* was not widely distributed; its size, cost, and limited availability put it beyond the reach of most independent astronomers and many observatories in the early twentieth century. Yet the *BD* was a very important part of the development and expansion of variable star observing with its unprecedented positional accuracies and magnitudes. It provided a standardized reference to fields of comparison stars for variable star observing.[18]

ARGELANDER STUDENTS AND THE FOUNDATION

Three Argelander students established the foundation for the study of variable stars for observers and astronomers in the early twentieth century. They were among the first astronomers who made long, continuous series of observations of variable stars and published their observations in a collected form.

Eduard Heis (1806–1877), a high-school mathematics teacher, became director of the Münster Observatory in 1852. His observations, begun in 1840 and continued over 27 years, were the basis of his *Atlas Coelestis Novus*, an atlas and catalog published in 1872 that was intended as the successor to Argelander's *Uranometria Nova*. The *Atlas* presents 5421 stars visible to the naked eye from central Europe. Due to his excellent eyesight, he accumulated 2153 more stars than in the *Uranometria*, including 14 variable stars.[19] The *Atlas*, like the *Uranometria*, gave right ascension in degrees, not hours, and the star positions for the 1855 epoch. The large chart format and its expense discouraged its use by observers. Nevertheless, like Argelander's work, the accuracy of positions and magnitudes in the *Atlas* made it a standard reference for anyone doing variable star work in later years.[20] J. G. Hagen of the Georgetown College Observatory compiled and published Heis' observations of variable stars in 1903. This collection comprised nearly 7000 observations of some 30 stars made from 1840 to 1877.[21]

Johann Friedrich Julius Schmidt (1825–1884) became Argelander's assistant at Bonn in 1846 and, later, director of the private Olmütz observatory in 1852 and the Athens Observatory in 1858. Although primarily known as a selenographer, his variable star observations (1845–1879) were sent to Potsdam Observatory after his death.[22]

Eduard Schönfeld (1828–1891) published several catalogs that became guides for early variable star observers and, later, variable star compilations. Schönfeld, who studied under Argelander at Bonn Observatory in 1851, became his assistant in 1853, helping with the *Bonner Durchmusterung* compilation. In 1859, Schönfeld became director of the Elector Palatine's observatory at Mannheim, where he observed variables and comparison stars over the next 10 years. Schönfeld published his own observations of variable stars made from 1853 to 1859.[23] He also compiled them in three catalogs: 113 variable stars in 1865; 119 variables in 1866, and 143 variable stars in 1875.

Schönfeld succeeded Argelander as director of the Bonn Observatory in 1875 and continued the *BD*, publishing the southern catalog in 1886 and the charts in 1887. He edited the *Astronomische Nachrichten (AN)* and prepared an annual catalog and ephemeris of variable stars printed in the *Vierteljahrsschrift der Astronomischen Gesellschaft*. Ernst Hartwig of Bamberg continued the annual publication of this variable star catalog and ephemeris after Schönfeld's death in 1891. W. Valentiner, Hartwig's successor at the Mannheim observatory, collected and prepared Schönfeld's variable star observations and published them in 1900. According to Caroline E. Furness, Schönfeld compared

117 variables with 1100 other stars and made 35 963 complete observations and at least 5000 observations with comparison stars.[24]

OUTSTANDING BRITISH OBSERVERS

In spite of John Herschel's 1833 admonition on the importance of variable stars, a decade passed before British astronomers became involved. Though there may have been some observing and discussions during the interim, it may have taken that long to realize that the techniques were well within their equipment and skills.

The first British astronomer to actively pursue variable star astronomy was John Russell Hind (1823–1895). He established himself as a serious observer by discovering two asteroids in 1847 while working at George Bishop's private South Villa Observatory in Regent's Park. Hind's initial variable star discoveries in 1848 were likely the consequence of his ongoing asteroid program, as was Carl Ludwig Harding's experience. Chandler credits Hind with discovering 9 variable stars in that first year and a total of 22 variable star discoveries by the time of his last discovery, 40 years later.[25]

Hind's early success, along with the Herschel and Argelander appeals, may have inspired three other British astronomers: Norman Robert Pogson (1829–1891), Joseph Baxendell (1815–1887), and George Knott (1835–1894). Their coordinated viewing programs represented some of the first such efforts and provided the impetus for the establishment of several organizations.

For example, an extended program of observations of U Geminorum involved all four astronomers after Hind's discovery of the star in 1855. Pogson first reported its flickering at minimum in 1856, and Baxendell reported similarly in 1858. Baxendell and Knott remained with the project for a number of years, comparing notes and exchanging telegraphic notices of U Gem outbursts. Knott's extremely valuable series of observations extended for 33 years.[26] Baxendell and Knott thought through the whole observing process in detail, publishing a manual, *Method of Observing Variable Stars*, printed by and distributed with Sanford Gorton's *Astronomical Register*.[27]

Pogson, who had an impressive lifetime total of 13 variable star discoveries, first began observing at the Radcliffe Observatory near Oxford University. He developed a catalog of 53 variable stars by 1854, including his 6 discoveries and comments on other suspected variables. Moving to the Hartwell House observatory, he continued observing variable stars, adding five more discoveries. During his tenure at Hartwell House (1856), Pogson proposed the adoption of a light ratio of 2.512 for two stars that differ in brightness by one magnitude. The Pogson ratio was later adopted by Edward C. Pickering for the Harvard Photometry and by Karl H. G. Müller for the Potsdam Photometry and eventually became an international standard.[28]

Pogson was regrettably isolated from his contemporaries when he became the Government astronomer in 1860 in Madras, India. He continued his work on variable stars, discovering four new variables and struggling to complete a massive but unfunded variable star atlas and catalog. Two confirmed variable stars were discovered by Pogson from Madras, bringing his lifetime total to 13 discoveries.[29] His catalog, near completion at the time of his death, included 134 stars, 106 confirmed variable stars, 21 suspected variable stars, and 7 temporary stars (likely novae or supernovae).[30]

Professional astronomer Herbert Hall Turner (1861–1930) edited and published the variable star observations of both Baxendell and Knott.[31] Turner's two publications constituted an appropriate cap on a very productive period in British variable star astronomy: Hind, Pogson, Baxendell, and Knott discovered nearly a third of all the variable stars known in 1871 and followed their discoveries with extended series of observations over time.[32] In addition to Turner's publication of the observations of Pogson, Knott, and Baxendell already mentioned, he also collected and published the extensive variable star observations of Charles Grover (1842–1921) at the Rousdon Observatory maintained by Sir Cuthbert Peek.

Turner stood out in the beginning of the twentieth century in his concern for the preservation and publication of collections of valuable variable star observations, but also for promoting the analysis of variable star observations to understand the patterns of variation and their significance. Turner was also responsible for turning Mary Adela Blagg (1858–1944) from her extensive cataloging of lunar craters and other features to the harmonic analysis of variable star light curve data.[33]

BRITISH ORGANIZATIONAL EFFORTS

Knott and Baxendell cooperated in other ways. For example, they attempted to form the first association of variable star observers. Their effort, described in detail by John Toone, was announced in 1863 in the *Monthly Notices of the Royal Astronomical Society* (*MNRAS*). Their intended Association for the Systematic Observation of Variable Stars (ASOVS) unfortunately failed to gain support from amateur astronomers. However, their efforts reflected a more general discontent among British amateur astronomers during this era. As British professional astronomers appropriated the Royal Astronomical Society as a purely scientific endeavor, amateurs looked for other venues to pursue their astronomical interests. Richard Baum has documented the Observing Astronomical Society (OAS) founded in 1869. Although the OAS counted in its membership a veritable *Who's Who* of British amateur astronomers of the period, it had apparently ceased to function as an organization by 1873.[34]

The Liverpool Astronomical Society (LAS) became the next association in 1881 with a Variable Star Section. The eventual demise of the LAS led to the formation of the British Astronomical Association (BAA) and its Variable Star Section (VSS). Initial members who had been active in the LAS included John Ellard Gore, BAA-VSS section leader; Ernst Elliott Markwick, BAA-VSS, second section leader; and John Birmingham, Thomas William Backhouse, and Thomas Henry Espinall Espin.[35] Between them, they added 38 variable stars to Chandler's *Third Catalogue of Variable Stars* and created several catalogs of red or unusual stars.

Variable star observing by the BAA-VSS observers continued sporadically, but, as Toone has shown, Gore's leadership was uninspired. It was not until Gore was replaced by Ernst Elliott Markwick that the VSS activities were systematized with data analysis and the development of observing programs. Following Turner's example and benefiting from Markwick's leadership, the rejuvenated section became a well-respected contributor to variable star astronomy.[36]

ASTRONOMERS IN THE SOUTHERN HEMISPHERE

Early British settlers in the southern hemisphere not only sought longitude and latitude determinations, but also wanted to explore and understand the rich astronomical resources in their night skies. In his history of southern hemisphere astronomy, David Evans has explored the work of both amateur and professional astronomers.[37]

In Australia, the earliest notable variable star observer was the farmer astronomer John Tebbutt (1834–1916), comet discoverer and a dedicated observer from his Paramatta, New South Wales, observatory. Tebbutt also appealed to his amateur colleagues to observe variable stars, asking, "Can no amateur be found in New South Wales to work in this field which promises to be so fruitful?"[38] Tebbutt's appeal, repeated in the 1880s, was almost completely ignored by active Australian amateurs.[39]

Tebbutt limited himself to a few known variable stars because they were not this comet discoverer's primary interest in astronomy.[40] Between 1880 and 1899, he observed 17 maxima over a total of 22 cycles of the Mira variable star R Carinae and published 30 papers on this star, η Carinae, and a few other southern hemisphere stars. Wayne Orchiston credits Tebbutt with the ex post facto detection of Nova V728 Scorpii in 1862, based on his scrupulously accurate observational records of a comet that year.[41] Tebbutt reconstructed his observation of the nova from those observations in 1877.

In South Africa, amateur astronomer Alexander William Roberts (1857–1938) made substantial contributions to variable star astronomy. Roberts was one of the early members of the BAA VSS, but his achievements in variable star astronomy began much earlier and went well beyond the scope of that section's activities. His observing, analysis, and publication established an ideal to which other amateurs might aspire, then and now.[42]

As a youth in Scotland, Roberts wanted to pursue a career in astronomy, but he later supported his avocation while a teacher and principal of the Lovedale School for Natives. Under the tutelage of David Gill (1843–1914), director of the Cape Observatory, Roberts discovered a number of variable stars and made astonishing, precise observations of eclipsing binary stars. Using his own observations, Roberts deduced accurate periods and determined the orbits for four binaries, showed that the stars were severely distorted oblate spheroids, and determined the approximate masses for each individual member of the four binary pairs. His data provided a clear indication that the four binary pairs were very low-density stars. His results published in the *Astrophysical Journal* were followed by Henry Norris Russell's computations for the combined masses of binary star systems.

ASTRONOMERS IN THE AMERICAS

In the United States, Benjamin Apthorp Gould (1824–1896) – considered to be an important influence on the development of astronomy in mid- to late-nineteenth century America[43] – recognized the valuable contributions of Argelander and other European astronomers. Throughout the 1850s and 1860s, Argelander had published observations and information about variable stars in *Astronomische Nachrichten* (*AN*) in Germany and in *Monthly Notices of the Royal Astronomical Society (MNRAS)* in England. These journals were widely read by professional and amateur astronomers.

Gould traveled to Europe in 1845 and met Argelander, Johann Franz Encke, Carl Friedrich Gauss, and Friedrich Georg Wilhelm Struve, as well as younger astronomers Friedrich Winnecke, Eduard Schönfeld, Arthur Auwers, and others. In 1847, Gauss accepted Gould as a student at the University of Göttingen. Gould became the first American astronomer to study in Europe and earn a Ph.D. there.[44]

While in Europe, Gould was especially impressed with the quality of work being published in Heinrich Christian Schumacher's *Astronomische Nachrichten.* It was regarded as an international astronomical publication rather than a German publication because it was contributed to and widely read all over Europe. Gould spent 4 months with Schumacher in 1848 in Altona on his way home to the United States. It was likely that Gould decided to publish his own journal during that time. When he returned to the United States in 1848, he founded *The Astronomical Journal* (*AJ*), with the first issue appearing in 1849.

Among the ideas that Gould transplanted from Germany to the Americas was Argelander's assertion of the importance of variable stars and refined techniques for observing them. In Gould's *AJ* of 1855 and 1856, Argelander himself published an appeal to American astronomers to observe variable stars, along with his own observations of Algol and T Cancri.[45] Noting Argelander's contributions, Gould stated: "[I]t is to be hoped that some interest may have been awakened for the investigation of the strange and striking phenomena of the variable stars; and that observers may be found wherever there is a love for astronomy, and a disposition to labor for its advancement."[46]

Figure 1.2. Benjamin Apthorp Gould.

When Gould returned to the United States, he was unable to secure work as an astronomer. Gould suspended publication of *AJ* in 1861 – partly because of the start of the Civil War and partly because of a lack of funds. Publication of *AJ* had become both a financial and emotional burden on Gould. He had intended to resume publication in August 1869, but this plan was interrupted when he was appointed Director of the Argentine National Observatory in Córdoba.[47]

On his arrival in 1870 at Córdoba, Gould and his assistants immediately began work on the *Uranometria Argentina* – a catalog of southern hemisphere stars completed in 1874 and published in 1879. Intended as an extension of Argelander's *Uranometria Nova*, Gould's catalog included 722 standard stars and magnitudes of 7730 stars south of declination +10° down to seventh magnitude. In the course of this work, he and his assistants discovered 12 new variables. Furthermore, their observations revealed many stars suspected of variability. Gould estimated that at least half, if not most, of the stars within the magnitude limits of his catalog would show some kind of variability over time and would have to be monitored.[48] This foreshadowed a problem that would weigh on Harvard College Observatory Director Pickering many times in the future.

While at Córdoba, Gould also compiled *The Argentine General Catalogue: Mean Positions of Southern Stars Determined at the National Observatory*. This work listed the locations and proper motions of 32 448 stars.[49] After 15 years at Córdoba, Gould returned to the United States and resumed publication of *AJ* in November 1886, after a 25-year break in its history.[50] From this base, Gould would next encourage a new generation of variable star observers to gather much-needed data.

2 · A need for observers

Those who have not tried it do not realize the growing interest in a systematic research and the satisfaction in feeling that by one's own labors the sum of human knowledge has been increased.

– Edward C. Pickering, 1882[1]

Variable star astronomy developed on the European continent, largely through the efforts of professional astronomers at observatories, while amateurs contributed in Great Britain. In the United States, variable star astronomy emerged with a mix of professionals and amateurs in a more complicated manner.

Initially, amateurs in the United States embraced variable star observing as a specialty and developed skills and programs that were consistent with good astronomical practice at that time. Many individuals were drawn into variable star observing through the appeals of Benjamin A. Gould and Friedrich W. A. Argelander in Gould's *Astronomical Journal* (*AJ*) and the work of Edward C. Pickering at the Harvard College Observatory (HCO).

In the first part of this chapter, we examine the impetus for variable star observing through some of its leading exponents. The second part deals with their concerns for accuracy and verification. The last part deals with the establishment of publications, organizations, and the evolution of standardized observing procedures.

APPEALS FOR OBSERVERS

American astronomer Gould returned from Europe with several astronomical agendas, particularly the promotion of variable star astronomy following the urgings of his German mentor Argelander (see Chapter 1). Gould encouraged his readers to observe variable stars in his *Astronomical Journal* and through personal communications. Among the earliest respondents was Stillman Masterman (1831–1863), who observed without instruments in the Minnesota Territory and Maine. Masterman reported naked-eye variable stars and their maxima and minima with great accuracy, but his untimely death at 32 years deprived astronomy of a skilled observer.[2]

The most prominent amateur astronomer to respond to Gould's appeal, Seth Carlo Chandler, Jr. (1846–1913), provided an important early link between amateur and professional variable star observing in the United States. Chandler was among the first American amateurs to publish his own observations; he was also the first to encourage other amateurs to take up the study of variable stars.[3]

Although his formal education was limited as a Boston English High School graduate, Chandler learned to make accurate longitude determinations from astronomical observations. He worked as Gould's aide with the US Coast Survey from 1863 to 1869, and Gould thought so highly of his capabilities that he invited him to his South America expedition in 1869. Chandler declined and remained in the United States, where he married in 1870, worked as an actuary, and continued observing as an amateur astronomer.[4] In 1874 and 1875, Chandler made observations of stars selected from Argelander's *Uranometria Nova*.[5]

Chandler soon became a recognized authority on variable stars through his publications in the United States and abroad. He lectured on variable star theory and discoveries at the Boston Scientific Society and elsewhere, and his lectures were frequently summarized or reprinted in the Boston newspapers in the late 1800s.[6]

In 1878, Chandler wrote four articles "On the Methods of Observing Variable Stars" in the *Science Observer*, which began as a journal for the Boston Scientific Society in 1877. Chandler's articles were reprinted in an *Astronomical Journal* booklet in 1887 with an introduction by Gould, encouraging individuals with small telescopes to take up observing, adding

Figure 2.1. Seth Carlo Chandler, Jr. (from *Popular Astronomy*, Vol. 22, No. 5, 1914, p. 271).

Figure 2.2. Backyard observatory of Paul S. Yendell, located in Dorchester (Boston), Massachusetts.

"To such, the study of the variable stars is especially to be recommended."

In this booklet, Chandler and Gould encouraged amateur astronomers to make accurate observations of variable stars on a regular basis, using the naked eye, a common opera-glass, spy-glass, or small telescope, "such as is within reach of almost every one." Chandler described Argelander's step method of recording observations. He also showed how to compute the brightness of a variable with a comparison star magnitude. Despite the somewhat mundane and modest character of these instructions, Chandler emphasized the value of such work:

> The importance of investigation relating to the phenomena presented by the variable stars cannot be over-estimated, for it is upon our knowledge of these phenomena that the solution of questions concerning the physical condition of the heavenly bodies will very largely depend.[7]

From 1881 to 1885, Chandler worked as a volunteer HCO observer and research assistant, receiving only a nominal salary.[8] His visual observations were correlated with those for the photometric catalog of HCO Director Edward C. Pickering.[9] Pickering also relied on Chandler's unpublished lists of suspected variables.

After 1885, Chandler continued as an independent amateur astronomer and made his most noteworthy accomplishment, a discovery called the Chandler Wobble. The description of the earth's variation in latitude was published in a series of papers from 1891 to 1894. He received two gold medals for his discovery; one from the US National Academy of Sciences in 1895 and another from the Royal Astronomical Society in 1896 for his discussion of the variation of latitude, his work on variable stars, and other astronomical investigations.[10]

Chandler was among the organizers, along with John Ritchie, Jr., of the Boston Scientific Society, which included amateur and professional scientists from many disciplines who met regularly to discuss the developments in their fields. This society became a nexus for amateur variable star astronomers in the Boston area.

Among those drawn into variable star observing through the encouragement of Chandler and Gould were Paul Sebastian Yendell (1844–1918) and Edwin Forrest Sawyer (1849–1937). Civil War veteran Yendell became a prolific observer using a 4.25-inch refractor from his home in Dorchester, Massachusetts, accumulating more than 30 000 variable star observations. He wrote extensively about variable star astronomy from 1894 until his death in 1918. During the last 3 years of his life, he prepared his observations for publication, but the work was not completed.[11]

Between 1882 and 1890, Sawyer determined the visual magnitudes of 3145 stars from Gould's

Figure 2.3. Edwin F. Sawyer (from the *Boston Sunday Herald*, August 7, 1898, p. 27).

Figure 2.4. Edward C. Pickering, Director, Harvard College Observatory.

Uranometria Argentina. He observed nearly every star three to four times or more for a total 13 654 observations and discovered 8 variable stars and 51 suspected variables. His results were published in 1893 as the "Catalogue of the magnitudes of southern stars."[12] Yendell praised Sawyer's catalog, calling it "one of the most valuable works of its kind in existence."[13]

COOPERATIVE EFFORTS

Edward Charles Pickering (1846–1919), director of the Harvard College Observatory, soon became one of the most important influences in variable star astronomy. According to Helen Lewis Thomas, Pickering "was more responsible than any other individual for shifting the center of variable star study from Europe to the United States."[14]

Pickering established HCO as the center of variable star astronomy. HCO staff and volunteer observers had collected tens of thousands of variable star observations and published them in the *Harvard Annals*. The staff had made thousands of variable star discoveries. HCO implemented and refined the photographic and spectroscopic methods in the discovery and analysis of variable stars and developed astrophysics as it applied to the study of variable stars. When Pickering became HCO Director in 1877, there were 180 known variable stars. By the time he died in 1919, his HCO staff and volunteer observers had discovered 3435 variable stars.[15]

Pickering wanted reliable photometric comparison star magnitudes for variable star observing. His interest in variable stars was a by-product of his attempt to obtain photometric magnitudes of north polar stars and establish standards for photometry. Like Gould in 1874, Pickering recognized the need to monitor stars thought to be variable if his standard star measurements were to be of any use. By 1881, he had developed a system by which variable stars could be classified.[16]

Pickering also recognized the opportunities and invested HCO resources in variable stars, although they were clearly secondary to the Observatory's larger programs. Pickering also encouraged participation by developing a corps of variable star observers, both amateur and professional astronomers, not associated with HCO. Thus, under his stewardship, variable star astronomy emerged as a specialty with a productive mix of amateur and professional engagement.

THE PICKERING PLAN

In 1882, Pickering published a pamphlet, *A Plan for Securing Observations of the Variable Stars*,[17] which appeared 4 years after Chandler's "Methods of observing variable stars" series in *The Science Observer*. This plan was directed toward other scientific institutions like HCO and proposed the enlistment of volunteers to make observations of variable stars. Pickering's plan simply involved soliciting help from amateur astronomers, providing instruction, and collecting their observations. The data collected would be reduced and analyzed by the professional astronomers. Pickering did not propose the establishment of a formal organization of amateur observers but only suggested that amateur observers were a resource that professional astronomers ought to take advantage of in the name of efficiency.

The Pickering Plan was noted by *The Sidereal Messenger* in 1882, the only astronomy publication in the United States at the time. The editor, William Wallace Payne (1837–1928), also the director of Carleton College Observatory in Northfield, Minnesota, published excerpts in 1883 and offered to send the complete pamphlet to anyone interested.[18] Pickering's proposal was also reprinted in *The Observatory* in Great Britain.[19] At the same time, Pickering published a "First circular of instructions for observers of variable stars," which he made available along with his "Plan" to anyone interested.

Pickering made similar appeals for "co-operation" between amateur and professional astronomers in *Popular Astronomy* in 1901, nearly 50 years after Gould's appeal for observers, and in *Harvard College Observatory Circulars* in 1906 and 1911.[20]

EFFICIENCY IN ASTRONOMY

While earlier appeals only called for cooperative efforts, Pickering proposed a more ambitious program in 1887. In his work, "The extension of astronomical research," published in the *Popular Science Monthly,* he proposed the establishment of a central astronomical agency to which research funds would be sent for distribution to observatories and researchers. Like his previous plan for efficiency in collecting variable star observations, this broader proposition sought the best use of observatory time, equipment, personnel, and other resources.[21] This idea of administrative efficiency is a recurring theme throughout Pickering's career and should be considered an important nonscientific motivation for his encouragement of amateur involvement in astronomy.[22]

The response to Pickering's appeals appeared to be negligible. He corresponded with Joseph P. Baxendell in 1892 and with Ernst Elliott Markwick in1895 at the latter's post at Gibraltar. But throughout the 1880s and the 1890s, Pickering was preoccupied with financial problems that plagued the HCO, the establishment of a foreign research station (eventually at Arequipa, Peru) under the terms of the Boyden Fund, and the extension and evolution of stellar spectral classification studies within the terms of the Anna Draper grant that emerged as the monumental *Henry Draper Catalogue* of spectral classifications.[23] Furthermore, Pickering was conducting his own intensive observing program of stellar photometry with the 4-inch meridian photometer.[24]

During this period, Pickering emerged as one of the intellectual leaders in the profession of astronomy as well.[25] Variable star astronomy was clearly just one of many lines of endeavor for the busy HCO director.

POPULAR INTEREST

A hint of popular interest surfaced in an article on observing variable stars in the May 1891 issue of *The Sidereal Messenger*.[26] However, nothing further appeared other than lists of ephemerides and the occasional report of a variable star. Further publications may have been deferred with the shift in the journal's editorial interest. In 1892, the journal changed its name to *Astronomy and Astro-Physics*, co-edited by George Ellery Hale (1868–1938) and Pickering. Technical articles by both Hale and Pickering featured stellar spectroscopy more than general astronomy. Hale wanted to attract the emerging astrophysicists who were unable to publish their work in Gould's *Astronomical Journal.* According to Donald Osterbrock, this new journal led to the creation of *The Astrophysical Journal* and *Popular Astronomy.* At first, Payne was a willing participant, but when the *Astronomy and Astro-physics* circulation fell in comparison with that of his previously well-established *Sidereal Messenger,* he established a new journal for amateur astronomers, *Popular Astronomy,* in 1893.[27] His "untechnical" publication soon attracted amateur astronomers like Paul S. Yendell, who wrote in September 1894 on "Suggestions to observers of variable stars" that the field was one in which an amateur with limited means "may most confidently hope to achieve results of value to science."[28]

Yendell followed with a four-part series, "On the variable stars of short period," beginning in the December 1894 *Popular Astronomy*.[29] Although aimed at an amateur audience, his articles contained considerable detail, such as the positions of 18 short-period variables, elements of variation, a finder chart, a table of comparison stars and light steps, and a table of mean values from his own reduced observations. Yendell also provided a brief observing history for each star from its discovery to its most recent observations. He followed with many *Popular Astronomy* articles on variable star astronomy.

VARIABLE STAR SPECTRA

The Harvard College Observatory continued its interest in variable stars. The spectrum for each star was being gathered in the Draper Spectral Classification program, which involved photographing the entire sky with an objective prism. Pickering assigned the problem of studying and classifying the stars based on these spectra to a trusted woman computer, Williamina Paton Stevens Fleming (1857–1911). She had worked for Pickering in 1881 as a copyist and later as a computer, a title for women performing the computational work in the period.[30] In 1885, Fleming worked with Pickering and developed a new, more detailed scheme for the mass classification of stars than originally devised by Angelo Secchi (1818–1878).[31] Secchi had originally shown that the variable star R Geminorum exhibited bright emission lines when it was near maximum brightness.[32] In her classifications, Fleming discriminated between the special spectral characteristics exhibited by known variable stars. She began finding other stars that exhibited these same characteristics but were not yet identified as variable stars. Suddenly, the discovery of variable stars was transformed from patient telescopic comparisons for a single star's brightness to comparison stars over time to an examination of a single stellar spectrum on a photographic plate.

CHANDLER'S VARIABLE STAR CATALOGS

Chandler and other established visual astronomers expressed their distrust of the growing reliance on the new astrophysical and photographic methods. They were especially skeptical about announcements of variable star discoveries that relied solely on photographic stellar spectra. It is hard to say when and how the seeds of this controversy were planted. It could have begun with Pickering's 1882 Plan for recruiting inexperienced amateurs in variable star work, or his 1886 *Investigation in Stellar Photography*, which included a description of obtaining photographic stellar spectra, or with the initiation of the *Henry Draper Catalogue* in 1890 and 1891. But it is clear that by the mid 1890s, there were two extremes of thought regarding what type of variable star observations really mattered.

From 1888 to 1896, Chandler published three variable star catalogs in *The Astronomical Journal*, which became the platform for his criticism against reliance on photographic spectra. For the most part, like their predecessors, the catalogs provided information about variables confirmed visually, including position, magnitude range and period, year of discovery, discoverer, cross-identifications, and notes, as well as suspected variables.

The first catalog was intended to fill an immediate need for current information about these stars, as 13 years had passed from Schönfeld's second catalog of 143 variable stars in 1875. Chandler's first catalog listed 225 variables; the second catalog, 260, and the third catalog (1896), 393 variables, a 42% increase from the first catalog.

Given the number of variable stars being discovered at HCO and elsewhere, the small percentage of increases in the variables in Chandler's catalogs is surprising. Yet, this appears to have been a point of pride for Chandler, who noted that the second catalog would have been much longer if announcements of "what might be regarded as reputable authority, had been admitted. But it would have ceased thereby to represent any exact knowledge."

In the second catalog, he stated his argument in three sentences: "It is a paramount consideration that our knowledge must be kept clear of confusion, even at the risk of an incomplete statement of it. No star should be inserted, no matter how high the authority on which its variability is declared, without independent verification on undoubted authority and evidence. Otherwise the result will be chaos."[33]

DEFENDING VISUAL OBSERVATIONS

Chandler stated that his purpose was "to bring order out of the chaos, by the careful and continuous observation necessary to discriminate the actual cases of variability from the numerous pseudo-variables with which the periodicals of the day are filled." Taking a critical but indirect stab at the increasing numbers of variables discovered photographically at HCO, he declared: "Considering it extremely desirable that no star should be placed in the list, no matter how high the authority on which its variability is asserted, without independent verification, I have had under observation a large number of stars during the last few years with this especial object in view."[34] Chandler continued this commentary in his second catalog introduction in the 1893 *AJ*. Here, he emphasized his imperative of "conscientious and critical care of construction" even more strongly, stressing "the labor involved in carrying out the design of making the elements of every star definitive, in the sense that every observation available up to date should be included in the calculation."

DEFENSE OF SPECTRAL CLASSIFICATIONS

Fleming responded to Chandler's "Second catalogue of variable stars" statements, stating they should more correctly be called the "Second Catalogue of Variable Stars discovered Visually," since "no weight is given to photographic observations further than is necessary to enable them to swell the list of stars discovered visually. Stars discovered photographically which have been announced as variables, and have been proved beyond doubt to be variables, are here credited as 'suspected.'" Her arguments were based on a paper she presented at the Congress of Astronomy and Astrophysics at the Chicago Columbian Exposition, which also appeared in an article, "A field for woman's work in astronomy," in the October 1893 *Astronomy and Astro-Physics Journal*, 2 months after Chandler's second catalog appeared in *AJ*. She then stated: "You can obtain in one night what would represent years of hard labor in visual observation" and went on to discuss the error in brightness and positions involved with visual observations, whereas the photographic record "can be consulted and compared with others years hence . . . "[35]

In his "Third catalogue of variable stars" of 1896, Chandler made a more pointed criticism of HCO, while defending "the degree of uniformity and completeness" of variable star observations made to date, noting that:

> this harmonious development has been attained without any concerted scheme of "cooperation," but by the free will and independently planned efforts of individual volunteers, each discriminatingly directing his work in accordance with his means and situation . . . [with] the enthusiasm which springs from the individual initiative of the observers themselves.[36]

In effect, Chandler was reflecting an abiding concern among scientists, both amateur and professional, in the late nineteenth century – specifically, the issue of what constitutes authoritative evidence in a scientific setting. Issues of authority in matters scientific tended to resonate with other issues surrounding the emergence of professions and their societies in this period. The tensions between the new professionals and the amateurs who stood to be left out or ignored grew substantially over just such issues.[37]

ARGUMENTS CONTINUE

These arguments reverberated in articles by others through the rest of the 1890s. Yendell, who frequently wrote articles, defended the traditional method, insisting "on the necessity of careful, deliberate, and conscientious work; and that the beginner remember that quality is of infinitely greater importance than quantity; that a single well and thoroughly determined fact is worth more than pages of hurried and half-digested observations."[38] Again in 1896, he questioned the determination of magnitudes by measuring the photographic images of stars in view of "the character of the images, the influence of the colors of the stars on their relative sizes, the many varying conditions of atmospheric disturbance, possible lack of uniformity in films and emulsions, and difference circumstances of development, the more one knows of these measures, the less he feels inclined to trust them."[39]

In the meantime, amateur astronomer (later turned professional) John Adelbert Parkhurst (1861–1925) of Marengo, Illinois, published eight articles in volume one (1893) of *Popular Astronomy*. These were the first in-depth discussions of variable star observing published in a general circulation journal for an audience of amateur astronomers. In the first article, Parkhurst stated, "I doubt whether there is another field in astronomy which offers the amateurs so good a chance to do work which will materially increase the sum of human knowledge." In the next articles, he provided general information, included Pickering's classification of variable star types first published in 1881, and reproduced Pickering's charts of circumpolar long-period variables.[40] Parkhurst also discussed Chandler's catalog and the step-method of estimating magnitudes and explained how to derive a light curve from observations. In his final article, he noted that "the current literature of variable stars indicates that interest in the subject is general and increasing."[41] Over the next 10 years, he published more than 20 articles and numerous notes on variable stars in *Popular Astronomy*, as well as in *AJ* and *AN*.

Parkhurst expressed the most objective view on the visual/photographic controversy. In his March 1894 *Popular Astronomy* article, he acknowledged the recent emergence of photographic observation methods and noted how photography can produce results for the fainter stars, especially when many plates of a region are made over a period of time. He also noted, however, that accurate photographic astronomy required specialized equipment and techniques beyond the reach of most amateurs. "Besides," he added, "the time required to expose and develop a plate and make the necessary measurements would suffice to observe a dozen stars visually."[42] Parkhurst went on to acknowledge objectively that any new method deserved careful scrutiny.[43]

Chandler and Yendell were justifiably proud of the accomplishments and discoveries that they and other skilled and dedicated amateur observers had made in the young history of variable stars. Yet Chandler himself went on to state in the "Third catalogue" that variable star observing "is a field affording ample room for more participants. In the southern hemisphere, particularly, the need for volunteers is pressing."[44] But Chandler undoubtedly was thinking in terms of highly skilled observers and computers like himself and his amateur associates.

Pickering's rationale for organization in variable star observing was also to avoid chaos: he saw that individual, nonstandardized efforts inevitably would result in confusion and errors because most observers had their own sequences of comparison stars, and many times an individual's star identifications would be questionable, rendering their observations useless.[45]

The visual/photographic controversy of the 1890s was merely a symptom of inevitable change in the way variable star observations would have to be conducted in the future. The key to involving many more amateurs in variable star observing – while ensuring the quality and consistency of measurements – would be to provide standard sequences of comparison stars that had assigned magnitudes. For the novice observer, this would make variable star measurement a much simpler activity than having to follow the cumbersome step method, and it would do away with the laborious reductions needed to derive a light curve. It is perhaps enough to say that Chandler and Pickering, and their respective camps, both were aware of the rapidly expanding field of variable star research. They both agreed that more observations of these stars would be vital toward understanding them, and they both agreed that variable star work could be done by amateurs with modest means.

AN INTERIM STANDARD

The *Atlas Stellarum Variabilium* (*ASV*) fulfilled the great need for reliable charts and lists of comparison stars for variable star observing between the years 1899 and 1908, before the standardized construction and distribution of variable star charts became a widespread practice at HCO.

Indeed, the *ASV* was the greatest improvement in charts for observers of variable stars since Friedrich W. A. Argelander's *Uranometria* of 1843 and its expansion by Eduard Heis in 1872. The *ASV* was a series of charts with corresponding catalogs of sequences prepared by the Reverend Johann Georg Hagen, S. J. (1847–1930). Hagen learned astronomy from Argelander and Eduard Schönfeld at Bonn and Heis in Münster.[46] Hagen was a careful observer who grounded his work in a thorough understanding of the history of variable star astronomy. After his ordination in 1880, Hagen was sent from the Jesuit seminary in England to Prairie du Chien, Wisconsin, as a mathematics teacher. There, he established a small observatory and began teaching his students to observe variable stars.[47] In 1888, Hagen became Director of Georgetown College Observatory in Washington, DC, where work on the *ASV* began and several series were published. In 1906, Hagen was appointed Director of the Vatican Observatory, where the work on subsequent series was completed.

In compiling the *ASV*, Hagen's intention was to provide charts and sequences that would be useful

Figure 2.5. Reverend Johann Georg Hagen, S. J., Director of Georgetown College Observatory, Washington, DC, where he began his work on the *Atlas Stellarum Variabilium*. He was appointed Director of the Vatican Observatory in 1906.

mainly to the amateur variable star observer, especially for observing stars having faint minima. The charts were designed so that the variable star and comparison stars could easily be located and the charts easily used at the telescope with chart gridlines invisible when viewed in red light. The need for such an atlas arose mainly because *Bonner Durchmusterung* (*BD*) charts were no longer available, but also because additional charts were needed that went below the limiting magnitude of the *BD* charts.

The Hagen *Atlas* in 1908 comprised six series. Series I through III (1899–1900) have a limiting magnitude of 10 and hence are suitable for larger telescopes. Series I contains 45 variables having minima below the *BD* limit, Series II contains 46 variables within the *BD* limit, and Series III contains 37 naked-eye variables. Series IV (1907) contains 101 variables visible at minimum in moderate-sized telescopes and that are within the *BD* declination and magnitude limits. Series V (1906) shows 48 variables having a minimum brighter than seventh magnitude, suitable for naked-eye or small-instrument observing. Series VI (1908), having 65 variables, is a supplement to Series I, II, and III. Eventually, the *ASV* would grow to nine series and supplements; the last two were prepared by Hagen's successors after his death in 1930.

Each series was published separately so an observer only needed to purchase the set for the range of the observing instrument. The price for each of the earliest series was about \$50, expensive – about \$1000 today – but within reach of many of the serious amateur astronomers of the day. Appended to an 1897 announcement by Hagen in *Popular Astronomy* was an enthusiastic endorsement by amateur variable star astronomer John A. Parkhurst, who later served as a professional at Yerkes Observatory and provided Hagen with many photographic charts of faint variables corresponding to those in the *ASV*. Parkhurst endorsed the project warmly, noting that many "would be glad to take part in this interesting work if only they were provided with means for conveniently identifying the variables."[48]

On each chart, variable star identifications, positions, color, spectral type, and maximum and minimum magnitude ranges were taken from Chandler's "Third catalogue." The charts in the first three series depicted a 1° square region of the sky, with the variable star in the center. An inner field, $30'$ square, shows stars that are as faint as the variable at minimum; the field outside of this inner square shows the same stars that are found in the *Durchmusterung*. The comparison stars listed in the catalogs are listed in grades, or steps, not magnitudes.

When the first three series were published, the only reliable standard magnitudes that could be used were those found in the *BD*. Magnitudes for the grades could be derived by a formula used with reference to the magnitudes of the *BD* stars appearing on the Hagen charts. Thus a uniform scale of magnitudes could be applied to all of the Hagen charts.[49] In 1900, a table for converting the Hagen comparison star grades for the first three series to the Harvard photometric scale was published in the *Harvard Annals*; a supplementary table was added in 1902.[50] When he prepared the fourth and subsequent series, Hagen included a column headed *HP* (Harvard Photometry) in the comparison star tables, which gave the photometric magnitudes as determined at HCO. In the case of the naked-eye stars of Series V, some comparison star magnitudes were taken instead from the *Potsdam Photometric Durchmusterung*, the *Uranometria Argentina*, or from magnitudes determined by A. W. Roberts of South Africa.[51]

FRUITS OF COOPERATION

The cooperation between amateurs and professionals was leading to a new era for variable star observers. This in part was bolstered by the interest in variable star research at Harvard College Observatory and the emergence of a new generation of variable star astronomers.

Between 1889 and 1899, a program of visual observation of 17 circumpolar variable stars was undertaken at HCO under Edward C. Pickering's direction. These observations were made mainly by Oliver C. Wendell, William M. Reed, and Annie J. Cannon of the HCO staff and by amateur astronomer Frank E. Seagrave of Providence, Rhode Island. What distinguished these observations from others made elsewhere, Pickering noted, was that the stars were observed "throughout their variations of light," and that all observations were reduced "to a uniform photometric scale." Also, each of the variables had its own sequence of comparison stars (published in an 1891 pamphlet by Pickering, "Variable stars of long period"), consisting of "stars differing from each other by about a third, or a half of a magnitude, the brightest being brighter than the variable at maximum, and the faintest fainter than the variable at minimum."[52]

The results of this survey, Pickering reported, "justified its extension to other stars." He soon followed it with a program of observations of 58 long-period variables made between 1890 and 1901. Although Pickering continued to expand his program of collection and publication of variable star data, he still had to rely on an informal network of amateur and professional astronomers. The observations again were made mainly by HCO staff Wendell, Reed, and Cannon, and Seagrave. Additional observations were made by Leon Campbell and other staff observers and also the amateurs John H. Eadie, Zaccheus Daniel, and Manoel Soares de Mello.[53]

In 1890, the British Astronomical Association formed a Variable Star Section (BAA-VSS). The BAA-VSS began an observing program of variable stars in 1899 using the scale of magnitudes from HCO's studies of 1889 through 1901.[54] This was the first organized group to use Pickering's standard sequences of comparison stars. This development, together with the publication of J. G. Hagen's first three series of variable star charts with magnitudes easily reducible to the Harvard scale and subsequent *ASV* series that included the HCO magnitudes in its catalogs, meant that Pickering's effort to encourage "co-operation" in the use of standardized magnitudes was gaining ground.

MORE URGENT CALL FOR OBSERVERS

In 1901, Pickering again called for "co-operation in observing variable stars," this time in the new *Popular Astronomy*. He repeated much of what he published in his 1891 pamphlet. However, this appeal had more urgency to it, as Pickering began: "The number of known variable stars of long period is now so great, and is increasing so rapidly, that the observation of many of them has been greatly neglected." Pickering acknowledged the usefulness of the Hagen series of charts and, following Hagen's lead, offered photographically reproduced charts – also derived from the *Durchmusterung* – for 72 variable star fields. Pickering praised Hagen's charts, but justified another, wider field series of charts for observing these same stars when they were brighter than ninth magnitude:

> Observations of nearly equal value [as those made by experienced observers] can be obtained by those unaccustomed to estimating intervals in grades. It is only necessary to enter on the charts the standard magnitudes of the comparison stars, and from these to estimate directly the magnitude of the variable.[55]

In 1903, HCO published a "provisional catalogue" listing 1227 variable stars. This was an outgrowth of a bibliographic card catalog begun at HCO by William M. Reed and continued by Annie J. Cannon. The cards and published catalog were created in an effort to distinguish new variable star discoveries from old. Cannon followed these compilations in 1907 with her "Second catalogue of variable stars."[56]

In the February 1906 *Harvard Circular* 112, Pickering made another appeal for observations of long-period variables, stating once again that this type of variable was ideal for amateur observers who could observe them at infrequent intervals with small telescopes; and, furthermore, because most of these stars had a large magnitude range, there was room for "moderate errors of observation." One important point being stressed more than in the 1901 circular is found in Pickering's explicit description of an alternate method of making observations, "being substituted here for that of Argelander." The magnitudes of a sequence of comparison stars were entered directly onto the HCO charts (either photographic enlargements made from the printed *BD* or sometimes using photographs taken directly from the sky). The observation was to be made by comparing the variable directly with a brighter and a fainter comparison star and matching or interpolating the variable's magnitude from the given comparison star values.[57] It is the method commonly in use today. This direct method of estimating magnitudes was of great significance to the future of variable star observing because: (1) it brought many more amateurs into variable star observing because the direct estimation method was easy to understand and to apply, enabling many more amateurs to contribute who would otherwise be put off by the more intricate requirements of the step method and the computational labor required to derive a magnitude; and (2) it required that the determination of comparison star magnitudes made from photometric measurements be a centralized activity – at HCO in this case – and it required that the preparation and distribution of charts imprinted with these standardized sequences be also controlled by a central authority.

Figure 2.6. Henrietta Swan Leavitt (from *Popular Astronomy*, Vol. 30, No. 4, 1922, p. 197).

By 1906, the discovery of a variable star had ceased to be a remarkable event. What was now remarkable was the number of variable stars being discovered, to the point of being almost overwhelming. In April 1905, for example, the number of variable stars discovered in the Small Magellanic Cloud by Henrietta Swan Leavitt (1868–1921) underwent a large revision: 11 months earlier, the number given was 57, but Leavitt's reexamination of the field using better plates revealed 843 new variable stars – that number increased to 970, and then even more, on further examination. This discovery of the unexpected number of variables in one region was remarkable enough in itself to prompt Leavitt to present her findings in a paper at the December 1905 meeting of the Astronomical and Astrophysical Society of America. "The question as to whether there are other regions containing such remarkable groups of variable stars," Leavitt concluded, "is of the greatest interest. Not less important is the general problem of the distribution of these fainter variables."[58]

PICKERING PARADIGM SHIFT

To have an understanding of the number of variables discovered – their distribution in a given region – was now at least as important as understanding the behavior of each star. From this point on, announcements of variable star discoveries occurred with monotonous regularity in the pages of the *Harvard Circulars*. Between January 1906 and July 1908, about 388 variable stars were discovered at HCO – an average of about 20 new variables per month.

This seemingly endless proliferation of discoveries led Pickering to emphasize a paradigm shift in how and why variable stars needed to be discovered and observed. In a May 1906 circular titled "A Durchmusterung of variable stars," Pickering announced that "certain regions of the sky have now been systematically examined for the discovery of variable stars, the number of which has already increased to more than three thousand." Rather than spending time in making close observations of each known variable to understand why it varies, Pickering – concurring with Leavitt's 1905 report – proposed that an all-sky survey be undertaken to understand the *distribution* of all variable stars "to the faintest magnitude possible." He further stated that careful observation of all variables discovered was no longer necessary because:

> they will tend to fall into a few well-defined classes, an examination only sufficient for classification will be necessary for the greater part of them, and

elaborate researches will need to be extended only to representative types, and to objects of special interest.[59]

One consequence of this shift in emphasis, Pickering noted, was that the age of the independent variable star observer was finished: "While we recognize the eminent service which has been rendered during the last century by visual observers, it is certain that photographic methods now yield vastly greater results for a given expenditure of time. This is especially true for densely crowded and faint stars." In other words, in Pickering's view, the value of visual observers of variable stars was no longer centered on the highly skilled individual's attention given to a single star, but on the recruitment of many observers of average ability willing to observe as many stars as possible.[60]

HARVARD CORPS OF OBSERVERS

From this point on, the number of astronomers – both amateur and professional – who were volunteering to make variable star observations for HCO was growing rapidly, along with a corresponding increase in the number of observations. Historian Howard Plotkin dubbed this group "The Harvard Corps of Observers." Never an organized group, individuals in the Corps of Observers likely suffered bursts of interest that waned over time; thus at any given time the composition of this group of volunteers varied. Undated address lists in HCO archives from about 1910 suggest that at that time about 32 individuals were active in the Corps of Observers, whereas an additional 14 individuals who had previously been active contributors had become inactive. Interestingly, only 14 of the observers on these two combined lists also appear on a list of observers compiled from other sources by Plotkin. Evidently the membership of Pickering's Corps of Observers was quite fluid.[61]

In his 1911 plea for cooperation, Pickering reported that about 17 000 observations of long-period variable stars were made by HCO staff and about 6000 observations were "kindly communicated by other astronomers" between 1906 and 1910. His list of volunteer observers included Professor Anne S. Young; Frank E. Seagrave, John H. Eadie, several astronomers, and assistants, and professors Mary Whitney and Caroline Furness and students at Vassar College. Among others were Helen Swartz, Ida Whiteside, and Psyche R. Sutton. Whiteside, Swartz, and Whitney contributed more than 3200 variable star observations from 1902 through October 1911, with Whiteside accounting for at least 2507 observations between 1904 and 1909.[62]

Figure 2.7. Caroline E. Furness, of Vassar College, one of E. C. Pickering's Corps of Observers and one of the AAVSO's founding members. In 1915, Furness published *An Introduction to the Study of Variable Stars.*

Observations in the thousands were added by HCO newcomers Henrietta Leavitt, who joined the staff in 1895 (1800 variables in the Magellanic Clouds by 1908); Leon Campbell, who joined in 1899; and other HCO staff astronomers, especially O. C. Wendell and Annie Jump Cannon. In addition to these, Pickering added observations from the BAA-VSS and by many individual observers. To avoid unnecessary duplication and to secure the best results, he noted that some form of cooperation seemed advisable. Pickering urged observers to make estimates on a star more than once each month, especially short-period variables. He also explained that some coordination of observers was necessary to ensure the continuity of observations otherwise interrupted by the problems caused by weather, moonlight, time of day, position of the Sun, and so on. Finally, he urged observers to "send to the Harvard Observatory on the first of each month, a copy of their observations giving for each star, the name or designation, the date, and the concluded magnitude. Forms will be furnished for this purpose, and charts will be given to observers of the regions of such variables as they will observe systematically." Thus, in addition to sets of standardized charts announced in 1906, Pickering further attempted to assert some controlling order by formalizing the reporting interval and format to be used. The date of this last Pickering commentary was June 29, 1911 – just 4 months before the founding of the American Association of Variable Star Observers.[63]

THE EMERGENCE OF ORGANIZATIONS

By 1910, the observation and study of variable stars was a line of astronomical work of recognized importance in its own right. A growing body of variable star observers and a growing number of observatories, professional astronomers, and associations devoted their attention to this work, and yet there was still a lack of coordinated effort.

Localized initiatives began taking shape as astronomical societies showed interest in variable stars or established their own sections of variable star observers, but the ideal of a centralized organization was never envisioned, let alone attempted. The most successful and productive of the early initiatives was the Variable Star Section of the British Astronomical Society. The Astronomical Society of France established a Variable Star Commission in 1901. The Astronomical and Astrophysical Society of America, formed in 1897, heard 21 papers on the subject of variable stars in its first 13 years, but as yet, there was no organized group devoted to variable star work in the Americas.

Some observatories, like HCO, relied on their own corps of observers and published their own results; at other observatories, variable star work was the domain of one observer/analyst – John A. Parkhurst at Yerkes, for example. Among the observers themselves, some were paid observatory staff; others were volunteers who contributed observations at the specific request of an observatory following that observatory's guidelines; still others preferred to keep working as unaffiliated, independent observers and reporting their observations in published journals.

In the case of individual observers, their observations, period determinations, and ephemerides, if published at all, were published mainly as single notes or as letters in *AN*, *AJ*, *The Observatory*, *MNRAS*, and similar journals. Compilations of historical observations by individual observers during their lifetime were rare: some observers considered their observations to be of secondary importance to their other observatory work; mainly due to space limitations, astronomical journals of the day did not regularly publish lists of variable star observations; preparations, reductions, and checking all took time; and publication of a volume of collected observations entailed considerable expense.[64]

Publications of collected observations were slow to emerge. Ten years of observations made by Henry M. Parkhurst, in collaboration with John H. Eadie, were published in *Harvard Annals* in 1893.[65] While traveling in Europe, Pickering obtained copies of the unpublished observations of Friedrich W. A. Argelander, William Herschel, and Julius Schmidt. Pickering published these in 1900, along with the previously printed but difficult to obtain observations of Eduard Schönfeld.[66] In 1901, Rev. J. G. Hagen dutifully published his own observations of variable stars from 1884 to 1890, and he compiled the variable star observations of Eduard Heis and Adalbert Krueger, which were published in 1903.

Still, Paul S. Yendell lamented the dearth of published collections of variable star observations. Reporting on the publication of Hagen's "The Heis-Krueger observations of variable stars" (1903), Yendell noted:

> The regrettable infrequency with which such valuable material comes within the reach of the private observer. In my own case, for instance, the observations of Argelander published in Vol. 7 of the Bonn Observations, those of Knott, published since his death, those of Plassmann, who is setting the good example of publishing his own observations during his life time, with a few single series by the Baxendells, Gore, Sawyer and one or two others constitute the entire accumulation of twenty-three years of constant work in this line.[67]

The numbers of variable star observers grew steadily from the time of Argelander's first appeal through those of Gould, Chandler, John A. Parkhurst, Yendell, and Pickering. The call for observations was being answered, but the growing number of observers and observations – coupled with the growing number of variables, new methods of discovery, and astrophysical investigation – brought an increase of activity that was only partly coordinated and, at best, suffered from a lack of standardization, bordering on confusion and chaos. The next great need was that of organization, and the earliest organizations to emerge were the short-lived Society for Practical Astronomy and the fledgling American Association of Variable Star Observers (AAVSO).

THE SOCIETY FOR PRACTICAL ASTRONOMY

Few realized that the inspiration for The Society for Practical Astronomy (SPA) in 1909 in Chicago came from two teenagers, Frederick Charles Leonard (1896–1960) and Horace Clifford Levinson (1895–1968), who had the potential to provide needed organization for variable star observers. With the guidance of a few professional astronomers, the SPA was organized on the same principles as the BAA, with observing sections for various topics of astronomical interest, including variable stars. Leonard, probably the main instigator, was well known to astronomers at Yerkes Observatory and the University of Chicago. The SPA journal, originally titled *The Monthly Register of the Leonard Observatory*, was typewritten and not widely distributed. The organization grew slowly as Leonard recruited members through his persistent and polished personal correspondence.[68]

In 1911, after publishing several technical articles in *Popular Astronomy*, Leonard advertised the SPA to the broader audience of *Popular Astronomy* readers and invited others to join. Although the society consisted primarily of amateur astronomers, Leonard welcomed professionals. The society's journal was now printed and distributed as the *Monthly Register of the SPA*. The officers of the SPA were Frederick C. Leonard, president; John E. Mellish, secretary; Horace C. Levinson, treasurer; and Ruel W. Roberts, organizer and lecturer.

Leonard described his intent to make the SPA "one of the strongest and largest amateur astronomical organizations in existence." The invitation went out to amateurs who are "so willing to do all they can to advance Practical Astronomy . . . so that their combined efforts may result in promoting this sublime science to even a still greater degree than formerly."[69] The beginnings of a technical organization emerged slowly, with sections devoted to the planets, variable stars, and other observational topics such as spectroscopy and instruments. Leonard maintained a steady flow of announcements in *Popular Astronomy*. He was successful in attracting more adults to the organization, although many may not have suspected that the founders/co-editors of the SPA journal were only teenagers.

Two early SPA members from Maine, Russell W. Porter and Robert H. Bowen, conceived a national SPA convention, which Porter hosted at his home at Lands End, Maine.[70] Only a few individuals participated in the August 1914 weekend, but the participants, including *Monthly Register* editor Levinson from Chicago, enjoyed sharing observing notes, and Porter regaled them with his arctic adventure tales as well as tips on telescope making. Those attending voted to do the same thing the following summer, and Bowen described the first "International Conference" glowingly in the *Monthly Register*.[71]

By the end of 1914, the SPA membership included several professional astronomers, including Forrest Ray Moulton, who helped organize the second SPA convention in August 1915 on the University of Chicago campus. The second convention attracted 30 or more attendees, mostly adults, and featured formal paper sessions and a banquet. The new slate of officers elected at the second meeting indicated that the SPA was evolving away from its status as the brainchild of two teenagers into a broader-based organization. Its new president, Latimer J. Wilson of Nashville, Tennessee, had demonstrated his ability as serious amateur planetary observer and would emerge as one of the pioneers in planetary photography in later years. The new SPA treasurer, Henry W. Vrooman of Kokomo, Indiana, a variable star observer and spectroscopist, was well respected by the staff at Yerkes Observatory. Secretary John E. Mellish was well known as discoverer of three comets as well as for his telescope making. Speakers at the second conference included professional astronomers Mary Byrd and Charles Pollard Olivier. In addition to Moulton, Byrd, and Olivier, other professional astronomers who had joined the ranks of the SPA included Homer Black, Edward A. Fath, and Francis P. Leavenworth. This is admittedly not a list of the leading professional astronomers in America, but it is important to note that these professionals saw this organization as one in which it was worth investing their effort. They saw the need for

Figure 2.8. Members present during the second annual conference of the Society for Practical Astronomy (SPA), held at the University of Chicago, August 16–18, 1915. From left: S. F. Maxwell, Miss M. Mueller, R. B. Potter, C. A. Mundstock, Father A. Petrajtys, F. C. Leonard, H. F. Black, H. C. Levinson, Rev. R. F. Bumpus, H. W. Vrooman, W. Henry (from *The Monthly Evening Sky Map*, Vol. 9, No. 106, 1915).

such an organization to facilitate amateur contributions to astronomy and believed that the SPA had potential to do just that with its technical observing sections.[72] Leonard remained an SPA director. Plans were made for the third convention in 1916 in Rochester, New York.[73]

SPA VARIABLE STAR SECTION

William Tyler Olcott of Norwich, Connecticut, served as the first leader of the SPA Variable Star Section upon Leonard's invitation. By January 1912, Olcott was writing notes on variable star observing in the *Monthly Register of the SPA.* The SPA-VSS had about 15 members in 1912. After about a year, Olcott resigned and was replaced as section leader by Edward Gray, who resigned a year later.[74] In late 1914, Leonard, an active variable star observer who contributed more than 1400 observations to Olcott's own fledgling AAVSO between 1913 and 1917, announced he would lead the SPA variable star observing effort.

Unfortunately, the VSS members had all resigned from the SPA. Their departure was a severe blow to the SPA, as many were involved in more than one SPA section. As a group, they represented a substantial loss in resources, both financially and in potential leadership. Their departure did not reflect any diminishment in their interest in the variable stars. Instead, they found another new organization with a more satisfying opportunity to contribute.

Leonard's interest in astronomy was undiminished; he completed a Ph.D. in astronomy and founded the astronomy department at University of California at Los Angeles, where he was chairman for many years. His scientific interests gradually focused on meteorites; he was a co-founder with Harvey H. Nininger of the Meteoretical Society in 1933.[75] That society eventually named its highest award, the Leonard Medal, in his honor.

In later years, Leonard continued to harbor resentment about the demise of the Variable Star Section of the SPA, even though that event likely played no role in the eventual collapse of the entire organization by 1917. Other factors that were likely more significant than the loss of a few variable star observers included the departure of Leonard and Levinson for college, withdrawal of financial support provided by Levinson's father, and the onset of World War I.[76]

These events have set the historical stage on which the AAVSO emerged, as yet another attempt to answer the need for an organization of variable star observers. We shift our attention now to William Tyler Olcott, his life, and the founding of the AAVSO.

Part II

THE FOUNDING OF THE AAVSO – THE WILLIAM TYLER OLCOTT ERA

3 · The amateur's amateur

There is an effort being made to organize a Variable Star section in this Country. Prof. Pickering of the Harvard College Observatory favors the plan.

– William Tyler Olcott, letter to Harriet W. Bigelow, Professor of Astronomy, Smith College, October 4, 1911[1]

William Tyler Olcott's resignation as the leader of the Variable Star Section of the Society for Practical Astronomy in 1913 coincided with his efforts to form another variable star organization.

In the previous 2 years, Olcott had formed the American Association of Variable Star Observers (AAVSO), officially begun on October 10, 1911.[2] Its existence inadvertently demolished the SPA's variable star section, whose members had left to join the AAVSO. Their migration may not have had anything to do with actions of the SPA's teenaged founders. What can be said, however, is that they remained with Olcott and the AAVSO after a year or more in correspondence with Olcott and recognition that Olcott and his AAVSO were connected in some way to the Harvard College Observatory (HCO).

This man of letters and science seemed ideally cast as the founder of a new organization. Even more so, he always seemed to have sufficient time and wherewithal to travel and support an international organization. This chapter follows Olcott's life leading up to his founding of the AAVSO and how he came to a privileged position.

TYLER OLCOTT'S PARENTS

Although he was born and raised in Chicago, Illinois, Tyler Olcott, as he preferred to be known, lived most of his life in the family's ancestral home, originally a church rectory, at 62 Church Street in Norwich, Connecticut. He was born January 11, 1873, to William Marvin and Elizabeth Olivia Tyler Olcott. They may have met through a mutual friend, Tyler Roath, a Norwich native whose family ties played a critical role for Tyler Olcott's later independence. Though William Olcott was born in Utica, New York, in 1839, his family moved to Chicago when he was a child. His father left farming to seek employment in the fastest growing city on the western frontier. William Olcott received his early education in the Chicago schools. He began his college education at Beloit College in Wisconsin, but after a year returned to Chicago, where he joined the famous military unit, the Ellsworth Zouave Cadets.[3]

The Cadets, a precision drill team, was nationally known for its exotic uniform styled after those of the French Algerian Legions. In total, 138 cadets were recruited for this unit.[4] After winning a drill team competition at the United States Agricultural Fair on September 15, 1859, the Cadets toured other states, challenging other drill teams to compete for their colors during July and August 1860. One of Olcott's fellow cadets, W. Tyler Roath, did not tour but perhaps briefed William Olcott on a proposed stop in Hartford, Connecticut. As a Norwich native, Tyler Roath might have invited William Olcott to visit his family in Connecticut, including Tyler Roath's father, Walstein Roath, a well-established watchmaker and jeweler in the region, and Tyler's uncle, Captain Edmund D. Roath, a respected merchant ship captain and successful investor. But since the Hartford performance was cancelled, it seems unlikely that William Olcott made it to Norwich.

Although the Cadets were called into service with the Civil War, neither William Olcott nor Tyler Roath went with them. Tyler Roath, who had developed some infirmity precluding his service in the Cadets, migrated to the Black Hills, Montana, and Denver, Colorado, before settling in Cheyenne, Wyoming. William Olcott served with the Chicago Board of Trade Battery of the Illinois Volunteer Artillery and was wounded at the battle of Peach Tree Creek in 1864. After the war, he rejoined a coal distribution business in partnership with his brother James F. Olcott, titled "Wm. & J.F. Olcott," in Chicago.[5]

Tyler Olcott's mother, Elizabeth Olivia Tyler, was born in Norwich, Connecticut, in 1841. Olivia, as she preferred to be known, was educated at Miss Edward's School in New Haven, Connecticut. After completing her education, she lived with her father William Samuel Tyler, a drug store owner, and her mother Olivia Ann Clarke in the Glebe House, an ancestral home leased to the Tyler family on a perpetual 999-year lease in 1829. Several Tyler generations lived at the Glebe House, built for their ancestor, John Tyler, the first rector of Christ Episcopal Church in Norwich. The family also occupied the Tyler house next door; the two buildings were connected with a passageway. The two women continued to live at the Glebe House even after William Samuel Tyler's death in 1864.

THE OLCOTT-TYLER CONNECTIONS

As long-term Norwich residents, the Tyler and Roath families were well acquainted,[6] and Tyler Roath and Elizabeth Olivia Tyler would have known each other well. It seems likely that Tyler Roath, William Olcott's friend, first introduced him to Elizabeth Olivia Tyler. They married November 9, 1870, and left for Chicago to establish their new family home there.[7] A year later in 1871, ties between the Tyler and the Roath families were further strengthened when Elizabeth's widowed mother, Olivia Ann Tyler, married Captain Edmund D. Roath. It was a first marriage for Captain Roath, who was 16 years older than Olivia Ann Tyler, and they continued to live at 62 Church Street.

FAMILY BUSINESSES IN CHICAGO

Numerous disasters in the next 5 years affected William Olcott's business in Chicago. The disastrous fire (October 8–9, 1871) destroyed the coal business.[8] The Olcott brothers quickly re-established their business by setting up three coal distribution yards around the central business district. Olivia Olcott's pregnancy paralleled the rebirth of the business, and in early 1873, William Tyler Olcott joined the Olcott family.

Like many other Chicago businesses, the Olcott coal business was devastated by the financial panic of 1873 and 1874, and late in 1874, the brothers declared bankruptcy.[9] Afterwards, William Olcott worked as a solicitor for a while before starting a new coal distribution company, this time known as "Wm. Olcott and Company." Fortunately, the Olcott coal company achieved a stability not enjoyed in the 1870s as the Chicago economy prospered with the addition of more railroads, stockyards, manufacturing plants, and other businesses. Throughout the late 1870s and 1880s, the family lived in the distinctly upper-middle class Douglas development, first at 1160 Prairie Avenue and then at 2550 Prairie Avenue when Tyler Olcott attended elementary school.

RETURN TO ANCESTRAL HOME

In 1888, Olivia Ann Roath, now 68 years old, invited her only daughter, Elizabeth Olivia Olcott, and her family to bring her only grandchild to live in the Glebe House. Captain Roath had become an invalid in 1881. No doubt the burden of his care as well as a growing sense of her own diminishing life expectancy played an important role in her invitation. William Olcott closed his coal business and the family moved to Norwich that summer.[10]

The Olcott family took the move from the bustling Chicago to a traditional New England town with some success. Life in Norwich likely exceeded William Olcott's expectations after leaving Chicago, the city that had served him well for nearly 40 years. Active in Christ Episcopal Church, William Olcott served on the vestry and as a volunteer in several church-related mission organizations. He also joined the local chapter of the Grand Army of the Republic, serving as a full colonel and assistant inspector general on the Commander-in-Chief's staff. He joined several fraternal lodges and was chosen for membership in the Arcanum Club, an exclusive organization of about 20 leading male citizens. For her part, Olivia Olcott re-joined social circles in which she had circulated in Norwich before her marriage and developed an interest in genealogy, tracing the Tyler family for more than 20 generations.[11]

WILLIAM TYLER'S EDUCATION

At the age of 15 years, William Tyler Olcott entered the town's public high school, Norwich Free Academy. His education likely followed a traditional college preparation course, with an emphasis on Greek, Latin, mathematics, and English. When he graduated in 1892, the Academy principal, Robert P. Keep, commented that Tyler Olcott was thoroughly prepared to matriculate to a college.[12]

While Tyler Olcott attended the Academy, the elderly Captain Roath passed away in February 1891. Apparently fond of his step-grandson, he bequeathed Tyler Olcott his gold watch. Captain Roath passed half of his estate to his wife Olivia Ann Roath, who had supported him during his later years, including 10 years as an invalid. The remaining half was divided between his sister and his two nephews, William Samuel Roath and W. Tyler Roath, who were to split the remaining share.[13]

COLLEGE LIFE IN THE GILDED AGE

By the early 1890s, graduation from a small college had become a rite of passage for the sons of most upper-middle class families.[14] Around 1890, a boom began in American collegiate education. Traceable to the formation of new universities like Stanford, Clark, and the second University of Chicago, the curriculum at most institutions shifted rapidly to accommodate innovations like Charles William Elliot's elective system and graduate schools at Harvard University.[15] Laurence R. Veysey has observed that there was a division in the student body. One group took the "the faculty, textbooks and debating more seriously." This contrasted with the upper-middle class sons who came to have a good time.[16]

Tyler Olcott's experience at Trinity College in the 1890s clearly fits Veysey's description of the less serious undergraduate. Although Trinity was slow to modernize its curriculum, it offered a full range of activities outside the classroom in which the good life was available. Olcott's academic record clearly reflected his intent to make the most of his undergraduate years in a social and athletic way. Olcott excelled in a few subjects (Themes, Ethics), but only scraped by in most other courses, including mathematics, chemistry, physics, metaphysics, Latin, German, and Spanish, graduating near the bottom of his class of 26 students.[17] As a member of Kappa Delta Epsilon, Olcott spent time in many activities other than studying, including intercollegiate lawn tennis; editing the school yearbook, *Trinity College Ivy for 1896*; secretary of his senior class; and president of Medusa, an exclusive senior honorary society that appeared to honor social standing more than academic performance.[18] He was an editorial board member of *The Trinity Tablet*, a campus newspaper published every 3 weeks. He also composed the lyrics of a song for which a fellow Trinity student, A. L. Ellis, composed the music. The pair sold the rights for their song to Broadway producer E. E. Rice for one of his musicals. Although Tyler enjoyed writing poetry in later life, he only ventured into the role of lyricist one other time.[19]

TRAVEL AND LAW SCHOOL

After graduation from Trinity in June 1896, Tyler followed another late nineteenth century practice of the upper-middle class and traveled "abroad," an experience that ranked along with a college education in the maturation process. Traveling on the continent in the first year after completing a college degree was considered de rigueur by many. During the summer of 1896, Tyler traveled but never recorded his experiences for public consumption. After his return to Norwich, Tyler took off a year to apply at a law school. Tyler spent the remainder of the year socializing with friends in Norwich, writing poetry, and taking one camping trip to Block Island, Rhode Island.

Tyler was well qualified to enter law school in 1897, as no law schools except Harvard required a college degree as a prerequisite to admission. With his B.S. degree from Trinity, he could enter any law school of his choice. In the fall of 1898, he entered the 2-year program at the New York Law School (NYLS).[20] A relatively new law school, NYLS had been founded by disgruntled faculty from Columbia University's School of Law around 1891. NYLS students could take the New York Bar examination in 2 years, as had been the case previously at Columbia.[21] NYLS had become the first independent institution chartered by the state of New York to confer degrees and quickly became the second largest law school in the United States.[22] In mid-town Manhattan, the NYLS was ideally situated for students who studied part time at night and worked full or part time in the law offices in the lower Manhattan business district during the day.

Being able to take the bar exams after 2 years of law school may have appeared especially attractive to Tyler Olcott, who never practiced law. During the

2 academic years beginning in 1898, Tyler lived at 107 East 45th Street in mid-town Manhattan. In addition to his proximity to the NYLS buildings, he was also close to the theater district and Manhattan's other cultural centers. It was an exciting time to be in Manhattan. With more than 3 million people, New York City was second only to London in size and greatly exceeded other European capitals in scale as well as population. With immigrants still flooding in and businesses forming, growing, collapsing, and dying on a daily basis, New York City certainly was one of the most exciting places to live and study law. It is no surprise that Tyler chose NYLS over the stodgy Yale Law School in nearby New Haven.

During his second year in New York, Tyler's maternal grandmother, Olivia Ann Roath, passed away. In her will, she bequeathed $10 000 directly to Tyler, but the remainder of her estate went to support her daughter, Olivia Olcott, for the rest of her life and then was to pass to Tyler Olcott.[23] Thus Tyler Olcott's financial security was assured even before he completed his legal training.

After completing the second year at NYLS, Tyler passed the New York State Bar examination and a year later was admitted to the Connecticut State Bar, also by examination. He returned to Norwich and opened a law office in his family's home at 62 Church Street.[24]

SOCIETY, MARRIAGE, AND RECREATION – THE GOOD LIFE

Norwich, a city that lingered in the nineteenth century, certainly contrasted with the rapidly growing cities (New York City, Boston, and Chicago) that strived to epitomize the modernity of the twentieth century.[25] Norwich thrived on the usual New England industrial base. Powered by coal as well as water at the confluence of three rivers, Norwich's textile mills provided a major source of employment. The city also served as a transportation and commercial hub for nearby agriculture and fishing.

As a freshly minted lawyer, young Tyler Olcott could surely be classified as one of the most eligible young men when he entered Norwich social circles in 1900. A social scrapbook reveals a calendar of whist games, dances, theater and concert engagements, tennis, and boating as his social activities. Almost all of those activities involved his eventual life partner, Clara Eunice Hyde.

Tyler Olcott was active in a number of clubs that apparently fitted his skills. He was elected secretary of the Chelsea Boating Club in 1901 and enjoyed membership in the Arcanum Club, an elite group limited to about 20 men. Although chartered for the improvement of the men's intellectual skills, the club's main function appears to have been the maintenance of a club house for billiards, bowling, chess and other board games, and social functions. Because his father had been elected to membership in the Arcanum Club some years previously, Tyler's early election may be regarded as a legacy selection, but it is evident that he fit into the social flow of the Arcanum Club and to Norwich society in general.[26]

Clara Eunice Hyde emerged as Tyler Olcott's constant companion in those days after he completed law school. A music teacher and more than 4 years older than Tyler, Clara was the oldest of three daughters and one son in the family of banker George Rodney Hyde and Katherine Rhoda Dickey Hyde. She enjoyed a privileged childhood, attending the Norwich Free Academy before graduating from Miss Dana's School, an elite finishing school for girls in Morristown, New Jersey. Miss Dana's School exposed Clara to a classical curriculum with required 4-year courses in Latin, English, and the Bible, and required courses in history, algebra, and geometry. These required courses were supplemented with electives from a broad list that included botany, physiology, astronomy, music, studio art, Greek, French, logic, chemistry, physics, psychology, and banking. Vassar College waived its entrance requirements for any applicant holding a diploma from Miss Dana's School, so Clara's education to that point was clearly well above average for a high school graduate.[27]

On June 16, 1902, Tyler Olcott and Clara Hyde were married in a quiet ceremony at her family home in Yantic, Connecticut, a separate town that was part of the Norwich Township. The impending death of Tyler Olcott's father dampened the couple's enthusiasm for a more elaborate affair. The event, attended by only a few of their friends, also drew family members from across the country. William Marvin Olcott died a week later, and his obituary on page 1 of the *Norwich Bulletin* gave ample testimony to his social importance since arriving in Norwich from Chicago only 14 years earlier.[28]

Figure 3.1. William Tyler Olcott and Clara Eunice Hyde Olcott on their wedding day, Yantic, Connecticut, June 16, 1902.

Figure 3.2. William T. Olcott with his 3-inch refractor, in his back yard, Norwich, Connecticut, about 1910.

NIGHT SKY REVEALED

The Hyde family owned a camp site on the South Bluffs of Block Island. Called *Camp As You Like It*, the site had been the scene of many happy family gatherings. Tyler Olcott purchased a lot near the Hyde camp. Shortly after their marriage, he and Clara built a cottage on the lot. It is easy to imagine the excitement in the Hyde and Olcott families watching the newlyweds' cottage, completed in the summer of 1905, emerge from the island's rocky soil.[29]

One early guest at the cottage made an incalculable contribution to this history; Helen McGregor Clarke introduced Olcott to the beauties of the night sky. As Clara's friend, she came to visit the Olcotts during the summer of 1905.[30] On one especially clear and dark evening, she produced a planisphere from her suitcase and announced that she was going out on the bluff to observe the stars. After about an hour outside, she returned to the cottage to invite Olcott to join her on the bluff, an invitation he gruffly refused in favor of a book on the conquest of Mexico. She went out again and spent some additional time under the stars before retiring for the night. This process was repeated for several evenings, until she decided not to invite him. On that evening, Olcott had decided to be a more polite host and had his coat ready when she appeared with planisphere in hand. Although he had taken an astronomy course at Trinity College and likely had some exposure to astronomy at the Norwich Free Academy, as well, Olcott professed no knowledge of astronomy as of that event.

After several evenings on the bluff with Miss Clarke and her planisphere, Olcott was hooked. He returned to the cottage frequently during their outings to record notes about what he was seeing. He quickly learned the constellations, the brightest stars, and the visible planets, so obviously enjoying the experience that when Miss Clarke returned to New York City she left her planisphere behind. Olcott went out every clear evening thereafter to study the stars. When he and Clara eventu ally returned to Norwich, he purchased a small telescope and avidly read books on astronomy. His orientation to the night sky and note-taking continued until late autumn, when he and Clara took a trip to New York City. On that visit, Olcott took his book of notes about the constellations and stars along when he called on the offices of publisher George P. Putnam. He showed Putnam his book of notes and asked if there was a chance the publisher would produce a similar book of astronomical information for beginners in astronomy. Putnam responded that such a book had long been sought by his firm for use in high schools and colleges. He asked that Olcott leave his notes and return in 2 weeks for further discussions and a decision on the matter.

On his return, Putnam agreed to publish the book if Olcott added a sky map and a few drawings of important constellations. While he completed his review of the seasonal constellations, he added descriptions and a few diagrams. *A Field Book of the Stars*, published in 1907, succeeded so admirably that Putnam reprinted it several times. Putnam urged Olcott to follow up on that success with another book at about the same time his mother gave him a 3-inch telescope, which led to his second book, *In Starland with a Three-Inch Telescope*, published by Putnam in 1909.

A career in writing now seemed possible and more attractive than Olcott's all but abandoned profession as an attorney. Olcott undertook another, far more successful writing project. Based on the success of his first two efforts at astronomy books, Olcott wrote a book to describe the history, myths, and legends related to the celestial constellations. He described his simplified diagrams of the constellations, each in one chapter, with the lore abstracted from sources that stretched into antiquity, including poetry and the translation of ancient texts. He used photographs of important artworks that illustrated the constellations in more material form in his book, *Star Lore of All Ages – A Collection of Myths, Legends, and Facts Concerning the Constellations of the Northern Hemisphere,* published in 1911 by Putnam. Its success followed his two previous books, well burnishing his credentials both as an astronomer and as a scholar.[31]

Figure 3.3. The Olcott residence, Norwich, Connecticut, the first, unofficial, home of the AAVSO. The dome of Olcott's octagonal observatory is visible between the chimneys. This photo was made probably in 1921 when the dome was built. (From an article by Olcott in *Popular Astronomy*, Vol. 38, No. 1, 1930).

VARIABLE STARS AND SCIENCE – AN OPTION?

Hoping to increase his understanding of astronomy, Olcott attended a year-end 1909 meeting of the American Association for the Advancement of Science (AAAS), held on the campuses of Harvard University and the Massachusetts Institute of Technology. At the meeting, Olcott observed a variable star astronomy presentation prepared by Harvard College Observatory Director Edward C. Pickering. The presentation discussed how amateur astronomers might make a significant contribution to that science by observing long-period variable stars. Olcott was fascinated by Pickering's light curves and charts and resolved to find out more about the subject.[32] He asked Pickering whether an amateur with a 3-inch telescope might make a contribution and how he should get started. Pickering responded immediately by sending a young night assistant, Leon Campbell, to Norwich to personally introduce Olcott to astronomical observation.

During the Campbell visit, Olcott learned about the delightful experience of finding and estimating the brightness of variable stars. After a long struggle to locate Mira (o Ceti) and make his first estimate on January 23, 1910, Olcott became a dedicated variable star observer, reporting his observations on a monthly basis as a member of Pickering's Corps of Observers.

Within a few months, Olcott reported his experience for the readers of *Popular Astronomy*, inviting others to take up variable star observing and share in producing scientifically useful results from their hours spent at their telescopes.[33] Olcott's article apparently attracted Frederick C. Leonard's attention and led to Leonard's invitation to Olcott to become "Director" of the Variable Star Section in Leonard's Society for Practical Astronomy. In his enthusiasm for the subject, so clearly expressed in the *Popular Astronomy* article, Olcott accepted Leonard's invitation, a hasty decision he likely soon regretted.

ORGANIZING OBSERVERS

At this point, then, it is appropriate to better explain what happened to Leonard's SPA Variable Star Section. To do so, we need to go back to the August 1911 issue of *Popular Astronomy*. After Leonard announced the SPA's formation, *Popular Astronomy* editor Herbert C. Wilson announced he would begin a new monthly article devoted to amateur astronomy. Wilson described the opportunities for amateurs to contribute to astronomy and decried the lack of any American organization to coordinate the scientific activities of amateurs. Although he did not mention the British Astronomical Association (BAA), it is clear from the text of the article that Wilson had in mind a nationwide organization made up of sections devoted to specific observational topics, similar to the BAA. Wilson appealed for the formation of such an organization and offered *Popular Astronomy*'s support by publicizing observations each month under the heading, "Notes for Observers," short lists of objects to be observed and suggestions on methods and forms of record. He asked: "Can we not have in America an association of observers with a 'Variable Star Section,' a 'Jupiter Section' etc.? We invite correspondence in regard to the matter."[34]

The disjunction between the Leonard and Wilson articles is evident in an examination of these pages. Wilson ignored Leonard's effort to organize exactly the type of society that Wilson so eloquently described, deliberately undermining Leonard's effort. He later made that intent explicit in a letter to Pickering:

> I have several letters from observers who would like to see the formation of such an association as was mentioned. I am chary, however, of the organization headed by Frederick C. Leonard who is only a boy with lots of enthusiasm but not very much knowledge of astronomical subjects. If there is any way we could head him off, it would be a good thing.[35]

The new national association of amateur astronomers that Wilson hoped would compete with the SPA never materialized. There was no one else with Leonard's energy willing to attempt such a broad organizational effort. However, two parts of the desired umbrella association, "sections" for variable stars and meteors, were organized in response to Wilson's appeal. In the September 1911 *Popular Astronomy,* Wilson quoted a letter from Pickering that strongly supported the formation of a variable star section like Wilson advocated. Pickering indicated that Harvard Observatory would provide technical guidance if someone else would form and administer such an organization.[36]

In his September 29, 1911, letter to Pickering, Olcott advised him that he had already arranged to announce the formation of a variable star section as proposed by Wilson and would handle correspondence necessary in connection with such an organization. "The sole idea in the formation of such a body is to supplement the splendid work you are doing and in no wise to interfere with your present cooperative plan," he continued. "All members of such a section would be required to send their observations to you." He also addressed several other concerns in recruiting members:

> It is my hope that the formation of such a section will attract recruits and awaken interest in a line of work that amateurs would do well to take up. I trust such a plan meets with your approval, as my endeavors in this direction are directed solely with a desire to call the attention of others to this work so that they may cooperate in your admirable plan of systematic observation.[37]

In the September 1911 *Scientific American*, Olcott announced he was forming "an association of variable star observers" and invited interested observers to write to him.[38]

Pickering agreed in broad terms with Olcott's plan, stating that HCO would do "anything we can to aid the work." He cautioned against use of the word *amateur* in the title of the organization and sent a stack of reporting forms for Olcott to distribute to members, setting up a monthly schedule for the arrival of reports at HCO. Finally, he suggested that Olcott contact Charles L. Brook, who had replaced Col. E. E. Markwick as the BAA-VSS Director in 1910, to establish a "much desired" cooperation between the two groups.[39]

RECRUITMENT COMPETITION

What actually happened in practice was slightly different than what Olcott expected when he wrote to Pickering announcing his intent to form "a variable star section." Olcott expected that Pickering's Corps of Observers would also become members of the new association. In an October 3rd letter to Pickering, Olcott outlined his understanding of the new arrangements:

> I think we should count on the group now cooperating with you to form the nucleus of this new association, which has for its aim the

furtherance of your plans. A word from you to them will accomplish this purpose. As I am merely acting as Corresponding Secretary in this matter I feel that my appeal to them would have little weight. I have requested these observers to send me by the 10th of the month their observations so that I can forward them to *Popular Astronomy* for publication. This will enable each member of the association to see what the other members are doing, will stimulate interest in the work by affording an opportunity of comparing observations, and should attract recruits for the work.[40]

Olcott was not only aggressively recruiting membership from among Pickering's Corps of Observers, he was also soliciting support from other professional astronomers. Olcott's October 4th letter to professional astronomer Harriett W. Bigelow at Smith College, Northampton, Massachusetts, demonstrates his activity:

Dear Madam: There is an effort being made to organize a Variable Star Section in this Country. Prof. Pickering of the Harvard College Observatory favors the plan and I thought possibly you might be able through some of your classes to cooperate in this work as Vassar, Mt. Holyoke, & Amherst are doing. Prof. Pickering furnishes the necessary charts, and I will gladly assist those who care to take up the work in any way that I can. I hope that you will give the matter your attention for your cooperation is earnestly desired.[41]

Pickering's response to Olcott's recruitment and organizational efforts was not exactly what Olcott expected. Pickering had already staked out the importance of primary review and control of the data at Harvard in his September letter to Wilson, noting that such matters "can not really be trusted to an amateur."[42] After suggesting that Olcott, or perhaps his entire organization, ought to become members of the Astronomical and Astrophysical Society of America, Pickering elaborated on his concerns:

There seems to be some misunderstanding with regard to the plans of *Popular Astronomy* in publishing the detailed observations . . . . If our observers are inclined to send you, on the tenth, a second copy of their observations, which they send us on the first, and Professor Wilson is inclined to publish them, I see no objection to your plan. Until this arrangement is definitely made, however, I hope nothing will be done to prevent our observers from sending their observations to this Observatory as heretofore, or to interfere with our monthly article in *Popular Astronomy*. Since your letter was received by our observers, there seems to be doubt in the minds of some of them as to where they should send their observations.[43]

Apparently some members of the HCO observing corps had already expressed their concern about the new organization. For example, Amherst Professor David Peck Todd indicated his willingness to join Olcott's efforts if that was what Pickering desired, though redundant reporting to both Harvard and Olcott's new organization could not have been very attractive to Todd. He asked Pickering if sending the reports to Olcott was necessary, agreeing to do so but only if Pickering wanted him to do so.[44] Pickering did not reveal to Olcott that on October 5th he had already advised Todd, and possibly other members of the HCO observer's corps, that they were not required to send observations to Olcott. In his response to Todd, Pickering clearly expressed his desire to keep the two efforts separate:

Mr. Olcott's plan is only to supplement ours, and he assures me that he will not attempt to interfere with it. I hope therefore that you will continue to send us the observations which have proved to be very valuable in the past. Of course, I have no objection to your sending him copies also, if you desire to do so, but he is apparently laboring under a misapprehension, as Mr. Wilson writes on September 11: – "We could not undertake to publish individual observations and computations as you have done and so the only thought would be to help to make the cooperation of more amateurs effective."[45]

With Pickering's reassurance that it was not required, Todd and his Amherst associate, amateur astronomer Charles J. Hudson, elected not to submit observations to the AAVSO.[46]

On October 29, 1911, Olcott wrote to Helen M. Swartz, a high school mathematics teacher in South Norwalk, Connecticut. Swartz, a Vassar graduate and astronomy student of Mary Whitney, had been submitting variable star observations directly to Pickering since 1902. In his letter, Olcott extends his gratitude to Swartz for volunteering to contribute variable star observations in answer to his call. Olcott's choice of words shows that he was sensitive to any objections Pickering might have had about an amateur calling for observations, and yet, that he was determined to organize a group of variable star observers that included both amateurs and professionals:

> Your very welcome letter recd. and I was so glad to learn that you would cooperate with us in this plan to popularize variable star observing, for that is what it amounts to. My sole aim is to call the attention of amateurs to this line of observational work and thereby secure for Prof. Pickering additional data.
>
> . . . Dr. Furness, Miss Young, Mr. Hunter and Mr. Jacobs have joined us and I have succeeded in interesting seven other amateurs to date. This is doing pretty well I think as the V.S.S. of the B.A.A. only contained 13 members last year.[47]

The envelope in which Olcott sent this letter displays the AAVSO seal, and the words "Office of the Secretary, 62 Church Street, Norwich, Conn., U.S.A."

In contrast to Todd, Wilson was apparently already promoting Olcott's plan. In spite of his assertion to Pickering that he would not do so, he published Olcott's first summary of individual variable star observations sent to him by his correspondents in the November 1911 *Popular Astronomy*. In addition to Olcott's observations, the first monthly report only included observations from one other regular contributor to the Harvard program, Professor Anne Sewell Young of Mt. Holyoke College. However, the AAVSO report was expanded in the December 1911 *Popular Astronomy* with observations from Stephen C. Hunter, M. W. Jacobs, and Henry W. Vrooman of the SPA variable star section; F. E. Hathorn of Iowa, a recruit to the program; and Swartz and professor Young, also regular contributors to the HCO variable star program. Significantly, the December report contained observations made by AAVSO members from October 10 to November 11, 1911, titled, "The First Monthly Report of the American Association of Variable Star Observers."

The battle was won, and AAVSO observations appeared regularly in *Popular Astronomy* for the next 25 years.[48] By that time, AAVSO would be well established as a regular participant in variable star astronomy.

4 · Amateurs in the service of science

I will be glad to act on any suggestions you may offer that will in your judgment render our observations of greater service to you.

– William Tyler Olcott to Edward C. Pickering, 1912[1]

William Tyler Olcott, who began married life as an attorney in Norwich, Connecticut, was drawn into viewing the sky one summer from Block Island, Rhode Island. His growing interest and success in publishing his notes about the sky and stars drew him into a more active involvement in variable star observing and the organization of the American Association of Variable Star Observers.

After his discovery of the skies, he was delighted to find that a small telescope could contribute to the science of variable star astronomy. Acting on this knowledge, Olcott established an organization to collect and publish the variable star observations made by himself and others, which he announced as the American Association of Variable Star Observers, or AAVSO.

The AAVSO grew under the influence of monthly appearances of AAVSO observing statistics and other news in *Popular Astronomy*. That growth and the advent of World War I had consequences for both Olcott's health and the AAVSO, which led to the incorporation of the AAVSO in 1918. All of these events are the subject of this chapter.

A NOVA OPPORTUNITY

The AAVSO had barely begun to function when an exciting opportunity was presented to participate in the broader work of variable star astronomy. On the night of March 12, 1912, a self-taught astronomer, Sigurd Enebo, discovered a nova in the constellation of Gemini from his private observatory near Dombaas, Norway.[2] Enebo reported the nova to the Keil University Observatory in Schleswig-Holstein, Germany. That observatory immediately cabled the news to the astronomical world so that resources could be redirected to observe what was now being called Nova Geminorum (2). Harvard immediately began observing the nova, as well as searching its photographic plates prior to the discovery. A March 10 plate provided no evidence of a star in the reported position. However, a March 11 plate showed the nova at approximately fifth magnitude.[3]

The speed with which the discovery was reported allowed astronomers to study the nova before it reached its maximum, a visual magnitude of 3.4 sometime on March 14 before starting its decline. HCO astronomer O. C. Wendell performed photometric observations of stars around the nova to form a sequence suitable for magnitude estimates during the nova's decline, but most professional observatories followed the rise and decline of the nova spectroscopically.[4] For example, John A. Parkhurst at Yerkes Observatory reported changes in the absorption spectral class of the nova from that of a normal F5 star near maximum light to the emission spectrum typical for a nova a few days after maximum. This was probably the first time such early spectral changes had been recorded for any nova.[5]

Although he was excited by the opportunity to observe this new object, Olcott lacked charts for doing so. He drafted a chart and submitted it to Pickering for review and verification of the magnitudes of comparison stars. Pickering returned the corrected chart to Olcott bearing Wendell's March 18 photometry. He allowed Olcott to reproduce copies for AAVSO observers and begin development of a light curve for the nova.[6] Observations of Enebo's nova by three AAVSO observers appeared in the March AAVSO report in *Popular Astronomy*. The AAVSO observations captured the sudden brightening of the nova by nearly a magnitude on March 22 before its continual decline.[7] Thus the AAVSO entered into the observation of transient events at a very early date.

Frederick C. Leonard diligently observed the nova on every clear night and reported his observations with equal fervor in *Popular Astronomy* in the next few issues. Leonard presented his results in both text and tables. His reports display an interesting approach to the problem. He determined that his contribution would be to report the magnitude at each observation, but also to comment in a coherent way on the color changes he observed in the nova over a period of time. As the only observer to do so, Leonard struggled with characterizing the subtle color differences he thought he detected. There is no mention of the SPA in Leonard's articles, as he apparently sought the spotlight for himself in this effort.[8]

Figure 4.1. Giovanni B. Lacchini of Faenza, Italy. With the AAVSO, almost from its inception, he would contribute more than 54 000 variable star observations over his lifetime.

GAINING CONTROL

While Olcott kept up with correspondence and wrote articles summarizing variable star work for the SPA's *Monthly Register* throughout 1912, he also recruited new members to the AAVSO and reported the new organization's observations in *Popular Astronomy's* monthly reports. The reports served their intended purpose well, as letters flowed in from all over the world. The sudden increase in inquiries about variable star observing caused Pickering and Olcott to formalize a process for handling them. Olcott served as the AAVSO's main correspondent, whereas Pickering and HCO sent a form letter and a package of information that included preliminary instructions and sequences for stars to observe, observing forms, and HCO circulars.[9] With the administrative kinks quickly worked out, the process soon functioned very smoothly. Giovanni Battista Lacchini (1884–1967) of Italy was one of the earliest international observers in the AAVSO, having transferred his allegiance from Leonard's SPA on Olcott's first notice. However, it is also clear that some of the overseas inquiries were from individuals who found the subject interesting and wanted to form their own association: Cooke in Western Australia, for example, and Raurich in Barcelona, Spain.[10]

To provide observing materials to new members, Olcott was frantically tracing Harvard's sequence charts, producing more than 6000 such charts for the AAVSO membership. Others realized the magnitude of the effort and began assisting with the creation of these charts. Although Olcott continued to copy charts by hand for each variable star, requiring an enormous investment of time and energy, Edward Gray, M.D., and Frederick Leonard began using blueprint charts in 1912, which could be easily reproduced in quantity.

At first, Pickering was cool to the blueprint charts, but he soon warmed up to the idea, suggesting "if Harvard magnitudes are entered on the blue-prints, I see no reason why they should not be used to advantage."[11] His early objections may have had more to do with Gray using the comparison star magnitudes from the first three series of Hagen's *Atlas*. Pickering had published tables to convert Hagen series I through III magnitudes to the Harvard photometric scale, and he also provided Hagen with magnitudes that were printed directly in the catalogs of Hagen's series IV, V, and VI – all of which Pickering found acceptable.

In a late 1912 letter to Olcott, Pickering acknowledged that the AAVSO observers were making progress. "The variable star observers seem to be accumulating a large amount of useful material," he wrote. "The

Figure 4.2. Edward Gray, M.D., with his granddaughter Alice, September 1912.

principal difficulty now seems to be that too many observations are obtained of some stars and too few of others." Olcott reacted sharply to this implied criticism and added a note to the monthly report of AAVSO observations in *Popular Astronomy*, suggesting that observers broaden their programs and observe more stars. He promised to publish a list of under-observed stars and began badgering Pickering for a list of variable stars needing more observations so that the AAVSO star repertoire could be expanded. Eventually Pickering produced a list and provided Olcott with the material for preparing charts.

Olcott finally had to admit that he could not support activities in both the AAVSO and the SPA. As founder of the AAVSO, his stake was obviously much higher in that organization than in SPA, and in 1913 he resigned from the SPA. In a January 1913 *Popular Astronomy* announcement, it was noted that Dr. Edward Gray, of Eldridge, California, had been appointed director of the Variable Star Section of the SPA to succeed W. T. Olcott, of Norwich, Connecticut, who resigned on account of the heavy work connected with his position as corresponding secretary of the American Association of Variable Star Observers. Furthermore, the notice stated that "Mr. Olcott will continue in this latter position, in which he is doing most efficient service, as shown by the great increase in the number of variable star observations which have been published in this department of Popular Astronomy during the past year."[12]

Throughout 1912, there were two AAVSO organizational strains that must have affected Olcott as he immersed himself in variable star observing and reporting. The heavy workload involved in this dual relationship with HCO imposed a burden on Olcott's usually sunny personality. In his letters to Pickering, Olcott's sensitivity to his position can be seen in such statements as: "I hope you approve of the work I am doing, and of my report in 'Popular Astronomy' as I desire to do nothing in this line that is not in accordance with your wishes"; "I hope my efforts to interest telescopists in variable stars is not causing you annoyance"; and "Do my reports on variable work in 'Popular Astronomy' and my endeavors to secure observers for the work meet with your approval?" Similar sentiments appear in the closing remarks of most of his letters in late 1911 and early 1912.[13] Pickering never responded directly to these questions and comments.

EXPANDING THE AAVSO PROGRAM

Just as Olcott terminated his SPA obligations, the situation in the AAVSO began to improve measurably. Assistance in drafting the blueprint charts was offered by a new member in 1913, Harry C. Bancroft, Jr. (1885–1935?), an architect living in Collingswood, New Jersey. With Olcott acting as a liaison with Pickering, Bancroft redrafted all of Olcott's rough sketches of variable stars and introduced a standard format for chart information. The new chart footers acknowledged both Olcott and Bancroft (with a stylized Bancroft monogram) as authors of the charts.

Bancroft's skills as a draftsman soon emerged in the new charts. His star diameters were uniformly circular and crisp, and the labeling of stars and header lettering were much more legible. Bancroft's charts became the standard and were prized by observers as the blueprint charts emerged as a valued AAVSO observing tool for the next 60 years.[14] Bancroft's skills and dedication to the chart-making program made it possible to promote the under-observed stars and upgrade the basic chart issues.

Bancroft brought yet another benefit to the program – lower-cost charts. Up to this point, Gray supplied the blueprint charts with his expenses reimbursed by Pickering. As an architect, Bancroft had excellent relationships with a print shop and arranged for a large supply of charts to be printed all at one time at very low cost. Charts purchased through Bancroft cost only 1.5 cents per copy in comparison to Gray's 3.5 cents per copy on the West Coast. The HCO charts were apparently even more expensive.

The expense factor shifted the center of chart work from Gray on the West Coast to Bancroft on the East Coast. Pickering provided Olcott with a budget for the charts from his observatory funds. When Olcott proposed that AAVSO charge a nominal fee for the charts, Pickering agreed readily. Harvard University, it seems, was nervous about their tax exemption even in those years before "Town and Gown" disputes became common in Cambridge. With the fledgling Massachusetts Institute of Technology planning to move across the Charles River to Cambridge, Harvard President Lowell fretted that the added burden of a fourth institution of higher learning could trigger a tax revolt. Thus any suggestion that the HCO might profit from the sale of merchandise raised concerns about a possible loss of property tax exemption and other implications quite apart from the effect of the expense-balancing revenue for the department involved. As a result, the AAVSO ended up in greater control of its destiny through complete control of its most important observing assets, the blueprint charts.

Pickering began to suggest improvements in the observing program as the wrinkles in the chart-making program and the new member administration were smoothed out. He observed that "many stars are receiving more observations than are needed" and suggested some alternatives. Short-period variables could be observed more frequently than long-period variables, and Algol variables were suggested as possible extensions of the observing program. Olcott expressed concern that more frequent observing sessions for the shorter period variables might not fit the plans of many observers. Also, he questioned why Algol variables might be of astrophysical interest.

These questions were discussed when Olcott arrived in Cambridge for a late November 1913 visit. The outcome of that discussion was never recorded in a letter. There was not, however, a marked change in the AAVSO observing program at this point, as neither short-period variables nor Algol variables were added to the program. Instead, the program was expanded to include more long-period variable stars that were under-observed and/or were at fainter magnitudes.[15]

BOOK PUBLICATIONS

Olcott remained a busy man in this period for other reasons, which can be seen in several events in 1914. Olcott's fourth book published by Putnam was *Sun Lore of All Ages: A Collection of Myths and Legends Concerning the Sun and its Worship*. Clearly modeled on *Star Lore of All Ages*, its successful predecessor, *Sun Lore* is a richly illustrated exploration of the extensive mythology associated with the Sun in many different cultures. Five of its 30 photographs were provided by Leon Campbell from Peru, the night assistant who oriented Olcott to variable star observing in 1910. Shortly after his visit to Norwich, Campbell and his family traveled to Arequipa, Peru, where he took up duties as the astronomer-in-charge of the HCO's Boyden Station.

Other aspects of Olcott's personal history are also reflected in his book, *Sun Lore of All Ages,* with short explorations of the presence and meaning of the Sun in the rituals of Freemasonry and the symbolism of social fraternity escutcheons. The publication of his fourth full-length book was no doubt a cause for celebrations in the Olcott family.

Olcott also prized time at the eyepiece of his telescope. His variable star observing expanded as HCO added more and more long-period and irregular variable stars to their program. Olcott's northern horizon was limited by a hill behind his house, so Pickering expanded his program to include more southerly variables.

As Olcott's confidence rose with the situation, he began to press Pickering for more material with which to make charts for the under-observed longer period variables. Clearly with AAVSO matters under control, Olcott was anxious to extend the association's and his own observing repertoire. His increased time at the eyepiece seemed to have paid off on December 17, 1914, when he notified HCO of discovering a possible nova. Pickering's letter to him the following day advised that he had, instead, discovered the minor planet Vesta. Pickering thanked Olcott for his prompt reporting of the observation and offered suggestions for avoiding such false alarms in the future. Clearly the opportunity to observe Nova Geminorum (2) before maximum had alerted everyone to the benefits of early notice.[16]

NEW STELLAR THEORY

Cepheid variables are a group of variable stars that have been observed for some time. Professional astronomers later came to consider them as we will see in this chapter, so amateurs have included them in their AAVSO observing programs. In December 1914, Mount Wilson Solar Observatory astronomer Harlow Shapley published a new argument in favor of the theory that the light variations of the Cepheid variable stars were caused by radial pulsations of the bodies of these stars. For several decades, a debate had raged over the cause of the variations, and most astronomers held that they were due to mutual eclipses in a pair of stars orbiting a common center of gravity. A smaller number of astronomers held that they were more likely caused by pulsations of a single star. The evidence seemed ambiguous, and there was no quantitative model, but observations of the Doppler shifts in spectral lines seemed to be consistent with the binary hypothesis. Shapley showed that new evidence adduced by Henry Norris Russell at Princeton and Ejnar Hertzsprung in Denmark proved that these stars were very luminous and therefore very large. In fact, they had to be so large that the idea of binary stars was ruled out. The relatively small orbits inferred from the velocities and periods of motion actually put the two stars inside each other. Thus the binary hypothesis led to an absurdity, and the pulsation theory seemed the only logical alternative. Shapley's conclusion, eventually supported by both theoretical and observational results, led to an important clarification of the causes of these types of variable stars.

A less happy event occurred far from Norwich that would have enormous consequences for Tyler and Clara Olcott. Although Norwich often seemed isolated from the rest of the world, the assassination of the Archduke Franz Ferdinand in Serbia plunged most of Europe into war. Seeking revenge, the Austrian army invaded Serbia, affording Germany the excuse to attack France, and all of Europe was engaged in the disastrous World War I. Although the response in the United States was at first limited, Olcott volunteered for the Connecticut Home Guard and drilled as a corporal in the Norwich battalion.

Thus observing variable stars and leading the AAVSO placed constant demands on Olcott's time. He had exhibited extraordinary commitment to the AAVSO with his routine correspondence, ledgering observations received each month from AAVSO members, and preparing those observations for publication each month in *Popular Astronomy*. Still, he also remained fully engaged in many other aspects of a busy life in Norwich.

A MEASURE OF SUCCESS

When Olcott resigned as the director of the SPA variable star section, Edward Gray agreed to replace him, but that arrangement lasted less than a year. Gray may have realized that the AAVSO would succeed because of its adult leadership and an independent means of publishing observations, neither of which had been secured for the SPA by Leonard. On the other hand, Leonard had refused to give up and became the director of the SPA variable star section himself.[17] In spite of his efforts, the SPA Variable Star Section was effectively defunct; by 1914 all of its members, including Leonard himself, had joined Olcott's AAVSO and were submitting observations to him. Thus Olcott relieved his dual allegiance by resigning from SPA, but the consequences of his resignation went far beyond relieving his own strain.

The same could not be said, however, about Olcott's difficult relationship with Pickering, which never realized the same relief. Although never acknowledged explicitly by either individual, the publication of the AAVSO observations in *Popular Astronomy* in spite of Pickering's strong resistance had created a strained relationship.

Pickering's coolness, evident in his letters at this time, must have concerned Olcott. Pickering's letters were crisp and to the point; he kept Olcott and the AAVSO, as well as the SPA, at arm's length, providing advice and technical support for chart development and not much else for the next 4 years.

Table 4.1. *A comparison of the total numbers of variable star observations made by AAVSO and HCO observers*

Tabulated year	Annual totals AAVSO	HCO Corps
1912–1913	13 199	2892
1913–1914	15 128	2420
Total by organization	28 327	5322
Total observations	33 769	
Percent by organization	84	16

Toward the end of 1914 – after the second full year of AAVSO operation – the facts were clear when the variable star observations for the year were totaled up at HCO. The tabulations, found in the Harvard Archives collection of papers from the Pickering era at HCO, report the annual observing contributions from each observer.[18] Table 4.1, extracted from these tabulations, shows the amazing progress made by AAVSO observers in comparison with the HCO Corps of Observers.

Pickering never conceded, however, that the HCO Corps of Observers was redundant and a diminishing resource to the observatory. In 1913, he continued to work enthusiastically with Gray on the blueprint charts, finally conceding to Olcott that the "new blue print charts are excellent" and that he planned to distribute them quite freely.[19] In something of an acknowledgment of the AAVSO's success, Pickering suggested that the Corps of Observers had shifted their priorities.

The larger telescopes at Harvard (Seagraves), Amherst (Todd), and Vassar (Furness) were concentrated on long-period variables that were at minimum brightness.[20] By the end of 1914, Pickering's problems in sustaining the Corps of Observers were exacerbated by the departure of his two top observers. David Todd stopped submitting observations from Amherst in June 1914. He was immersed in controversy over his proposed schemes for radio communication with life on Mars, and perhaps in the early stages of the mental disorders that eventually led to his institutionalization.[21] Todd never resumed submitting variable star observations. Another Corps stalwart, Frank Evans Seagraves, became embroiled in a bitter lawsuit with Harvard in which he demanded compensation for the variable star observations he had already submitted. Seagraves could no longer be considered an effective contributor to the Harvard program.[22]

PICKERING'S EARLY SUPPORT

Correspondence between Olcott and Pickering in the 4 years following the AAVSO founding in 1911 suggests that Pickering and his staff dealt with Olcott cautiously during that time. That Pickering would hold Olcott and his AAVSO at arm's length during this period is entirely understandable in several contexts. One factor contributing to this distancing may have been the early confusion over the publishing of AAVSO observations by *Popular Astronomy*, previously mentioned. Given the strength of Pickering's assertion that variable star data should only be published after assembly and analysis at Harvard, he may have felt Wilson violated an understanding in supporting Olcott's desire to publicize the new organization. If Pickering saw this as a mild threat to the integrity of the HCO variable star program, it would be natural for him to take a "wait and see" attitude toward the new organization.

A second reason for the apparent distancing recognizes that Pickering maintained relationships with three variable star groups that had separate and competing demands for HCO resources. At least in the early years, this distinction may have been important to Pickering because of his desire to maintain the HCO Corps of Observers. In his letters, Pickering kept Olcott informed about the SPA and also referred separately to the efforts of the HCO Corps of Observers (Seagraves, Swartz, Todd, Young, Furness, and others).[23]

Because the AAVSO was Olcott's separate effort, it was up to him to bring it to maturity with only that technical support that Pickering and the HCO provided to all three groups. Until the SPA failed, and its corps of observers disintegrated, Olcott's AAVSO was just one of several resources, all of which demanded technical support from Pickering and the HCO staff.

Most importantly, however, it is unlikely that Pickering had time to devote to the fledgling AAVSO. From a scientific perspective, Pickering was engaged in numerous projects, the results of which were to become standards in their respective fields. For example, the Harvard systems of photographic magnitudes and spectral classification of stars were accepted as international standards in 1913.[24]

The massive *Henry Draper Catalogue of Stellar Spectral Classifications*, which was initiated under Pickering's direction by Annie Jump Cannon in 1911, would surely have involved almost daily discussion with the director. Publication of the *Draper Catalogue* was not completed until several years after Pickering's death.[25]

Through this work, and his cooperation with international projects during this period – for example, photometry for Kapteyn's Plan of Selected Areas – Pickering established the HCO as one of the preeminent astrophysical observatories of the era. Furthermore, from 1909 until his death in 1919, Pickering actively led the professional astronomical society in the United States. As its president for 10 years, Pickering was involved in every aspect of the Society's affairs. His early years as president included a contentious debate that resulted in changing the name of the Society from the Astronomical and Astrophysical Society of America to the American Astronomical Society (AAS) in 1914.[26] At the peak of his power professionally during this entire period, and particularly from 1913, Pickering dispensed funding and other forms of patronage to the astronomical community in America.[27] The substantial scientific and professional workload described above was added to the routine administration and fundraising for the observatory and its Boyden Station in Arequipa, Peru, and maintenance of relations with Harvard University. It is not difficult to believe that Pickering would have found it impossible to spend much time on Olcott and his variable star observers.[28]

CLOSING THE GAP

In his letters to Olcott, Pickering frequently referred to the AAVSO as "your observers." Although Pickering's letters dealt nearly exclusively with charts, sequences, and related technical matters, he occasionally encouraged Olcott by complimenting him on his effort with the AAVSO. In a 1913 letter, for example, Pickering commented, "The variable star association, largely owing to your supervision, appears to be doing its work in a very satisfactory manner."[29]

In his annual report for the observatory, Pickering was more guarded. In the 1912 annual report published in 1913, he stated, "The organization for the observation of variable stars has greatly increased during the last year." He then tabulated the results in terms of charts prepared, variable stars observed, and the number of observations from each observer detailed by name and location, without regard to whether the observer's results were from AAVSO members, SPA members, or members of the HCO observing corps. He closed this section of his report with the following statement:

> The important aims are to secure observations, at short intervals, of the principal variables of long period, and to obtain useful results from large numbers of owners of small telescopes whose work, otherwise, might be of but little value. Both of these conditions seem now to be fulfilled in a highly satisfactory manner.[30]

Pickering mentioned Olcott's observations, but not the name of the association; the reader was left to understand that the "organization" that Pickering mentioned is a creature of HCO in some vague way.

In April 1914, Olcott and some of his AAVSO colleagues began holding informal meetings in New York City and had assembled there on three occasions by mid 1915.[31] Pickering rejected, without specifying any reason, an invitation to meet with AAVSO members at one of these sessions, though as outlined above, his busy schedule at HCO would surely have precluded travel to New York for this sole purpose. For their fourth meeting in the fall of 1915, the informal group agreed to meet in Boston to attend a Harvard–Yale football game and discuss AAVSO matters over dinner. Olcott again made an effort to draw Pickering a bit closer to the AAVSO by inviting him to participate in their Boston dinner meeting. In response, Pickering invited the group to come to the observatory for a tour.[32]

At the last minute, Olcott nervously inquired whether or not the football game would be played. Olcott's question was not without merit, as Harvard football had become a bit controversial. Pickering replied that he knew only what the newspaper reported, wooden stands were being constructed at the stadium, and it appeared the game would proceed. The variable star observers enjoyed the game, and the following day assembled at the observatory for a tour.[33]

After the observatory tour, Pickering and Campbell joined the group for the evening dinner at the Copley Square Hotel in Boston and were quite impressed with the variable star observers. Olcott had obviously invested a lot of effort in making the dinner as memorable as possible. He even prepared a special menu for

Figure 4.3. The AAVSO's first look at Harvard College Observatory, November 1915. (Top row, from left) Alan B. Burbeck, David B. Pickering, Solon I. Bailey, William T. Olcott. (Middle row, from left) Forrest H. Spinney, Charles Y. McAteer, Edward C. Pickering (HCO Director), George F. Nolte. (Bottom row, from left) Rev. Tilton C. H. Bouton, Leon Campbell, J. L. Stewart.

the dinner presented in the form of a blueprint variable star chart with North at the top and the lines of declination and right ascension centered on the location in Boston. The food served in the fixed menu was all characterized in pseudo variable star language. For example, the broiled halibut steak was identified as 012502 OK Piscium, whereas the beef plate was identified as OU Lardi Tauri. The chart authority was identified as "From H.C.O Chart by W.T.O." All participants in the meeting signed a picture postcard – a picture of HCO and its domes. Dated November 21, 1915, and mailed to H. C. Bancroft, Jr., at his home in Collingswood, New Jersey, the postcard confirms that the participants in this meeting included Olcott, David B. Pickering, S. C. Hunter, C. Y. McAteer, G. F. Nolte, Leon Campbell, James L. Stewart, A. B. Burbeck, and E. C. Pickering.

Pickering's first exposure to the members of the AAVSO elicited a remarkably positive response. His thank you letter to Olcott radiates a new enthusiasm for Olcott's association. In his letter Pickering mentioned the possibility of an AAVSO affiliation with the AAS and even suggested that some of the AAVSO members could be nominated for membership in AAS if they wished. He further suggested that Harvard publications related to variable star work could be made available to AAVSO members if Olcott thought it worthwhile.[34]

As late as 1916 though, Pickering still did not refer to the American Association of Variable Star Observers by its title. In his correspondence with Olcott, he addressed the group as "your observers" or "your Association." If at first it was not evident to Pickering that the AAVSO would succeed, by late 1916 it must have been abundantly clear that Olcott and "his observers" were solid contributors to variable star astronomy. At a meeting of the association in the fall of 1916, discussions among the members made the observant Pickering aware that Harvard's financial assistance to the fledgling association might be appreciated. Pickering asked Olcott for a summary of the organization's finances. In response to Olcott's report, Pickering proffered Harvard's first financial aid to Olcott's operation.[35] It was in this late 1916 letter advising Olcott of that decision that Pickering used the title American Association of Variable Star Observers for the first time.[36] This new support, the first support not related directly to the technical operation of the association, occurred 5 years after Olcott founded the AAVSO.

By the end of 1915, the remainder of the SPA Variable Star Section, with the exception of Leonard, had resigned from the SPA and joined Olcott in the AAVSO. In his studies of the later career of Frederick C. Leonard, Smithsonian Curator Emeritus Roy S. Clarke, Jr., noted that in the years after the SPA collapse, Leonard seemed to reflect a distinct bitterness about the AAVSO. For example, Leonard refused to allow Henry W. Vrooman to dispose of the residual funds in the SPA treasury by donating them to the AAVSO for the Pickering Memorial Fund. This somewhat sour outlook was evident in letters Leonard wrote to both Lick Observatory Director W. W. Campbell and Harvard College Observatory Director Harlow Shapley.[37]

STEADY AS SHE GOES?

Never enjoying robust health, Olcott nevertheless maintained his busy schedule, as his social scrapbook showed in earlier years. He and Clara traveled to Hartford, New Haven, and occasionally to New York City for theater and other forms of entertainment. In early 1916, a respiratory illness that was to plague Olcott in later years forced him into a period of inactivity. Although it was not long-lasting, the illness was serious enough that Olcott commented about it to correspondents. In a letter asking Olcott to endorse some membership applications for the American Astronomical Society, E. C. Pickering inquired: "I hear with regret that you have a cough and are going to a milder climate." Pickering suggested that the Olcotts visit his brother William Henry Pickering, who maintained a comparatively well-equipped observatory in Mandeville, Jamaica. Olcott replied quickly that he was well enough recovered from a mild attack of grippe to travel to Florida in a few weeks but had not resumed observing. He appreciated Pickering's suggestion though and wished some day to travel to Mandeville and meet Pickering's younger brother.[38]

The monthly reports in *Popular Astronomy* had exactly the effect that Olcott and *PA* editor Wilson had originally intended. The AAVSO and the observations its members reported grew slowly but continuously. By the end of 1916, when the membership met again informally at HCO, the cumulative total AAVSO variable star observations had grown to nearly 70 000. Olcott seemed well within his capacity to organize and report the growing numbers in the *Popular Astronomy* each month.

In its November 1916 meeting, AAVSO members discussed the various problems faced by observers with respect to charts, telescopes, and the weather. These issues concerned the participants, who had little opportunity to discuss such matters outside these informal AAVSO gatherings. The question of the accuracy of visual observations provoked those present to test the matter, and 19 observers gathered at a 12-inch telescope to each privately make estimates of the brightness of a variable star. The results were later reported by the HCO Committee of Visitors (COV) in their Annual Report to the Harvard College Board of Overseers.

In this report, for the first time the AAVSO had been mentioned in the COV annual reports, the committee reported that the magnitudes estimated by the 19 observers varied "by less than one-tenth of a magnitude – an experiment altogether unprecedented." committee chairman Joel H. Metcalf, himself a distinguished amateur astronomer, appreciated the

Figure 4.4. AAVSO 1917 Spring Meeting at the home of David B. Pickering, East Orange, New Jersey. Front row, from left: Alan B. Burbeck, William T. Olcott, Charles Y. McAteer; back row: David B. Pickering, Helen M. Swartz, Forest H. Spinney, Lillian Pickering, Silas Wright Pickering, Leon Campbell. Olcott holds a loving cup presented to him by an appreciative membership for his service to the AAVSO.

significance of this demonstration of both the visual acuity of the volunteer observers involved and of the implication of a combined result that deviated over so narrow a range for the longer range utility of AAVSO observations.[39]

Some members attending the November 1916 meeting must have felt the organization had reached another point of maturity. In May 1917, David B. Pickering, a manufacturing jeweler, invited members to meet at his home in East Orange, New Jersey. At the meeting, there was a substantial discussion of how the AAVSO should move toward a formal organization. Leon Campbell reported that the meeting "accomplished much" in this regard. In addition, the assembled group had presented a cup to Olcott in recognition of his services to the organization and as a testimony of the membership's regard for him.[40]

WORLD WAR I REACHES AMERICA

In this same time period, other events complicated Olcott's life and service to the AAVSO. Early in 1917, Germany extended its aggression to the United States. Congress declared war and the United States began formal mobilization to join the Allies in Europe. By May, Congress had enacted the Selective Service Act and 4000 local exemption boards were formed across the United States. In total, 23 million men were registered for the draft, and slightly more than 10 percent of those were actually inducted. In Norwich,

Connecticut, Olcott was selected to serve as the secretary of the local Exemption Board. The title "Exemption Board" is deceptive because the Board was in charge of registering all eligible males, issuing draft cards (numbered by hand in red ink) to each eligible individual, and administering the lottery selection process. The Board also conducted hearings for those who sought exemptions on various grounds, including religious conscientious objection.

Olcott was occupied full time in the work of conducting the draft lottery, publicizing the results, handling applications for exemption, and dealing with those who failed to comply, all spelled out with rigid Federal schedules for compliance.[41] It must have interfered with his ability to handle the equally tedious tasks of ledgering variable star observations and preparing monthly reports for publication in *Popular Astronomy*. Clara was not available to help with these astronomical tasks since she had volunteered full time as chairwoman of the American Red Cross Volunteers. Olcott needed help from other observers. Campbell, who was already seeking volunteers to assist with the calculation of 10-day means for HCO, agreed to help ledger observations and prepare reports.[42]

AAVSO'S INCORPORATION

By July 1917, Olcott had ordered letterhead stationery displaying the now-familiar AAVSO seal and drafted a constitution and by-laws.[43] At the November meeting in Cambridge, the association adopted the constitution and by-laws and took the first steps to incorporate the AAVSO. Officers elected at that important meeting included David B. Pickering as president; Harry C. Bancroft, Jr., vice president, Olcott, secretary, and Allen B. Burbeck, treasurer.[44]

One interesting business item at the meeting was the appointment of a committee to develop a plan to improve the productivity of the organization.[45] At an uneventful first formal Council meeting, the members mainly honored the astronomers they regarded as the leaders in variable star astronomy, namely Edward C. Pickering, Fr. Johann Georg Hagen, and John A. Parkhurst.[46] In these as well as in both earlier and later minutes by Olcott, his legal training is evident in that the minutes are sanitized of any discussion that might be misunderstood or misinterpreted and limited for the most part to actions actually taken by the meeting.

The decision to incorporate in Massachusetts rather than Olcott's home state of Connecticut was likely influenced by the warm relationship developing between AAVSO members and HCO, but also partly by the proximity of many active AAVSO members, including Allan B. Burbeck, Ida E. Woods, William H. Reardon, and Michael J. Jordan, in addition to Campbell and Pickering. The task of incorporating the organization was led by Jordan, a Cambridge Justice of the Peace. On October 31, 1918, this group assembled at HCO with HCO staff members Solon I. Bailey and Dorothy Block of the HCO staff joining the group to meet the minimum number of incorporators.

When all were assembled, Jordan explained the process, and then Campbell was chosen to chair the organizational meeting. Ida Woods was sworn in by Jordan as the first AAVSO clerk to take the meeting minutes. The by-laws were adopted, after which the necessary officers for incorporation purposes were elected, with Campbell as president, Jordan as vice president, Burbeck as treasurer, and Woods as clerk. Following the by-laws, a council was then elected and included Reardon, Pickering, Bailey, and Block. The incorporation council adopted the resolution that all Association members would become members of the corporation, having been enrolled on a list of members by the clerk. The list included not only these officers and council members, but also others present at the November 1917 meeting when the decision was made to incorporate. At that earlier meeting, those signing the incorporation certificate included Annie Jump Cannon, Anne Sewell Young, William Tyler Olcott, David B. Pickering, Stephen C. Hunter, and other AAVSO regulars who could not be present for the formal incorporation meeting. The incorporators then signed the articles of incorporation, which Jordan subsequently presented to the Secretary of the Commonwealth of Massachusetts, together with the Certificate of Incorporation for official recognition and sealing.[47]

The next Annual meeting of the newly incorporated AAVSO celebrated the incorporation of the organization on November 23, 1918, at HCO. It was necessary for the incorporation officers to resign so that a new permanent set of officers could be elected. At the general membership meeting, several committees – already engaged in some work – reported on their efforts: Leon Campbell, chair of a telescope committee, a position he held for many years, reported that the

American Association for the Advancement of Science (AAAS) had granted funds to the AAVSO to purchase a telescope. The money ($300) had been used to purchase a 5-inch refractor, which had immediately been loaned to Charles Y. McAteer of Pittsburgh. On a recommendation from Solon I. Bailey, a committee was appointed to acquire, organize a catalog, and oversee distribution of a lending library of lantern slides about astronomy. As an informal AAVSO historian, Henry Schulmaier reported on his efforts to gather biographical information and pictures of all AAVSO members.[48] Campbell reported for the Plan of Work Committee since David Pickering was chairing the meeting and had originally suggested the Plan of Work idea. The Plan, which was read and approved without comment, proposed the formation of a Chart Committee to prepare standardized variable star observing charts and handle their distribution. The original chart committee included David B. Pickering as chairman, John J. Crane, and J. Ernest G. Yalden.[49] Several additional elections to honorary membership and patron status were approved, and the organization began to function formally with its second election. Both Harry C. Bancroft, Jr., and Steven Crasco Hunter were nominated for the presidency. The second set of elected officers included Bancroft as president, Charles Y. McAteer, vice president, Burbeck, treasurer, and Olcott, secretary.[50]

Strangely though, Bancroft dropped out of sight at that point. In addition to the meeting minutes, a blank ballot from the election indicated the candidates and a letter from Solon I. Bailey addressing Bancroft as "Dear Mr. President" are the only evidence that Bancroft was elected. He apparently did not attend any meetings the following year and was not in any way a presence in the AAVSO.

The Association apparently suffered little from Bancroft's absence with Olcott active as secretary, Burbeck managing as treasurer, and Campbell watching out informally for the AAVSO on the technical side from his vantage point at HCO. In a February 1925 letter to Campbell, Bancroft revealed he had been swamped with work, family problems, and other matters that had precluded any active role as an observer for a number of years, but he was dedicated to returning to the observation of variable stars. The overload mentioned by Bancroft likely accounts for his disappearance from the AAVSO after his election as president in 1918.

By 1919, the continuing AAVSO workload for Olcott had also become too great a burden and seriously effected his health, already compromised by the demands of the wartime Exemption Board activities and previous illness. Campbell and the HCO staff stepped in to temporarily handle the monthly observation reports to *Popular Astronomy*. Various individuals, including Ida Woods, Arville D. Walker, and Florence Cushman of the HCO staff, signed the monthly *Popular Astronomy* reports. Woods was identified as "acting secretary," but Walker and Cushman were identified as "recording secretary." It appears from member correspondence that monthly observations reports were addressed to Campbell at HCO.

GROWING CLOSER TOGETHER

By late 1918, Campbell had become an important figure in the AAVSO. He played a significant role in its affairs after 1915, but little has been said about him. Therefore, at this important juncture, we depart from a purely chronological march through the AAVSO's history to recover Campbell's background before continuing beyond the end of 1918. It will suffice to state that in the 3 years after the first informal AAVSO dinner in Boston in 1915, Pickering and Harvard College Observatory drew much closer to Olcott and the AAVSO through Campbell's voluntary efforts.

Part III

RECORDING AND CLASSIFICATION – THE LEON CAMPBELL ERA

5 · Leon Campbell to the rescue

Late in 1909 along came Olcott (better known as William Tyler) . . . And what a find he proved to be! How well I recall that first trip to Norwich when I attempted to impart to him some of my own practical ideas on variable star observations with a small telescope.

– Leon Campbell, 1930, reflecting on the history of the AAVSO after 20 years[1]

William Tyler Olcott founded and nurtured the American Association of Variable Star Observers (AAVSO), seeing its membership and observations grow in its first 6 years. But as the United States became involved in World War I, his appointment to the Norwich, Connecticut, Exemption Board created additional demands on his time and energies that compromised his weak health condition.

Other members of the Association recognized that it would no longer be possible to operate so informally in view of Olcott's physical condition. They began taking steps toward a more formal organization, a plan they accelerated as Olcott's need for help became ever clearer. In addition to the help he needed in restructuring the work of the Association in some more formal way, Olcott needed help handling the increasing numbers of observations from observers around the world and preparing monthly reports for publication in *Popular Astronomy*.

After 1916, Harvard College Observatory astronomer Leon Campbell began to provide some of that assistance in an informal way by organizing and ledgering AAVSO observations at his Cambridge, Massachusetts, home. This chapter provides background on Campbell's life prior to 1919 and his special background, which made him eminently eligible to assume an ever greater role in the AAVSO.

EARLY LIFE IN CAMBRIDGE, MASSACHUSETTS

Leon Campbell was the son of immigrant parents. His Irish father, William James Campbell (1857–1911), also a Boston native, established a separate residence in Cambridge in about 1877. Employment opportunities as well as inexpensive quarters in boarding houses on the Cambridge side of the Charles River attracted the immigrants. The 1880 census dramatically reflected this flow; about half of the Cambridge population of 52 669 described themselves as natives of Ireland or of Irish descent.[2]

For an Irish immigrant such as William James Campbell, Cambridge offered many different opportunities, especially highly skilled work for mechanics in its fabric mills, book printing and binding factories, glass factories, and in firms like the Alvan Clark and Sons telescope manufacturing shops.[3] After attempting employment as a soap maker and a clerk, William Campbell landed a job in 1880 at the Mason & Hamlin factory. The firm, which had been manufacturing organs since 1854 in Cambridge, had a successful national business exploiting Hamlin's discovery of a unique technique of voicing reeds, the key element of tone creation in the cabinet organ. Initially, William Campbell worked as a mechanic, blacksmith, and then a piano maker, since the firm added pianos to their line in 1882. Eventually he rose to foreman of the reed department, after mastering the technique of making and voicing the reeds.

William Campbell married Leonora Rawding from Nova Scotia, Canada, on May 12, 1880. She gave birth prematurely to twin boys, Leon and William Ellory Campbell, on January 20, 1881, and died the next day. Additional tragedy stuck when William Ellory died of diphtheria in 1884. Leon survived to adulthood, but his small physical stature may have remained as testimony to the difficult circumstances of his birth. In August 1882, William Campbell married May Lillian Towle, who gave birth to three more children and cared for Leon. After several moves, the family settled in the northern section of Cambridgeport at 66 Tremont Street, the address from which Leon attended Cambridge High School.[4]

In his senior year, he studied astronomy with Miss Esther Dodge, a university graduate and science teacher, who found a 4-inch refractor in a high school closet. After it was refurbished by Alvan Clark and Sons, Leon viewed celestial objects and especially enjoyed tracking sunspots across the solar disk. But the family's meager resources meant that Leon sought employment after high school in June 1898 and became a grocery store clerk.[5]

But his lifetime opportunities changed when Professor Oliver C. Wendell of the Harvard College Observatory (HCO) visited Miss Dodge seeking an assistant to record data at the HCO 15-inch Equatorial refractor used for astronomical photometry. Wendell had various observing assignments, including the timing of eclipses of binary stars, changes in brightness of Jupiter's satellites when eclipsed by the planet's shadow, and precisely measuring variations in the brightness of Saturn's satellites, minor planets, and variable stars.[6] Miss Dodge unhesitatingly recommended Campbell. After interviews with Professor Wendell and Edward Charles Pickering, the Observatory director, Campbell was hired.[7]

EMPLOYMENT AT HARVARD COLLEGE OBSERVATORY

Beginning in January 3, 1899, Campbell assisted Wendell by recording data while the astronomer observed at the eyepiece. A conscientious observer, Wendell waited at the observatory for clear moments, no matter how stormy the night. Campbell recorded instrumental settings, the exact times of observation, and the sky conditions at the time of the observation using a desk in the observatory library that eventually became his desk. When Campbell's immediate attention was not required during long periods, such as an eclipse of one of Jupiter's satellites, Campbell learned and practiced visual astronomy from the observatory roof adjoining the Great Refractor dome.[8]

During 1899, Campbell made his first variable star observations from the observatory roof, at first by naked eye following maxima of o Ceti (Mira) and minima of β Lyrae. He soon supplemented these observations using field glasses and then the observatory's 5-inch portable refractor. Campbell became an accomplished observer before the end of his first year and mastered the art of estimating the brightness of variable stars using the Argelander step method and reducing the observations graphically.[9]

An anonymous gift to the observatory made it possible to procure a 24-inch reflecting telescope in 1903.[10] By the time the telescope was in service in late 1904, Wendell and Pickering were so pleased with young Campbell's enthusiasm and accurate work as Wendell's dome assistant after 5 years that they placed him in charge of the new 24-inch telescope. The change reflected high praise for a young astronomer who had no formal training in astronomy. With the new telescope, Campbell's work involved, almost exclusively, observing a growing list of variable stars and developing light curves showing their changes in brightness over time. For this work, he also developed the sequences of comparison stars needed to make visual observations of these stars, especially those near minimum in brightness, again using the Argelander step method.[11]

With the seeming employment stability, Campbell made a major change in his life. He proposed to Frederica J. Thompson of Columbia, Connecticut, and they married on June 15, 1905 in Columbia. Leon moved from his father's Cambridgeport residence to Old Cambridge, much closer to Observatory Hill. The couple's first child, Leon Campbell, Jr., was born in August 1906, and was joined by a sister, Florence, in January 1908, and brother Malcom in January 1909.[12]

Perhaps due to his responsibilities with his growing family, Campbell worked very hard as an observational astronomer. Additionally, he worked with the ledgering and discussion of the variable star observations of the Harvard Corps of Observers and eventually became the HCO variable star astronomer.[13] In 1909 Pickering dispatched Campbell to Norwich, Connecticut, to instruct Olcott on finding variable stars and estimating their brightness. The visit marked the beginning of a life-long friendship, as Campbell noted that Olcott in his first observation of Mira "displayed an interest in the work rarely found among amateurs."[14]

ENTERPRISE IN PERU

With characteristic vision, Pickering recognized that his standards for measuring the brightness and spectral characteristics of stars should be extended uniformly to the stars in the southern skies. With funding from the Uriah A. Boyden endowment, Pickering surveyed three regions before settling on southwestern Peru.[15] Solon I. Bailey, Pickering's survey team leader in Peru, selected a station at a slightly greater than 8000-foot elevation, 3 miles northwest of Arequipa, and demonstrated the value of the new site with photometric and photographic surveys of the southern skies.[16] Then, Pickering's brother, W. H. Pickering (1858–1938), and a second team completed permanent physical facilities for the Boyden Station at Arequipa. Bailey returned to establish stable and highly productive astronomical operations from 1893 to 1905.[17]

A succession of station supervisors after Bailey continued the operation with varying degrees of success until 1911, when Pickering sent Campbell with his wife and 3 children to replace the station supervisor. The young family traveled by ship from New York City through the Panama Canal to the port of Mollendo, Peru, and then by rail to Arequipa, arriving at the Boyden Station on May 15, 1911.[18]

By mid-September, Campbell expressed his concern about the weather and other conditions, such as the decline in the production of photographic plates, the principal product of the Arequipa operation. The three problems that plagued Campbell most were the shortage of manpower to operate cameras and develop plates, unfavorable weather at night, and the condition of the equipment itself, which limited the number of plates that might be taken. In addition, Campbell became increasingly aware of the logistical, communications, and especially the financial difficulties associated with the long distance to Cambridge. After Campbell appealed for more manpower several times, Pickering finally revealed the economic issues facing the observatory. The Boyden Station had to be operated with far fewer operators than previously available, and Bailey had convinced Pickering that Campbell could maintain its productivity at the lower staffing levels.[19]

By the time Campbell arrived, the Boyden station had been in operation for 20 years of fairly continuous and demanding service with equipment that was, in most cases, not new when it arrived in Peru. Its first priority was to produce reasonable quality photographic plates for Pickering's astrophysical "factory" in Cambridge. Compromises were necessary, and temporary solutions were frequently improvised to keep equipment in service. During cloudy or rainy weather, the staff repaired the equipment to the best of their abilities with tools available to the station. Performance problems with the Bache camera, in particular, were severe enough that Campbell suggested shipping it back to Cambridge for a complete overhaul in 1912. Instead, Pickering suggested a series of detailed remedies, apparently receiving suggestions from Bailey and others.

Not all astronomical work at Boyden Station was a part of a frustrating routine. For example, Campbell observed Nova Geminorum 1912, several eclipsing binary stars discovered on Arequipa plates, and discovered a comet. If these excursions from the routine pleased Campbell, one wonders if he was enthusiastic about Pickering's other projects, a campaign to photograph the 1912 opposition of the minor planet Eros and a photographic search for a new planet in July 1914 with rather vague instructions on what was needed.[20]

On April 1, 1914, Frederica gave birth to their fourth child, Ruth Evelyn Campbell. Living with a baby and three other small children (ages 8, 6, and 5) under the crowded circumstances of the Boyden Station's staff residence created problems for the family and exacerbated tensions, already high, among the resident staff. At the same time, the Peruvian government's instability created additional logistical and financial problems for the Boyden Station operation. Photographic plate production, Pickering's ultimate measure of the success of the operations, sagged to the lowest level in more than 20 years of operation. By the end of 1914, Pickering had decided to change supervision of the station, a change that Campbell welcomed. However, by the time Campbell packed up his family and turned the entire operation over to his replacement on July 11, 1915, he did so with mixed feelings. He was relieved to leave an unpleasant situation at Boyden Station and, at the same time, regretted leaving the variable star work to southern hemisphere observers.

Pickering measured the success of Arequipa by the total number of photographic plates produced as raw material for his computers in Cambridge. On that basis, Campbell's tenure as the Boyden Station supervisor (1911–1915) could not be judged a huge success. But there were conditions mitigating Campbell's production. Before his arrival, there were four assistants to help with the equipment and operations, but during his tenure, there were only two, and for several periods, he was alone at the station. For example, Frank E. Hinkley left before February 1, 1912, whereas his replacement, Harold I. Peckham, arrived May 17, 1912. Thus Campbell was alone for more than 4 months. In addition, both Peckham and Campbell had extended periods of illness, and there was no one to pick up the slack in those periods. It is unclear whether Pickering considered these circumstances because his letters never showed any sympathy for the young assistant he had placed in such difficult circumstances.

After deciding Campbell should return to Cambridge, Pickering adopted a slightly more conciliatory language for the first time in several years. It can't be known how he assuaged Campbell's bruised ego in their first conversations after the young astronomer arrived back at HCO. It would not be surprising if Campbell left Boyden Station with a different view of himself and his potential than when he first arrived there. Subsequent developments indicate that when he returned home he was determined to specialize in variable star astronomy and draw closer to the AAVSO.

Campbell's tenure in Peru interrupted his emergence as the variable star astronomer at HCO. He had hoped to do more variable star work at Arequipa to reinforce that specialization. His regrets over opportunities to observe the southern variable stars may have been mollified by other developments. For example, with Pickering's support on his return, he continued to ledger the observations of the HCO Corps of Observers. In connection with that work, Campbell eventually convinced Pickering to consolidate the Corps' observations with those from the AAVSO observers that Olcott was reporting separately in *Popular Astronomy*.[21] His recommendation may have seemed reasonable, as the AAVSO observers' totals quickly and substantially surpassed those of the Corps' observations (see Chapter 4, Table 4.1). Campbell also communicated with Olcott, receiving news of the Association's activities.

JOINING THE AAVSO

When the Campbell family returned to Cambridge, they rented a home much closer to HCO than previously at 11 Bellevue Avenue in the Old Cambridge district north and west of the Harvard Campus. They

Figure 5.1. Harvard College Observatory's Building A (right foreground). On the first floor to the right of the stairs was Leon Campbell's office and AAVSO Headquarters.

re-established relations with their families, the church parish, and related organizations after a 4-year absence. A fifth child, Eleanor Beatrice Campbell, was born in 1916, completing the family but putting further strains on their requirements.

When Campbell returned to HCO, he resumed the photometric work that had been dormant since Professor Oliver C. Wendell died in 1912, less than a year after Campbell departed for Peru. Pickering had taken care not to replace Wendell, even writing to *Popular Astronomy* editor H. C. Wilson to correct the journal's misstatement about the matter, obviously holding the position for Campbell on his return.[22] Campbell's primary observing program was photometry with the 15-inch Great Refractor, including variable stars, light curves of minor planets, and eclipses of Jupiter's satellites. As had been the case with Wendell, Campbell settled into this routine, quickly mastering the large refractor and re-establishing his role as the variable star astronomer at Harvard. It is clear, though, that after his experience at Arequipa, his long-term interest focused more strongly on variable star astronomy.

After his return, Campbell joined the AAVSO and attended his first meeting with Pickering in the fall of 1915 in Cambridge (Chapter 4). Campbell's role at HCO was sufficiently ambiguous as to promote his communication with AAVSO observers as both an HCO staff representative and an AAVSO member. The Association welcomed Campbell's participation, as might be recognized by the fact that David B. Pickering (no relation to Edward C. Pickering) invited Campbell to participate in the first AAVSO Spring Meeting in 1916 at his home in East Orange, New Jersey.

As HCO's variable star astronomer, Campbell discussed the variable star observations. To do so, Campbell had to interpret a staggering amount of accumulated data. First, there were observations available not only from the Harvard Corps of Observers, but also those of the AAVSO and the British Astronomical Association Variable Star Section (BAA-VSS). The BAA-VSS had gathered data for 20 years, over 10 years based on Harvard photometry of comparison stars.[23] Second, Campbell conceived the utility of preparing light curves for long-period variables using 10-day means of the available data rather than individual observations. He recognized that volunteers could calculate those means, simplifying his workload and giving him more time for understanding the light curves and working on a classification system.[24] Several volunteers assisted, including Henry C. Vrooman of Indiana, Forrest H. Spinney of New Hampshire, and Charles Brook of the BAA-VSS.[25] The 10-day means for individual stars were undertaken by John J. Crane for R Coronae Borealis and by Ida E. Woods for R Cygni and the variable stars in Ursae Majoris. The volunteers calculated these means for about 50 stars, enabling Campbell to propose the publication of a book that eventually emerged as his *Studies of Long Period Variables*.[26]

In late 1916, Campbell exchanged letters with several members on other topics as well. For example, Campbell proposed a program to evaluate the variability of red sensitivity in the night vision of AAVSO observers with Vrooman, Spinney, Bancroft, and perhaps others.[27] This was not an HCO program, and no formal AAVSO organization existed to initiate such technical investigations. Using his home address on his Cambridge letters, it appears that this initiative was his own interest with the long-term goal of correcting the observations of certain observers who routinely overestimated the brightness of red stars. Although Campbell was able to confirm that, with a correct procedure, the variability of these stars in red light was substantially less than that observed in white or unfiltered light, the program was not rigorous enough or long enough to draw valid scientific conclusions about the variation of observer sensitivity to red light. The program was soon dropped, but the issue did not go away, and there were renewed efforts to study the problem in later years.[28]

These two programs, the 10-day mean light curves of long-period variables and the red stars variability studies, allowed Campbell to open direct correspondence with AAVSO observers and helped solidify not only their identification of him as Harvard's variable star astronomer, but also the feeling that they were, in some vague way, a part of Harvard College Observatory.

Campbell's growing correspondence with AAVSO members focused on their problems with charts used at the telescope for making observations. The available charts had severe problems in some cases, and in other cases, members were eager to expand their observing program as HCO announced the discovery of new variable stars. With the problem charts, at least, the solution seemed to be to acquire a complete set of the Hagen charts, although those were very expensive. J. Ernest G. Yalden solved that problem by buying a complete but lightly used set from a London bookseller, where he typically purchased mathematics texts not available in the United States. The Hagen charts were ideal for use in variable star fields and for identifying the variable star itself, but suffered the deficiency – in at least the Hagen Series I, II, and III – that the comparison star magnitudes were not taken from the "Revised Harvard Photometry," which was not available when Hagen started his project. Campbell agreed to add the correct magnitudes to Yalden's charts as a favor.[29]

Instrumental measurement emerged as a recurring theme in AAVSO observers' questions to Campbell. An early AAVSO observer, Howard O. Eaton, began submitting visual observations in 1914 with a 6-inch refractor. He asked Campbell for advice on using a photometer in 1916 for viewing variable stars. In responding, Campbell assured him that his photometer was identical to Harvard's photometer R that had been used for some years, provided references to the *Harvard Annals*, reminded him that a clock drive was needed for a telescope during photometric observing, and made several other useful comments about photometry in general.[30] Whether Eaton was successful with his photometer is not known, but he reported observations for 7 years and was active in AAVSO leadership. Campbell also encouraged Forrest H. Spinney in the use of a photometer with Spinney's telescope, also emphasizing the need for a robust clock drive. Instrumental measurement of variable star brightness was an accepted practice among professional astronomers by 1916, although a variety of techniques were applied.

Figure 5.2. AAVSO Annual Meeting, November 1918. Left to right: Leon Campbell, William H. Reardon, Dorothy W. Block, L. Brewster, F. L. Ducharme, William T. Olcott, Annie J. Cannon, Edward C. Pickering, Anne S. Young, Henrietta S. Leavitt, David. B. Pickering, Michael J. Jordan, Ida E. Woods, C. T. Whitehorn, Alan B. Burbeck, Mrs. Gould, V. Francis, E. L. Gould, Stephen C. Hunter, Solon I. Bailey.

DEATH OF PICKERING

As a participant in the November 1918 AAVSO meeting, the first after incorporation, Edward C. Pickering no doubt relished the maturation of Olcott's association. The organization was functioning with an established leadership and was accomplishing valuable work in a normal and productive manner. Unfortunately, it was the last meeting Edward C. Pickering would ever attend. Two months later, he died on February 3, 1919, in Cambridge. There are no minutes for the 1919 Spring Meeting, once again at David B. Pickering's residence in East Orange, New Jersey. However, dialogue about a Pickering memorial had begun. Before the meeting, Campbell requested that Anne S. Young, who had known Pickering as well as any of the members, take an active role in the matter. He suggested a memorial edition, "to be printed uniformly with the Presidential Address of Mr. D. B. Pickering, be issued, which will contain the biography of the late Director, with special reference to his activities in the line of variable stars and of his close association with the A. A. V. S. O."

Young accepted Campbell's nomination but immediately wrote to Olcott and requested his help, which he eventually provided. Her description to Campbell of her efforts to involve Olcott in this regard is a bit ambiguous:

> I am glad you agree with me that Mr. Olcott should still be expected to do his part. I cannot believe he will fail. Of course I will do all I can to help. I wrote him suggestions right after the meeting, and perhaps he thought I did not mean to urge him to do it. I surely hope he will.[31]

To prepare the Pickering memorial discussed in Campbell's letter to Anne Young, a committee was appointed May 1919, including Young, Annie Jump Cannon, Charles Edward Barns, and Olcott. Whether or not Olcott, Young, and Cannon met while those three were together at the meeting in East Orange is unclear. However, Barns' location in California made it difficult for him to participate in such an activity. Already deeply involved in printing various bulletins and other materials for the AAVSO, Barns' appointment to the committee reflected his desire to print the memorial booklet.

CAMPBELL TAKES CHARGE

Other than the reading of the Pickering Memorial Committee's report, Olcott's minutes for the November 1919 annual meeting are strangely silent about Pickering's passing. The committee report (November 8, 1919) signed by Cannon, Young, and Barns identified Edward Charles Pickering as "a founder and the first Patron and Honorary Member of our Association." The report continued:

> His unfailing confidence in our ability to do work of value has been a constant inspiration. His interest in our individual efforts has added a touch of personal friendship, ennobling us by contact with a master mind, and enriching our lives by recollections of a delightful companionship. Generous with his time and wealth of knowledge, he placed the recourses [sic] of the Harvard Observatory at our disposal. We shall always cherish his memory in loving gratitude, while we realize that the best and most fitting Memorial will be a continuation of the work in which he was so deeply interested.[32]

Although he was present at the meeting, it is remarkable that Olcott did not sign the report. Instead, his minutes state simply that the Committee's report was accepted with no further discussion and the report was neither incorporated in the minutes, nor adopted as a resolution by the Council. At the same meeting, with C. Y. McAteer presiding, Leon Campbell was elected president, replacing Bancroft, who had ceased being an active member after his election to the presidency at the 1918 Annual meeting.

After his election as the third AAVSO president, Campbell asked members several questions, such as "What can we do for you?" and "What can you do for us?" This survey may have been discussed at the 1919 Spring Meeting at David B. Pickering's residence, but the fact that Campbell initiated it so rapidly suggests that a vacuum had developed in AAVSO leadership with Bancroft's failure to serve the presidency. The results of Campbell's survey were never presented to the Council or membership and were not discussed in any way in the available historical record.[33]

Having served as the informal chairman of a Telescope Committee, Campbell quickly formalized that role by appointing Anne S. Young to replace him as the chair of the Telescope Committee. The AAVSO soon acquired an impressive inventory of telescopes to loan to observers as members upgraded from the entry-level telescopes (typically a 3-inch refractor) to larger instruments that could follow long-period variables through their minima. A recurring theme in the earliest available member correspondence reflects their disappointment in not being able to observe stars at minimum with their first telescopes. Many donated their smaller telescopes to the AAVSO, such as McAteer, who had obtained a 5-inch American Academy of Arts and Sciences telescope.

When Young later replaced Campbell as President, he resumed his previous role as Telescope Committee chairman. At that time, Campbell documented the association's holdings in a letter to Young.[34] His list reflects the association's ownership of 21 telescopes, of which half were 3-inch or smaller refractors. Two additional telescopes were scheduled to be donated to the AAVSO, including a 6-inch Clark refractor from Princeton University. Decisions on where to place instruments were carefully discussed between the Telescope Committee chair and the president. In the case of the Princeton instrument, not yet received but shown in Campbell's inventory, Campbell had already made a recommendation to send this to an aspiring young observer, Leslie Peltier, who had observed very productively with a much smaller telescope from a family farm near Delphos, Ohio, since 1918.

It seems important to discuss at this point the strangely anomalous position Campbell then found himself in with respect to his roles at HCO and within the AAVSO. For example, although he was no longer the AAVSO Telescope Committee chair, Campbell still fielded inquiries from aspiring young astronomers in this period, some that would emerge as lifetime relationships. In May 1920, for example, at Solon I. Bailey's request, Campbell responded to a letter from Donald H. Menzel of Denver, Colorado, explaining in general terms how variable star observers worked. He referred the boy to the work of Henry Norris Russell, Harlow Shapley, Theodore Phillips, and Herbert Turner, who did visual or photometric work that young Menzel could "take up to advantage."[35] Menzel's reply indicated his desire to begin work immediately, without waiting for charts. By August, Menzel reported he had observed some eclipses of binary stars with the 6-inch Grubb refractor of the University of Denver.[36] Within a year, Menzel won an appointment to graduate study in astronomy at Princeton University. He attended his first AAVSO meeting, the 1922 Spring Meeting hosted by the Olcotts in Norwich, and then traveled on to Cambridge for research in the HCO plate stacks. During his stay in Cambridge, Menzel roomed with the Campbell family.[37]

THE PICKERING MEMORIAL

By May 1920, after 4 years in a row in which David B. Pickering hosted an informal spring meeting for members of AAVSO in his home in East Orange, New Jersey, the AAVSO had matured to the point that a formal spring meeting could be arranged. Held at the Uptown Club in New York City – David B. Pickering having made the arrangements – the activities of the meeting failed to inspire much in the way of minutes. David B. Pickering Nova Discovery Medals were presented to Ida E. Woods and Johanna C. S. Mackie. Formal papers were presented for the first time by Annie Jump Cannon, Caroline Furness, J. Ernest G. Yalden, Irvin C. Murray, Leon Campbell, Stephen C. Hunter, Charles W. Elmer, and David B. Pickering. None of this would likely have been possible in Pickering's home, but it seems likely, too, that the popularity of the spring meetings had grown such that the number of attendees simply overwhelmed the space available there. Again, nothing is said in the minutes for this assembly about the Pickering memorial booklet project.

At the 1920 Annual Meeting, there is again no indication in the minutes of any report from the Pickering Memorial Committee or approval of the proposed booklet text. Instead, it appears that the Pickering Memorial Committee simply agreed upon the text and format. Barns printed and assembled the booklets, which represented the AAVSO in a very attractive manner. The title page indicates the document was "Published by the Association: William Tyler Olcott, Secy., Norwich, Conn." The frontispiece is a portrait of Pickering seated in his academic regalia. On the back of the title page, additional information that bears discussion includes the fact that the memorial was "Authorized by Council at the Annual Meeting of the Association held at Harvard College Observatory Cambridge, Mass. Nov. 8, 1919," with a listing of the committee members. This inscription is interesting in that, although the preparation of the memorial was perhaps authorized, it is unclear from Olcott's minutes that the booklet text was even available to the 1919 Annual Council or general membership meetings for approval. As noted above, the minutes record only that a committee report was accepted.

The text of the memorial booklet merits further discussion. After a paragraph of complimentary introductory remarks about Pickering, the Memorial asserts:

> At its annual meeting at Harvard College Observatory on November 8, 1919, the following resolution was passed by the Council and endorsed by the Association:
>
> The members of the American Association of Variable Star Observers record with deep sorrow the death on February 3, 1919, of Edward Charles Pickering, a founder, and the first Patron and Honorary Member of our Association.

What is remarkable here is that no such resolution was ever recorded in the minutes of the association for these meetings, which were written by Olcott himself. With his attention to the legal niceties involved, Olcott would surely have recognized the significance of such a resolution and recorded it in the minutes of whichever meeting was charged with approving the resolution.

The memorial booklet goes on for another three and one-half pages of text that might be described as a scientific biography of Pickering, with special emphasis on his contributions to variable star astronomy and the AASVO. This is followed without interruption by the following text:

> Our Secretary, Mr. Olcott, has written the following appreciation of Professor Pickering's connection with the Association.
>
> In addition to the many great contributions to Astronomical Science made by the late Professor Edward C. Pickering, it is of particular interest for us to recall that he was the first to place before the amateur astronomers of this country the

> opportunity afforded them for accomplishing valuable scientific work in the observation of Variable Stars. Through the medium of an appeal published by him in "Popular Astronomy" in 1901, he called for volunteers to cooperate in this valuable line of observational work, thus carrying out the great Argelander's wish expressed at the conclusion of his wonderful career.
>
> It was but natural therefore that Professor Pickering should have taken a great and personal interest in the action that was to result in the formation of our Association.

Olcott is very careful to not ascribe the status of founder to Pickering in these words or in what follows in his writing. Instead, he heaps praise on Pickering for his unstinting technical support and encouragement in the earliest days and his eventual warm welcome of AAVSO observers into his home during meetings eventually held at HCO.

The memorial booklet goes on for another three pages of text, but it is unclear in all of this how much of the remainder should be ascribed to Olcott and how much to the rest of the committee. Although quotation marks are used carefully in most places in the text, they are not used to set off either the beginning or the end of Olcott's text. The implication is left that the remainder of the booklet was written by Olcott, including a quotation of David B. Pickering's speech in connection with presentation of a gift envelope opener to Edward Pickering only months before his death. The fact that the title page reflects Olcott's role as Secretary of the association also implies that he was personally responsible for the final version of the text for the entire booklet.

The actual text quoted above and ascribed to Olcott, in fact, sounds very much like his minority report on the matter of Pickering's status as a founder of the AAVSO. It is no stretch of the imagination to believe that Olcott's modesty prevented his active objection to the resolution presented to and apparently approved by the Council. That he might *not* have warmly endorsed the description of Pickering as "a founder," easily understood in light of available information to the contrary, is further supported by Campbell's stilted language in asserting that Olcott and David B. Pickering "fight shy of such a task" when he requested that Anne S. Young write the memorial.

THE OLD GUARD STEPS IN

Because the constitution and by-laws provided only for a secretary and mention neither a corresponding secretary nor a recording secretary, it is clear that the AAVSO Council had to improvise in order to formalize relief for the overloaded founder/secretary of the association. In another action taken at the November 1919 Council meeting, the role of secretary was divided, with Howard O. Eaton, a graduate student at the University of Wisconsin, assuming the role of recording secretary, while Olcott continued in his role as corresponding secretary. Others assisting with the work as Council members included Leon Campbell, who was helping as a volunteer on the work of the recording secretary, John J. Crane, Charles Y. McAteer, Helen Swartz, and Anne S. Young. All had known each other from previous meetings; it was an agreeable transition in leadership and promised 2 additional years of stability for the organization. The minutes for that 1919 Annual Council Meeting were recorded by Olcott, but the minutes of the AAVSO Council's 1920 Spring Meeting, recorded by Allan B. Burbeck, are very sketchy and do not record who was present; it seems likely that Olcott was not in attendance, confirming that Olcott was on an extended medical leave.

After the election of officers and the division of the secretary's job took place at the 1919 Annual Meeting, Olcott appears to have dropped out of sight. Clara had resigned from her post as the director of the Norwich Red Cross earlier in the year and no doubt sensed that Olcott had reached the limit of his physical endurance and insisted that he back away for a while. In any event, there is no evidence that he was contacted when a short-term crisis emerged just a few days after the 1920 Annual Meeting of AAVSO.

The minutes book reveals that AAVSO Treasurer Allan Burbeck died unexpectedly on November 29 at his home in North Abington, Massachusetts. On December 23, a Council quorum consisting of David Pickering, J. Ernest G. Yalden, and Charles Whitehorne met in the offices of the Harriman Bank in New York City, with Yalden taking the minutes. Resolutions were adopted memorializing Burbeck and electing Michael J. Jordan as the new treasurer. This would insure that the AAVSO retained full control over its bank accounts. This quick action by these Old Guard members was essential in the event and demonstrated that the association had reached a new plateau in maturity.[38] Interestingly, at this truncated Council meeting, a resolution was also adopted to memorialize Dr. Edward Gray, who died suddenly while waiting for a trolley in Oakland, California.[39]

With the death of Edward C. Pickering, the AAVSO and Campbell had both lost the support of a strong friend. Campbell no doubt took some solace in the fact that the memorial booklet would serve as a reminder to history of what he imagined Pickering's role to be in the founding of the AAVSO, though Campbell was in Peru at the time. After hearing the news of Burbeck's sudden death and the actions taken by the New York contingent of the Old Guard to safeguard the AAVSO's funds, Campbell no doubt reflected on the uncertainty that he faced as president of the AAVSO.

A list of the risks that faced the fragile new organization Campbell now led was neither short nor insubstantial. Several aspects of Olcott's personal situation remained uncertain; Campbell could not reasonably expect that Olcott's future support to the AAVSO would remain the same. For example, health issues clouded the status of the founder and previously stalwart supporter. There were, as outlined above, the questions that had arisen on the early history of the organization that detracted from Olcott's unique position in that history. These issues made it unlikely that Campbell could count on as much support from Olcott as had previously been available. Reliance on volunteer officers like Harry C. Bancroft, Jr., to lead the organization created yet another set of issues that must have been sobering to Campbell. Then too, with the first death of a member of the Old Guard, Treasurer Allan Burbeck, Campbell must have considered the fact that he (Campbell) was the youngest of that distinguished group and that their continued support could not last forever.

Finally, Campbell must have reflected on his own personal situation with respect to his employment at HCO. Having returned from Peru under a cloud as far as his performance at the Boyden Station was concerned, Campbell dedicated himself to variable star astronomy, both as an observatory employee and as a leader in the AAVSO. Solon Bailey was surely also committed to variable star astronomy, but his status as acting director was of limited consolation. Who would be chosen as the new HCO director? Variable star astronomy had not gained a lot of traction, and observational astronomy seemed headed toward distant galaxies. Campbell's own concerns thus placed him in an anomalous position, although his dedication to both the HCO and AAVSO at this point were obvious. As Campbell faced his second year in office as president of the AAVSO, the risks for both him and for the fledgling organization were large. With many other questions in the air, there was much for Campbell to ponder.

6 · Formalizing relationships

Systematic visual observations of variable stars have been continued by Mr. Campbell, who also supervised the observations received from the members of the American Association of Variable Star Observers, and others in foreign countries.

– Solon I. Bailey, 1921, on Leon Campbell's role at Harvard College Observatory[1]

A few years after the end of World War I, the status of the AAVSO changed from a small gentleman's club of amateur astronomers loosely organized around the observation of variable stars to a formal organization of both professional and amateur astronomers, maintaining some vague connections with the Harvard College Observatory (HCO).

During the last years of Edward C. Pickering's tenure as the HCO director, the AAVSO gradually emerged with its own identity in his annual reports. Pickering's 1911 report, as might be expected, makes no mention of the AAVSO's founding, as the reporting year ended before William Tyler Olcott's October announcement. The report does recognize, however, that Olcott had undertaken the effort to organize such an association. The following year, Pickering only notes that "the organization of variable star observing has greatly increased during the last year," an indirect reference at best to Olcott's new association.

What is obvious is that the AAVSO had become a formidable organization; its observers were increasing their observations at a phenomenal rate. In the 1913 report, Pickering noted that observations had increased from 10 117 the previous year to 16 830 observations in 1913. "The great increase," he wrote, "is mainly due to the growth of the American Association of Variable Star Observers under the careful supervision of Mr. Olcott." This is the first attribution to the AAVSO by name in these annual reports. Similarly, in 1914, Pickering recognizes "The organization for the observation of variable stars, under the efficient supervision of Mr. W. T. Olcott" and similarly in the 1915 Annual Report.

Over three decades, variable stars had grown in importance in the HCO program. The annual reports reflect both the change in the importance of variable star observations to HCO and the evolution of the AAVSO's relationship with HCO.[2] This chapter and the next trace the intertwined and ambiguous relationship between the two organizations, attempt to clarify Leon Campbell's role at HCO and in the AAVSO during the Association's maturation in the 1920s, and finally explore his relations with the younger Harlow Shapley, the new HCO director after Pickering's death in 1919.

EVOLVING HCO INVOLVEMENT

An interesting change took place in the 1916 HCO Annual Report: Pickering again referred to the Variable Star Section of the British Astronomical Association (BAA-VSS), whose observations "under the supervision of Mr. C. L. Brook" numbered 671 for the year and 10-day means of 28 long-period variable stars (LPVs) observed over the previous 5 years. Pickering accorded similar status to the AAVSO reporting "11,284 observations under the supervision of Mr. Olcott." For 1917, Pickering added, as separate cooperating organizations, the contributions of the Goodsell and McCormick observatories, in addition to the BAA-VSS and the AAVSO, again under the supervision of Messrs. Brooks and Olcott, respectively. The 1918 Report added southern hemisphere observers J. F. Skjellerup and the Cape Astronomical Society in South Africa, J. M. Baldwin at the Melbourne Observatory in Australia, and Bernard Dawson of the La Plata Observatory in Argentina. Until his death, Pickering clearly identified the AAVSO as a separate organization contributing to the variable star program at HCO.

The scope of Campbell's work at the Observatory may be seen in subsequent publications. He assembled and "discussed" variable star observations from all these sources, including the AAVSO, in addition to his own observing program. In the 1919 Annual Report, Solon I. Bailey, now the acting director upon Pickering's death, credits Campbell as having "supervised the observations received from the members of the American Association of Variable Star Observers, and others in foreign countries." In the 1920 Annual Report, Bailey noted that about 40 percent of Campbell's observing was on the minor planets and the eclipses of Jupiter's satellites. However, this report also names AAVSO members and others reporting, with all observations "supervised" by Campbell.[3] In 1920, 18 437 observations came from the AAVSO, as well as from observers in Australia and South Africa. Bailey's report for 1921 was similarly detailed, with Campbell's role described as being "in charge of this department, attending to the correspondence with Variable Star Observers and others outside the observatory," while also handling 24 018 observations from 83 observers, 40 of whom submitted more than 100 observations and were listed by name. In these 3 reports by Bailey, we see him identifying Campbell as the astronomer in charge of the HCO variable star program and handling the observatory correspondence with variable star observers outside the observatory, but there is still no clear appropriation of the AAVSO into the HCO organization.

With the apparent approval of HCO's acting director Solon I. Bailey, Campbell served as the AAVSO president from 1919 to 1922 and for two terms as a Council member. Thus Campbell maintained a presence in the AAVSO, even though it was not a formal part of his HCO role. His HCO assignments extended to many areas outside variable star astronomy, such as timing eclipses of Jupiter's satellites and producing photometric light curves for minor planets. He observed with both the 12-inch polar photometer and the 15-inch Great Refractor; his observational routines involved either visual or instrumental photometry, LPVs at minimum, other variable stars, and the development of comparison star sequences.

AAVSO'S SCIENTIFIC FUNCTIONS

AAVSO's activities were reorganized in 1917 when Olcott's workload was divided, and the work of organizing and reporting observations on a monthly basis was delegated to a recording secretary. There were, of course, several ongoing problems, including the management of the ever-increasing flow of observations on a monthly basis, the standardization and distribution of charts, availability of the Julian Day calendar, availability of telescopes, broadening the observing program, and instrumentalizing variable star observing.

Standardization and distribution of charts

In Campbell's 1919 informal survey of AAVSO members, one overwhelming concern was the "sorry" condition of the variable star charts used by observers. It was often difficult to resolve problems promptly because of the length of time for mail delivery, less than 3 days in the northeastern states and more internationally for an observer such as Radha Gobinda Chandra in Bagchar, Jessore, India.[4] The checking of problematic observations frequently involved a month or more of tedious and irritating delays in problem resolution. These problems were not just limited to the AAVSO. Chart improvements constituted a high priority for the American Astronomical Society's (AAS) Committee on Variable Stars and the International Astronomical Union's (IAU) Commission on Variable Stars; Campbell was appointed to both.[5]

The AAVSO Council appointed a committee in 1918 to work on standardizing and reproducing variable star charts. David B. Pickering agreed to chair this important committee, initially supported by John J. Crane, J. Ernest G. Yalden, and Olcott.[6] The workload grew in the 1920s with the need for more new charts and the problem of standardizing existing charts. New members were added to the committee, including Dean Potter, Arthur C. Perry, and finally, Dalmiro Francis Brocchi in 1922. The addition of Brocchi deserves special recognition. A railroad civil engineering draftsman by profession, Brocchi's carefully drawn charts, initialed DFB, were valued by observers for many years to come. He eventually replaced Pickering as the Committee chairman in 1934, serving until near his death in 1955. Among his accomplishments were a series of useful finder charts and the first AAVSO *Star Atlas*, which showed LPVs that reached naked eye brightness.

A continuing challenge for the Chart Committee, however, involved finding the best comparison stars and their magnitudes. Although Harvard provided their existing sequences, they were not always ideal, and

Figure 6.1. Dalmiro F. Brocchi of Seattle. He produced more than 800 charts during his 30 years with the AAVSO.

Figure 6.2. At Amherst College Observatory in 1929 with J. Ernest G. Yalden; Alice H. Farnsworth of Mt. Holyoke College; Warren K. Green, Observatory Director; and David B. Pickering.

Campbell actively sought assistance externally. Professor Warren K. Green volunteered to work on checking and improving sequences shortly after taking charge of the Amherst College Observatory. Green needed a photometer and placed an order for one shortly after first inquiring about Campbell's efforts to construct an inexpensive device.[7] In the meantime, Green had many problems reactivating the Observatory's 12-inch Clark refractor after too many years of neglect. Eventually, Green made measurements with a new photometer, but never achieved a productive level of support for the AAVSO chart sequence work.[8] Well intentioned though he was, the actual work at the necessary detail level exceeded Green's capacity to contribute after his regular teaching and other responsibilities.

Drafting the charts accurately and neatly – a major issue in itself – constituted only part of the Association's problems with charts. Another major issue centered on the distribution of the charts to observers all over the world. At first the Chart Committee chairman handled this distribution problem, but that soon became impractical. As awareness of the issue overwhelmed David Pickering, he appointed Arthur C. Perry, who handled distributions until his health failed. William F. H. Waterfield, who served temporarily, followed Perry. In 1928 Pickering appointed Helen Sawyer, a graduate student in astronomy at Harvard, to assist as Chart Curator, but she asked to be replaced a year later when she married another HCO astronomy graduate student, Canadian Frank Hogg. The Council then formalized the position of Chart Curator and appointed HCO astronomer Margaret Mayall to the position. This was Mayall's first formal involvement with the AAVSO, which she had supported informally as a member of the HCO staff.[9]

Julian Day calendars

Another important tool provided to observers by the AAVSO, in addition to charts, was an annual Julian Day calendar.[10] John H. Skaggs an Oakland, California, printer and variable star observer, originated the idea of printing Julian Day calendars for AAVSO members in 1921. He printed them for the year 1922 and sent them to Campbell to include in mailings to members along with a paper by Paul Merrill. Skaggs hoped the members found the calendar to be a useful Christmas present. As it turned out, they were hugely popular and remained a valued addition to the variable star observer's tool kit for the next 80 years.[11]

Availability of telescopes

During the Olcott era, most AAVSO observers used 3-inch refractors as a starting telescope. Before long, however, "aperture fever," a common ailment among astronomers over the history of the telescope, took its toll as many LPVs sank out of sight at minimum in such small instruments. Frequently, those upgrading to larger telescopes donated their old refractors to the AAVSO or HCO. Among those were several larger telescopes and a number of mirrors for reflecting telescopes.

AAVSO's inventory of small telescopes soon served as a source of loaners for those unable to purchase their own telescopes. Campbell established a telescope business, which he operated in addition to his full-time employment by HCO, likely to supplement his meager HCO income and to support a growing family. When HCO's telescope inventory accumulated, Campbell purchased the excess instruments and sold them as the opportunities arose. Thus, for a period of time, Campbell not only operated a small job-printing business on the side, but also a telescope resale business.

Broadening the AAVSO observing program

When Pickering first suggested that the AAVSO broaden its observing program in November 1913, Olcott rejected his suggestion. Perhaps Olcott appreciated the extant membership and the limitations of their small telescopes. The observation of either the eclipsing binary stars or shorter LPVs suggested by Pickering involved more frequent observation, more attention to small variations in brightness, and more attention to accurate recording of times for each observation. All of these new requirements were perhaps seen by Olcott as inconsistent with gentlemanly observing as a leisure activity.

When Campbell became AAVSO president, he had considerable experience observing eclipsing binaries (EBs) and shorter period variable stars. Although capable of broadening the observing agenda, he did not. When Shapley came upon the HCO scene he focused on both of these classifications.[12] Furthermore, if HCO was to maintain its position on the cutting edge of astronomy, it would seem natural to encourage the AAVSO to support the determination of distances and masses of stars through documentation of their light curves, periods, and apparent brightness. But unaccountably, AAVSO members had never pursued E. C. Pickering's suggestion and instead during the Shapley era extended their observing program in other directions.

The reports of AAVSO observations and other notes in *Popular Astronomy* attracted members to the organization. The monthly totals of observations – listed with the names of observers – and the number of observers reporting observations grew steadily over the years through World War I. Many capable astronomical observers joined the AAVSO's ranks; they were individuals committed to contributing to the advancement of astronomical science. As the reputation of AAVSO observers for conscientious and valuable observing efforts continued to grow, professional astronomers became interested in AAVSO observers as a resource for projects not related directly to variable star observing. The first of many such distractions came, not unexpectedly, from professional astronomer Charles Pollard Olivier, who recruited amateur astronomers as meteor observers in 1911 for the Society for Practical Astronomy (SPA). While serving as the SPA's director of meteor observation, Olivier also founded his own organization of meteor enthusiasts, the American Meteor Society (AMS). Olivier proposed that the AAVSO undertake a Telescopic Meteor reporting program. It was seen as a minimal distraction for an observer of a telescopic meteor to simply record the variable star under observation, the time of the observation, and the general path of the meteor through the field. He proposed that observers record this information on a separate form mailed to the AAVSO along with their monthly variable star observation reports. The AAVSO could consolidate those reports and mail them all to Olivier. The Council agreed to adopt the telescopic meteor program in November 1920 on behalf of the AMS.[13]

Another distraction, similar to observing telescopic meteors but demanding more member time and energy, was the lunar occultation program. On invitation from either Campbell – or, more likely, Shapley – Yale Professor Ernest W. Brown (1866–1938), an acknowledged authority in the field of celestial mechanics, encouraged AAVSO members to time lunar occultations at its October 1926 meeting at HCO. The logic, if any, was that lunar occultations could be observed at times when the light from the Moon interfered with variable star observing. Brown's proposal was tentatively accepted by the Council, so his letter to the AAVSO membership in 1927 in *Variable Comments* explained how the timings could be done with the more limited equipment that might be available for most amateur astronomers.[14] Several AAVSO observers, especially J. Ernest G. Yalden, took up Brown's challenge, and the AAVSO Occultation Committee was formed. AAVSO members eventually mastered the full range of occultation activities, from prediction to observation to reduction, all as a service to lunar theory, but not variable star astronomy. AAVSO observer's timing of lunar occultations was a valuable contribution to celestial mechanics and the improvement of the lunar equations.

Instrumentalizing variable star observing

Since the beginning of the eighteenth century, philosophers and scientists have increasingly favored the substitution of a mechanical or electromechanical device with some kind of numerical readout for the perception in a human brain of what is seen visually or felt through the tactile senses as providing for better science.[15] That idea manifested itself in variable star astronomy in two ways. The first technique was to insert a device in the path of the light gathered by the telescope so that a star of known brightness or an artificial star could be compared with an unknown object and then dimmed until the brightness of both were sensibly equal. The technique relies on the principle that the human eye is more capable of detecting very small differences in brightness qualitatively than in measuring them quantitatively. Thus matching the unknown star to one of known brightness or a standardized but variable light source allows the observer to "measure" the unknown star's brightness. That approach, known as comparison photometry, dominated stellar photometry practices at HCO during the Pickering era. Such apparatus could be standardized and operated to produce high volumes of fairly accurate data to a probable error of only 0.1 magnitude and was ideally suited to Charles Pickering's drive to produce large volumes of useful data.[16] A second technique operates on a different principle in that the light collected by the telescope is never seen or judged by the human eye and is instead passed directly into a photoelectric cell. This device converts light energy into an electrical signal – a voltage difference or a current – that can be measured with common laboratory meters.[17]

Joel Stebbins at the University of Illinois discovered ways to improve the sensitivity and reduce the noise levels of his selenium cells by 1910. Using a 12-inch telescope, he gained sufficient experience and confidence in his results to publish a detailed light curve for the eclipsing binary Algol.[18]

With that remarkable demonstration, researchers identified more suitable photosensitive materials than selenium, improved envelopes within which to contain the cell, and made many electrical measurement improvements. Astronomers elsewhere rapidly accepted and applied the new astrophysical technique of photoelectric photometry (PEP). That was not the case at HCO, where collection of basic data on many stars rather than study of single objects defined the mode of operation.

In his anomalous role, Campbell seemed to represent professional astronomy to many AAVSO members, who looked to him as a representative of HCO more than for any role in the AAVSO. Thus, although it involved technology not being practiced at HCO, Campbell became involved in one of the earliest amateur efforts in the United States to apply the selenium photoelectric cell to the measurement of variable star brightness. In 1919, Lewis J. Boss, the 21-year-old nephew of Dudley Observatory Director Lewis Boss, constructed and experimented with a selenium cell photometer. Using Frank E. Seagrave's 6-inch Clark refractor in Rhode Island, Boss applied his primitive photometer to making experimental brightness measurements on α Ori and α Her. He discussed his work in two articles in *Popular Astronomy* at that time, but his efforts apparently attracted little comment from professional astronomers.[19] In 1921 he joined the AAVSO, and in 1922, at Campbell's request, Boss sent him a copy of a paper he had delivered at an earlier AAVSO meeting. Boss described the telescope, selenium cell, and galvanometer setup and asked Campbell how to improve his technique so the results could be used for variable star observation. Boss' crude observational data consisted only of readings from the galvanometer; he apparently did not know how to convert them to stellar magnitudes. After consulting with a Harvard physics professor, Campbell responded with a method of calibration, a procedure for reducing the current readings to magnitudes, and suggestions for the use of a nearby star of known constant brightness as a comparison star, making alternate measures of the variable and the comparison star and then reducing the measurements to the actual magnitudes involved.[20] Later correspondence between Campbell and Boss did not mention photometry, but eventually Boss negotiated with Campbell about a mount for his photometer.[21]

Eventually Campbell undertook an effort to develop a comparison photometer that could be reproduced inexpensively for AAVSO observers. He solicited the help of several AAVSO members: J. Ernest G. Yalden, who prepared drawings according to Campbell's "design for a wedge photometer," and Morgan Cilley, a civil engineer returned from the Philippine Islands, who examined Campbell's photometer but preferred his own modified design that would be simpler and less expensive. Cilley's design could be easily manufactured from simple castings and commercially available parts.[22]

Campbell's wedge photometer used a neutral-density gradient filter that slid across one half of the telescopic field of view to extinguish the light of the brighter star. The gradient filter would be advanced with a micrometer head mechanism until the brighter star matched the brightness seen in the other half of the field without a filter. Each density gradient filter would be calibrated independently against stars of known brightness to understand the extinction of light along the filter.[23]

Anne S. Young expressed interest in the project, both as an active AAVSO leader and in teaching at the John Payson Williston Observatory at Mount Holyoke College. She was concerned that the sky background would affect the light level through the photometric wedge to the extent that it would make it difficult to match the two stars.[24]

Campbell then fabricated a photometer body and searched for a source of an inexpensive photometric wedge. Eastman Kodak suggested a neutral tint-dyed gelatine wedge they manufactured and used in their own photometry applications as an alternate to a photographically prepared wedge.[25] When Campbell tested one of the Kodak gelatin wedges, he found it required frequent recalibration because its density changed over a comparatively short period of time. Kodak finally conceded that photographically prepared wedges would be more stable. Campbell apparently lost interest in the project because there is no indication that any such device ever reached practical use in the hands of AAVSO observers.[26] Thus a second approach to instrumentalizing variable star observing was shelved.

In his anomalous role at HCO and within AAVSO, Campbell had neither the time nor the charter to effectively alter the dominant practice of visual observing. It is also true, however, that most AAVSO members remained gentlemen observers who enjoyed this approach to leisure and were not necessarily inclined to seek the next level of accuracy if their visual observations remained scientifically useful. The same notion probably applied for the small college professors, like Young, for whom the instructional value of the measurement was as important as its scientific value. Simplicity – for teaching purposes – was a necessity, not just a virtue.

MANAGING THE OBSERVATIONS

As Olcott's health failed at the end of World War I, a series of stopgap measures filled the need to organize observations for publication on a monthly basis in *Popular Astronomy*. At the 1919 Annual Meeting in Cambridge, the AAVSO Council acted to reduce the organizational strain and regularize the work by splitting the job of the Secretary into two parts, thereby redefining the Secretary's role. Olcott assumed the role of Corresponding Secretary, whereas Harvard graduate student Howard O. Eaton accepted responsibility for a new position of Recording Secretary and the work of assembling observations for monthly publication. In spite of this change, Olcott continued to be more absent than present, and Campbell shouldered Olcott's new role as well.

Campbell matured and grew into the more complex workload, which still included his work with HCO – both his own observing and data handling – and a huge and growing workload for the AAVSO as the number of observers and the monthly observations continued to grow. His influence in the Association as well as in variable star astronomy grew rapidly as he became known as the variable star expert at HCO, at that time the leading observatory in the field.

PROGRESS IN VARIABLE STAR ASTRONOMY

At the time of Edward Pickering's death, progress in theoretical variable star astronomy had been limited. Indeed, there was little one could identify as a recognized field of variable star astronomy other than Henrietta Leavitt's period-luminosity relationship for Cepheid variable stars and Harlow Shapley's research: (1) recognition in 1914 that the light variation in Cepheid variable stars could not be explained by eclipsing phenomena, and (2) calibration and use of Leavitt's relationship in defining the structure of the Milky Way Galaxy in 1918. Thus it is more than a little surprising that AAVSO members faithfully observed and reported data for what could only be considered a speculative field that had just begun to mature.

Henry Norris Russell recognized that this was the case in a 1919 address to the American Association for the Advancement of Science. He assembled his talk hastily as he wound down his war effort and returned to Princeton and teaching. In his talk, Russell focused exclusively on variable stars, identifying what was known, and what was not, and recognized the growing body of historical observations available to astronomers. His address, later published in *Science Magazine*, extolled the virtues of the data being collected by AAVSO observers and further urged effort on the part of theoretical astronomers to understand the data already available to them.[27]

That picture began to change gradually with progress in the theoretical understanding of stellar evolution. In two seminal papers in 1918 and 1919, A. S. Eddington began to elaborate on Shapley's 1914 deduction that pulsation of some sort was involved in Cepheid light variation. Eddington followed those papers with his publication in 1924 of the stellar mass-luminosity relationship, which paved the way for the realization that the evolutionary history of a star is a function of its mass, age, and initial composition, as described in his *Internal Constitution of the Stars* in 1926.[28]

Following this, and the purely theoretical work of Meghnad Saha, Edward Milne, and William Fowler, Cecilia H. Payne published her dissertation on *Stellar Atmospheres*, which began the process of combining laboratory, observational, and theoretical considerations in modeling the interior of actual stars. Although it initially misstated the abundance of hydrogen in stellar atmospheres, within 5 years the additional accumulation of observational data and analysis led to the publication of a corrected, rearranged, and greatly expanded version as Harvard Observatory Monograph No. 3, *The Stars of High Luminosity*.[29]

In contrast to the rather slow pace of theoretical development in variable star astronomy, observational results in the form of photometry and light curves had been appearing for more than half a century. By the time of Pickering's death in 1919, HCO had published three compilations of variable star observations in the *Harvard Annals* (*HA*), volumes 37, 57, and 63. Campbell's first publication of his own variable star work under Pickering's direction in 1918 appeared as *HA*, volume 79, part 1 (*HA* 79.1).

When he arrived at HCO, Shapley (like Pickering before him) recognized the substantial value of unpublished observations that were already archived, as well as those accumulating through the efforts of AAVSO, the BAA-VSS, the AFOEV, and several other sources. Determined to reinforce HCO's position as a center for variable star astronomy, Shapley immediately set about publishing the available observations in *HA*. By 1925, Campbell had his next contribution prepared for publication. Whereas *HA* 79.1 consisted only of a listing of the unpublished observations of 323 variable stars from 1911 to 1916, *HA* 79.2 reported times of the maxima and minima of 272 LPVs from 1900 to 1920. Campbell also qualitatively described the differences in the light curve shapes but offered no criterion on which a variable star was to be considered "long period."

Recognizing both Campbell's fundamental technical limitations and limited time, Shapley brought in Sidney D. Townley from Stanford University to lead the next analysis of variable star data. With Annie Jump Cannon and her spectral classifications of stars and Campbell's phenomenological descriptions, Townley assembled a considerably more complete catalog and discussion of what were for the first time defined with specified criteria as long-period variable stars (LPVs).[30] Townley's sources were the earlier catalogs of Karl E. G. Müller and Carl E. A. Hartwig, and those of Richard Prager, in addition to Harvard spectroscopy and photometry. His summary gives a good picture of variable star astronomy in 1926 – of the 1760 stars in the catalog, probably about 50 are not typical LPVs. Of the total, 59% have periods determined, 33% have known spectra, and approximately 27% have been sufficiently studied to permit the derivation of values of M-m, the interval from maximum to preceding minimum.[31] His "not typical" stars included U Geminorum, R Coronae Borealis, and RV Tauri type stars; apparently they fit the criteria of period and magnitude range but were otherwise poorly understood.

AAVSO'S SOCIAL AND ADMINISTRATIVE FUNCTIONS

For the first decade of the AAVSO's existence, the organization remained small, and for the most part, members knew each other from meetings, especially those who lived in the Northeastern United States. In that mode – letter writing with 1-day delivery common – interpersonal communications could accommodate most of the administrative needs of the association. After incorporation, a more formal structure and designated officers with clearer lines of authority and responsibility made it necessary to begin formalizing the necessary administrative functions of the AAVSO.

Member communications

Both the AAVSO leadership and the membership recognized that communication of events and news about the organization were missing elements critically needed as the organization grew. In 1922, the Council took a tentative step in this direction with two resolutions at their spring meeting. The first provided a modest expenditure from the AAVSO General Fund for additional type fonts for Campbell's basement "printery," whereas the second authorized the formation of a publications committee. Both could be connected to the development of some internal communication for which certain new fonts would lend an air of authority, as did the new Association seal.[32] West Coast observer John H. Skaggs offered to print a quarterly newsletter, suggesting the title *Variable Comments*.[33] When AAVSO President Young visited Lick Observatory in September 1923, Skaggs presented his idea to her and a few other AAVSO members. The publication was launched after Young presented the idea to the Council at the annual meeting. David Pickering wrote a lengthy essay "Proceedings and Portraits," filling the entire first issue and describing its benefit for members who had never attended a meeting. In the process, Young and Campbell agreed that it would not be necessary for Olcott to review everything as the AAVSO Editor because that role had already been ignored and to do so would simply hold up the publication unnecessarily.[34] *Variable Comments* proved to be very popular with the membership. By 1929, the first "volume" of issues had been bound and made available for libraries and other collections.[35]

Meetings

Annual meetings, held in the fall every year at HCO, became rather stylized by the time the records really provide much detail. The annual meetings attracted members who valued the opportunity to visit the observatory and hobnob with well-known professional

Figure 6.3. Charles E. Barns, Anne S. Young, and John H. Skaggs, Morgan Hill, California, 1928.

astronomers for a few days. It seems natural, therefore, that after incorporation, the annual meetings would convene at HCO. With facilities that accommodated the relatively small gatherings at a short distance from the Harvard campus, the annual events fell into a comfortable pattern, with a Council meeting on Friday afternoon or evening, followed by a day-long meeting at the observatory on Saturday. The afternoon sessions were typically broken for a tea at the HCO Director's residence. The final banquets, fairly informal at first, were held in local church halls. As the meetings grew in popularity, the number of members and guests exceeded local facilities so the banquet was shifted to hotels. Annual meeting banquets usually featured short presentations by both popular members and guests. A "Toastmaster" appointed by the president called upon Olcott and other members of the Old Guard for a few words, usually a poem by Olcott and humorous remarks by others. From the reports mainly recorded in *Variable Comments*, a good time was had by all.

Receptions at the Observatory Director's residence soon took on status as a cherished feature of the annual meeting. Assisted by some of the senior women staff at the observatory, the HCO Director's wife, Martha Betz Shapley, entertained the AAVSO members every year at their residence on the Observatory grounds. AAVSO Council meetings were also held there on occasion. Clearly, Shapley's view of the AAVSO enlarged greatly upon that held by Pickering, who looked at the AAVSO mainly as a technical organization and resource for variable star observations. Shapley, on the other hand, seems to have seen the AAVSO gatherings as a social function as much as a scientific meeting.

Figure 6.4. At the AAVSO 1922 Spring Meeting at the Olcott home, Norwich, Connecticut. From left: George Waldo, Caroline E. Furness, Charles H. Godfrey, Mrs. C. H. Godfrey, E. Macaughey, Lillian Pickering, David B. Pickering, Mrs. George C. Waldo, Helen M. Swartz, Clara Olcott, William T. Olcott, and Leon Campbell.

Spring meetings, on the other hand, offered the AAVSO Council and officers the opportunity to meet in different venues, providing members in more distant locations an opportunity to participate at lower cost and less inconvenience. From 1920 to 1931, the locations of spring meetings were quite diverse, with three meetings early in that period held at the residences of members, like the pre-incorporation meetings at the home of David Pickering. Olcott hosted one such spring meeting in Norwich, Connecticut, whereas Yalden offered his home for the May 1925 meeting in Leonia, New Jersey. This was the last AAVSO meeting to be held at a private residence, as the spring meetings had grown in participation. Afterwards, spring meetings were frequently held at college and university observatories.

The 1923 Spring Meeting deserves some special attention. Court recorder Charles W. Elmer hosted the meeting, which included a clambake, at his home in Southold, Long Island, New York. It was a measure of his gregarious nature, as well as his love for astronomy, that within 5 years after joining in 1919 he hosted an AAVSO meeting at his home. The Southold meeting was memorialized in one of Olcott's more memorable poems in the style of the then-popular poet Robert Service, "Settin' Elmer's Glass – The Ridicule Method," a pun on "the reticule method" of aligning a telescope.[36]

To a casual observer, Elmer's Southold property was a hotbed of astronomy, with a number of small observatory domes for his larger instruments (a 6-inch refractor). For all this hardware, however, Margaret Mayall's later characterization of Elmer as 99 percent enthusiasm and 1 percent experience contained more than a grain of truth.[37] Although his enthusiasm for the accoutrements of astronomy was greater than making useful variable star observations, he became one of the very strong insiders on whom the organization relied as a foundation. The AAVSO has traditionally relied on more than a few such non-observers who served in other important ways.

PASSING OF THE OLD GUARD BEGINS

For nearly a decade, the AAVSO founding members and those who joined shortly afterwards dominated

Figure 6.5. Clambake held during the May 1923 AAVSO meeting at the home of Charles W. Elmer, Southold, Long Island, New York. From left: Caroline Furness, Morgan Cilley, George Waldo, David Pickering, Leon Barritt, Leon Campbell, unidentified server, A. W. Butler, Arthur C. Perry, Inez Clough, J. E. G. Yalden, Donald Menzel, W. T. Olcott, Charles Elmer, and two unidentified servers.

Figure 6.6. William T. Olcott, William Henry, and Leon Campbell, at Charles Elmer's, Southold, Long Island, for the 1923 Spring Meeting.

Figure 6.7. At Mount Holyoke College, South Hadley, Massachusetts, for the AAVSO's 1924 Spring Meeting. On the left are William T. Olcott, Alice H. Farnsworth, Anne S. Young, Helen M. Swartz, and Harriet W. Bigelow; on the right are Caroline E. Furness and Leon Campbell.

the association. Sometime after the incorporation, that group became known informally as the Old Guard, and although the group's membership was somewhat fluid, the name stuck and was in common usage in *Variable Comments*, as well as in their personal letters.[38] The first Old Guard member to pass was Dr. Edward Gray (1851–1920). His main contribution was in suggesting and preparing the first blueprint comparison star charts, a substantial contribution in the first years of the AAVSO's history.[39] The death of Treasurer Allan Beal Burbeck (1882–1920) came as a shock to the Old Guard. No contingency plan existed for the Treasurer's death, but when it occurred, the Old Guard stepped in quickly to appoint a replacement (Old Guard member Michael Jordan) to insure continuity of control of important bank accounts and records. There was no warning, either, with the unexpected death of railroad engineer Charles Young McAteer (1865–1924).

Figure 6.8. AAVSO meeting at Vassar College, June 1928. From left: Rev. John A. Ingham, Rev. E. C. Phillips of Georgetown College Observatory, William T. Olcott, David B. Pickering, and Leon Campbell.

Figure 6.9. AAVSO meeting at Vassar College, June 1928. Caroline Furness is seated on left, next to David Pickering, and Mrs. Yalden next to him. Standing second from left is Clinton B. Ford, just elected to AAVSO membership at age 15. To the right of him are J. E. G. Yalden and Alice Farnsworth.

In a hurry to catch a streetcar to get home and back to variable star observing after completing his work one night, he carelessly ran in front of a truck and was killed.[40] McAteer's death, though acutely felt as the loss of a great old friend, did not create the same risks to the AAVSO as Burbeck's unexpected death. Similar risks are evident in the circumstances with the death of Dr. Charles Cartlidge Godfrey (1855–1927), an autodidact in the sciences in general and longtime telescope maker. The Connecticut physician and surgeon, who had a noteworthy career in medicine and was the personal physician to showman P. T. Barnum, was especially involved in the amateur telescope-making movement. He also discovered the joys of variable star observing through the efforts of David Pickering in 1919. A natural leader in all that he attempted, he was elected president of the Association in late 1926. After chairing only one meeting, Godfrey died suddenly in August 1927 at his home in Bridgeport.[41]

Dr. Godfrey's unexpected death brought a professional astronomer, Alice Farnsworth of Mount Holyoke College, to the presidency. Perhaps uncomfortable with her comparative lack of such leadership experience, Farnsworth declined the honor, acceding to Campbell's request to allow David Pickering to take up the President's post for a second time.[42] This agreement would keep the AAVSO's affairs running smoothly, allowing Farnsworth to continue to serve as vice president and replace Pickering as president in 2 years.[43]

The death of AAVSO Treasurer Michael J. Jordan (1867–1929) on December 22, 1929, also occurred suddenly and had not been anticipated. Campbell suggested to Farnsworth, who had succeeded Pickering as AAVSO president, that she appoint HCO Professor Willard J. Fisher as the interim treasurer. Fisher served out Jordan's unexpired term and continued as treasurer for the Pickering Fund.[44] Thus there was less of a problem at Jordan's sudden passing than created by the death of Burbeck nearly a decade earlier. One of the earliest members, Jordan had served faithfully for a number of years in various capacities, including a lead role in the 1918 incorporation of the Association. By the spring meeting at Maria Mitchell Observatory on Nantucket Island, Fisher had completed his audit of Jordan's records, and a normal meeting transpired.

The lack of any complications resulting from Jordan's death illustrates that the AAVSO had begun to mature as an organization. In October 1929, David B. Pickering chaired his last annual meeting and Council meeting. The record of the general meeting prepared by John A. Ingham for *Variable Comments* reflects significant progress in the Association. There was plenty of good news in which the membership could be justly proud. Farnsworth accepted the reins of the organization from Pickering in a normal transfer of leadership. Excellent technical talks by a number of professional astronomers told of progress on a wide front in the advancement of both observational and theoretical astronomy. All remaining members of the Old Guard were present or accounted for.[45]

The Nantucket Council meeting presented modest modifications of the AAVSO constitution to members for their approval. The proposed amendments provided for the election of officers by the Council each year after the Annual meeting rather than by the general membership, added a second vice president, and established positions to be appointed by the president each year, including a recording secretary to supervise the publication of observations, a chart curator, and a librarian, in addition to two official auditors. For the 1930 Annual Meeting at Harvard, arrangements had been made for Percy W. Witherell to attend as a guest, apparently to meet and observe the Council members in action in anticipation of his election to replace Jordan as the treasurer. Some familiar faces were missing from that critical meeting. Farnsworth was on a sabbatical leave visiting California observatories, so Harriett Bigelow of Smith College, attending her first Council meeting in any capacity, presided over the meeting as the newly elected vice president. Olcott was still ill in Tucson, Arizona, so Campbell filled in for him as secretary. Fortunately, Young, Pickering, Yalden, and Elmer, who at this point was entering his third year on the Council, were present to stabilize matters and continue the Association's forward progress. Among the problems discussed and dealt with at this Council meeting was the apparent need to replace or supplement an ailing Olcott as a mentor for new members, who were joining the AAVSO in increasing numbers. At the general meeting, the membership elected Witherell as treasurer to replace the deceased Jordan. Fisher agreed to continue as the treasurer for the Pickering Fund until the related issues with HCO were resolved.[46]

Figure 6.10. Anne S. Young of Mount Holyoke College Observatory, with S. A. Mitchell of Leander McCormick Observatory.

INTERNATIONAL CONSIDERATIONS

From the beginning, the AAVSO had an international membership with Giovanni Battista Lacchini, a significant contributor from several observatories in Italy. One of the consequences of World War I for the United States involved a greater openness to international involvement, and the AAVSO and its members reflected some of that new openness. On his grand tour, Pickering visited not only Lacchini, but also Mario Ancarani and Domenico Benini in April 1926. His evangelism for amateur astronomy, especially the AAVSO, is described in *Popular Astronomy*.[47]

Although amateur astronomers in many countries formed their own national associations and variable star sections, the AAVSO continued to attract a small but steady number of observations outside North America. This continuing influence no doubt reflects the widespread availability of *Popular Astronomy* and perhaps the HCO publications as well.

The opportunity for international involvement manifested itself in Council actions as well. In November 1920, the AAVSO Council recognized the strong support rendered by several professional astronomers by electing Canadian-born S. A. Mitchell, director of the Leander McCormick Observatory, as a patron, and the British astronomer H. H. Turner as an honorary member. In addition, the Council considered a suggestion from Indian member and regular contributor R. G. Chandra that the AAVSO undertake a statistical

Figure 6.11. AAVSO members at Yerkes Observatory, May 1930. Front row, Kyoto Observatory Director Issei Yamamoto, Mrs. Yamamoto, John A. Parkhurst, H. L. Ealin; back row, Ralph Buckstaff, Anne S. Young, and Yerkes astronomer Georges Van Biesbroeck.

Figure 6.12. Radha Gobinda Chandra of Bagchar, Jessore, India, in 1937, with telescopes loaned to him by the AAVSO. Between 1919 and 1954, he made more than 49 000 variable star observations.

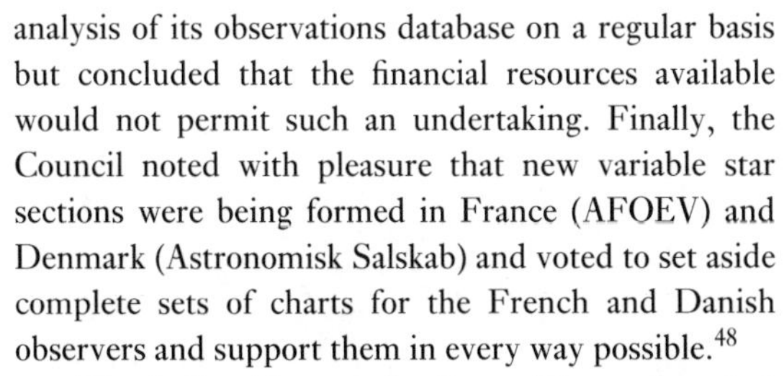

analysis of its observations database on a regular basis but concluded that the financial resources available would not permit such an undertaking. Finally, the Council noted with pleasure that new variable star sections were being formed in France (AFOEV) and Denmark (Astronomisk Salskab) and voted to set aside complete sets of charts for the French and Danish observers and support them in every way possible.[48]

Henri Grouiller of the Lyon Observatory had apparently expressed an interest in forming an organization like the AAVSO for French amateur variable star observers. Campbell was happy to provide Grouiller with any information he desired concerning the formation of such an organization, based on the experience of the AAVSO.[49] Campbell also realized the importance of providing Grouiller with samples of standard charts and comparison star sequences, lists of variables in need of observation, instructions for observers, and other relevant material. Cooperation in these vital areas would help to insure that each new observer could produce observations that were compatible with those being generated by extant observers.

Stephen Crasco Hunter was an AAVSO founding member whose professional career as an underwriter of insurance for the marine shipping industry provided him with both the wealth and opportunities for international travel; he used them to AAVSO's advantage in 1921. Hunter volunteered to promote variable star observing in Europe, where he met Grouiller and Lyon Observatory Director Jean Mascart, and Henri Deslandres, astrophysicist and director of the Meudon Observatory. In the last case, Hunter encouraged French professional astronomers to support formation of the French Association of Variable Star Observers (AFOEV). In June 1921, Hunter reported back to Campbell that they had started something that properly nurtured, "may grow into almost anything, to the greater and broader glory of our modest but persistent efforts here."[50] Hunter's prediction was more accurate than even he could have expected, for in that same month, June 1921, AFOEV organized under the leadership of Grouiller, Mascart, and amateur astronomer Antoine Brun.[51] The chart donations and Hunter's visit formed the foundation of a lasting cooperative relationship between AAVSO and AFOEV.

The case of Radha G. Chandra of Bagchar, Jessore, India, illustrates a growing international influence in this period. A clerk in the tax collector's office by day, he enjoyed watching the stars at night. When Chandra made an independent discovery of Nova Aquila 1918 and reported his results to HCO, Campbell recommended him as a member of the AAVSO.[52] Chandra carried on a lengthy correspondence with the AAVSO through Campbell, conveying his enthusiasm and pride in his association with the AAVSO. Recognizing that more observations were needed from his hemisphere, he suggested recruiting more Indian members and asked permission to translate instructions for observers into Bengali. Campbell encouraged Chandra's recruitment efforts and loaned Chandra several telescopes, including the Elmer 6-inch refractor and mounting.[53]

European and South American observers also joined the effort in greater numbers in the 1920s, including Max Beyer and Paul Ahnert in Germany, Rastislav Rajchl in the "new country" of Czechoslovakia, and Martin Dartayet in Argentina. Recognizing that cooperation was as useful as membership, the AAVSO Council made Felix de Roy of Belgium and A. N. Brown of the BAA-VSS honorary members.[54] Italy's Giovanni B. Lacchini earned recognition as the AAVSO's top observer with 4500 observations during fiscal year 1932–1933 and was elected by the membership to the AAVSO Council in 1930.[55] At the 1930 Annual Meeting, the Council took the unusual step of electing Lacchini, who attended the meeting, as a second vice president, even though he would be unable to participate regularly in Council meetings from his home in Italy.[56] All things considered, the members of the Association could be proud they belonged to a growing international effort to produce valuable scientific observations.

SO FAR, SO GOOD

Looking back from the perspective of the general membership of the AAVSO, the decade of the 1920s could generally be summarized as one of substantial progress. The Association's leadership oversaw the organization's normal activities but allowed Campbell increasing latitude in his informal role of providing HCO's support for the Association. Growth in the Association's membership increased in the United States and abroad as more observers learned about the opportunities to contribute to scientific progress in variable star astronomy.

7 · The Pickering Memorial Endowment

The members of the Association and the many friends and colleagues of Professor Pickering have reason to be proud of their achievement. It is not only a fitting memorial to a distinguished astronomer, but it affords a greater opportunity of extending our knowledge of Variable Stars.

– William Tyler Olcott, 1931[1]

The American Association of Variable Star Observers grew in membership during the 1920s and added substantially to its archive of variable star observations and, in addition, initiated several projects not related to variable stars, such as telescopic meteors and lunar occultation timings. By the end of the decade, the AAVSO had acquired a good reputation for its service to the science of variable star astronomy, though primarily in the narrow, and not particularly exciting, area of long-period variable stars.

In parallel with AAVSO activity that might be considered normal, however, were a series of events taking place around the AAVSO that held out both promise and peril for the still maturing Association. This chapter discusses how the young new Harvard College Observatory Director Harlow Shapley actively involved the AAVSO in HCO's affairs while attempting to rebuild the aging though justly famed observatory and its staff.

AN AMBIGUOUS RELATIONSHIP

Through the years, it is understandable that relationships between the AAVSO and HCO were unclear to observers. There had been assistance from HCO staff members, such as Leon Campbell and the HCO Director Edward C. Pickering, even though there was no formal relationship. Although HCO gave limited technical support to the Association since 1911, Pickering only became involved with the organization after 1916, but thereafter his involvement and support were increasingly active.

Leon Campbell, who worked at HCO, was committed to building his career in variable star astronomy. In his role at the observatory, he was able to assist many variable star observers. When he returned from a 4-year stint as the supervisor of HCO's Boyden Station in Arequipa, Peru, he joined the AAVSO and participated vigorously in the Association's activities, establishing strong relationships with several members, both at meetings and in correspondence, regarding several projects he initiated. He also helped incorporate the association in the Commonwealth of Massachusetts (see Chapter 5).

The death of HCO Director Edward C. Pickering soon after that incorporation strengthened Campbell's position as the liaison between the Association and the observatory. Campbell's election as the third president in late 1919 is therefore not surprising; the election further enhanced his roles as both a leader and a liaison. By offering a room for the Association to use at the observatory, acting HCO Director Solon I. Bailey further strengthened ties with the Association.

But a statement in the Pickering memorial booklet further complicated the understanding of the AAVSO's founding. The Memorial Committee, including Anne Sewell Young, Annie Jump Cannon, and the ever-modest William Tyler Olcott, identified Pickering as "a founder" of the Association. This statement diminished Olcott's crucial role as the founder, starting and nursing the organization for 9 years with his own energies and financial support (see Chapters 3 and 4). Although the minutes do not reveal the discussion on the Memorial statement, Olcott did not speak up to claim his own position, and the AAVSO Council followed the Committee's statement.

More than anyone else, David B. Pickering (not related to E. C. Pickering) who served as the first AAVSO president, was aware of Olcott's problems and his tireless support through the organization's earliest and difficult years. It is not surprising that neither he nor Olcott insisted on crediting Edward C. Pickering as the "founder," a role that Pickering clearly had not played. Olcott was generous in his comments; praising E. C. Pickering in everything except in a role as founder.

Although Campbell was deeply involved in AAVSO activities by 1920, the support he provided to the AAVSO was still not a formal part of his role at Harvard College Observatory. The records of the AAVSO in the 6-year period from Campbell's election as president in late 1919 until about 1925 are disappointingly sparse, being limited to the sterile minutes of council and general membership meetings and a few letters from Campbell's correspondence. However, as the relationship between the AAVSO and HCO evolved during that period, that evolution can be followed to some extent in the annual reports of the HCO Director and the contents of the Harvard Archives.

AAVSO PICKERING MEMORIAL FUND

Shortly after the Pickering memorial booklet was published in 1920, Olcott proposed to Campbell a physical memorial to Pickering in the form of an observatory that would also serve as the AAVSO Headquarters. Campbell was taken with Olcott's idea because he then presented it in correspondence to both Anne S. Young and Stephen C. Hunter. Although Young had reservations about the financial commitment, Campbell concurred but felt the project "could be put across if we all put our shoulder to the wheel" and noted that Olcott was thinking of a $5,000 donation to move the project along.[2]

An unexpected contribution of a telescope from the estate of a wealthy New York amateur astronomer in January 1921 may have galvanized the Council to take action on a Pickering Memorial Observatory. C. A. Post, who passed away in New York City, had a well-publicized observatory at Bayport, Long Island.[3] A serious amateur astronomer and astrophotographer since before the beginning of the century, Post housed his $6^1/_2$-inch Gregg refracting telescope with a 5-inch photographic refractor on the same mounting in a unique counterweighted swing-off roof observatory.[4]

Hearing of Post's death, someone from the AAVSO, possibly Olcott or Charles W. Elmer, who lived near the Post observatory on Long Island, contacted Post's widow about donating the entire telescopic outfit to the AAVSO.[5] As the nominal custodian of small telescopes for both the AAVSO and the HCO, when Mrs. Post agreed, Campbell picked up the telescope and its mounting from Post's estate.[6] Along the way he visited Olcott's new observatory in Norwich and stayed with AAVSO member George Waldo. Once acquired, Olcott and the Council recognized that the Post telescope would make an ideal centerpiece for a Pickering Memorial Observatory.

At the 1921 Annual Meeting of the AAVSO Council, Olcott made a motion to establish a fund honoring the memory of Edward C. Pickering, thus formalizing the idea he had discussed 10 months earlier with Campbell. Olcott proposed that the AAVSO establish an endowment fund of $100 000 to support the construction of a traditional observatory with a telescope and rooms to house AAVSO Headquarters. After some discussion, the Council approved Olcott's motion, and Campbell appointed Olcott to chair the Endowment Fund Raising Committee and select its members.[7]

Olcott's motivation in making his initial proposal must be considered. One view, of course, is that Olcott simply wished to honor the memory of the deceased HCO director. A more persuasive argument of Olcott's motivation, however, is that his proposal was an effort to move the AAVSO away from HCO by moving it off the Harvard campus onto its own property. Possibly, Olcott feared that the AAVSO would be absorbed into the fabric of HCO and disappear as a separate entity, losing its identity as a social as well as a scientific organization. He may have feared that the gentleman amateurs would lose control to the scientists. We don't have the documentation to prove Olcott's motivation, but subsequent events point to this as his primary motivation in the matter.

Olcott and his committee set out to raise the money necessary to fund the endowment, but 1921 and 1922 were poor years for fund-raising. Just recovering from the effects of World War I, the economy faced problems absorbing returning veterans after their discharge from the services. Prices had risen dramatically during and shortly after the war, followed by a sharp deflation in 1921, and then stabilized at levels nearly 50 percent greater than pre-war levels. Under these difficult economic conditions, the members of the AAVSO and others still exhibited some enthusiasm for the project. A high percentage of members contributed to the fund, but the sizes of the gifts were well below what would be required to raise Olcott's goal of $100 000.[8]

A NEW HCO DIRECTOR

For many years before his death, Edward C. Pickering had informally designated Solon I. Bailey as his deputy, and Bailey served during Pickering's extended absences. Pickering requested the Harvard Corporation to appoint Bailey formally as the assistant director of the observatory, but Harvard President Abbott Lawrence Lowell would not support Pickering in this appeal, and the request was denied. Even after Pickering's death, President Lowell and the Harvard Corporation would not appoint the 65-year-old Bailey to replace Pickering as the director of Harvard College Observatory. Instead, Lowell designated Bailey as acting director and initiated a nationwide search for Pickering's replacement.

The idea of an external search was unusual at Harvard. With similar independent departments at Harvard, for example, the Botanical Gardens and the Anthropology Museum, appointments had been made from within so that institutions tended to be inbred to the point of being refractory.[9] Another issue facing the Corporation in connection with the Observatory was a separate teaching department of astronomy with pressures to confer degrees in astronomy.[10] Appointment of a new director from within the observatory would likely make it more difficult to consolidate the observatory and the teaching department in the future. Thus it is understandable, in this context, that an external search was mounted for a younger astronomer to lead the HCO. A fresh approach would be required to extend Harvard's distinguished record of research and service in astronomy accrued during Pickering's lengthy tenure.[11]

Among the candidates considered, one stood out – Harlow Shapley (1885–1972).[12] During his graduate years at Princeton, he studied theory with Henry Norris Russell (1877–1957) and practical work with observational astronomer Raymond Smith Dugan (1878–1940). In both cases, Shapley studied eclipsing binary stars, and his dissertation involved his theoretical analysis of the orbits of 87 eclipsing binary systems. These eclipsing binaries were thought to be Cepheid variables in those early days, but Shapley and Russell demonstrated that they were individual pulsating stars. The analytical methodology developed by Russell became the foundation for the evolution of this important field of eclipsing binary astronomy.[13]

After receiving his Ph.D. in 1913 and spending a few months in Europe, Shapley joined the Mount Wilson Observatory staff.[14] At the suggestion of Harvard's Bailey, Shapley launched a study of Cepheid variables in globular clusters. Though a comparatively young astronomer, Shapley was privileged to observe with the 60-inch reflecting telescope, the largest functioning telescope in the world at that time.

In 1918, some of Shapley's conclusions about the approximate size and shape of the Milky Way Galaxy and the position of the solar system in the outer portions of the Galaxy drew national attention. He mapped the distance of cluster Cepheids (now known as RR Lyrae stars) from the Sun and found that the globular clusters, concentrated mainly in the directions of the constellations of Sagittarius and Scorpio, were distributed in roughly a spherical shell at vastly different distances from the Sun.[15] His work was based on Henrietta Leavitt's discovery of the relationship between a Cepheid's period of variation and its absolute brightness and his own calibration of that relationship using the known distances for a few Cepheid variables.

ASTRONOMY'S GREAT DEBATE

With Shapley's finding, a debate ensued, called Astronomy's Great Debate, about the distance scale of the universe. George Ellery Hale, director of the Mount Wilson Observatory, suggested that Shapley debate his most vocal critic, Heber Doust Curtis of the Lick Observatory, the most important astronomical center apart from Mount Wilson. The event was held in April 1920 at the National Academy of Sciences in Washington, DC. President Lowell of Harvard sent two representatives, George Russell Agassiz, an important university benefactor and member of the HCO Visitors Committee, and Theodore Lyman, the UV spectroscopist from Harvard's Physics Department, to hear the NAS discussion and comment on Shapley as a candidate for the director's job. Much had been written about Shapley's view that the debate would make or break his chances for the job. That concern grew so strong that his performance in the debate was adversely affected.[16] The history of this debate contains interesting insights into the scientific process in that both astronomers presented information that illustrated their points, and in both cases, some, but not all of their conclusions, were later accepted broadly in the astronomical community.[17]

The Harvard representatives were more impressed, however, with a member of the audience, Henry Norris Russell, Shapley's Princeton dissertation advisor, who asked questions during the presentations.[18] They later were even more impressed after a meal with Russell. On their recommendation, Lowell first offered the HCO director's position to Russell. However, after a visit to Cambridge and further negotiations with the Princeton administration, Russell declined the position.[19] Then, Lowell offered Shapley a temporary position, which he accepted. He began working in 1921, and 6 months later his appointment was made permanent. Once Shapley was on campus, it seems likely that Lowell was strongly influenced by Shapley's naturally relaxed manner, gift for conversation, and the breadth of his interests in the classics and other non-astronomical subjects.

Figure 7.1. Leon Campbell, AAVSO Recording Secretary 1924–1931; AAVSO Recorder and Pickering Memorial Astronomer at Harvard 1931–1949.

LEON CAMPBELL AND VARIABLE STAR ASTRONOMY

With Shapley's appointment, Campbell may have stopped to ponder how his life at HCO was likely to change. He would be working for a director, 4 years younger, instead of one who was nearly 35 years his senior, or an acting director, 27 years older. His uneasiness may have increased on his first meetings with the boyish-looking new director. The differences in their experiences and intellectual activities may have also loomed large when one considers their different education and Shapley's previous work at the foremost astrophysical observatory in the world. Their shared interest in variable stars may have been comforting, but was limited as Shapley's main interest was astronomy on a galactic scale.

Campbell did have a reputation as a variable star observer since serving as the Observatory's contact person for variable star communication with amateur astronomers. Even professional astronomers such as Paul W. Merrill at Mount Wilson, as early as 1920, requested light curves and predictions of maxima and minima on a regular basis from Campbell. His correspondence with Merrill was helpful in placing the value of Campbell's work in a clearer perspective; it enhanced Campbell's understanding of the evolution of variable star astronomy. For example, Merrill pointed out the special characteristics of the long-period variable X Ophiuchi, a recently discovered double star in which one was the variable, sometimes brighter than the other star and sometimes much fainter.[20] Although perhaps an isolated example, X Oph provided a rare opportunity for a more direct measure of the distance to the binary system and therefore of mass determinations for both of the stars. Thus X Oph provided important astrophysical information not likely available for other well-understood long-period variables. Such insights were not easily achieved in those earliest days of both astrophysics and variable star astronomy.

Campbell received a boost in his reputation as a professional astronomer when he was appointed to the International Astronomical Union (IAU) Commission 16 (Planets) for his HCO work on Jupiter's satellites and asteroid light curves, and as the AAVSO's representative to IAU Commission 27 (Variable Stars). He returned from the 1925 IAU meeting in Cambridge, England, full of vigor and enthusiasm for his position in the field of variable star astronomy.[21] Campbell's position after 1925 was well recognized beyond the confines of HCO and the IAU as well.[22]

THE PICKERING MEMORIAL CHANGES

From Chapter 6, one can see that the AAVSO had stabilized and by 1931 functioned as a viable organization making a good contribution to astronomy with accumulating archives of variable star observations. Professional astronomers participated in the leadership of the association and contributed to both the stability and to modest technical progress over time.

The same could not be said of HCO, where the appointment of Harlow Shapley as director in 1921 resulted in rapid changes in program and in the culture. Shapley immersed himself in the mainstream of rapidly evolving issues in stellar, galactic, and cosmological astronomy during his employment at Mount Wilson Observatory. By 1920, or only 6 years after he joined the Mount Wilson staff, that involvement constituted something of a maelstrom of activity, which he brought with him to HCO. On his arrival at HCO, Shapley discovered that the program there verged on stagnation; aged physical facilities failed to meet the needs of the staff, which for its own part continued to practice mainly late-nineteenth century astronomy. As a matter of Pickering's intent, for example, the HCO program included no significant theoretical work.

With AAVSO activities increasingly centered on HCO, the Association was not immune to the effect of Shapley's arrival and the changes he began making at the observatory. It is not clear how or why the matter was elevated to Shapley's attention to such a degree that he decided to act in the matter of Olcott's idea for a small observatory that would double as the AAVSO Headquarters. Perhaps this issue had been raised to him by the previous Acting Director Bailey. After Pickering's death, Bailey had gone out of his way to make the AAVSO feel they had a space to use at the observatory and therefore a reason to be more strongly affiliated with HCO. Bailey would certainly have recognized the value of the progressively increasing flow of variable star observations coming to the observatory on a monthly basis. In any event, shortly after he arrived at HCO, Shapley suggested that Olcott survey professional observatory directors, asking for their opinions about Olcott's ideas for an Edward C. Pickering Memorial observatory.

Olcott did undertake a survey like that suggested by Shapley. He wrote to the directors of a number of observatories, both large and small, in the United States and Canada describing the project and asking for their opinion of the idea.[23] He apparently also sent his letter to a few other individuals prominently connected to astronomy, such as Garrett P. Serviss, author and astronomy popularizer. The result of that survey could not have been comforting to Olcott, for although professional astronomers endorsed the idea of a memorial to the late director and an endowment for that purpose, most apparently recommended against the idea of investing that endowment in a small, private observatory building to house the association. Instead, most recommended that such an endowment should be directed to the expansion of research in variable star astronomy.[24]

The response Olcott received from Edwin Brant Frost at Yerkes Observatory may be representative for our purposes. Olcott invited Frost to serve on the endowment advisory committee. Although supporting the idea of an endowment as a memorial to Pickering, Frost explained frankly that he did not agree with Olcott's goal for an endowment, claiming $100 000 fell far short of what would be needed. Citing construction costs for the basic observatory building and the need for more facilities than were being planned, Frost concluded:

> Thus it seems to me that we would have another un-endowed observatory on our hands; and you would have no salary for the director or the janitor, or funds for the upkeep of the institution or its library. . . . Without going into more details, you see that I would regard it as highly unfortunate that another struggling observatory should be started, liable to periods of strong interest and to periods of diminished interest.[25]

Lick Observatory Director William Wallace Campbell was very explicit about the difference. As he later described it to Shapley, his strong recommendation, and that of other astronomers at Lick Observatory, was that the memorial endowment be established at and managed by Harvard University for the support of variable star research.[26] In their final wording of the appeal brochure, Olcott and the Pickering Memorial Committee had not been so direct, stating instead that the memorial "should be, in some way, closely associated with the Harvard College Observatory."[27]

Although the individual letters from the other observatory directors are not available in the Harvard or AAVSO archives, it appears they also recommended that such an endowment be managed by Harvard, and not by the AAVSO. Olcott, already suffering from chronic health problems that caused Clara and him to later spend substantial time in the warmer climates of South Florida and Tucson, Arizona, reluctantly accepted the defeat evident in the observatory directors' responses. When taken together, it seems, those responses discouraged Olcott, and – as noted earlier – he did not pursue the fund-raising effort with his usual enthusiasm. Instead, Olcott wrote to Shapley, transmitting the letters he received from the observatory directors, and suggested that AAVSO would raise such an endowment, but that the money for the endowment might be managed by the HCO director.[28] From Shapley's reply to W. W. Campbell, we can assume that the letters from other observatory directors laid somewhere between the strongly negative letter from W. W. Campbell and a very positive letter from Dominion Observatory Director J. S. Plaskett.[29]

Figure 7.2. David B. Pickering and Morgan Cilley examine the AAVSO's C. A. Post Memorial Telescope in the dome on HCO's Building A, where it was installed in 1922 as part of the AAVSO's E. C. Pickering Memorial.

When he received Olcott's letter, Shapley immediately sent it to President Lowell. In addition to his draft reply to Olcott, Shapley included only the letter to Olcott from Lick Observatory Director W. W. Campbell and added a two-page essay titled "The Edward C. Pickering Memorial."[30] In his letter to Lowell, Shapley explained the situation briefly, advised Lowell that he had offered AAVSO the use of a small dome and an office, neither of which had "been used to speak of for some ten or twenty years." Shapley was referring to the dome on HCO's Building A, where the Post telescope would be installed in 1922, and to a library room and Leon Campbell's own office.

In his draft reply to Olcott, for which he requested Lowell's comments, Shapley proposed that if the AAVSO would raise as much as $60 000, Harvard would undertake to complete the endowment. A week later, Lowell replied to Shapley and agreed with Shapley's proposal that Harvard accept the endowment money raised by the AAVSO. Lowell gave his tacit agreement that Harvard would complete the endowment if insufficient funds were raised by the AAVSO. He warned, however, that it was not a good idea to imply Harvard would put any money into the arrangement:

> The only question I would suggest in regard to your draft of a letter, which I return herewith, is that I am not sure that it is wise to say we will undertake what is proposed if they will raise $60 000. I suspect that this is likely to have the effect of relaxing their efforts to that point. Before making a definite agreement, the matter ought to be approved by the Corporation.[31]

We do not have Shapley's actual response to Olcott in writing, and it is unclear from Shapley's files that he ever did respond formally.

What actually happened may indicate what passed between Shapley and Olcott as a result of Shapley's exchange with Lowell. From 1921 to 1924, Olcott, Tilton C. H. Bouton, Shapley, Bailey, David B. Pickering, and possibly others, made independent Edward C. Pickering Memorial Lecture tours to raise money.[32] Formal elements of this effort included advertising of the endowment fund drive in such national and regional publications as the *New York Times*, *Scientific American*, and *Science*. The AAVSO also solicited subscriptions to the memorial fund from members.[33] Membership response was good in terms of the number of

Figure 7.3. Harlow Shapley and William Tyler Olcott, about 1925.

members who donated, but donations from members and others came slowly and in lesser sums than originally hoped. The campaign suffered from the fact that Olcott's health, never strong, caused him to travel with some frequency to warmer climates.

By 1924, the fundraising effort had clearly stalled. Olcott himself did not make a large gift to the fund, a gift of a size for which he clearly had the financial capacity and which he had previously alluded to in his communication with Leon Campbell.[34] That he did not do so is the clearest indication we have at this point of Olcott's gradually changing attitude. As it sank in that his role as the founder of the AAVSO had been diminished by others who, at the same time, wished to strengthen ties between the AAVSO and HCO, Olcott's enthusiasm for fund-raising for the memorial was proportionally diminished.

At the 1925 Annual Meeting, Shapley apparently made an effort to rejuvenate the fund-raising drive by proposing a renewed effort through a speaker's bureau, although the exact nature of what he proposed is unclear. In an excited letter to Shapley after the meeting of the Pickering Memorial Endowment Committee, Olcott described discussions he had with Godfrey and Yalden on the way home from the meeting:

> . . . we thought that the best plan would be to work through the Harvard Clubs in N.Y., Boston, and Chicago. In this way we could reach the most influential of the Harvard Alumnae. If we get the Presidents of these Clubs to appoint committees to canvas the Clubs, I think we could put it over. Have also written to [S. Lynn] Rhorer to see what he could do in Miami. I am confident of success.[35]

Despite of this burst of enthusiasm, however, the drive for funds continued to falter. It seems likely that Harvard would have rejected the proposal to work through their alumni clubs if that proposal had ever been put forward formally.

HARVARD TAKES CONTROL

With Leon Campbell's time increasingly devoted to the AAVSO as well as to variable stars, Shapley might have seen that the AAVSO could provide a vehicle for funding Campbell's salary and at the same time remove Campbell from the mainstream of HCO science by establishing a secure niche for him in the AAVSO. Even by 1922, it was clear that stellar astronomy, including variable star astronomy, was evolving much more rapidly than Campbell, with his limited scientific background, could assimilate.[36] At the same time, however, Campbell had gained substantial recognition and trust among AAVSO members by his diligent effort and easy personality. Campbell was a popular figure in the AAVSO at the time Shapley arrived in Cambridge, so the possibility of simply sliding Campbell aside in this manner would have been an attractive option for the new director as he considered how to upgrade his scientific staff.

As early as 1923, it is possible to see Shapley's plan for Campbell by using the AAVSO in this manner emerging in his published works. In his first annual report to the President of Harvard for work completed through September 30, 1922, Shapley devotes a half-page of text to long-period variable stars, extolling the virtues of the AAVSO and its observers. By stating explicitly that no satisfactory explanation of the cause of the variation of these stars existed, Shapley laid the groundwork of the importance of continuing the AAVSO's work. Noting that the AAVSO Council "is now endeavoring to raise a fund as a memorial to the former Director of the Harvard Observatory, Professor

Edward C. Pickering, under whose guidance the Association was developed. It is proposed to operate this memorial fund in connection with the Harvard College Observatory," Shapley had, in effect, started this process.[37]

In his annual report for the year ending September 1923, Shapley continued to lay the groundwork for a longer-term plan for the AAVSO. The study of variable stars was, at that point, a major activity at the observatory. It was appropriate in that regard for Shapley to devote a full two pages of his report to the subject. In his discussion of long-period variables, he emphasized that more than two-thirds of the observations in observatory files were provided by AAVSO observers and were being studied by Campbell. He then observed that "The close affiliation of the Harvard Observatory with the Variable Star Association continues to be very fruitful in the study of this important class of stars."[38]

By the time of his September 1924 Annual Report, Shapley felt confident enough of his position to begin articulating a grand plan for the overall evolution of the observatory. One sees distinct elements of what Shapley later described as an Institute for Cosmogony at HCO, although he did not use that explicit term to identify his intent at that time. Instead, he began to rely on terms like "the physical chemistry of the stars" to describe the work of graduate students in astrophysics. Those students, Donald H. Menzel at Princeton and Cecilia Payne at HCO, worked with the archives of plates at HCO for their dissertation research. Their work impressed Shapley with the potential of their approach. He thus described in these general terms the more detailed and specific work of astrophysicists in understanding stellar evolution, including, of course, stellar variability.[39] As the HCO staff assigned to variable star work increased considerably in the process, Campbell and the AAVSO gradually shrank into the background in Shapley's annual reports.[40]

By 1925, Shapley felt he had all the pieces in place with staff and justification for a large investment in the observatory. Furthermore, he had apparently gained the confidence of Harvard president Lowell, who directed Shapley to Wickliffe Rose of the International Education Board (IEB), a subsidiary of the Rockefeller Foundation.[41] Shapley approached Rose with a proposal for a large grant of approximately $1 000 000. It would be inappropriate to map out Shapley's total plan in the context of this history; suffice to state that it provided for a substantial upgrading of both facilities and program at the observatory. Included in Shapley's list of seven major elements of his proposal as item C, however, is Shapley's request for $100 000 for the Pickering Memorial Endowment, described as follows:

> C. *Support for the direction of the scientific work of the hundred amateur observers and calculators*, whose present contributions to knowledge of the stars that vary in light are assisting in the chemical analysis of matter under extreme conditions of temperature, pressure, and radiation, and whose work on the position of the Moon bears directly on the problem of the constancy of the Earth's rotation and therefore on our basic unit for measuring time.[42]

Shapley's overall proposal for the observatory improvements was apparently persuasive. It was hand carried to Rose on March 27, 1925. On October 16, 1926, Shapley announced to Rose that the IEB grant had been matched by Harvard University and plans were already underway for a new telescope mounting.

However, convincing Rose of the general value of the Pickering Endowment as part of Shapley's whole project did not end with this announcement. The undated copy of Shapley's final report "The Harvard Observatory Development" found in the AAVSO archives contains references to events that occurred as late as 1929. It appears that IEB's demand for additional justification and progress reports for the whole proposal from Shapley resulted in a significant evolution of Shapley's proposal document over time.

One example of such later evolution of Shapley's proposal is found in the inclusion of lunar occultation timing by AAVSO observers in Shapley's justification for Part C. It appears that Shapley had discussed the possibility with Ernest Brown of Yale University before October 30, 1926, when Shapley first mentioned that as a possible extension of the AAVSO program to Rose. As a consequence, Brown, likely at Shapley's urging, amended a request for more occultation timings that he had sent to professional astronomers. In his amended version dated in December 1926, Brown included instructions suitable for amateur observers. Brown's letter appeared in the next issue of *Variable Comments* in April 1927, after which the AAVSO Spring Meeting occurred at Yale University in May 1927. Thus the extension of the AAVSO program to include lunar occultations only occurred 2 years after Shapley's original grant application for the Pickering endowment and to all appearances represented a Shapley initiative designed to help justify the Pickering Memorial Endowment as well as to aid the Yale astronomers.[43] Shapley also included more refined cost estimates for the major capital investments in this version of the project proposal. At that point the requested funds totaled $1 800 000. Harvard would not likely have undertaken the expense of preparing such estimates unless the university had been assured by the IEB that those major portions of the funds requested would be approved.

What is critical to discuss in this context is Shapley's proposal on behalf of the AAVSO as part of that request. From the beginning, his overall proposal included $100 000 to endow the AAVSO Recorder's salary and certain expenses, and that amount apparently never varied. The IEB, initially unwilling to accept that aspect of the grant proposal, apparently asked for clarification of what Shapley intended. Shapley spent the next month working with and reassuring Rose in a series of letters that the money was essential, the science valuable, and the AAVSO trustworthy. Furthermore, Shapley took care to note that observations of variable stars were being provided to the AAVSO from a number of members and other variable star associations around the world to "internationalize" his proposal. Thus the endowment funds were not formally requested in AAVSO's name, although the AAVSO's activity was used to justify the endowment. Instead, the fund was designated as an endowment of the Edward Charles Pickering Memorial Astronomer of Harvard University.

Shapley's presentation of information about the AAVSO in the process of persuading Rose to support that last part of the large grant request left a great deal to be desired. In a series of letters and presentations to Rose, Shapley not only misrepresented his authority to speak for the association (he had no such authority), but he also distorted the history of the founding of the organization and the early role that Harvard had played in sustaining the volunteer observing effort. Furthermore, he deliberately broadened the role of the AAVSO to include, by implication, meteors, occultations, and "such other astronomical subjects as are now studied or will hereafter be studied by the American Association of Variable Star Observers."[44]

The AAVSO Council had never authorized Shapley to speak for the association. Shapley had been present in only one AAVSO Council meeting as a guest prior to the date of his initial proposal to the IEB, and the minutes of that meeting reflect that no such authorization was ever discussed. In his letter to Rose dated November 16, 1926, Shapley stated: "The central office of the American Association of Variable Star Observers has been at Harvard Observatory from the beginning and from this office practically all managerial work of a scientific nature is done." The first part of this statement is clearly not true, no matter whom one regards as the founder of AAVSO. The central office of the AAVSO was, for the first seven years of its existence, located in Olcott's home at 26 Church Street, Norwich, Connecticut. Indeed, the formal mailing address for the AAVSO was not changed to HCO until 1919, and even then only for reports of observations.[45] No one then or now could argue against the second part of Shapley's statement; clearly the scientific leadership for the AAVSO came mainly from HCO.

Figure 7.4. AAVSO member Charles C. Godfrey, Raymond S. Dugan of Princeton University, and William T. Olcott at the October 1926 Annual Meeting of the AAVSO, held at Harvard College Observatory.

For the next several years, Shapley understandably occupied himself with matters other than the AAVSO as he struggled with the substantial investment underway at both the existing HCO site at Oak Ridge, Massachusetts, the new site for the Boyden Station on the Harvard Kopje, Bloemfontein, South Africa, and the transfer of assets and personnel from Arequipa, Peru, to Bloemfontein. At the June 1930 Council meeting, Shapley, by now playing a more aggressive role in the AAVSO, suggested that the funds raised by the AAVSO be turned over to the HCO and that they would be augmented by the Harvard Corporation. Apparently Shapley did not reveal to the AAVSO Council that Harvard had already received final agreement for the IEB funds that were to be devoted to the Pickering Memorial Endowment. He also did not say that Harvard had not yet provided the necessary matching funds that IEB required as Harvard's own share. Thus his presentation made it appear that the endowment funds from IEB were somehow or other contingent on AAVSO's willingness to turn over their funds to Harvard.[46]

The AAVSO's Pickering endowment fund treasurer, Willard J. Fisher, presented a draft letter of transmittal for the funds, but Anne S. Young of Mt. Holyoke College and businessman David B. Pickering raised mild objections, expressing concern about "loose wording" in this letter. In June 1930, the Council went ahead and approved a formal motion to transfer the funds in the AAVSO's bank accounts to Harvard University. With this approval by the Council, Fisher withdrew funds from each of five separate savings accounts after annual interest payments had been credited and transferred the money to Harvard in five separate payments, the final payment being made on May 25, 1931. The funds actually transferred totaled $6356.76.[47]

Preparation of a draft wording of the terms of the agreement between the AAVSO and Harvard dragged on until the following spring, when it was discussed at the Council meeting. Young, now joined by Harriet Bigelow of Smith College, again objected to the wording, this time more aggressively than before. But Olcott, defeated at every turn, weakened by illness, and perhaps – like others on the committee – concerned that the offer might be withdrawn, did not object, and in fact wrote to Shapley to declare that he "heartily approve[d]".[48]

From the early 1920s, a Janus-like Shapley had used the AAVSO's Pickering Memorial Endowment fund as an important indication of HCO outreach and education efforts in petitioning the IEB for funds to modernize and update the observatory. Shapley deliberately distorted Harvard's role in the founding of the AAVSO and its early involvement and, as well, overstated Harvard's control over the AAVSO to the IEB. His goal: to convince the IEB to approve the $100 000 endowment for the Pickering Memorial Astronomer as part of his million-dollar package for his Cosmogony Institute. The endowment was the final step in his gradual displacement of Campbell from the mainstream of HCO's variable star effort. The AAVSO Council minutes reflect that the Council agreed to turn over the AAVSO's money raised for the Pickering Fund and that the decision in the matter was unanimous. However, Campbell's wording of the minutes would indicate otherwise: for example, the entire meeting was consumed by only this one matter – the Pickering Endowment – and the Treasurer was *directed* to turn over the money.[49] Time for the completion of the Council's business at this meeting expired, and the meeting was adjourned to be resumed when convenient.

When the Council reconvened, it was in an unusual January meeting held in New Haven, Connecticut, where the American Astronomical Society was also conducting its annual meeting. Council members present for this meeting on January 2, 1931, included Vice-President Harriet Bigelow, William B. Henry, David B. Pickering, and Ernest W. Brown, with Helen Sawyer Hogg, Shapley, Anne S. Young, and Charles Elmer sitting in by invitation. A quorum existed for purposes of executing the association's business legally, but missing from the meeting were President Alice Farnsworth, Secretary Olcott, Treasurer Willard J. Fisher, and councilors Leah B. Allen, Arthur W. Butler, Giovanni B. Lacchini, and Percy W. Witherell, the latter elected to the Council but not yet formally elected Treasurer. Thus Harriet Bigelow was the only elected officer present. Campbell acted as Secretary Pro Tem.

As the first item of business handled, Young read a report from a special committee on the *AAVSO Bulletin*. Their recommendation for the Council to accept an offer from Shapley for the observatory to handle, partially or fully, the publication and distribution costs for a bulletin was approved. Next, the Council addressed a motion to establish a system of Regional Advisors. For nearly two decades, one of William Tyler Olcott's major contributions to the AAVSO consisted of acting as a mentor to new members. As members joined, their applications to the AAVSO reached Olcott in Norwich. He typically responded in 2 days or less with charts and simplified observing instructions and encouragement to contact him if questions or problems arose. As Olcott's health failed him, more and more of this load fell to the staff at Harvard College Observatory. By 1930, the situation had deteriorated to the point that some alternative mechanism to shift that load off the HCO staff had to be developed. In discussions during several meetings, the Council came up with a plan to appoint regional advisors to handle the workload that had previously been handled by Olcott alone.

The Council's initial appointments included Sigeru Kanda, Japan; Dalmiro F. Brocchi, Washington; John H. Skaggs, California; Leslie C. Peltier, Ohio; Oscar E. Monnig, Texas; Sigmund K. Proctor, Michigan; Harold B. Webb, New York and New Jersey; Hans D. Gaebler, Wisconsin; Giovanni B. Lacchini, Italy; Walker, New England; and J. Ernest G. Yalden, New York and New Jersey (occultations). Each new member was to be vetted by one of these Regional Advisors before election to the Association. The Council further voted that the Curator of Charts should certify new members before any of the free charts were sent out. A third Council action authorized the Chart Committee chair to bring the supply of charts in her hands up to an efficient level, meaning Hogg was authorized to buy more charts to hold in inventory. Pleased with this development, Campbell touched up a document that he had used training observers in the past and sent a copy to Olcott for his comments.[50]

At first, the Council's ability to appoint regional advisors was limited by the availability of qualified individuals to handle the task. We are unclear, for example, who Walker, the Regional Advisor appointed for New England, might have been. No individual by that name submitted observations of variable stars to AAVSO's database during that period; it would appear that Walker later demurred from the role.

A final action by the Council was to authorize Shapley to apply to the Permanent Committee of the American Academy of Arts & Sciences in the name of the AAVSO for a grant of $200 for improvements in the chart distribution. In spite of the fact that the New Haven Council meeting was very productive, nothing was done to clarify Campbell's new role as far as the Council was concerned, nor was any action taken regarding Olcott's serious health problems, which were creating a serious overload for Campbell. It is clear, however, that Shapley dominated the two-part meeting and got exactly what he wanted from the AAVSO Council at that point, and as yet he was unwilling to reveal exactly what he had in mind for Campbell.[51]

It is unclear exactly when Shapley knew with certainty that the Harvard Observatory Development project, including the funds for the Pickering Memorial Endowment, were completely approved by the IEB, but likely he knew well before this point. This question is clouded by Shapley's need to continue the evolution of the proposal statement with more details as they became available, but also by changes that occurred at the Rockefeller Foundation. Shapley's prolonged effort to upgrade his request for funds to meet the evolving demands of the IEB after Rose's preliminary agreement, occurred at a time when the Rockefeller Foundation and the IEB were in organizational turmoil internally. Historian W. A. Nielsen identified the period from 1924 to 1928 at the Rockefeller Foundation and its constituent organizations as a "period of transition." By 1928 the IEB had been liquidated, and its scientific activities, including no doubt Shapley's proposal for development of the Harvard College Observatory, were absorbed into the Rockefeller Foundation itself. By 1928, Rose and many other key leaders from the early years of the Foundation had retired.[52]

Shapley, in the meantime, had been pursuing the necessary matching funds that were a condition of the Rockefeller Foundation, but with no success. By now, President Lowell supported Shapley and his project very strongly. Between them, they apparently came up with the idea of matching the IEB grant with funds from an unrestricted bequest from Stuart Wyeth. The idea had already been proposed by Harvard's Biological Endowment, also in search of funds for expansion, but the IEB rejected Biology's proposal on the grounds that the matching funds had to be from actual new fund-raising efforts, not from a bequest, under the terms of the IEB grant. Still, in early 1930, Lowell, now firmly in Shapley's camp, thought it worth a try, so he wrote not once but twice to Max Mason, who had replaced Rose at the Rockefeller Foundation. Lowell's second letter pointed out the damage that the October 1929 market crash had inflicted on the Harvard donor base. Mason rejected both requests tactfully but firmly.[53] Toward the end of 1930, Shapley tried again with a note to Harvard's corporate secretary, F. W. Hunnewell, only to be rebuffed based on the logic above.[54]

In the end, though, it appears that Lowell and Shapley found a way to have the Wyeth Gift used temporarily to fund the Pickering Memorial Endowment. Harvard's files contain documentation of the fact that at a later date, an undated charge of $93 554.62 against the Rockefeller funds repaid a temporary loan from the Stuart Wyeth Bequest. Perhaps Lowell and Shapley, at that point, felt Harvard had to make good on their end of the matching funds commitment. The Harvard Corporation took its final action to approve the creation of the Edward C. Pickering Memorial Endowment in the amount of $100 000 at its meeting on June 17, 1931.[55]

Finally, in late 1931, all the pieces were apparently in place in Shapley's mind, so Campbell's appointment to an endowed chair as the Pickering Memorial Astronomer was ready to be announced. For purposes of this history, though, what is more interesting is the manner in which the announcement was made. First Shapley met with the AAVSO's Pickering Memorial Fund Committee and reported to them that the University had appointed Campbell to this endowed chair as a fait accompli. Shapley then reported to the AAVSO Council that an advisory committee (Farnsworth, Olcott, with Shapley as an ex officio member) recommended the following actions be taken by the AAVSO Council:

> That Mr. Leon Campbell receive the title of "Pickering Memorial Astronomer," and a clerk be appointed to assist him. / That his duties should involve the entire supervision, direction and discussion of the Variable Star work of the Association. / That his title of "Recording Secretary," be changed to "Recorder". / That the title of Mr. Olcott, Corresponding Secretary be changed to Secretary. / That a room in the Observatory be equipped and set apart as Headquarters of the Association. / That a bronze tablet suitably inscribed be placed in a conspicuous place in the Observatory attesting to the fact that the Association had raised the sum of $100 000 as a memorial to Professor Edward C. Pickering, the income derived therefrom to be used in promoting Variable Star work.[56]

The report of the Committee was adopted, apparently without discussion. The Council approved a suggestion made by Campbell that "his room in the Observatory be equipped as the Headquarters of the Association." The Council also approved a motion that an Advisory Committee of the Pickering Memorial Fund be elected annually, the membership of the Committee to include the Director of the Harvard College Observatory; and finally that the committee consisting of Shapley, Farnsworth, and Olcott be continued with David B. Pickering as an alternate for Olcott. This last motion, apparently intended to establish the composition of the first Pickering Memorial Fund Advisory Committee, confuses more than it clarifies but also clearly illustrates how quickly and thoroughly Shapley insinuated himself into the leadership of the AAVSO once his arrangements for the Pickering Memorial Endowment were finalized. Campbell was nominally released from all his other duties at the observatory (although as we will see, that full separation never materialized).[57]

WILLIAM TYLER OLCOTT AFTER 1920

If Campbell found his position with the AAVSO and HCO comfortable in 1931, the same could clearly not be said of Olcott. One thing that is apparent in the broad perspective provided by formal histories of the AAVSO, minutes of meetings, and the extant correspondence files is that Olcott suffered diminished health capacity. He and his wife Clara spent less and less time in New England and more on the road to Florida, Arizona, California, and, to the extent his health permitted, some international travel. Olcott maintained a fairly steady presence in meetings, although the formal record of meeting minutes sometimes indicated that he concurred with minutes taken by someone else rather than directly by him. Eventually, his attendance at meetings became spotty and for reasons that were not always clear.

Although his health might explain some changes in Olcott, it is also apparent that Shapley's aggressive assimilation of the AAVSO into the HCO orbit as a satellite had a great deal to do with changes in Olcott's attitude. He had attempted, and failed, in his effort to maintain the association as a club of gentlemen astronomers. His plan for an independent headquarters building and observatory was, from at least his perspective, rudely rejected by the HCO director and his staff, as Shapley, in particular, saw the AAVSO as a secure home for one of his longer-term staff members who clearly could not compete in the mainstream of an emerging new scientific discipline. Although the evidence for this outlook on Olcott's part is only circumstantial, when taken in the whole, this appears to explain a change in his attitude that clearly took place over the course of the 1920s. Thus science prevailed as the *AAVSO's raison d'être*, and as a consequence, the AAVSO evolved more and more away from the gentleman's club of those interested in variable stars as an empirical curiosity and more and more toward a rational search for scientific understanding.

Things really came to a head in 1929 when Campbell in Cambridge typed a letter to Olcott in Norwich asking him to provide a few addresses for members. It was a routine matter not different in any way from hundreds of such brief notes the two exchanged over the years. The letter was forwarded to Olcott where he had been vacationing in the Berkshire Mountains of Massachusetts. Olcott, as it turned out, had fallen ill and visited a doctor in Stockbridge, who immediately confined him to a hotel room. In a note Clara scrawled across the top of Campbell's letter and returned to him in Cambridge, she expressed concern that Tyler should have been hospitalized as he could neither speak nor read and was not getting better. She suggested that Campbell had to get the addresses he had requested for himself, for it would likely be a long time before Tyler was ready to function in any way. The Olcotts were still at the historic Red Lion Inn in Stockbridge when Clara wrote a follow-up note to Campbell 2 weeks later to report that Olcott's health appeared to be improving; although they would be confined to the room for some additional time, she was unclear how long. In his reply to Clara at the Inn, Campbell expressed relief to know that his friend was on the road to recovery.[58]

From that point on, it is clear that Campbell took charge and asked very little of Olcott, fulfilling Olcott's entire role as Secretary in addition to his duties at the HCO and other AAVSO responsibilities. Olcott's role, at Clara's insistence, boiled down to that of a corresponding secretary and recruiter – perhaps cheerleader might be a viable description – but not really as a leader of the AAVSO. In a 1931 letter to Olcott in Tucson, Arizona, where he and Clara had gone to recuperate from the illness that struck Tyler in Stockbridge, Campbell revealed that the entire secretary's role had been assigned to him by the Council, although he would gladly relinquish those duties again to Olcott whenever the latter felt up to performing the tasks.[59]

Figure 7.5. Harvard astronomer Cecilia Payne-Gaposchkin with J. E. G. Yalden of the AAVSO.

LOOKING BACKWARD (AND FORWARD)

Leon Campbell had good reason to feel he had arrived in the spring of 1931. That he felt supremely confident of his position at both Harvard College Observatory, and within the AAVSO, is understandable. After more than 15 years of effort as an observer for HCO, and in a variety of leadership and service roles for the AAVSO, he found himself in the best position he had enjoyed in some years. Whether or not Shapley had advised him that he would soon be awarded a securely endowed chair as the Edward Charles Pickering Memorial Astronomer, he had been formally identified as Recorder, the senior professional astronomer in the AAVSO organization. His recognition in the field of variable star astronomy grew annually as he gathered and reported variable star observations and answered questions for professional astronomers and AAVSO members with both fragmentary light curves and observations. He provided quick and jaunty but serviceable replies to members and attended professional meetings of astronomers. Self-confident and relaxed, Campbell took on the persona that would serve him well for nearly two decades more in this position.

Campbell's barely concealed pleasure with his position can be easily detected in his summary of the first

20 years of AAVSO history, which he delivered as a talk at the 1931 spring banquet and then published in *Variable Comments*. The style in which he did so, familiar to all in the AAVSO by that point, mentioned as many names of members as possible as he wended his way through AAVSO history. After grandiosely summarizing the history of variable star astronomy up to 1911, with a nearly singular focus on Edward C. Pickering and HCO, Campbell described the founding of the AAVSO in terms that shifted the credit to Pickering, and H. C. Wilson, editor of *Popular Astronomy*, in effect displacing Olcott to a secondary role as handling all the secretarial part of the effort. Campbell's intent, no doubt honorable, appears to have been to ascribe and draw into his story the broadest number of participants possible in any role he could fashion. Instead, when viewed in the context of what has been presented in this history, Campbell's talk and later essay undeservedly deflects to others credit rightly owed to Olcott in the matter. Small wonder that Olcott might have felt slighted by the presentation, in spite of numerous flowery and humorous other references to him as Campbell completed his historical review. Campbell's self-confidence in this spring presentation makes it evident that he knew the Pickering Memorial would be funded by fall and that would protect his position at the HCO and the AAVSO thereafter.[60]

Campbell could not have appreciated at that point the rapidity with which Shapley altered and upgraded the HCO staff around him, hiring young leaders in astrophysics like Cecilia Payne and Donald Menzel and expanding opportunities at HCO in astrophysical programs as well as traditional astronomy, while managing to shuffle Campbell to the side in a field not yet quite ready for that talent. The HCO continued as one of the leading institutions in the burgeoning fields of astronomy and astrophysics, employing staff and expanding capability with instruments and observatories, but variable star astronomy still stood as an empirical scientific field without an accepted theoretical grounding.

8 · Fading of the Old Guard

. . . A fellow ought to be thankful to have anything these parlous times.

– William Tyler Olcott, 1934[1]

The American Association of Variable Star Observers (AAVSO) enjoyed a decade of organizational prosperity in the midst of the Depression after the firming of relationships with the Harvard College Observatory and the new ties established by its director, Harlow Shapley. The Association settled into a routine of annual meetings, increased observing, and a continued growth in its membership.

After Campbell was appointed Pickering Memorial Astronomer and AAVSO Recorder, he still remained in his offices at Harvard College Observatory (HCO). Even though Campbell had been performing tasks for the AAVSO in previous years, and even though he shared his space in the HCO library with the AAVSO space granted by Solon I. Bailey in 1919, his office had never been identified as the organization's headquarters because of his anomalous position with the AAVSO. By 1932, with formalization of the several relationships, Campbell's office and AAVSO Headquarters became firmly identified as being one and the same.

In following the long process to complete the Pickering Memorial Endowment, which funded Campbell's staff position at HCO, little notice has been taken of the other circumstances that confronted those endeavors. The national impact of the financial crash of October 1929 is well understood and documented. From a peak in September 1929, the stock market lost over half its value by the middle of November. Most economic historians accept that the Great Depression began in October 1929 and did not end until the United States entered World War II in 1941.

The ways in which the Great Depression affected the AAVSO and its members provides an appropriate setting for this chapter, which focuses on progress in the Association during the decade that followed that crash and slightly beyond. Astronomy, it seems, offered a relief from the troubles of this world with a comparatively inexpensive ticket to another world, at least for amateur astronomers. Their efforts resulted in a more expansive program for the AAVSO in the international arena, in developing charts, and in establishing committees dealing with topical areas such as novae, red stars, and occultations.

THE GREAT DEPRESSION AND THE AAVSO

The Great Depression was not a remote event for Campbell and others living in Cambridge, Norwich, and most other industrially based cities. Cambridge suffered a considerable loss of jobs and employment at the same time its population had reached a peak of 113 643 by 1930. Shapley employed young women for various observatory tasks with funding he received through donations.[2] Others on the HCO staff who enjoyed the benefit of continued employment also contributed their time to assist those suffering the effects of the Depression.

The Depression, barely noted in the cheerful pages of *Variable Comments*, had a marked effect on a few AAVSO leaders, especially David B. Pickering, everyone's friend in the AAVSO and a member of the Old Guard. Pickering, a jeweler, suffered considerably. In his correspondence with Campbell, William Tyler Olcott noted that Pickering was "on the rocks financially," having to sell his home and his "glass," concluding: "A fellow ought to be thankful to have anything in these parlous times."[3] J. Ernest G. Yalden and Olcott both suffered health problems during the early years of the Depression and significant financial difficulties. Olcott's correspondence during the rest of his life is sprinkled with references to being tight on cash and avoiding the extended travel that he and Clara had enjoyed for two previous decades.[4]

Figure 8.1. Four South African observers: A. W. Long, G. E. Ensor, H. E. Houghton, and W. H. Smith. They made well over a combined 48 000 observations for the AAVSO.

On the other hand, the Depression appears to have enhanced all things astronomical, an upsurge of public interest that generally benefited the AAVSO in the 1920s and early 1930s. Well-publicized discoveries seemed to emerge regularly from the great telescopes on Mount Wilson and elsewhere. The completion of George Ellery Hale's dream of an even bigger telescope seemed imminent when Corning successfully cast the 200-inch Pyrex disk for the unique new telescope on Palomar Mountain in Southern California.[5]

The discovery of Pluto at Lowell Observatory in 1930 brought fame to Clyde Tombaugh, a young man barely out of high school. Perhaps not unexpectedly, AAVSO member George Gray inquired whether Tombaugh was an AAVSO member (he was not) to be added to his list of members who had made other contributions to science. Campbell identified Leslie Peltier of Delphos, Ohio, and Max Beyer of Hamburg, Germany, who had discovered comets.[6] Within a year, he added two more variable star observers who had comet discoveries – George E. Ensor and H. E. Houghton of South Africa. Gray's list, it seems, offered practical testimony to the importance of amateur variable star observers in interesting ways.[7]

Public interest in astronomy grew with the fanfare surrounding the openings of new observatories and planetariums and media reports of astronomical discoveries and understandings. Investments undertaken in the 1920s came to fruition in the 1930s with the opening of five major planetariums featuring elaborate Zeiss projectors, including the Adler Planetarium in connection with the World's Fair in Chicago in 1930.[8] All this stimulated rapid growth in the number of astronomy clubs.

The amateur telescope-making (ATM) movement that blossomed during the Great Depression provided the vehicle for many individuals without scientific training to participate in activities at least vaguely related to the scientific discoveries that appeared regularly in the news. Most members of the AAVSO regarded their telescopes as a means to an end, not the end itself. The availability of inexpensive but serviceable reflecting telescopes, in most cases with a much larger aperture than could otherwise have been afforded, provided a few individuals access to variable star astronomy through the AAVSO. However, the small growth in the AAVSO membership, though important, was not proportional to the widespread interest in the ATM movement and recreational astronomy in general. As will be seen, the ATM movement was important to the AAVSO for other reasons. First, though, let us consider a decade of other progress in the AAVSO.

AAVSO LEADERSHIP TRENDS

When considering how the AAVSO leadership changed during this period, an important consideration needs emphasis. Despite his long experience as an astronomer, Campbell had little direct supervisory experience. His skills as a supervisor, demonstrated in his handling of the limited resources available to the Boyden Station at Arequipa, were not impressive. One could make the case that his success with the AAVSO depended far less on his ability as a supervisor than on his well-refined sense of interpersonal relationships, and in many respects, not much more would be involved in his new dual assignment as the AAVSO Recorder and the Pickering Memorial Astronomer.

However, it is fair to say that the formal establishment of the position of AAVSO Recorder, and Campbell's role as the first incumbent in that position and as the Pickering Memorial Astronomer, changed the entire way the Association was managed. The importance of two points cannot be overemphasized here: first, Campbell's responsibility for AAVSO matters was at last pretty clearly defined, and second, it is critically important to recognize Campbell's dual-reporting relationship to the HCO Director and to the AAVSO

Council – the former on a daily basis as an HCO staff member and the latter primarily through correspondence and in twice-yearly meetings as the AAVSO Recorder. Nothing in Campbell's prior experience prepared him to deal with the difficulties in lines of authority and responsibility. Indeed, although those problems existed by nature of the position's dual-reporting relationship, no one, with the possible exception of Shapley, likely thought much about it.

The one saving grace in this situation may be found in the comparatively late arrival of Charles W. Elmer, a well-accepted Old Guard member who had been around the AAVSO for 15 years. Although Campbell's previous experience prepared him for his role as the Pickering Memorial Astronomer in a technical sense, it did not in a managerial sense; Elmer's experience, on the other hand, had almost no technical content but could be said to be directly relevant to the managerial side of Campbell's new position.

Figure 8.2. AAVSO officers 1939: Leon Campbell, Percy Wetherell, Helen Sawyer Hogg, David W. Rosebrugh, Charles W. Elmer.

Figure 8.3. Leon Campbell, Harlow Shapley, Helen Sawyer Hogg, J. E. G. Yalden, and Charles Elmer, Philadelphia, May 1934.

Campbell and Elmer formed a deep bond similar to the one that existed between Campbell and Olcott in earlier years. But Olcott's more senior status in the organization as its founder made it more difficult for Campbell to confer with him when confronted with a decision involving the association. That is especially true in those situations involving Campbell's dual role, a situation for which Olcott could offer no real advice from experience. Elmer, on the other hand, had run his own business and had worked in other business organizations. His daily work involved him in the details of business organizations and business relationship issues, in deal-making, and in getting things done as a practical matter.

Elmer joined the AAVSO in 1918 and soon enrolled as a life member. He and his wife, May, began making substantial donations to HCO through the Association; for example, a 1921 AAVSO gift of $1000 to HCO to support the publication of the *Henry Draper Catalogue*. Elmer served in a number of positions on the AAVSO Council before Olcott died. He served as secretary in 1936 and subsequently as president from 1937 to 1939, which represented an important consolidation of power for Campbell.

The close friendship between Elmer and Campbell during the 1920s provided many opportunities to visit about AAVSO matters, especially when Campbell assisted with the installation of many telescopes at Elmer's Long Island residence and helped him understand how to use those instruments. Campbell finally took a well-deserved vacation in 1930, during which he and his family spent 5 days with the Elmers. When Elmer and Campbell attended the 1935 International Astronomical Union (IAU) meeting in Paris, they roomed together for the period of the meeting.[9]

Thus Elmer's relationship with Campbell and his role as a leader in the AAVSO were well established by the fall of 1936, when Elmer ran into an

Figure 8.4. Charles Elmer and Leon Campbell.

acquaintance from the Amateur Astronomers Association of New York. Richard S. Perkin also attended the Astronomical Symposium that was a part of Harvard University's Tercentenary celebration. During the Symposium, their conversations stimulated a strong friendship; the two agreed to form a partnership in optics that eventually led to the founding of Perkin-Elmer Corporation. Elmer held the position of Secretary-Treasurer of Perkin-Elmer until he retired from the firm in 1948.[10]

During Elmer's terms, the Council included more professional astronomers than had previously been the case. This no doubt reflects several parallel changes that occurred simultaneously, including increased recognition of the possible importance of variable stars theoretically, mixed perhaps with some reconciliation of professional attitudes that the AAVSO could be considered a scientific organization, rather than an amateur astronomy club. The shift may also reflect some recruitment effort on Shapley's part, and finally, it also no doubt reflected professional recognition that the AAVSO's resources might be used opportunistically for studies other than variable stars. In any event, it is clear that the Association's credibility in the professional community improved when the AAVSO drew considerably closer to HCO. In 1933, when the Council elected Ernest Brown as AAVSO's president and Shapley as the vice president, the Council consisted of eight amateur and seven professional astronomers.[11]

Rectifying administrative committees

Over many years, the AAVSO maintained two functions that the Council had considered essential "benefits" for the membership: a circulating library of books and a similar library of lantern slides that illustrated astronomical subjects, including variable star astronomy. The AAVSO Lantern Slide Collection was created at the suggestion of Solon I. Bailey at the November 1918 Council meeting.[12] The Council approved his motion, and HCO astronomer Willard J. Fisher volunteered for a decade or more as slide curator to periodically inventory, catalog, and maintain the slide collection. George Ellery Hale at Mount Wilson donated slides, as did the Yerkes Observatory and other sources willing to help the fledgling organization with a worthwhile service to the community.[13]

Although the slide collection was used in technical presentations and by those members who engaged in public speaking, especially Olcott, its value to the AAVSO and variable star astronomy seems questionable. The routine management of the lantern slide collection demanded a lot of time and energy over the years to periodically organize and update the collection and circulate the slides to members on demand.

The AAVSO library was established in 1921 at the suggestion of Charles Y. McAteer, who donated 200 books from his personal library.[14] He had received the books from the estate of his friend, John A. Brashear, a famed telescope and instrument maker who died in 1920. After McAteer's untimely death, the AAVSO Council named the library in his honor.[15] Other deaths led to the library's rapid growth, and by 1940, when Campbell recognized that he had 40 to 60 valuable books from the eighteenth and nineteenth centuries on a shelf in his office, he transferred those books to Harvard's Widener Library for safekeeping.[16]

Several other committees had been created over the years, but in 1932 the Council reduced them to three standing committees: Charts, Occultations, and Publications. Other functions were delegated to the recorder for the previous committees: Slide Curator, Library, Telescope, and Telescope Making[17] Although this restructuring simplified matters for the Council, the problems associated with cataloging, maintaining, and circulating materials from the library and slide collection plagued Campbell for many years. Campbell had gotten a handle on the telescope problems when he established "The Telescope Mart" as a side business dealing in used telescopes and accessories for a number of years.

Table 8.1. *AAVSO Regional Advisors, 1936*

Region	Advisor	Region	Advisor
Arizona	J. W. Meek	California	H. Raphael
Canada	H. S. Hogg	Florida	T. C. H. Bouton
Illinois	W. Callum	Maryland	P. S. Watson
New York City and vicinity	H. B. Webb	New York – Eastern	D. W. Rosebrugh
New York – Central	C. R. Gregory	Ohio	L C. Peltier
Oregon	R. E. Millard	Pennsylvania – Western	L. J. Scanlon
Texas	O. E. Monnig	Virginia and West Virginia	M. Cilley
Washington	D. F. Brocchi	Wisconsin	L. E. Armfield; then E. A. Halbach

Vetting and training observers

After the Council established a panel of Regional Advisors (Chapter 7), Campbell asked for their assistance in mentoring young applicants. The Regional Advisors, where they existed, helped young observers come onboard with the needed skills. Harold B. Webb agreed to mentor young David W. Rosebrugh of Poughkeepsie, New York, a candidate whom Campbell believed would be "a worthwhile member."[18] His forecast was good; Rosebrugh eventually served 7 years as secretary, a term as president, and a number of years as a Council member. Thus, the Regional Advisors helped spread the leadership burden, but also identified potential candidates for the Council and elected officer positions. In his later years, long-time Council member and benefactor Clinton B. Ford fondly recalled his advisor and mentor, Sigmund K. Proctor.

By 1933, Olcott expressed his satisfaction with the system but identified a growing need for more volunteers and identified some who were not working out.[19] The appointment and supervision of Regional Advisors was originally delegated to the recorder, but the Council later opted to become more actively involved. Typically, these shifting processes will occur in a growing organization. Questions are bound to arise regarding what to delegate to the staff and what to retain within the authority of the elected Council.[20]

By 1936, the number of Regional Advisors had grown and most of the original advisors had been replaced so that the array was as shown in Table 8.1. [21]

Valued as it was, the Regional Advisor program dissolved during World War II, and the Council never reactivated the program afterwards.

Figure 8.5. Eugene Jones, Charles Elmer, and Edward Halbach, Harvard College Observatory, 1938.

Other leadership issues

After Campbell was appointed Pickering Memorial Astronomer, the AAVSO Council faced an entirely new relationship with HCO. Because Campbell's office never changed from 1919 to 1931, it might be expected that there would be no change in Campbell's status. But there were other agreements and understandings that were expressed or implied in that arrangement, and it is clear that the atmosphere changed in a measurable way. In 1932, when the editors of *Popular Astronomy* sought an increased level of support from the AAVSO for the publication of the monthly reports of variable star observations in its pages, they unwittingly created a test case of the relationship. One might have assumed this request would be a routine matter with the members of the Council, but after much discussion it was referred to the Pickering Memorial Committee and through it to Shapley.[22] The Council placed considerable importance on the increasing number of pages of variable star observations in a popular journal as both feedback and recognition for members and as advertising to attract other observers. Observations of variable stars made by AAVSO observers first appeared in the December 1911 *Popular Astronomy* and only covered two pages. By comparison, the list of observations published in the December 1935 *Popular Astronomy* took up 34 printed pages. The number of observations per year had tripled over the previous 10 years. The Council discussed the question far too long, but an agreement to increase the funding and how the funds were to be provided never emerged from these discussions.

Popular Astronomy Editor Curvin H. Gingrich again noted the problem in a letter to Campbell on September 30, 1935, adding that this was an inordinate amount of space to be used in a general interest astronomy publication. Furthermore, at a total cost of $6.00 per manuscript page, *Popular Astronomy* could no longer carry the cost of composing these lists for publication with Campbell submitting 150 to 200 manuscript pages each month.[23] The problem was again discussed at the AAVSO's October 1935 Council meeting, and the final list of observations published in *Popular Astronomy* appeared in December 1935. Without going into the issue of cost, Campbell simply stated that a "change in the manner of publication has been made necessary." And, as though to further mollify the amateur observers who were used to seeing their names and observations in print, Campbell added that the future reports "will be in the same publication [i.e., *Harvard Annals*] with similar observations made by the noted pioneer variable star observers – Argelander, Schönfeld, Schmidt, and others."[24] AAVSO observations began with *HA* volume 104 and continued in volumes 107 and 110. The quarterly installments of these volumes covered observations made from 1935 through 1942.[25]

The handling of the *Popular Astronomy* cost issue undoubtedly suffered because of confusion associated with the death of AAVSO Treasurer Michael Jordan. HCO astronomer Willard J. Fisher agreed to fill in as a temporary basis while the Council identified an individual to take the job on a more permanent basis. For the fiscal year 1930, Fisher was the sole individual involved, but his work was complicated by the special handling required for the Edward C. Pickering Memorial Endowment and the eventual transfer of those funds to Harvard University. At the 1930 Annual Meeting, Percy W. Witherell attended the Council meeting as a guest, and apparently, conversations with the Council and his introduction were sufficient to bring him in as new AAVSO Treasurer.[26] Witherell, who had training in accounting, was treasurer for his family-owned business and lived near headquarters as a Boston resident. Even so, the transition could not have been an easy one. Fisher and Witherell worked for several months to examine and understand Jordan's books as well as the steps taken by Fisher in the interim. For example, there were problems with the way that Jordan had managed funds for the Chart Committee. Fisher's interests were best served by not changing such relationships. However, Witherell soon recognized the problem and took steps to regularize the cash handling and accounting for the chart work. The Chart Committee chairman was authorized to maintain a separate account for the Chart Committee and to ask for additional funds from time to time as needed, but directly from the Council, not the Treasurer.[27]

Seemingly simple problems, when managed by just one person, emerged as more difficult coordination problems when the Administration gradually shifted from Olcott to Campbell. For example, the maintenance of a coherent and accurate list of the paid members consumed substantial time and created frustrating – indeed, at times irritating – problems in the correspondence between Olcott and Campbell. Olcott complained, for example, that he received a ballot from an R. A. Seely of New York City, but his membership list did not reflect any such name.[28] After the Council finally clarified a policy on dropping members for nonpayment of dues, Campbell and Olcott worked to create a list of paid-up members. That project dragged on for a period of time, no doubt extended because of Olcott's illness and travel, as well as Campbell's increasingly heavy workload.[29]

Meetings

Meetings provide another example of the growing complexity of AAVSO administration and leadership issues. The AAVSO annual meetings at HCO fell into a familiar pattern in the 1920s that changed very little for several decades. Much of the work of conducting those meetings was shared by members of the HCO staff, who welcomed, for the most part, the opportunity to renew acquaintances with friends among the AAVSO members. Such meetings provided the faculty and staff with an opportunity to present their work in technical sessions always appreciated by AAVSO audiences, hungry for astronomical science. Shapley's annual state of astronomy talk at the banquet became an eagerly awaited feast of new information and understandings.[30] For the AAVSO membership, these meetings at HCO, especially Shapley's summaries, became the annual highlight.

Shapley's influence soon became visible at the Association's spring meetings in several ways. First, Shapley's well-known interest in the popularization of science led to an invitation to hold the 1929 AAVSO Spring Meeting at the offices of Science Service, Inc. in Washington, DC. Shapley was a member of the Science Service Board and exhibited great pride in the founding of the Science Fairs project for high school students.[31] Obviously the HCO director was anxious to burnish his influence as a popularizer of science in the nation's capital. As the Association's first meeting outside New England, the meeting was an important one for him, if only for that reason. Shapley's presence was again significant at the 1931 meeting at the American Museum of Natural History (AMNH) in New York City, when Olcott created considerable excitement by announcing that the final funding of the Pickering Memorial Endowment had been secured: "Thanks to the interest of Dr. Shapley and the generosity of Harvard University that $100 000 has been set aside for the study of variable stars."[32]

Spring meetings in large public planetaria scheduled at a time when such facilities had captured public attention helped increase public awareness of variable star astronomy. An invitation came from the Franklin Institute in Philadelphia, which had opened the Fels Planetarium in November 1933. The AAVSO meeting there in 1934 was the first meeting of a scientific organization in the Franklin Institute. With the opening of the Hayden Planetarium at AMNH in October 1935, a second spring meeting was scheduled there for 1936.

The next big step involved holding an AAVSO meeting outside the United States – it came with the membership of Helen Sawyer Hogg among the many Canadian observers. Hogg was the only Canadian member who participated actively in the AAVSO leadership at that time. A joint meeting between the Royal Astronomical Society of Canada (RASC) and the AAVSO seemed inevitable, given Hogg's dedication to variable stars and to the AAVSO. She served as President of the AAVSO for the years 1939 and 1940 and was active in RASC leadership.

The 1940 Spring Meeting held in Toronto represents the first time the AAVSO conducted a meeting outside the US borders. The organization of the RASC into geographical centers, rather than topical sections, likely made it difficult for purely topical associations like the AAVSO to receive much attention in RASC leadership circles, concerned instead with servicing the needs of the centers while publishing a fine journal and a widely used annual celestial handbook. At the 1940 Spring Meeting, the AAVSO Council meetings took place at the University of Toronto, whereas the general meetings were conducted at the David Dunlap Observatory. Other than the welcoming speeches, only two other Canadian astronomers were named in the minutes for the meeting. In the Aurora Committee report, Ed Halbach identified Bert Topham of Toronto as a co-author and a co-chairman of the committee, along with himself and Louise Ballhausen of Milwaukee. Quebec resident Paul Nadeau joined the AAVSO and was elected to annual membership. Nadeau would later be elected to the AAVSO Council, only the second Canadian to serve in that capacity. At the traditional banquet, RASC Founder C. A. Chant spoke warmly of the AAVSO's efforts.[33] Several of the Canadian speakers hoped the AAVSO would soon return, but that did not occur for another 17 years.

VARIABLE STAR PROGRAM EVOLUTION

New observers displayed increasing dedication, resulting in the rapid growth of reported observations. Campbell continued to plot observations and published reports focused largely on the long-period variable stars (LPVs). New stars were added to the AAVSO program as more LPVs were discovered on Harvard patrol plates and from other sources and sequences developed by the Chart Committee.

It is probably appropriate to comment at this point on how, in the 1930s, LPVs were defined. With insufficient theoretical tools and understanding, many of the

Figure 8.6. Eugene Jones, Helen Sawyer Hogg, Margaret Mayall, Martha and Harlow Shapley, R. Newton Mayall, Frank Hogg and son David, Clinton B. Ford, and Leon Campbell, at the June 1940 meeting of the AAVSO held in Toronto.

program stars from that period were misclassified; those stars still in the program today are classified in other categories. For example, R Coronae Borealis, SS Cygni, and U Geminorum were all considered LPVs, or probably irregular variables. But in emphasizing that the AAVSO program focused on LPVs, including these few exceptions, we intend to emphasize the fact that there were at that time many shorter period stars now classified as Cepheid, RR Lyrae, RV Tauri, eclipsing binary, and other classes, all of which could be usefully observed by AAVSO observers with small to moderate aperture instruments. Professional astronomers were, in fact, more interested in these shorter period stars than LPVs, so an incentive existed to add them to the AAVSO program. A different approach may be required to accurately observe these shorter period variables: more frequent observations, more accurate recording of times, making and recording brightness estimates more carefully, and so on. But as the passage of time has shown, when those constraints are properly understood, AAVSO observers were as capable and willing as anyone to comply.

Theoretical interest in variable stars stalled with the mathematical models of A. S. Eddington in the middle and late 1920s. For a period of time, professional interest drifted away from phenomenological interest in LPVs to other emerging topics in astrophysics and observational cosmology. However, interest in the subject among amateurs never faltered: for instance, J. Wesley Simpson of Webster Groves, Missouri, read his paper on why variable stars vary at the 1933 Spring Meeting.[34] The fact that his paper was credible enough to be accepted for presentation at the meeting also lends support to the likely openness of AAVSO observers to expand the observing program to shorter period stars. The lack of an appropriate theoretical framework within which to work on LPVs, perhaps, accounts for the willingness of Shapley and others to push AAVSO observers into other observing projects like lunar occultations.

What seems to have been missing in all this was much attention to the AAVSO's growing database of observations. Campbell routinely ledgered the observations, calculated 10-day means, and hand-plotted mean light curves for his use in making predictions of maxima and minima. Campbell's predictions were of great value to professional spectroscopists like Paul Merrill. The data accumulated rapidly in the files, but, in essence, no one really used them to support further theoretical studies.

The power of the accumulated data for at least phenomenological studies might have gone unnoticed, except for the surprising discovery made by journalist and amateur astronomer, Felix de Roy, in Belgium. Looking at his own data and that available to him as Director of the British Astronomical Association-Variable Star Section (BAA-VSS), de Roy discovered a class of stars that should be separated from the LPVs and, perhaps, observed more frequently. This new class of variable stars is now known by the name of one of the two stars de Roy studied to reach his conclusion, Z Camelopardalis, and the other being RX Andromedæ.[35]

The Chart Committee

Variable star workers from around the world convened to discuss chart and sequence problems during the IAU's triennial gathering in September 1932 in Cambridge, Massachusetts. The meeting provided a convenient forum to consider the sensitive issue of sequence standardization. S. A. Mitchell at the Leander McCormick Observatory had been providing sequences to the AAVSO Chart Committee for the fainter stars needed for observations of LPV minima. When this became known, he received similar requests from the BAA, AFOEV, and other groups also involved in making charts. At the IAU meeting, astronomers agreed that the Harvard sequences would be adopted as the international standard. This signal achievement, a relief to all concerned, facilitated the sharing of variable star data internationally with confidence that the observations were consistent.[36]

With steady progress over the previous decade and more, the AAVSO Chart Committee issued a catalog of the charts available in the new format drafted by D. F. Brocchi. Without complaint, the Committee continued its work as new stars were added to the program and new corrections identified. By 1934, David Pickering felt his work was complete, or at least it was time to resign after nearly 20 years of involvement. He and Campbell persuaded Brocchi to accept the Committee chairmanship, for which Brocchi felt honored. Brocchi contributed not only to the regularization of the chart drafting, but also completed an "all-sky" AAVSO atlas showing the position of program stars relative to each other on a scale useful for planning both nightly and long-term observing programs and developed a series of finder charts for individual stars and/or groups of stars that facilitated locating variables that were difficult to find. His contributions to this effort were already very substantial before assuming chairmanship of the Chart Committee.[37]

Brocchi's dedication typified that of many volunteers on this and other committees at the time. It was easy to take that dedication for granted, and that resulted in some miscalculations as well. Chart Curator Helen Sawyer Hogg, recently graduated, married, and starting a family, asked to be compensated fully for all the expenses involved in performing the role – postage, mailing, and other necessary office expenses. The older, more settled members had never asked for compensation for these expenses as Chart Curator, only for the actual cost of making large chart purchases. In an undated letter to the Council, Olcott warned the Council that it needed to accede to Hogg's request, but the warning apparently went unheeded. Hogg resigned in a huff so the Council was faced with changing Chart Curators again.[38] The change was not a trivial one, involving the transfer of a large inventory of charts and records. The Council accepted Hogg's resignation with regrets. After further discussion, they appointed Ferdinand Hartmann as the Chart Curator, and replaced David Pickering with Harold Webb on the Chart Committee.

Webb traveled from his home on Long Island to the Pickering home in New Jersey to learn more about the AAVSO chart work and pick up files that Pickering had not already shipped to Brocchi. In a letter to Campbell, Webb noted that Brocchi was in favor of "printing our regular charts in black and white on 8 x 10 paper the same method as our practice charts."[39] The fact that Brocchi's recommendation, which would have been a savings, was not accepted provides an indication of the popularity of the AAVSO blueprint charts at that point.

By 1937, Brocchi and the Chart Committee announced the completion of the first round of the Leander McCormick sequence revision work. As always, the Chart Committee's work was never done. New stars, mainly LPVs, were slowly but continually added to the AAVSO program.

Nova search

While still in Arequipa, Leon Campbell had proposed a formal effort for searching for novae as early as 1912, following Siguard Enebo's discovery of Nova Geminorum in Norway.[40] Olcott remarked on the discovery in his monthly reports in *Popular Astronomy*. However, the AAVSO was still a young organization; no one responded and there was no formal mechanism for creating a committee.[41] Olcott, who was preoccupied with recruiting members and formalizing charts, had no time and little interest in taking on more activities for the fledgling organization.

Interest in a dedicated search for novae first appears in AAVSO files in the early 1920s, when Issei Yamamoto in Japan joined the AAVSO and began a systematic effort to discover bright novae. However, no formal structure yet existed in the AAVSO to support that effort.[42] Such searches had been the topic of discussion for many years, especially when David Pickering designed and crafted a Nova Discovery Medal that was adopted as an AAVSO award by the Council in 1928. J. P. M. Prentice of Needham Market, Suffolk, England, received the first AAVSO Pickering Nova Discovery Medal for Nova Herculis on December 13, 1934. AAVSO President Ernest W. Brown presented the medal to Prentice at the June 1935 BAA meeting.[43]

A Nova Search Committee was formed after Nova Herculis 1934 stimulated widespread attention to variable star astronomy. Luverne Armfield of the Milwaukee Astronomical Society took an early lead role. After that effort faltered, the Council appointed a new Nova Search Committee consisting of Roy A. Seely as chairman, Grace S. Rademacher, and David Rosebrugh. The new committee developed a plan for searching selected areas of the sky in which a nova could be expected and then prepared both an overall map of the search area and more detailed maps of each search field. When it was discussed at the 1940 Annual Meeting in Toronto, David Pickering evidently thought their plan a good one for he complained humorously, in opposition to the plan, that the larger number of discoveries flowing from this new approach would result in more AAVSO Nova Discovery Medals.[44]

Red Stars Program

The Red Stars Program involved a small group of observers who were given special charts and special red filters through which to observe the stars frequently throughout their light cycle to demonstrate whether observational scatter could be reduced using that technique. The selected observers included Rosebrugh of New York, Peltier of Ohio, Halbach and Arthur Peck from Milwaukee, Mark Howarth, Rosina Dafter, and another observer from the BAA New South Wales VSS in Australia. A longstanding interest existed in red stars, starting with the early catalogs of such objects by the Rev. T. H. E. C. Espin, John Birmingham, and later, A. F. Kohlman.[45] The difficulties of observing very red variable stars without a filter introduced more scatter in the AAVSO light curves than for ordinary LPVs. As Rosebrugh and Clinton Ford later described the results, too few observations were gathered before the committee members lost interest and the program faltered. The reason, they said, was that the filters blocked so much light from all the stars in the eyepiece field of view that only the brightest part of the light curve of the program variable stars could be observed.[46] One thing that was learned was that the amplitude of variation observed in red light appeared to be less than in visual light. Later, Leon Campbell told Ed Halbach that the recent developments in red-light photography would allow more accurate light curves to be observed photographically, so there were no regrets regarding the eventual failure of the Red Stars program.[47]

Photographic program

In the 1930s, there were many individual attempts among AAVSO observers to apply photography to variable star astronomy. This was understandable with the interest among scientists in the application of photography to all sorts of problems for which a permanent record was invaluable. The issue was, so to speak, "in the air" because Eastman Kodak advertised the achievements in their own laboratories and encouraged others with their successes. Kodak chemist C. E. K. Mees promoted the application of photography to astronomy and developed an excellent reputation among astronomers who valued his recommendations.[48]

The newly established Milwaukee Astronomical Society (MAS) in 1935 promoted photography in its observing program. In Connecticut, George Waldo was a proponent of photovisual photometry as a substitute for standard visual observing or observing with visual photometers. In 1933, he still held that photovisual observing and calibration of charts should be accepted.[49] Others agreed. By 1935, MAS member Lynn H. Matthias practiced photographic work from a rooftop observatory he set up on an apartment building in Milwaukee.[50] In addition to taking photographs of the night sky, Matthias borrowed photographic plates from the Yerkes Observatory and studied a few Cepheid variables, including SU Cygni. He wrote Campbell asking for comments on his magnitude estimates. Encouraged by his photographic research, Campbell replied in a lengthy letter 2 months later. Campbell described other photographic work by AAVSO members; his suggestions were vague with respect to taking pictures, but very explicit on photographic plates. In effect, he held out the prospect that Matthias' photographs could be used and that the HCO was willing to send Matthias plates from their archives from which to extract additional historical observations. A precedent had already been established with AAVSO member Noah McLeod of Christine, North Dakota, although McLeod's work actually occurred outside the AAVSO framework. Campbell then asked if Matthias would be willing to take charge of such an activity if the Council encouraged a program of work.[51] Matthias must have answered affirmatively. In October he reported to Campbell on his recruitment of members for the committee. However, a year later he reported that not much progress had been made due to "the present uncertain observing program. This I have every hope will be rectified after a discussion of the matter with Dr. Shapley and yourself."[52] By the end of the year Campbell had proposed, with Shapley, a program published in *Popular Astronomy* for 10 regions of the sky, each about 100 square degrees in size, as the target areas for the AAVSO Photographic Committee. Furthermore, Campbell let Matthias know that AAVSO had arranged with HCO to loan one of its spare patrol cameras to Matthias to get started on the program.[53] Within 2 months Matthias sent Campbell prints made from plates taken of several fields and reported he was corresponding with other interested observers. Seemingly, the program gained momentum in these exchanges.[54] Campbell acknowledged his "very excellent photographs of the AAVSO fields. Dr. Shapley quite agrees with me that they deserve nothing but the highest commendation. We are having the variables on the fields identified and on the charts."[55]

By 1938 Campbell cited a number of individual efforts by AAVSO members to apply photography to variable star astronomy[56] and the purchase of six cameras with 3-inch objectives and focal lengths of about 30-inches.[57] But Matthias had nothing to report for 1937–1938; work pressures and other obligations had overwhelmed his best intentions. He offered to resign, but vowed he would continue his own photographic program.

The Council discussed the Photographic Committee in 1938 at Brown University's Ladd Observatory in Providence, Rhode Island, citing the difficulty and expense of this work. Measuring the plates with a microphotometer seemed to have limited interest to only a few individuals like Matthias with specialized skills. However, there were others interested in a photographic program, including Leo J. Scanlon, Joseph L. Woods, Claude B. Carpenter, Clifford Seauvageau, and Albert E. Navez, so President Elmer appointed a committee consisting of Woods as chair with Dr. Navez and Scanlon as committee members.[58]

Even under new leadership, the Photographic Committee never really functioned effectively. As Ford and Rosebrugh reflected, "apparently the calibration of the plates, and the general expense involved, were greater burdens than an ordinary member could bear." Matthias may also have had constraints on his time. Every volunteer organization like the AAVSO experiences such problems.

OCCULTATION COMMITTEE

The first successful observing program organized as an AAVSO committee was the Occultation Committee – formed in 1926 by Ernest Brown from Yale University. The enthusiastic chairman, J. Ernest G. Yalden, a fanatical perfectionist, obviously enjoyed both the observing experience and reducing the observations to a form usable in correcting the Lunar Equation. In his 1929 Occultation Committee report, Yalden was encouraged because 11 of the 19 observers reporting occultations in the *Astronomical Journal* were AAVSO members, who made 200 of the 277 observations, a satisfying result for 1 year's activity.[59] The following year, Yalden reported that 302 reductions had been completed by 13 computers, many of whom were professional women astronomers: Arville D. Walker, Alice H. Farnsworth, Marjorie Williams, Harriet W. Bigelow, M. A. Hawes, and Caroline E. Furness. Yalden with his friend in Leonia, New Jersey, John Ingham, and five other males, accounted for the remainder.

Brown, who conducted the celestial mechanics program at Yale University, assumed a strong presence in the Council and impressed its members to such an extent that after only 3 years he was elected AAVSO president in 1933. The timing of lunar occultations may have been satisfying because it was in a high state of mathematical refinement and involved a substantially different type of work compared with variable star observing, which seemed to lack any theoretical basis.

Figure 8.7. Walter Scott Houston, Joseph W. Meek, and Leslie C. Peltier, AAVSO Spring Meeting at Williams Observatory, Hood College, Frederick, Maryland, May 1932.

RECOGNIZING CONTRIBUTIONS

Nevertheless, variable stars were the Association's *raison d'être*. They formed the main attraction that brought amateur astronomers and their telescopes into the Association in the first place. Observing the ever-changing nature of the heavens signaled by the varying brightness of individual stars over observable periods constituted the work on several hundred backyard and rooftop observatories around the world nightly. Stephen C. Hunter put it nicely in his letter to Olcott: "The very thought of locating and studying the changeful phenomena of such an object is indeed enough to stimulate the most sluggish imagination."[60]

The international effort was seen in the increasing observations – numbers scribbled on sheets of paper – flowing into the AAVSO Headquarters at Harvard College Observatory month after month. Although the monthly reports initially only included a dozen observations, Campbell noticed that the sheets had expanded to hundreds of observations by the 1920s. A few individuals were recognizable for both their regularity and for the volume of observations, particularly Leslie C. Peltier, a young observer on a farm in Ohio. By 1930, Campbell was frequently citing Peltier as he described the work of the Association, such as in this *Harvard Bulletin* article: "One of the best observers is a young stock-clerk in a mid-western town, who, incidentally, finds time between regular variable star observing to hunt for comets, with more than average success." Campbell also cited Giovanni B. Lacchini in Italy, whose variable star observing led to his appointment as a professional astronomer there.[61]

By 1931, Campbell wanted to do something to recognize Peltier and his observations. In a letter to AAVSO President Harriet Bigelow, he suggested that the Council appoint Peltier as an honorary vice president.[62] The Council considered the idea without any action. But Campbell persisted, and finally the Council agreed with the idea of recognizing an outstanding performance. Shortly afterward, Peltier observed an outburst of the recurrent nova RS Ophiuchi when it was still brightening. Such observations were rare and extremely valuable because they allow a spectrum to be taken, which in this case confirmed the status of the star as a recurrent nova.[63] When Campbell raised the idea of an award for Peltier at the 1933 Annual Council meeting, a committee was appointed to consider an AAVSO prize award, and those present thought it should be something other than money, possibly a book accompanied by a specially engraved memorial, but not necessarily awarded every year.[64]

The creation of the AAVSO Merit Award was finally approved by the Council, with agreement that the award should consist of a certificate, accompanied by a prize. The first award, a copy of the Flamsteed star atlas *Atlas Célestis*, published in 1781, was presented to Leslie C. Peltier of Delphos, Ohio, at the 1934 Annual Meeting, one of the few he ever attended. An elegantly decorated parchment was inscribed to Peltier, "whose faithful and untiring service has placed him in the front ranks of variable star observers and his discoveries have won him international fame."[65]

There was no Merit Award given in 1935, but the Council soon took advantage of this new award to honor other faithful members. The Council honored the Rev. Tilton C. H. Bouton in 1936 with the second AAVSO Merit Award at the last Council meeting that Olcott ever attended. Campbell then apparently nominated Eugene Jones, David Pickering, and Eppe Loreta for AAVSO Merit Awards. After discussion, the Council sent a letter of commendation to Loreta.[66] Eugene Jones, admired by many for his prodigious observing efforts, received the fourth AAVSO Merit Award in 1937.

Initially, the Merit awards went to the leading observers (Peltier, Bouton, and Jones), seemingly ignoring the contribution of several others. Without Olcott's determination and dedication, the organization might not have existed or survived. His contributions were recognized when he was named the recipient of the third AAVSO Merit Award at the 1936 Annual Meeting, after his death. David Pickering had performed yeoman service for the Association over that same period in many different ways, but was not among the most productive observers. That long overdue recognition came to Pickering in 1938 as the recipient of the fifth AAVSO Merit Award.[67]

EVOLUTION OF AMATEUR ASTRONOMY IN THE 1930S

As mentioned in the opening of this chapter, the economic strictures imposed on personal expenditures by the Great Depression, as well as frequent radio and newspaper references to progress in astronomy and astrophysics, led to a substantial increase in avocational participation in astronomy. To a large extent, that added participation was recreational in nature, but the AAVSO also experienced several favorable trends that can be traced to these same influences.

The AAVSO would eventually benefit from the widespread interest in telescope making, known as the amateur telescope-making (ATM) movement, and a substantial growth in the number of local astronomy clubs, some of which encouraged scientific research rather than recreational observing. More importantly, the desire for better telescopes and the abilities to create them without substantial expenditures encouraged individuals to obtain larger apertures than previously available in small refractors. This section will discuss these separate but related phenomena and their impact on the AAVSO.

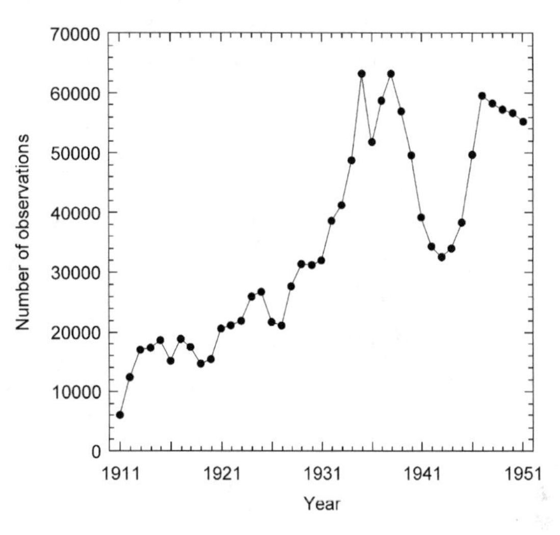

Figure 8.8. Annual totals of variable star observations received by the AAVSO, 1911–1951.

The amateur telescope-making movement

A sudden upsurge of interest in telescope making occurred during the 1930s and is easily identified, at least in the United States, with the monthly columns of Albert G. Ingalls in *Scientific American*. Through Ingalls' evangelistic approach to the hobby, the number of telescopes and the average size of telescopes available for astronomical observing increased dramatically between 1926 and the beginning of World War II.

What is important, for our purposes, is the impact that the ATM movement had on variable star observing and the AAVSO. One noticeable effect that might be predicted with the availability of the larger and less expensive telescopes would be a growth in AAVSO membership and a corresponding growth in the number of observations. The number of observations reported to the AAVSO on an annual basis during the 1930s rose rapidly compared with the two previous decades (Figure 8.8).

There are several other likely explanations besides growth in the number of observers, including the fact that more variable stars were being discovered and charted for AAVSO observers, along with better comparison star sequences for the fainter phases of the long-observed LPVs. However, having more telescopes in the hands of individuals who wanted to contribute to the scientific progress of astronomy was undoubtedly a contributing factor. Over a decade or so, more than a few amateurs completed 8- and 10-inch mirrors that became their primary variable star observing tools.

Organizations for amateur astronomers

Astronomy clubs were being formed in many cities and towns with the interest in astronomy during the early 1930s and the ATM movement. The major planetaria, for example, became the meeting places for a local society once each planetarium opened its doors.

However, the character of these clubs varied, with some emphasizing telescope making and others regarding telescope making as a means to an end, specifically as a way to develop tools for doing scientific work as amateur astronomers.[68] One particular example that illustrates this latter approach and had a significant impact on the AAVSO is the Milwaukee Astronomical Society (MAS), founded in 1932 by Luverne E. Armfield and E. A. De La Ruelle. Initially, some among the 12 founders appeared to be attracted to telescope making. Others, responding to the leadership of professional physicist Edward A. Halbach, soon concluded that telescope making was essentially a means to an end; the ATM hobby never preoccupied the MAS membership. Instead, the MAS met in Armfield's backyard and adopted regular programs of scientifically oriented observing, including variable stars, meteors, and occultation timing.

The MAS membership's contributions of variable star observations to the AAVSO in the decades since the MAS was founded stand as a monument to a club-oriented dedication to science.[69] By 1937, the MAS had developed stature in the community and acquired a dark observatory site near New Berlin, Wisconsin, a Milwaukee suburb. Over time they developed the site with several telescopes in shelters. One of those telescopes, a Newtonian reflector, featured a 13-inch plate glass mirror donated to the MAS by the AAVSO. As the MAS gained recognition nationally, AAVSO member Ralph Buckstaff of Oshkosh, Wisconsin, donated a full telescope and mounting, a Newtonian of 12.5-inch aperture, to the MAS observatory.

Two other clubs were formed that became a part of the threesome that would change the scene for national organizations. One, the Missouri-Southern Illinois Observers – or MSIO – was formed in Webster Groves in the St. Louis area, under the leadership of teens J. Wesley Simpson and Donald Zahner. MSIO published a newsletter, *The Astronomical Discourse*. The other club was the Madison Astronomical Society in Madison, Wisconsin, which organized and eventually published the *Madison Bulletin* and attracted the editing and production skills of a budding journalist, University of Wisconsin student Walter Scott Houston. Their efforts led to a national organization with a first-class publication, but the effort could not be sustained for more than 5 years. What never faltered, however, was the MAS commitment to the AAVSO in this period from 1934 onwards. Nearly every AAVSO meeting attracted at least one automobile-full of participants from Milwaukee. Armfield attended regularly and was elected to the AAVSO Council in 1934, serving 4 years, including 3 years as a vice president. Halbach also served on the Council in the same decade. More importantly, the MAS observers always supported AAVSO observing programs. In the early 1930s, that meant primarily observing LPVs and other variable stars and timing lunar occultations. As other AAVSO programs were added, MAS members provided some leadership, as well as active participation.

Individual contributors

Although many individuals in the ATM movement initially observed enthusiastically with their telescopes, inevitably, many telescopes ended up in closets, and the fledgling recreational observers moved on to other avocational pursuits. Of the few who continued to observe, there were some who tired of recreational observing and wanted to contribute to science.[70] Those with scientific aspirations who gravitated to the AAVSO eventually became strong supporters and substantial observers, like those from the Milwaukee group. Lancaster Hiett, a young junior high school science teacher in Virginia, worked with guidance from long-time AAVSO member Morgan Cilley. Hiett crafted several mirrors that he mounted, the largest being an 8-inch Newtonian telescope. Hiett joined the AAVSO in 1936 and steadily observed the regular LPV program stars, gradually adding other stars to his program. Rarely missing a monthly report, Hiett became one of the more reliable contributors to the AAVSO database, and from 1936 to 1998 he submitted 115 918 observations. He paid careful attention to the morning stars when the Moon was in the evening sky and generally followed textbook planning and observing practices.

Cyrus Fernald of Wilton, Maine, came into membership indirectly through the ATM movement. An unmarried CPA, he returned to Wilton to assist his parents after a period of employment as an efficiency expert for E. I. DuPont de Nemours and Company in Wilmington, Delaware. Fernald began a correspondence with Campbell, exploring such topics as what was involved in observing variable stars and the necessary equipment. He then acquired an 8-inch f/9.4 Newtonian telescope in a Springfield mounting from an amateur telescope maker who lived nearby.[71] After being satisfied that he was "ready," Fernald applied for membership in the AAVSO in 1937, received charts, and began observing with a passion. His enrollment signaled a very different approach to variable star observing. His telescope and its mounting and other aspects of his observing set-up were designed to be ideal for variable star observing.[72] Setting out to become the AAVSO's leading observer, he clearly placed himself in competition with Leslie Peltier, who had a long lead in total observations. Applying the lessons he learned as an efficiency expert, Fernald maximized his observations on a monthly basis. Besides mastering the Springfield mounting, he observed LPV stars that would be near minimum when the Moon was absent, leaving brighter stars to the times when moonlight would not interfere. Using the setting circles built into the Springfield mounting and visible from the fixed eyepiece, Fernald navigated the skies with speed and precision on a plan designed to minimize the time spent traversing from one star to another. By 1939, the perfectionist Fernald was also calling Campbell's attention to errors in the published results in the *Harvard Annals*.[73]

In other cases, the interest in variable star observing waned when new observers found themselves confronted with the charts, the telescope, and the night sky. It would be easy to dismiss all these as frivolous, but a few like Leo J. Scanlon seriously tried it before finally losing interest in observing and returned to telescope making and recreational astronomy. Scanlon's case is worth considering briefly because he made a substantial commitment to the AAVSO in time and energy before giving up. He joined the AAVSO in 1933 and began submitting observations to Campbell. After attempting to systematize his observing procedures, Scanlon wrote a thoughtful paper on how to observe variable stars seriously and with the greatest efficiency.[74] His efforts were reliable enough in Campbell's view to merit appointing Scanlon as a Regional Advisor for Western Pennsylvania. After 4 years of variable star observing, Scanlon stopped submitting observations to the AAVSO. He remained active in Pittsburgh amateur circles for a number of decades, but mainly as a planetarium demonstrator and ATM instructor.[75]

PROGRESS IN VARIABLE STAR ASTRONOMY

The keys to a coherent theory of Cepheid variability remained inextricably tied to the successful development of an overall theory of stellar evolution. That process, only just beginning by the early 1930s, heightened with the publication of A.S. Eddington's *Internal Constitution of Stars*. After Eddington, the more mathematical forms of astrophysics assumed importance as various areas of physical theory were applied to understand possible mechanisms for radial pulsation and later nonradial pulsation in terms of not only thermodynamics, but also fluid dynamics and magnetic field theory.[76]

Observational astronomy continued to produce fruitful results that enhanced understanding of stellar evolution as time passed. Theoreticians waited for data while observational astronomers struggled to provide increasing numbers of spectroscopic parallaxes, the measurement of some stellar diameters, and rotation rates for nonvariable stars. From these data, theoreticians could infer stellar masses and densities, as well as temperatures. Paul Merrill's patient studies of LPVs at Mount Wilson continued to produce their fundamental properties in the same vein. At Harvard, the Rockefeller grants resulted in the establishment of the Milton Bureau under the directorship of Cecilia Payne-Gaposchkin and Sergei Gaposchkin to mine as much useful information as possible about the variable stars captured on the hundreds of thousands of photographic plates in the HCO archives. All this information was grist for the theoretician's mill.

Inquiries to the AAVSO offices indicate the growing importance of observational data for various purposes. In answering a query, Campbell noted several

Figure 8.9. A view of AAVSO Headquarters – Leon Campbell's office in HCO's Building A. From left: Campbell, unidentified, and Campbell's assistant, Helen Thomas, 1936.

astronomers outside of Harvard relied exclusively on AAVSO data, including Hans Ludendorff in Germany and Albert Nijland in the Netherlands.[77] Jesse Greenstein at Yerkes Observatory reflected a growing recognition of the value of the AAVSO and the light curves of LPVs that were the primary product of its work. As part of the preparations for a meeting of the American Astronomical Society (AAS) at Yerkes to mark the observatory's 50th anniversary, Greenstein asked Campbell to provide an exhibit of AAVSO light curves.[78]

NEAL J. HEINES AND SOLAR ASTRONOMY

The AAVSO's involvement with solar observing was first evidenced in a letter to Campbell of November 5, 1937, in which member Neal J. Heines (1892–1955) stated that he had sent a copy of his Zurich solar observation reports to Donald Menzel at HCO "for his criticism and recommendations, with the request that it be shown to you and that this might be considered as an A.A.V.S.O. activity."[79] Although no action was taken on his suggestion, Campbell encouraged Heines to continue with his solar and variable star observations in a September 9, 1939, letter noting the need for more US amateur observer involvement, as the war had interrupted observations received from Europe. In his response, Heines regretted that he was unable to contribute any variable star observations because he was making solar observations and reductions for contribution to Zurich, adding "But I love it . . . it is all so interesting." Heines, who joined in 1934, first turned to variable star observing, but his interests soon were consumed by solar observing.[80] After contacting Dr. William Brunner at the Federal Observatory, Zurich, Switzerland, on July 9, 1934, he had been sending his solar observations there independently. The IAU had appointed the Federal Observatory as the central office for the collection and dissemination of solar data.[81]

EDWARD A. HALBACH AND THE AURORAE

Although the AAVSO did not have any program for aurora observing, the efforts of one member, Halbach, and MAS brought the AAVSO into a program sponsored by the National Geographic Society. The Association had never responded to previous member interest in the aurorae.[82] When Cornell University physicist Carl W. Gartlein requested auroral observations, Halbach responded with an ambitious plan.[83] From two separate sites, MAS would attempt to photograph an aurora simultaneously to yield information regarding the lowest altitude at which the aurorae occur and possibly their overall height. Halbach's plan called for ham radio stations at each site so that the photographers could make synchronized photographic exposures with identical cameras.[84]

Halbach first introduced the involvement of the Milwaukee astronomers at the 1939 AAVSO Spring Meeting in Ann Arbor, Michigan.[85] The Milwaukee group presented two more papers on aurorae at the 1939 AAVSO Annual Meeting. At that meeting, the Council established an AAVSO Auroral Committee and appointed Halbach as chairman.[86] Halbach's first report to the Council demonstrated the amazing progress in this program at the 1940 Spring Meeting at Toronto. Joined by Bert Topham of Canada and Louise

Ballhaussen of Milwaukee, Halbach reported that the two sites were fully equipped and operating successfully.[87] The aurorae reports enriched the agenda of the 1941 annual AAVSO meeting as well. Halbach reported on AAVSO's observing program, whereas Donald Kimball exhibited a painting of an aurora. Clearly, Halbach's program succeeded, whereas Neal Heines' efforts to start a solar observing program continued to falter.[88]

A NEW PRESIDENT FOR HARVARD

The relationship between the AAVSO and HCO was, by this time, so intertwined that we pause, briefly, to note that Harvard President Abbott Lawrence Lowell retired in 1933 after 36 years of service to the University.[89] Over his 24 years as president of Harvard, Lowell had a great influence on HCO, especially through his selection of Harlow Shapley as Edward C. Pickering's successor, a choice that had a substantial impact on the AAVSO. For that reason, it seems important to take more than a passing interest in his replacement during this critical period.

The Harvard Corporation plucked Lowell's replacement, James Bryant Conant, from Harvard's chemistry faculty. Conant's first preoccupation, even before his inauguration, consisted of planning for the Harvard Tercentennial celebration in 1937. The 2-week long celebration brought distinguished academics from many fields to campus to lecture on the latest developments, including Henry Norris Russell's speculations on the recent discovery of cosmological expansion, which he termed "the Cosmic New Deal."[90] Conant showed little interest in astronomy in the first years after the Tercentennial. Instead, he focused on academic matters in other sciences. In contrast to Lowell, Conant was interested in graduate education, especially in chemistry, physics, and biology, and on improving related campus assets – for example, the laboratories, museums, galleries, and the botanical gardens.[91] His interest in resolving the difficult relations between those separate institutions and the academic departments they represented to the public foreshadowed his later interest in the HCO and the astronomy department.

The Story of Variable Stars

As the new observatory programs matured, Shapley and his senior colleague, Bart Bok (1906–1983), planned for a new series of books in 1936, which they styled the *Harvard Books on Astronomy*, to fill a niche in the books available to informed readers with an interest in astronomy. In contrast to the available popular astronomy books, the Harvard series would emphasize the work of the astronomer (i.e., the tools and practical aspects of astronomy), as well as provide technically competent but nonmathematical explanations of astronomical objects and phenomena.[92] Astronomical topics were divided among the various faculty and staff members of the observatory. With agreement on the content of the series, a contract to publish the books was executed with The Blakiston Company of Philadelphia. The series of nine books were scheduled for publication in June and September 1941, but when the schedule was not met, Shapley and Bok turned to Campbell. Recognizing both his heavy normal workload and limited theoretical qualifications, they invited Italian refugee Luigi Jacchia to join with Campbell to write *The Story of Variable Stars*. Jacchia, who received a Ph.D. in physics from the University of Bologna in 1932, had published an Italian book on variable stars in 1933.[93] After fleeing the anti-Semitic political regime in Italy, he was employed in 1939 at Harvard as a research assistant. Jacchia provided a comforting level of technical expertise, whereas Campbell's authorship almost guaranteed the popularity of the volume among the growing AAVSO membership and other amateur astronomers. Together, their presence would appeal to exactly the audience Shapley and Bok most wanted to reach. The task of co-authoring was not easy, as Campbell revealed to his close friend Charles Elmer.[94] The book was published nearly on schedule in October 1941, sold well, and was reprinted in February 1945 and January 1946.[95]

SIGNS OF COMING CHANGES

The last few years of the 1930s gave ample evidence of coming changes in the international situation, which had mixed effects on the astronomical community. Campbell became aware of the effects of the war. George Ensor reported that a sequence of mishaps and the onset of World War II delayed construction of the Radcliffe Observatory in Pretoria. Ensor also reported that his work supervising an air force hospital laboratory drained all his energy, leaving none for variable star observing or occultation work.[96]

Figure 8.10. At the home of Giovanni B. Lacchini, Faenza, Italy, June 23, 1928. From left: Mario Ancarani, Bianca Rosa Lacchini, G. B. Lacchini, Tyler and Clara Olcott, Emma Lacchini, Domenico Benini.

Figure 8.11. Edwin F. Carpenter of the Steward Observatory, and his wife, Emily, with William Tyler Olcott, Arizona, fall 1931.

Refugee scientists emigrated to the United States throughout the 1930s. Shapley hired several at the HCO, including Jacchia, Martin Schwarzschild from Oslo, Richard Prager, and Sergei Gaposchkin from Berlin, and Zdeněk Kopal from Prague. Shapley invested a great deal of energy in the refugee scientist problem otherwise – a point that could not be missed by those around him.

Even in the United States, there were signs pointing to eventual American involvement in the war. In the ATM community, closely connected to amateur astronomy, there was talk of the coming need for optical workers and optical parts. Fred E. Ferson began to hone his skills in the production of specialized optics like the roof prisms for binoculars, sure to be a high demand item in the event of a war. Ferson did so after reading of this possible need described by Carl L. Bausch of Bausch and Lomb Optical.[97]

OLCOTT'S LATER LIFE

After William Tyler Olcott's health deteriorated in June 1929, he never fully recovered his previous vigor. His wife, Clara Olcott, determined to preserve what was left of Tyler's life as well as possible, signaled Campbell that she would not allow her husband to participate in the AAVSO with the same intensity he had for the previous two decades.[98] A brief return to AAVSO affairs in May 1930 only led to another decline in September 1930. The Olcotts took refuge from the New England winter in Tucson, Arizona, where Olcott regained some of his vigor, allowing him to explore his astronomical interests in different ways. Edwin F. Carpenter (1898–1963), an astronomer who managed the Steward Observatory at the University of Arizona in Tucson, and his wife Emily befriended the Olcotts after learning of his long involvement in variable star astronomy and his illness.[99] Carpenter gave Olcott a key to use the facilities at the Steward Observatory for some observing.

Upon returning to New England in the spring of 1931, Olcott attended the critical AAVSO meeting in New York City, when the Council reached an agreement to transfer the assets for the Edward C. Pickering Memorial Endowment from AAVSO to Harvard University. After the meeting, Campbell recognized that the Olcotts were "looking in better health."[100] His continued improvement seemed apparent when he wrote in November 1931 from Tucson describing his work at the observatory, helping Carpenter.[101] By December 1931, Olcott was well enough to observe variable stars again.[102]

During January 1932, Olcott became involved in a debate of a Pickering memorial tablet that would hang on the wall in Campbell's office. Olcott complained that the tablet was inadequate without ever spelling out his objections, which seem to boil down to the language focusing too much on Pickering's scientific achievements and not enough on his contributions to the AAVSO. Then, after reminding Campbell that the memorial had originally been proposed as an observatory that would double as the Association's headquarters, Olcott seemed resolved that the bronze tablet would have to serve as that physical memorial "Because we have wisely chosen to embody the Memorial in a Fund is no reason why the facts should not be as widely known." Without reading too much into Olcott's formal language, his statement to Campbell expresses a concern that the tablet "in a dignified way express the actual facts. They are facts that all who contributed may well be proud of. Such an expression would be an enduring reminder that the Association founded by Professor Pickering was not lacking in gratitude, and not unmindful of their obligation to him."[103]

We take all this to mean that in Olcott's view, and likely in contrast to the originally proposed wording, the tablet should simply, and with dignity, state that Pickering founded the AAVSO, that the AAVSO was grateful for his support over the years after founding, and that the AAVSO created the memorial in his honor. In Olcott's opinion, the rest should be left to others – including historians.

Campbell's response implied that he believed that Olcott was crediting the AAVSO as the source of the money for the endowment. Campbell reminded Olcott that the AAVSO contributed only one-sixteenth of the necessary funds. In his view, therefore, the AAVSO should not claim that the endowment represented the AAVSO's memorial to Pickering. "On the whole," Campbell wrote, "I believe that those who have discussed the matter with us feel that the tablet should contain just enough to make things clear as to what the Memorial is."[104] The discussion continued in a number of letters between Olcott, Campbell, and others. At the 1932 Annual Meeting, the full Council approved the concept, wording, and expenditure to produce the tablet.[105] After more than 3 months of discussion, the actual tablet, as cast and mounted in Campbell's office, contained final wording very similar to that which Olcott had proposed:

> In Memory of / Edward Charles Pickering / Founder of the American Association / of Variable Star Observers / Attesting the establishment in 1931 / of the Pickering Memorial Fund / to promote the study of / Variable Stars.[106]

More than anything else, the extended exchange over the wording of the Pickering memorial tablet highlights the communication problems that had emerged in the relationship between Campbell and Olcott.[107]

Olcott's winter in Tucson passed quickly in 1932 with the use of the telescope at Stewart Observatory and daily work on the Hartmann photometer. In addition, Olcott continued to recruit and train members for the AAVSO, including J. D. Williams of Pennsylvania, who wintered in Tucson. Olcott commented on how quickly Williams "picked up 10 fields by sweeping. This is a record and no one has taken hold so rapidly in my experience . . . . Williams is a find and will do good work."[108] This letter, more than recent ones, sounded a great deal like the Olcott of 10 years previously.

In April 1932, the Olcotts returned to Norwich for the summer. When asked by Campbell whether or not he would participate in the spring meeting at Hood College in Frederick, Maryland, Olcott cited his difficulties with colitis and going slow on his expenses.[109] Perhaps Olcott's concern for expenses was only normal prudence in the face of a bleak economic environment, but the response also signals a somewhat reduced commitment to the AAVSO.

Although Olcott could not be present, Campbell discussed his problems on a regular basis with Olcott. With Olcott's absences, Campbell was fielding many problems with AAVSO observers. For example, several members wondered how Eugene Jones achieved his reported high monthly totals. Justified or not, Campbell responded to those comments, and his correspondence with Olcott indicated that he (Campbell) had seen that

Figure 8.12. Clara Olcott, Annie Jump Cannon, William Tyler Olcott, and Edwin F. Carpenter, at the Olcott residence, Norwich, Connecticut, about 1933.

his "circles are now adjusted about as accurately as any I have known . . . and in spite of what some say about Jones' accidental errors I am pretty well convinced that his observations are honestly made and as accurate as the general run of observations can be."[110] Campbell's role was to keep everyone happy and working productively and to prevent backbiting from infecting the whole organization, a task at which he succeeded admirably. Those who were more deliberate in making observations or less dedicated to spending time at the eyepiece found it difficult to understand either the dedication or the facile way that the outstanding observers could move from star to star, make an observation in an already familiar field, and quickly move on effortlessly to the next star.

The cold Norwich winter drove Olcott to Miami in November 1933, where he joined other astronomical enthusiasts in frequent nighttime viewing.[111] Although weak and sickly in Norwich, Olcott was vigorous in Miami – a difference that suggests more was involved in his attitudes and illness in Norwich than just colitis. While in Miami, for example, he was strong enough to help re-energize the Southern Stars Astronomical Society that had been founded by AAVSO member Lynn Rhorer.[112]

Unfortunately, Olcott's health remained marginal in spite of his renewed energy. The following April, Campbell wrote to Olcott, still in Miami, that "I am mighty glad to get your card and to know that you are able to write, even though it must be done in bed."[113] Back in Norwich, Olcott sent his regrets about not attending a meeting: "It will be a keen disappointment to me not to be with you all but I am pretty well shot & lucky to be alive at all."[114] Here is another case in which Olcott, seemingly fine in Florida, was suddenly leveled by health problems. His schedule in Florida had been an ambitious one, with public talks, observing, and mentoring younger observers.

Olcott's relationship with Campbell began to deteriorate, but then that seems true about his relationships with other members of the Old Guard. By 1935 Olcott had missed many AAVSO meetings, and his demands for assistance from Campbell had become more insistent.[115]

Olcott's illness and perhaps his growing pettiness, even with old friends from the AAVSO, may have drawn him back to those aspects of astronomy he loved most, namely writing about it for a popular audience and giving telescope demonstrations and lectures about the wonders of the sky. And perhaps most indicative of his changed outlook, Olcott again turned to authoring books about astronomy. Leaning on Campbell for help on a number of points, he wrote and published a booklet in 1935 titled *The Romance of the Southern Cross*.[116] Olcott dedicated this eight-page work to the memory of S. Lynn Rhorer, founder of The Southern Cross Observatory, Miami, Florida. Olcott also began gathering materials for a book on the constellations of the southern hemisphere to complement his original best-selling *A Field Book of the Skies*. In June 1936, Olcott sent Campbell a draft planisphere he had assembled for the southern stars. Olcott recognized his own confusion about the reversal of handedness that is essential as one crosses the equator. In a postcard he asked Campbell for help straightening out the problems with the device.[117] Olcott must have mailed it to Campbell as he and Clara left their home in Norwich to drive to Lake Sunapee, New Hampshire, Olcott's favorite summer retreat. As he had done many times before, Olcott agreed to give an informal star lecture on the night of July 6, 1936.

Figure 8.13. William Tyler Olcott. From *Popular Astronomy*, Vol. 44, No. 8, 1936.

He was in the middle of delivering his familiar talk to a small group gathered in the Georges Mills Methodist Church when he suddenly stopped talking and fell to the floor. He had suffered a massive heart attack and passed quietly away, according to David Pickering. At a funeral at the Christ Episcopal Church in Norwich, where he had served as a senior warden for many years, Olcott was honored for his service to that church as well as to the community.[118]

It seems appropriate, at this point, to pause and consider Olcott's later years. As the founder of the AAVSO, he invested his time and energy heavily to make the association a success. He did so willingly and as a labor of love for the variable stars and for the people who gradually surrounded him as the AAVSO's Old Guard. The cost of that dedication came not only in time and energy, but also in lost opportunities to engage in other pursuits he might have found equally satisfying, such as authoring additional books about astronomy, lecturing about astronomy, writing poetry, and reading.

During the decade from 1910 to 1920, Olcott's dedication can be seen as unequivocal, moderated only by considerations of his health and diversions due to his sense of civic duty during World War I. As the Association grew, and more members shared responsibility for the Association's affairs, his self-effacing approach to the details of the founding allowed the history to become gradually distorted. First, some felt the need to embrace Edward C. Pickering as a co-founder. Then, Campbell inserted *Popular Astronomy* Editor Curvin Gingrich as yet a third founder. These early alterations of the AAVSO's founding must have weighed on Olcott, and over the next decade, roughly 1921 to 1932, it is evident that his attitude changed. By the time the Edward C. Pickering Memorial had changed his proposal for an observatory and AAVSO Headquarters separate from Harvard to an endowed Harvard staff position with only an informal or "understood" relationship to the AAVSO, Olcott's own role in the founding had become obscured in the minds of most members. The relationship with the prestigious Harvard College Observatory began to replace the camaraderie of the Old Guard as the glue that held the AAVSO together.

With Harvard's embrace of the AAVSO and the coming formalization of the endowment, Olcott's acceptance of the events of these two decades can be found in a poem he wrote at that time:

Fait Accompli
As husbandmen watch tenderly the soil,
With dreams of happy reaping days to be,
And count the hours of labor and toil
A harvest gleaned through dint of industry;
As they rejoice, when, fructified by showers,
The seeds reach up their fingers to the sun,
And strengthened by a wealth of kindly powers
Proclaim a service rendered and well done;
So we, who almost twenty years ago,
Sowed goodly seed far better than we knew,
Behold the fruit, – the A-A-V-S-O,
That now with pride and happiness we view.
To you who have been faithful to the cause
I send to all my greetings and applause.

– William Tyler Olcott, Tucson, Arizona / May 21, 1930[119]

We see in this poem the roots of Olcott's acquiescence to Shapley's proposal to absorb the Pickering Memorial as a Harvard endowment 18 months later. Furthermore, it was with a sense of deep satisfaction, and not necessarily disaffection, that Olcott backed somewhat away from the Association and returned to those activities that attracted him to astronomy in the first place: writing, lecturing, and viewing the stars with wonder.

9 · Growing pains and distractions

The more fields of endeavor we can open to the amateur the better, even if their work may not be so thorough or extensive. We never know to what ends such work may lead, or what may result from such a program.

– Leon Campbell, 1945, from a memo to Harlow Shapley[1]

Conditions in the United States changed dramatically on December 7, 1941: the Japanese attacked Pearl Harbor, resulting in enormous losses in US lives and warships. The US Congress declared war, and the subtle preparations that had been evident to a careful observer for some time suddenly became overt and intense.

Everyone's life would be affected by the events that followed, and they would have a share in the burden of the war that had, heretofore, been remote for most Americans. A few, such as Leon Campbell, had been aware of the war's impact on observers. George Ensor, a supervisor of a hospital laboratory, had already stopped observing in 1940, acknowledging that at the end of a day, "The work is stiff so I am too tired for anything but bed after my dinner. . . . Even occultation work has been dropped."[2]

There seemed to be little doubt; this war would not end as quickly as World War I after the United States entered that conflict, nor with as little engagement from ordinary US citizens. The AAVSO would feel the demands of the war, particularly on its skilled members who were called to assist in war efforts. Yet, having grown to some maturity, the AAVSO still maintained its program throughout the war, though with lower meeting attendance. This chapter is devoted to the wartime efforts of AAVSO observers and the subsequent activities undertaken by the Association.

WAR COMES TO THE AAVSO

Many aspects of the earlier programs for the AAVSO would be changing during the war years. With the deaths of William Tyler Olcott, J. Ernest G. Yalden, and many others, the organization had lost members of the Old Guard who had been so active in the 1930s. Their work could be considered complete, and well done at that. The transition was being handled with a few remaining stalwarts like David Pickering, and Campbell as Recorder and Pickering Memorial Astronomer.

The AAVSO, whose relationship had been stabilized in its affiliation with Harvard College Observatory (HCO), represented the efforts of many reliable astronomers, both amateur and professional. With the contributions of its Director Harlow Shapley, HCO continued to operate as one of the leading, but perhaps aging, astronomical institutions in the world. The science of variable star astronomy showed signs of maturing and, together with the changing of the guard, signaled that the gentleman's club era of Olcott's AAVSO had definitely drawn to a close.

But now its members were facing the crisis of being called into active military duty; the nation's demand for manpower would definitely affect the programs of the Association. Cyrus Fernald's concerns no doubt echoed those of many other AAVSO members. In his letter to Campbell in February 1942, Fernald reported a "drastic drop in number of active observers in the last three months. Presume this may be worse before long. It will not be surprising to me if I am called into the service. Because of my father and mother I do not plan to try to get in until I am called. Such a call will very likely put an end to my variable star observing for a time."[3] Fernald, kept in suspense most of the year, was finally rejected by his local Selective Service Board because his eyes were not good enough to serve in the Army. Campbell seemed to enjoy the irony of the Board's explanation, "As far as I have ever been able to tell, your eyes seem perfectly satisfactory for AAVSO work, whatever the Army may think of them. So perhaps you will be able to continue with your excellent work."[4]

Another member, Clinton B. Ford, was not as fortunate as Fernald. In the midst of his dissertation research, Ford was concerned he would be drafted into the Army. To avoid that possibility, Ford asked Campbell for a letter of recommendation to the Navy as an officer candidate; Campbell wrote his recommendation immediately.[5] After a special physical and eye examination in which a sympathetic medical officer agreed to overlook a muscular deviation in his right eye, Ford qualified for a commission as a reserve naval officer and reported to duty at Cornell University in July 1942 for his basic officer's training. Over the next 4 years, Ford trained in several fields, including cryptography in Washington, DC, and shipboard damage control in South Carolina. He remained in Washington until June 1943 as a communications officer. Then, he taught navigation and other naval science courses in duty tours at Rensselaer Polytechnic Institute in Troy, New York, and Pennsylvania State University in State College, Pennsylvania, where he also served as commanding officer for the Naval Reserve Officers Training Corp unit. Discharged as a lieutenant, Ford's military experience was not typical for the war years, but served his purposes well.[6]

The Harvard College Observatory suffered serious setbacks with the onset of the war. The professional staff members were divided in their attitudes toward serving or not serving actively; some left Cambridge for assignments in the military or war-related work. Among those supporting the war effort were Donald Menzel, Fred Whipple, Fletcher Watson, James Baker, Dorrit Hoffleit, and Margaret Mayall. For example, Menzel, Whipple, and Watson accepted commissions in the military services and were assigned to posts away from Cambridge. Others remained in Cambridge but devoted themselves full time to the war effort. For example, James Baker's well-known optical design work on wide-field aerial cameras was critical to the success of the aerial reconnaissance and bombardment campaigns so characteristic of World War II.[7] Others, including Shapley, Bart Bok, and Cecilia Payne-Gaposchkin, resisted direct involvement and remained at the observatory. In his autobiography, Shapley emphasized the positive aspects of staff involvement, including assistance to refugees. Shapley's wife, Martha Betz, computed firing tables and other mathematical applications for the War Department at the Massachusetts Institute of Technology (MIT).[8]

Many astronomers brought vital skills into roles important for success during the war. Lt. Cmdr. Donald Menzel (US Navy Reserve), who later became the HCO Director, spent most of World War II in Washington, DC.[9] His work there involved the effect of the Sun on radio propagation and preparation of radio transmitter frequency tables to guide operations when solar conditions changed. He wrote to Campbell that he was the "key man" in his section.[10] This information was particularly important for naval operations but was important in other military branches as well. After the war, Menzel published some of this wartime work for use in all types of radio operations.[11]

Campbell's own family was also involved in the war effort. Leon Campbell, Jr., served in the US Navy as a seaman aboard USS *Missouri* and was on board September 2, 1945, the day the armistice with Japan was signed on the deck of that famous warship. Campbell reported in a letter to Menzel that Leon, Jr., would be returning to New York City Harbor on *Missouri* on Navy Day, that same year.[12]

Overall, the actual involvement of all AAVSO members in the war effort is difficult to assess. No roster of members serving in the Armed Forces exists. It is possible to glean from Campbell's correspondence with members and from other sources some sense of the involvement. For example, J. Russell Smith and Y. William Fisher landed on the beaches at Normandy. Foster Brunton, C. E. Depperman, and Elmer Canfield were prisoners of war, and in the latter case, part of that time was spent studying astronomy. Ed Halbach and Luverne Armfield tried their hand at manufacturing roof prisms to the US Army's exacting standards, whereas John Ball, Jr., also of Milwaukee, and J. Wesley Simpson of Webster Groves, Missouri, were navigators for the Air Transport Command, and Walter Scott Houston, similarly for the Army Air Force. Radio operators, artillery forward observers, navy psychiatrists, and ambulance drivers are a few of the other military occupations that appear in these letters.

One of the more tragic losses to variable star astronomy came near the end of the conflict with the death of the outstanding Italian variable star observer Eppe Loreta. Two stories exist regarding the circumstances: one story claims that it was possibly retribution for his efforts to support the Allies as the Axis forces retreated, whereas the other claims he died after the end of the war, but with no explanation of the cause of his death.[13] Felix de Roy continued to observe variable stars under the adverse conditions of the occupation of Antwerp during the severe winter of 1941 to 1942. He contracted a cold and, because of the inadequate medical treatment, died at age 59 years – another tragic loss.[14] Italian contributor Giovanni Battista Lacchini, a charter AAVSO member, survived the war, but his home and observatory were destroyed by aerial bombing.[15] Miraculously, he continued to submit observations to the AAVSO until he was 79 years old in 1963.

Maintaining normalcy

In spite of the war and attendant travel problems, the AAVSO held nearly every one of the regular semiannual meetings that the membership had come to expect. Some shortcuts were taken. For example, in the spring of 1942, after several delays and relocations, the Council meeting convened at the home of Council member Donald Kimball in New Haven, Connecticut. A quorum was present so normal business could be conducted, and a few papers were presented at Yale University. There was no meeting in the spring of 1944, but the fall meetings were held on schedule. These are the meetings in which the Council met to elect new members and officers, hear committee reports, and consider the Association's general status. The main content of the general business meeting included technical presentations by the faculty remaining at HCO and the nearby universities in the Northeastern United States. Continuing a tradition from the previous two decades, Martha Shapley hosted an afternoon tea at the Director's residence. The social highlights continued to be the annual banquets with a toastmaster and impromptu short speeches by members. To a casual observer, it would appear that it was business as usual for the AAVSO. The appearance of normality was maintained in minutes and descriptions of the meetings, which rarely mentioned the momentous events taking place in the world outside variable star astronomy or previously active members drawn away into military service.

Part of its continuing business included recognition of the long-time contributions of members. At the 1942 Annual Meeting, the Council honored Dalmiro Francis Brocchi with its sixth AAVSO Merit Award, citing not only his work as a chart draftsman, but also his contribution of more than 8000 variable star observations in over 20 years of observing. The seventh AAVSO Merit Award the next year honored Charles W. Elmer, who could hardly be considered an observer with his 33 observations between 1923 and 1937. Still, Elmer had enlivened meetings with his humor, but more importantly, the Council, and especially Campbell, came to appreciate his steady, reasoned wisdom and advice. Elmer was a power to be reckoned with behind the scenes of the AAVSO, and the Council recognized that fact with this award. At the 1944 Annual Meeting, the eighth AAVSO Merit Award recognized the years of steady and dedicated service of its recorder. Campbell, who had not been an observer, had been surprised by the award. Fernald later commented to Campbell: "Charlie Elmer & Co. sure did put one over on you. The reward was richly deserved, but I was rather sorry that they ruffled up your fur a little before letting the cat out of the bag."[16] While Campbell disparaged about having any "deserving members" for that year, Elmer, Pickering, Eugene Jones, and his brother Fred, who designed and hand-lettered the award, conspired with AAVSO President Roy Seely to seal the deal.[17]

For many members who didn't attend the meetings in the Northeast, the main connection to the AAVSO was through its publications. Twice each year, *Variable Comments* reported the meeting minutes and announcements, which alternated with Campbell's reports mentioning many members by name and helping to maintain relationships as well as an understanding of how Headquarters functioned. Campbell also prepared a few technical summaries on variable star topics, for example, a description of "Forty Years of SS Cygni," and occasionally discussed his never-ending efforts to establish correlations between the light curves and other known factors regarding the LPV stars.

AAVSO SOLAR DIVISION

While Neal J. Heines worked in defense at the Propeller Division of the Curtiss-Wright aircraft company, his dedication to observing sunspots grew to greater importance to the military. A call had been issued for western hemisphere observers to make up for the delay and difficulty in communicating with the Zurich Observatory. Heines may have been among the few sunspot counters (probably two to four) contributing information to the Department of Terrestrial Magnetism (DTM) of the Carnegie Institution, Washington, DC.

Heines had suggested affiliating his solar-observing activity with the AAVSO in some way, and in October 1943, Campbell finally suggested that Heines propose a solar section to the AAVSO Council. This urging to get the AAVSO involved in solar observing may have come from Shapley, or from Donald Menzel, who had corresponded with Heines since the mid-1930s.[18] Subsequently, Campbell urged Heines to act in the matter, saying, "the idea of a Solar Division . . . still seems to be quite worthwhile."[19]

On October 6, 1944, the AAVSO Council accepted Heines' six-page single-spaced proposal and adopted a resolution to sponsor a Solar Division. In his proposal, Heines gave the following reasons for solar observations in general: "to improve the continuity of the present observations . . . by extending the area in which the observations are made and to supplement observations, which have been made impossible because of weather conditions. To supplement observations for educational institutions equipped for solar work but who do not function during vacation periods throughout the year. To widen the scope of American Association of Variable Star Observers by opening new avenues of activity for the amateur astronomer."[20]

At the time of its formation, the Solar Division Committee included Heines of Paterson, New Jersey, chairman; Menzel of HCO but in Washington, DC; Harry B. Rumrill of Berwyn, Pennsylvania; Rev. William Kearons and Mrs. Winifred Kearons of West Bridgewater, Massachussetts; and H. Helm Clayton of Canton, Massachussetts.[21] The Kearons and Rumrill were all experienced solar observers from a previous decade; in fact, Rumrill's solar observing dates back to his introduction in about 1905 by Rev. Alden Walker Quimby, also of Berwyn. The Kearons had developed their interest in the subject through contact with Menzel during the 1930s. Rev. Kearons photographed the Sun for Menzel on every clear day for a period of years and provided the plates to HCO.[22]

In his proposal to the AAVSO Council, Heines did not specifically cite the wartime need for data on solar phenomena, nor did he mention any disruption of the flow of information to or from Zurich because of the war.[23] In the following months Heines did, however, speak of them as if they were the primary reasons for forming the Solar Division. Writing to *Popular Astronomy* Editor Curvin H. Gingrich, in an effort to encourage solar observations by amateur astronomers, Heines stated the need had arisen because of the war effort and the inability to obtain the sunspot numbers from the Zurich Observatory.[24] Heines published similar notices in the *Texas Observers Bulletin*, *Astronomical Information Sheets*, *JRASC*, and *PASP*. He did not acknowledge any such "voluntary notices" in *Sky and Telescope*, though he thanked the magazine "for keeping the reading public sun-spot minded."[25]

With the formation of the AAVSO Solar Division in 1944, the reports by 20 to 30 additional observers were sent to the DTM of the Carnegie Institution. With the increase in data, DTM began issuing monthly summaries in December 1944 of the AAVSO's sunspot reports, thus asserting an active, public role in the matter, perhaps as a cover for the vital nature of the information for military purposes.[26] On January 15, 1945, DTM Director J. A. Fleming wrote appreciatively to Heines that "these observations have an application of considerable importance in the war effort through work done by the Dept. of Terrestrial Magnetism and by the Interservice Radio Propagation Laboratory . . . under contract with the Army and Navy . . . . Your initiative in organizing . . . produced such prompt results and is gratefully acknowledged."[27]

The Solar Division's "contribution to the war effort" was also recognized in a note published in the *American Association for the Advancement of Science Bulletin*, noting, "The excellent and useful work of the American Association of Variable Star Observers is well known, but its contribution to the war effort is not so widely appreciated." Further, the reports, which were analyzed by DTM in the Carnegie Institution of Washington, were being used in collaboration with the Interservice Radio Propagation Laboratory under contracts of the Army and Navy.[28] The DTM supported this effort by providing forms and instruction leaflets to the AAVSO, as well as making grants to cover the Association's out-of-pocket expenses.[29]

The first manual for solar observers was issued in a seven-page pamphlet on May 1, 1945. In his first report as chairman of the Solar Division, Heines noted June 17, 1945, that the reports had been used for the war effort. The DTM analyzed, reduced, and distributed the submitted observations. The reductions took the form of a Preliminary Relative Sun-spot Number, which came from Dr. Brunner at Zurich. Dr. Alan H. Shapley, son of HCO Director Shapley, supervised the reduction work at DTM. Alan Shapley also handled the correspondence with the Solar Division. By mid 1945, the AAVSO Solar Division had grown to 53 members submitting monthly reports.[30] To communicate news and summaries of observations, Heines began compiling and publishing a monthly *AAVSO Solar Division Bulletin* in September 1945.

Campbell was very pleased with the progress Heines made with the Solar Division and commended him in letters, as well as at the May 1946 Council Meeting, when Campbell recommended that Heines be elected to patron membership in the Association. In September 1946, Campbell notified Heines that the International Astronomical Union (IAU) had appointed him to its Commission 10 (Sunspots). Heines' IAU membership and commission appointment were apparently sponsored by HCO.[31]

Still, Campbell harbored some doubts about the proper place for Heines' solar group. A few months later, past AAVSO President Dirk Brouwer of the Yale Astronomy Department questioned Harlow Shapley on the necessity for amateur solar observations, as the solar work done by professional observatories was better quality and more thorough. Campbell vigorously defended the value of the amateurs' work, adding that "the more fields of endeavor we can open to the amateur the better, even if their work may not be so thorough or extensive. We never know to what ends such work may lead, or what may result from such a program." Then Campbell added: "As to having the Solar work under the auspices of the AAVSO, I am not so sure. If the N.A.A. [National Amateur Astronomers] should get going, that might be the answer. How about the Dept. of Terrestrial Magnetism? . . . To be sure, the solar division has increased the financial liabilities of the Association, with not all of them offset by the addition of new members."[32]

By May 1947, the National Bureau of Standards (NBS) of the US Department of Commerce had taken on the responsibility for receiving and reducing American sunspot data to a Preliminary Relative Sunspot Number from the Carnegie Institution. Dr. A. G. McNish was the department head, and Alan H. Shapley at NBS by 1948 compiled the sunspot data, assisted by J. Virginia Lincoln.[33] On June 30, 1947, Heines reported to Campbell that he had a contract "for signature in hand for execution" with NBS for support of the AAVSO's Solar Division operations "and myself and assistant." Heines was seeking $4354, of which he earmarked $2700 as the "chairman's salary," $500 for a part-time assistant, and $1154 for Solar Division expenses. This is a dramatic increase in the Division's estimated expenses compared with the $184.93 reported for the October 1946 to May 1947 period. Heines ended: "It is needless to say that the AAVSO will not be forgotten for its sponsorship of the Solar Division."[34] Heines' appropriation of most of the funds as payment for services rendered apparently raised no eyebrows at AAVSO Headquarters. Campbell's letters indicate he was comfortable with the arrangement because it was not even mentioned in the correspondence. This may reflect the fact that Campbell was only too happy to have the help and any part of the funding. Indeed, as David Rosebrugh, a solar observer, later wrote, "This was good as it enabled Mr. Heines to devote his full time to building up the SD on a firm basis."[35] Regarding Solar Division expenses, Heines provided no breakdown in his May 1948 report, only stating that "the chairman of the Solar Division has been able to negotiate contracts with the government, through the NBS, which defray all the expenses incurred by the Solar Division and in addition include a salary for the chairman which now enables him to devote all his [time] to the work of said division."[36] Subsequently, Solar Division reports submitted by Heines did not mention financial matters at all.

On October 7, 1948, J. Virginia Lincoln at NBS sent Leon Campbell a letter commending the AAVSO Solar Division observers, stating that their reports, used to compute the American Relative Sunspot Numbers, "are vital to the prediction of radio propagation phenomena. Their importance would be greatly increased in the event that prompt and assured communication with Zurich were not possible."[37]

Although the primary function of the Solar Division was to obtain sunspot counts for the NBS, Heines had, from the start, actively sought to provide other services to solar researchers. In October 1948, Heines reported, under the heading "Solar Division Activity," three projects undertaken at the suggestion of professional scientists: Granular Solar Surface and Color in Sunspots, Dr. James C. Bartlett; Unusual Spot Configuration and Colors, Dr. Walter Orr Roberts; and Foreshortening of Sunspot-Groups, Prof. W. Gleissberg. In addition, the Solar Division observers reported seasonal observations of high-flying migratory birds crossing the solar disc to Heines at the request of Prof. George N. Lowery, ornithologist, and Prof. William A. Rense, physicist, both at the Louisiana State University.[38]

Figure 9.1. At the AAVSO Annual Meeting in 1949. From left: Fred Jones, Cyrus Fernald, David Rosebrugh, Leon Campbell, and Neal Heines.

THE AAVSO AURORA COMMITTEE

Edward A. Halbach reported on the Aurora Committee in 1944 and then resigned because of the press of other work. He emphasized the longer term nature of the study, including plans to add continuous monitoring equipment in Ithaca and Milwaukee. At this point in the war, Halbach also noted that military applications of the aurora work helped justify the effort. Professor Carl W. Gartlein added comments reflecting the early results of the studies, including demonstrated correlations of the magnitude of the aurora with "the K magnetic figure," and also that sudden outbursts of an aurora, as arcs into rays, coincide in time with distinctive changes in the earth's magnetic field. On Halbach's resignation, Donald Kimball accepted responsibility for the Aurora Committee, but the absence of solar activity, combined with the war, effectively closed down this work for a period.

As solar activity for the next cycle began, Solar Division Chairman Heines encouraged aurora observing. Active participants in the aurora program included Seely and Rosebrugh, in addition to Donald Kimball. At the 1949 Fall Meeting, Menzel also urged greater effort on the part of the Aurora Committee.

THE NATIONAL AMATEUR ASTRONOMERS

The Solar Division, as shown above, expanded the scope of AAVSO activities quite substantially, taking in areas of science and scientists not directly involved with variable star astronomy. The precedent for this expansion was established in the 1920s with the addition of the lunar occultation work and telescopic meteor reporting. Further expansions occurred in the early 1940s with the addition of the Aurora Committee, and finally, the Solar Division. Thus it seems that to Shapley and Campbell at least, the AAVSO represented a resource of dedicated amateur astronomers who worked both night and day to support the scientific work of professional colleagues. Perhaps in Shapley's and Campbell's minds, the fact that the AAVSO program remained limited mainly to long period variables could justify these extensions. That is, at least, the context within which we consider the following developments that directly or indirectly involved the AAVSO Council in the mid 1940s.

At the October 1946 Council meeting, Campbell reported that a sponsor had been found for the Solar Division but gave no details.[39] The AAVSO may have been involved in discussions with another national organization of amateur astronomers, which would have separated the Solar Division from the AAVSO. After investigating such a separation, Heines apprised Campbell on October 31, 1946, that he found "no grounds" for such a change. Campbell then concurred that he, like Heines, found no grounds for separating solar work from the AAVSO, adding, "I have felt this way all along." Neither Heines nor Campbell gave any details of the "investigation" or any reasons to justify their conclusion. Campbell's concurrence here is surprising, given his earlier opinion that the Solar Division ought to be part of some other organization to relieve the financial strain on the AAVSO.[40] What did Campbell have in mind in making these comments?

With the hundreds of avocational astronomy clubs springing up during the 1920s and 1930s, several individuals, most notably Charles A. Federer, Jr., and Helen Spence Federer, the founders of *Sky and Telescope* magazine, worked to establish a national organization for the mutual benefit of all such clubs and individual members. Several national conventions led to the preparation of a proposed constitution and by-laws for the Amateur Astronomical League of America (AALA), but with the advent of World War II, the AALA's organizers suspended their activity.

There is plenty of evidence that Campbell was sensitive to much more than the progress of variable star observing as it might be measured by the AAVSO's observing totals. His recognition of amateurs in connection with the AAVSO's 30th anniversary stretched to include a few amateurs who, though variable star observers, were exemplars of amateur astronomy for other reasons. In his 10th Annual Report as the Pickering Memorial Astronomer, Campbell recognized individuals like Avery Hildom of California, Robert Millard of Oregon, and Oscar Monnig of Texas, none of whom were outstanding variable star observers, but each had made substantial contributions in other ways to amateur astronomy and astronomy in general.[41]

Campbell recognized that professional astronomers who came to him for assistance were frequently attempting to tap into this network of skilled variable star observers for other purposes. He recognized that these requests reflected a long-standing need for a national organization for amateur astronomers, perhaps one organized into technical sections like the former Society for Practical Astronomy and the British Astronomical Association. Such an organization would relieve the pressure on the AAVSO to provide access to its large corps of skilled observers. Furthermore, a new organization would reduce demands on Campbell's time and leave him free to pursue his main love, variable star astronomy. It must also be said that in the mid 1940s, the HCO staff faced constant reminders of the need to reduce the observatory's expenses; it was standard practice to cut corners in every way possible.

At the 1944 fall Council meeting after the formation of the Solar Division, Campbell suggested that the AAVSO take the lead in forming a new national organization for scientifically inclined amateur astronomers. The discussion continued with the support of Charles H. Smiley from an afternoon meeting until the following morning. Among those present for the morning meeting were four professional astronomers (Campbell, Dirk Brouwer, Marjorie Williams, and Shapley) and six amateur astronomers (Seely, Percy W. Witherell, Lewis Boss, Elmer, Fred E. Jones, and Heines).[42]

A consideration in this discussion was, of course, the pending formation of Charles Federer's AALA, discussed briefly above. It was noted, however, that Federer's purpose was to unite astronomy clubs. Instead, the AAVSO Council members agreed their purpose was to unite "working amateur astronomers as they are united in the British Astronomical Association." Even further, Heines suggested that the AAVSO constitute itself as the sponsoring body of an organization similar to the BAA in scope and suggested that perhaps lunar, cometary, binary star, planetary, and spectroscopic groups might be formed from its members. The Council authorized a committee to consider the matter, and in a later full Council meeting, Campbell proposed that Shapley be asked to chair the committee.[43] A few weeks later, Shapley agreed to chair this committee with members suggested by Seely and Campbell (Brouwer, Elmer, Federer, Gartlein, Halbach, Heines, Helen Hogg, Albert G. Ingalls, Oscar Monnig, Harvey H. Nininger, Charles Olivier). They represented the AAVSO's activities, but included other amateur interests, such as meteors, meteorites, telescope making, and popularization. The suggested committee was a balanced mix of five professional astronomers and six well-known amateurs, representing, as nearly as possible, a diverse geographic mix as well as astronomical interests.[44]

Shapley asked Campbell to draft a more formal proposal for how such an association might be organized. After surveying somewhat similar organizations, Campbell presented a lengthy proposal to Shapley. Campbell's proposal, reshaped in interesting ways, emerged as a formal letter from Shapley to the committee, which had, in the meantime, changed considerably in composition, now including Shapley, Brouwer, Federer, Olivier, Nininger, Elmer, Hogg, Heines, Halbach, Earle Linsley, George V. Plachy, Scanlon, Wagner Schlesinger, and Smiley.[45] The membership had not only been broadened substantially, but the balance of power had shifted toward its professional membership (10 professionals, 8 amateur astronomers). The substitution of members represented the selection of several individuals known for their interests in popularizing astronomy: Plachy, a telescope maker from the powerful Optical Division of the New York Amateur Astronomers Association; Scanlon of Pittsburgh, primarily as a telescope maker, and F. Wagner Schlesinger, the newly appointed director of the Adler Planetarium in Chicago.[46]

Shapley acknowledged in his April 15 letter that both Campbell and Federer had assisted him in drafting the proposal. Federer was no doubt concerned that if the new organization, styled the National Astronomical Association (NAA) by its AAVSO supporters, were successful, there would be little room for yet a third organization like the association of clubs he was actively promoting. Thus it seems natural that Federer would do all that he could to influence the shape of the new organization. In doing so, Federer could ensure that the NAA encompassed his ideals as well as those of the AAVSO Council members, even if those ideals were clearly divergent.

But the path for the proposed NAA, an association to be patterned after the BAA, proved to be tenuous. At the next AAVSO Council meeting, AAVSO President Smiley of Ladd Observatory, an otherwise strong supporter of amateur astronomy, "took up the cudgels against the NAA, not necessarily because he opposed it, but to bring its weaknesses into the open." Interestingly, Helen Federer, an active member of the AAVSO Council, who worked with her husband to organize the AALA, expressed the hope that the NAA would be delayed until after the war to see what progress could be made with the AALA. In effect changing horses in mid-stream, as David Rosebrugh reported in the minutes, Campbell then opined that: "there was not enough astronomical interest in the country to support both the proposed NAA and the AALA." Campbell opposed an attempt to organize two nationwide associations: "There had never been any cooperation between the A.A.L.A. and professional astronomers. In practice the A.A.L.A. had consisted entirely of non-observing amateur groups." Campbell once again expressed hope that if the NAA was formed, it would be possible for AAVSO to spin off its Solar Division and perhaps other activities as separate but co-equal societies with the AAVSO.[47]

Council members who supported the formation of the NAA (Campbell, Rosebrugh, and others) were unable to bring the matter to a head in the face of opposition from the Federers and Shapley. As the president of AAVSO, professional astronomer Smiley was unwilling to buck the interests of Shapley, the respected HCO director. Instead, a select committee consisting of Shapley, Campbell, and Federer was asked to study the matter and make further recommendations to the Council. At the May 3, 1946, Council meeting, Smiley reported that Shapley had advised the AAVSO to delay the formation of the NAA until after the AALA meeting in Detroit. Obviously, the power on the subcommittee and the interests of Shapley and Federer finally prevailed. Campbell had been forced to sit and watch as Federer was freed to pursue his dream of a league of recreationally oriented clubs.[48]

After subsequent Council meetings and a member survey, the Council dropped the idea of the NAA. Instead of shifting the Solar Division and several other activities to the NAA and then discharging that organization to stand on its own, the AAVSO continued to support much more than simply variable star observers.

RETURN TO PRE-WAR NORMALITY

During the war years, Campbell demonstrated his finesse in dealing with problems with observers. His efforts to maintain a level of normalcy would later pay dividends to the AAVSO. For example, a controversy surfaced over Fernald's prodigious variable star observations. He had amassed a substantial record of achievements, leading the annual list of total observations for 5 years in a row (1940–1945), but drew criticism from another AAVSO member, James L. Russell, who finally complained that Fernald's performance was impossible.[49] Campbell sent his letter of complaint to Fernald, who defended his observing practices, citing Campbell's ability as the AAVSO Recorder, "for it stands to reason that he would be fully satisfied of the reliability of the source of 15 to 20% of the observations of the last five years."[50] Offering forgiveness to his critic, Fernald went on to explain the advantages of the Springfield mounting, referencing his years of experience as an efficiency expert, and 8 years as an AAVSO observer. His ending claim that "an observing rate of 120 per hour is within the realm of possibility" must have seemed ludicrous to experienced variable star observers. As he noted, "An observer maintaining that rate would, of course, run out of stars to observe, but he could easily make 1000 useful observations per month. Perhaps when he comes along Prof. Campbell will put more stars on the lists for him to work on. Then, given good sky conditions, we might see 20 to 25 000 observations per year from one observer."[51] Fernald sent Campbell a copy of his letter to Russell, with advice that he should handle it with asbestos gloves, but Campbell only added his own support to Fernald's claims.[52]

Campbell's relaxed and easy style and the extent to which that encouraged the best in contributions from individual members can be seen in his handling of the membership of Lawrence N. Upjohn, M.D., Chairman of the Board, The Upjohn Company, Kalamazoo, Michigan. In May 1944, Upjohn had been referred to Campbell's Telescope Mart business rather than the AAVSO by Perkin-Elmer, likely Elmer. Campbell easily accommodated his need for some cleaning and repair

Figure 9.2. Lawrence N. Upjohn, M.D., with his wife, Gratia, May 1947.

of a couple of eyepieces. Apparently encouraged by the easy wartime access and reliable service, Upjohn continued to work with his telescope in his spare time and by May 1946 felt confident enough to apply for AAVSO membership. As he expressed it, he was interested in membership mainly for the privilege of purchasing star maps.[53] Campbell quickly added Upjohn's name to the roster and replied that Upjohn would soon receive his first set of blueprint variable star charts. In July 1946, Upjohn sent a check for 50 dollars to pay for a life membership, which was approved at the Council's annual meeting that same year.[54]

By July 1947, Upjohn had accumulated enough observations in routine monthly reports that he began to wonder how he was doing. It seemed to him that his observations differed systematically from the average run of observations plotted on the light curves; thus, in effect, asking how his results compared with those of others. It is in this kind of delicate situation that Campbell's long experience dealing with observers paid dividends. First, he allayed Upjohn's concerns:

> Although there seems to be a slight systematic difference between your individual observations and the mean of them all, I do not think you need feel discouraged, because the difference is no larger than is to be found for many other observers. I can imagine that perhaps the main difficulty has been the matter of color, which affects different eyes differently in different instruments. In fact, I think you should be congratulated on having obtained such a fine, smooth curve.[55]

In ending his letter, Campbell dropped back into a classic Old Guard approach, asking, "Am I correct in assuming that the framed pictures which I see in some of the drugstores here are of Lawrence N. Upjohn? If so, I am much pleased to have a better mental picture of what our Kalamazoo observer looks like."

Campbell's approach might have backfired, for as the chairman of the board at a large pharmaceutical corporation, Upjohn likely endured requests for donations from strangers on a regular basis. But by couching the inquiry in just the right personal – you are one of us – approach, Campbell made his point in a harmless enough way. Upjohn responded cheerfully, apparently pleased with the allusion to his membership, "regarding the picture you see in drug store windows, this is a gallery painting of an unknown model, procured for the purpose by our advertising agents. I enclose some snapshots in order that I may not be masquerading further under false colors."[56]

LOSS OF A PRESIDENT

By spring 1946, activities around HCO returned to a more normal state, but the clock could not be turned back. David Pickering, one of the last members of the Old Guard, suffered from declining health that reduced his ability to correspond with Campbell. Diagnosed with a tumor on his lung, Pickering received palliative treatment and care at his home. Heines visited Pickering and reported his conversation to Campbell and Rosebrugh, both close friends of Pickering. Heines' report is moderately cheerful, though it cannot have been easy for him to type out his recollections of the conversation, interrupted on occasion by a nurse who interjected care instructions from the next room. Pickering had reconciled himself to his condition and ultimate fate, referring to his good friend and councilor Dr. Godfrey, who had "told his family that there was nothing unusual about such a condition, we are all born into this world and must of necessity leave it, so, leave it I shall."[57] Pickering was proud that his son James Sayre Pickering was writing his first book on astronomy, noting that MacMillan and Company would likely publish it.[58]

Despite a brief recovery, Pickering declined rapidly and died on June 13, 1946, having been in a morphine-induced coma to relieve his pain for many weeks.[59]

Campbell was now one of the few surviving members of the AAVSO's Old Guard.

A WEAKENING BASE

Time was evidently running out at the Harvard College Observatory as well. Shapley decided that the end of the war signaled a new opportunity for fundraising. Money, or rather the shortage of money, permeated every event, every detail of HCO operations. Having invested the Rockefeller grant money in new but not really state-of-the-art facilities in the late 1920s and early 1930s, Shapley faced a bleak financial picture by 1946. Overextended in terms of both physical facilities and program, HCO simply did not have enough funds to cover everything.

Shapley organized a centennial celebration for the observatory, perhaps in recognition of this need for additional funds. As part of that effort, HCO published a booklet, *Harvard College Observatory – The First Century*.[60] Its brief historical summary proclaims proudly: "There is hardly a single field of research which is not included in the present-day programs of Harvard astronomers. Among them are variable stars and novae, the Milky Way, nebulae and galaxies, meteors and comets, stellar spectra and the constitution of stars, and solar phenomena."[61] Omitted here, of course, were planetary astronomy and astrometry, long since considered passé by astronomers at Harvard, but both due for a second life in the future at other observatories. The history further credited Shapley with expanding HCO activities more than during any preceding time since he took over in 1921: "improved and extended instrumental equipment; enlarging the number of astronomers, research workers, and students; and initiating a graduate school in astronomy, whose graduates hold important places in the astronomical observatories of this country and abroad."[62] Still, Shapley's boast to have it all covered conceals the thinness of that cover in most places. The equipment at HCO was for the most part second-rate, and the best of astronomers would have had problems achieving much with those resources.

The costs of the HCO operation are apparent in the extent of its far-flung facilities. HCO's outlying observatories included the general astronomical research facilities at the Oak Ridge station, Harvard, Massachusetts; the high-altitude solar observing station near the mining village of Climax with administrative offices in Boulder, Colorado; solar and meteor tracking stations in Alamogordo and Las Cruces, New Mexico; and the Boyden station, Bloemfontein, South Africa. Also, there was the new astrophysical research building on the HCO grounds in Cambridge to house the glass plate collection, the library, and additional offices.[63]

The Shapley administration had done all these things. And indeed, from a strictly scientific perspective, HCO was, as Owen Gingerich characterized it, "a special and exciting place in the three decades of [Shapley's] directorship."[64] But an economic depression followed by wartime turmoil combined to leave their mark on HCO's older physical plant. Shapley had complicated his problems by dispersing resources and staff to distant locations. Support for those remote operations weighed on HCO's expense budget as much as its capital budget. Comments in the day-to-day memos at HCO demonstrate the bleak working conditions that existed for most of the Observatory staff. Most of these messages pertain to routine office matters – all hinging on not having enough money. Correspondence and interoffice memos in the AAVSO archives give witness to the almost daily concerns about telephones (the number of telephones allowed, bills and usage), the number and reliability of mimeograph machines, physical infrastructure and absence of janitorial support, and general unsuitability of the provided facilities.

Shapley's managerial style inevitably contributed to the bleak atmosphere. He reacted viscerally to problems, dashing off notes on his typewriter, typing multiple notes per page, and then tearing the page into separate pieces to conserve paper. Invariably, most of these messages contain the same implied questions: "Do we really have to spend this money; Is there not some way to do this on a shoestring so we will not have to spend it?" and "What's in it for the Harvard College Observatory?" These are the questions always asked by prudent managers on occasion, especially in straitened circumstances, but as a relentless drumbeat that influenced every task in the observatory, they were a dispiriting distraction for the administrative staff. In his earliest years at Harvard, when Shapley had choices, he had consistently favored the expansion of facilities and program and then hoped he could manage to scrape by on the funds left. In part, Shapley's plans and the HCO fell victim to the reduced place of astronomy in the Harvard hierarchy of priorities for the natural sciences. The giant departments of chemistry, physics, and biology were favored with post-war corporate largess. By the time Shapley realized he needed to increase the observatory endowment, it was too late; university support had moved in other directions. The HCO, with its overstretched program, was left mainly to fend for itself.[65]

Embedded as it was in HCO, the AAVSO shared the same concerns about expenses. Much of the AAVSO's work involved administrative tasks, but HCO's acute shortage of administrative help spilled over on Campbell as well. For example, Campbell, Shapley, and Menzel all shared the employment time of Helen M. Stephansky, the AAVSO/Pickering Memorial administrative assistant.[66] The publication of the fast-growing numbers of AAVSO observations in the *Harvard Annals* became a concern. Volumes 104, 107, and 110 in numbered quarterly installments covered observations from 1935 through 1942. However, publication was delayed during World War II, particularly with the difficulty in obtaining observations from abroad. After the war, observations from 1943 through 1946 were published in *HA* volume 116; each of its four quarterly installments contained about 72 pages of observations. It is easy to see how, with HCO's circumstance, the AAVSO came to be seen as an expensive burden.

Comments alluding to the expense of publishing AAVSO reports, including Campbell's proposed *Studies of Long Period Variables,* in AAVSO newsletters did have one positive effect. In June 1949, Campbell received a letter from Lawrence Upjohn, who asked, "In one of your recent announcements you spoke about funds for completing publication of one of the variable star reports. I wonder if these have been forthcoming, and how much would be involved."[67] Campbell immediately passed the letter to Shapley, who advised him that the general cost of overhead and the preparation of the volume would be in the vicinity of $1500. "Tell him," Shapley wrote, "that publications have been extremely expensive in late years but that we are now successfully using the verityper and the offset process and it is possible to continue with some of our publication work. I believe it remains for you to communicate with Mr. Upjohn, expressing your appreciation of his interest in this problem of finishing off your big labor on the long-period variable stars."[68]

Although Upjohn's inquiry was about publications in general, Shapley, in his note to Campbell, deftly steered Upjohn's inquiry in the direction of Campbell's work on the LPV book. Shapley included a lot of other ruminations about the expense problem and publications in his note to Campbell. Campbell's reply to Upjohn, dated June 16, 1949, conveyed Shapley's message without some of these details. H. B. Allen, Upjohn's secretary, notified Campbell that they had "found it possible to appropriate $1500 toward the project, which amount is available as needed to meet the cost of publication. We would like to have the records indicate the gift as from an anonymous source."[69] Campbell thanked Upjohn with a copy to Allen: "Little did I expect such a generous donation from any one person!"[70] Thus the funds for the publication of the long-awaited *Studies of Long Period Variables* were made available to the AAVSO from an AAVSO member, with likely the largest donation the Association ever received from one person at that time.

Other cost-control measures were taken for the necessary AAVSO publications. At the AAVSO Council meeting held on May 27, 1949, Recorder Campbell proposed a new method for publishing AAVSO observations. Instead of publishing them in *Harvard Annals* at a cost to the AAVSO of $5 per page, they would be published as mimeographed sheets at $2 per page (the 10-day means would be published in the *Annals* at another time). This, Campbell pointed out, would bring the publication of observations – backlogged during the war years – up to date. The AAVSO Council approved Campbell's proposal at the October 1949 meeting.[71] The mimeographed observation lists would be sent to all active observers, as with the *HA* published reports. *AAVSO Quarterly Report #1* contained the observations for the 100-day interval from March 20 to June 27, 1946.[72]

THE NEW ERA BEGINS FOR OBSERVERS

As World War II eventually concluded, ace observer Fernald, after providing Campbell with his usual list of corrections to the printed record of his observations, hoped that the cessation of the war would allow them to bring the observing reports up within a few months behind the actual observations. "Then," he wrote, "if there are any unusual circumstances connected with questionable observations I may remember

Figure 9.3. Reginald P. de Kock at the 6-inch refractor of the Royal Observatory, Observatory Cape, South Africa.

them and thereby improve my technique."[73] Campbell commended Fernald for a fine year of observing, but recommended adding new stars rather than observing the same stars more frequently. But Fernald, who had pestered Campbell to add more stars to the list, had missed the opportunity to pick them up for his own use.[74]

With the efforts of Peltier, Fernald, De Kock, and others to maintain observing totals during the war, the average total annual observations for the decade of the 1940s amounted to about 41 000 observations, with a peak year at about 56 000 observations in 1947. The more important statistic for the decade, however, was established March 3, 1946, when William L. Holt, M.D., of Scarboro, Maine, observed X Ceti. After due consideration of the exact Julian Date and time of Holt's observation compared with all the observations made by others on the same evening, Campbell decided that Holt's observation was the one-millionth observation added to the AAVSO database. Unfortunately, his fame was short-lived, for Holt died in October of the same year. Fernald of Wilton, Maine, agreed to serve out Holt's unexpired term as a member of the AAVSO Council.[75]

As amateur astronomers returned to observing in 1946, seemingly anxious to put the war behind them, they brought some positive experiences. The advance of technology visible in the war evidenced itself in increased interest in instrumentalizing variable star observing and in applying mechanical and especially electronic technology. A good example of this is the interest in photometer design. John Grinch of the Bergen County Astronomical Society in New Jersey requested Campbell's advice on a useful design for a photometer they intended to construct for their serious work in observing variable stars.[76] Campbell described the polarizing photometer used at HCO, the most effective type of instrument, but expensive to construct (mainly because of its elaborate prisms) and difficult to use. Campbell then described the simpler form of the wedge photometer, as described in Chapter 6.[77] Two things emerge from this exchange: first, Campbell's interest in instrumentalizing observations, that is, of finding a way to read a meter dial, scale, or gauge, rather than simply comparing and recording visual impressions of the stars, had never disappeared, even a quarter century after Campbell's first work with Yalden and Cilley on a simplified version of a wedge photometer. Second, Campbell, probably reflecting the views of the professional astronomers around him, never gave up on the idea that such an improvement in amateur observing practice was desirable. He may have hoped, recognizing the innumerable other examples of innovation in the ATM movement, that some enterprising amateur would solve the challenge of building such a photometer at a reasonable cost in money and time. Regrettably, the problem remained intractable at the amateur level, at least for the wedge photometer.[78]

A more successful example of the increased interest in applying recent technology came to Campbell the next year in a letter from John J. Ruiz, an electrical engineer in Dannemora, New York.[79] He was interested in a photoelectric photometer described by professional astronomer Gerald Kron in 1947 in a *Sky and Telescope* article intended to interest amateur astronomers in applying the new technology of photoelectric photometry to astronomy with their small telescopes. A year earlier Kron had published a more technical article on the subject in *Harvard Circular 151*.[80] The photo-emissive detector (originally selenium) had been replaced with the RCA 1P21 photomultiplier tube developed and applied widely during the war. Ruiz wondered if such a device could be applied to variable star astronomy, and Campbell assured him that "once you have your photometer in working order, we shall be pleased to advise you on the type of work to undertake."[81] It would be several years before Ruiz completed a working photometer and demonstrated its utility by sending light curves of eclipsing binary stars to HCO.

Encouraged by this evidence of amateur interest in the application of photoelectric photometry (PEP) to variable star observing, Shapley expressed his desire that some AAVSO members be trained to use PEP methods. Shapley justified this interest on the possibility of observing faint stars in standard color bands using appropriate filters to AAVSO President Rosebrugh.[82] Later, Rosebrugh summarized his own views on the state of PEP in a memorandum to Margaret Mayall. Rosebrugh concluded that the technology was not ready for amateur application, because "it appears that better mountings and better electronic tubes are required before the average AAVSO member can do much with electronic photometers."[83]

POST-WAR SUPPORT TO VARIABLE STAR ASTRONOMY

By 1947, professional astronomers were fully engaged in astronomy for the most part. Having spent the war doing optical work at Yerkes Observatory, Jesse Greenstein returned to spectroscopy with the 82-inch telescope at McDonald Observatory, Fort Davis, Texas. In November of that year, Greenstein reported changes in the spectrum of ρ Cassiopeiae and wondered in a letter to Campbell whether there had been any observations of the brightness of this irregular variable star.[84] Campbell, of course, did not have the observations to match the dates of Greenstein's spectrographic observations until after the end of the month. By January 1948, Campbell had plotted the available observations of ρ Cas and provided the light curve to Greenstein, who had returned to his home base in Wisconsin. Greenstein responded that his own visual estimates of the star indicated there might be large brightness changes from night to night, "because I am certain that spectroscopic changes of great interest do occur within two nights," and suggested ρ Cas would be worth watching for that reason.[85]

The exchange indicates both the growing importance of variable stars to professional astronomers and their recognition of the AAVSO as a possible resource. The exchange also illustrates how difficult it was for Campbell to respond immediately to such inquiries. His ability to detect night-to-night variations in one variable among 400 was not easy since it began with a lengthy chain of events: backyard variable star observers reporting their observations to Cambridge, typically a month after the observations were made, followed by Campbell's patient entering all such observations in the Headquarters ledgers each month and then manually plotting the light curves at an even later date. Recognizing these difficulties in data handling and communication, Paul W. Merrill's longer term studies held out more promise for effective use of AAVSO data by professional astronomers.[86]

With data collected on about 400 of more than 1700 LPVs that had been cataloged at that point, it was important to find some rationale for selecting one or more of these interesting stars for closer study. As Paul Merrill later observed:

> Mean light curves are important for many statistical studies. Because the possibility of progressive changes has been illustrated by R Hydrae and R Aquilae, we ought to draw up reliable light curves every decade for scores, or preferably for hundreds, of long period variables.
>
> For some of the brighter variables, the AAVSO should distribute predicted curves for six-year intervals, with a new set every four years. Such curves would be most useful to spectroscopists and others in planning detailed observing programs; and they would aid early recognition of any unusual behavior of a variable.[87]

CAMPBELL'S BROAD COMMISSION

Looking back over the years, one sees an enormously expanded commission, one which Campbell assumed, in addition to such duties delegated by Shapley as "the care of buildings, grounds, and furnishings in the observatory" and other menial tasks.[88] Campbell was so accommodating to all requests to answer questions or to provide help that it became far too easy for everyone to rely on him for things they could not be bothered to do themselves. As Margaret Harwood aptly described the problem in her tribute to his career: "His fund of practical knowledge and thorough knowledge of his subject render him a valuable and active member of the Harvard staff and of these societies [AAVSO, Bond Club, ATMoB]. 'Ask Mr. Campbell,' is the common solution to many problems which arise in the observatory itself – he is never too busy to be interrupted by a call for information and assistance."[89]

For many years, one of Campbell's additional duties had been to serve as the Liaison Officer from the observatory to the AAVSO, the Bond Astronomy Club, and the Amateur Telescope Makers of Boston. Shapley and the HCO sponsored the Bond Astronomy Club and supported it strongly. Many from the HCO staff were members. Over the years, Campbell had provided its members with small telescopes. Many needed Campbell's patient help getting started with their new telescopes; Campbell occasionally spoke at Bond Club meetings or otherwise arranged for speakers for their meetings. In a similar way, Shapley and HCO also supported the Amateur Telescope Makers of Boston (ATMoB). Campbell was a familiar figure to the members of the ATMoB, who for many years met and did part of their mirror making in the HCO optical shops.[90]

The opportunity to meet young people and introduce them to astronomy constituted one of the pleasurable aspects of Campbell's employment at HCO, although those interviews frequently had nothing to do with variable stars or the AAVSO, along with many other duties completely unrelated to the AAVSO. Still, Campbell took pleasure in processing the applications of newcomers and welcoming them to membership with some counseling on how to get started. The exceptional applications, such as that of a North Newton, Kansas, youth, Owen Gingerich, attracted special interest from Campbell, who wrote that he should "be able to find observing variables very profitable as well as interesting and inspiring." When Gingerich asked about the fees for a lifetime membership, Campbell notified him that for a life membership he would have to pay the maximum $50 since he was under 20 years of age, but over 16 years. "It is the other end of the scale that is the lowest, that is, the older you are when you join, the less it costs. I think that if you care to pay the life membership fee in two parts, that would be quite agreeable to the Council," he explained. "Of course, he added, it would be understood that actual membership would not take place until the total fee was paid."[91] As young Gingerich made his second and final payment to achieve his Life Member standing at the very outset of his membership, he proudly announced the completion of his second telescope, an 8-inch Newtonian reflector. As it later developed, a larger telescope did not simplify finding variable stars when only an AAVSO d-scale chart was available.[92]

STUDIES OF LPV STARS

Campbell, likely more than anyone in the world at that point, had immersed himself in the data of LPVs over several decades. From 1942 on, he struggled with the problem of "what to discuss?" Lacking theoretical tools to address the problem, Campbell was forced to address the data strictly as a problem of morphology. He could only hope the geometry of the light curves would later yield to theoretical tools in the hands of others. For now, though, the problem remained one of sorting and classifying, of finding the right bins into which to logically sort the light curves in the hope that those bins would later be meaningful physically. Was it important to sort by such characteristics as the symmetry of the rising and declining branches of the light curve? Or sort by the rate of change in brightness on the rising or declining branch? Or by the breadth of the maximum compared with the breadth of the minimum? Or by the periods? Or by the spectral class? By 1942 Campbell had identified 15 types of long-period variability, with a representative star for each of these types. Eventually these were boiled down to just four classes.

An equally intractable problem, at least for Campbell, lay in how to present, as well as what to present, in terms of the enormous volume of data he had before him. He favored presenting the data as mean light curves based on the visual magnitudes, but others preferred using computed intensities instead.[93] There was no agreement on this question among the staff at HCO, with Menzel and likely Shapley favoring intensities, whereas Richard Prager, Cecilia Payne-Gaposchkin, and Dorrit Hoffleit all favored the use of visual magnitudes. Given the enormous amount of additional computing that would be required to convert the light curves to intensity curves, the pragmatists favored getting actual data out sooner rather than the computed data later.

FIFTY YEARS OF SERVICE

By 1948, Campbell saw his time at HCO drawing to a close. In a note to Willard P. Gerrish, an emeritus professor retired and living in Ashland, Maine, Campbell reflected on his career at HCO with evident satisfaction:

> Do you realize that I am now in my 50th year on the Observatory staff? I well remember that morning in January 1899 when I sat in your office and you had

Figure 9.4. At a special banquet hosted by Harvard College Observatory honoring Leon Campbell's 50 years of service. Seated next to Campbell is his high school teacher, Miss Esther Dodge. Donald Menzel (standing) is master of ceremonies. On the left are Cecilia Payne-Gaposchkin and Albert E. Navez. On the right are Harlow Shapley and Mrs. Frederica Campbell.

> me sign the agreement with the Observatory that I would start working here at $300 for that first year. I little thought then that I would stay on for 50 years, much interested as I was in astronomy.[94]

On January 29, 1949, Harvard College Observatory hosted a special banquet to honor Leon Campbell's 50th year of service. Those seated at the head table with Campbell included Miss Esther Dodge, the high school teacher who originally recommended Campbell to O. C. Wendell in 1899, and of course Frederica Campbell, but also Shapley and HCO astronomers Cecilia Payne-Gaposchkin, Albert E. Navez, Donald Menzel, and others. As chairman of the banquet committee, Menzel was charged with organizing the event.

Strangely absent from the invitees were any formal representatives of the AAVSO, the association that was the center of Campbell's life for so many years. At the last minute, largely at the importuning of former AAVSO President Rosebrugh, the oversight was corrected, and past president Smiley attended for the AAVSO. Invidious as the lack of a formal invitation might have seemed, at the time, it can certainly be understood to the extent that the AAVSO, through its recorder, had become embedded into HCO. In spite of the fact that it had been incorporated as a separate organization and held many meetings away from HCO, it seemed to most people at the observatory that Campbell was the AAVSO.

Other well-earned honors flowed to Campbell as his remaining time as the Pickering Memorial Astronomer lessened. He had been elected to Sigma Xi, the scientific honor society, in 1934 and received an honorary Master of Science degree in recognition of his years of service to Harvard and contributions to variable star astronomy at Harvard's June 1949 commencement. As important as that honorary degree was to Campbell, he was no doubt equally pleased with the congratulatory letters that he received from professional astronomers around the world. The extent to which all this pleased him is evident in his response to a former graduate student at HCO, Leo Goldberg, one of Harvard's outstanding graduates who became director of the McMath-Hulbert Observatory and chairman of the Astronomy Department at the University of Michigan. Campbell explained: "I am certainly delighted to be so honored and to be classed among such other alumni as yourself." Writing to Boston resident Charlotte Greene, Campbell recalled that it had been "a great thrill to be on the platform to receive the honorary degree from Harvard University."[95]

Letters of congratulations flowed into HCO from a number of AAVSO members.[96] Fernald's admiration for his friend and long-time correspondent stood out in one letter. Writing after the AAVSO's Annual Meeting, when the membership honored Campbell with a special ceremony, Fernald said:

> I was a little afraid that our general sadness because of your retirement would have a depressing effect, but everyone cooperated marvelously to cover up. We are all glad that the retirement is from the recordership only, for your work on the analysis of the million observations will be even more important, if anything, than their collection has been.
>
> When I went into this work, and all thru [sic] the years, the thought has been in my mind that sometime in the future some genius would use our observational records as the basis for developments that would lead to a great improvement in living conditions for all humanity. Just as Tycho's observations laid the groundwork for Kepler and Newton, who were the forerunners of the mechanical revolution that makes our lives so much better than those of our ancestors of only four or five generations back, so our observations as

Figure 9.5. Leon Campbell. This photograph was used as the frontispiece in *Studies of Long Period Variables*, published after his death.

> analyzed by you may be the forerunner of developments which we cannot now imagine, but which will mean much to our descendants. You can see the importance I place on the work you are now undertaking.[97]

Fernald's comments about the importance he assigned to Campbell's completion of his work on LPV stars were echoed by many AAVSO members who had contributed to the million-plus observations yet to be analyzed and by professional astronomers who needed guidance on what to observe in detail spectroscopically. Campbell told many of his friends how he looked forward to finally discussing the LPV light curves contained in the AAVSO records – a retirement project.

After Campbell retired on November 1, 1949, he came to his office in the HCO regularly and worked with the data ledgers, assembling the tables to be used in his report. For most people, this would amount to incredible drudgery, but for Campbell, it was a labor of love, and at last the work could receive his undivided attention. Tragically, his health, compromised by years of smoking and long hours at night as well as during regular office hours, declined more quickly than anyone expected. Campbell died of kidney failure on May 10, 1951, scarcely 19 months after his retirement.

While the observatory staff mourned the death of their old friend, the word quickly spread throughout the variable star community. An article appeared on the first page of the July issue of *Sky and Telescope*. Margaret Mayall prepared a longer memorial obituary that appeared in *Popular Astronomy* in December 1951.[98]

LAST OF THE OLD GUARD

With the death of Leon Campbell, an era surely ended. As one of the last of the AAVSO's Old Guard, he represented far more to the Association, and to variable star observers around the world, than to the Harvard College Observatory. Revered by most AAVSO members who came into contact with him, his simple but elegant style and constant willingness to help anyone who needed his help won him many friends. Those same character traits won many new friends to astronomy as well; Campbell worked hard to popularize the science he loved by conducting public nights at the observatory, talks to local clubs, and periodic radio broadcasts.

As it turned out, Clara Olcott had died on the same day as Campbell. The issue of *Popular Astronomy* that contained Margaret Mayall's obituary for Campbell also contained an obituary for that publication's editor, Curvin Gingrich. At the end of that year, *Popular Astronomy* ended its 58-year run as well. The Goodsell Observatory of Carleton College, Northfield, Minnesota, concluded that, after publishing astronomical journals since March 1882 – first, the *Sidereal Messenger* and then, *Astronomy and Astro-Physics,* prior to *Popular Astronomy* – they would not replace Gingrich and instead would end their involvement in astronomical publishing. So Campbell's death, significant as it was to the AAVSO and to variable star astronomy, was only one of several markers for the end of an era.

Part IV

THE SERVICE BUREAU – THE MARGARET MAYALL ERA

10 · Learning about independence

It is human nature always to strive to reach fainter and more distant objects. . . .

– Margaret W. Mayall, 1950, from a radio talk about Annie Jump Cannon[1]

With the retirement of Leon Campbell in 1949, the AAVSO waited expectantly for an announcement of who would replace him as the second AAVSO Recorder. The choice was important: knowledgeable members recognized the change would come at an obvious inflection point in the history of variable star astronomy.

Although well established and recognized in both the scientific community and by amateur observers, the AAVSO was in a state of change. The Old Guard, so instrumental in its survival, had faded away, and the organization was looking for new leaders. Its affiliation with HCO had been facilitated by Campbell, who served as AAVSO Recorder and Pickering Memorial Astronomer. Whether he would be replaced by Harvard College was a question of vital importance to the organization. At the same time, HCO and its Director Harlow Shapley were fighting their own struggle to replace deteriorating equipment and to support their far-flung operations in Africa. This chapter provides background on Campbell's successor, Margaret W. Mayall, explores her early career at HCO and her first years as AAVSO Recorder, and shows how she responded to the emerging crisis that threatened the continued existence of the AAVSO.

THE QUESTION OF A SUCCESSOR

Expecting to retire in 1949, Campbell himself raised the question of his successor as Pickering Memorial Astronomer and AAVSO Recorder. In a confidential memo to Shapley dated January 19, 1949, Campbell offered "a suggestion as to a possible successor for one or both of the titles" and outlined the necessary qualifications: familiarity "with the variable star field, historically and otherwise"; a willingness "to do what might be called drudgery – plotting, ledgering, acknowledging, preparing for printer, one-day and ten-day means, etc, etc."; and an ability to "get along with AAVSO contributors of observations. In other words," he concluded, "one who is well liked by folks in general. Perhaps the two posts, Pickering Memorial Astronomer and Recorder are, for the present at least, to be separated."[2]

The one person he had in mind was Mayall, who was "familiar with the variable star problem, having worked for years under Miss [Annie Jump] Cannon. With her spectral background she should be especially fitted, not to mention her experience with photographic observations." He added: "I have *not* even hinted all this to her, so do not know if she would even be interested. She has a way of getting along with folks, and at her age she should be able to carry on for some years yet."

Having gotten the ball rolling on his successor, Campbell wanted to make sure that it would keep moving. Seven weeks later he suggested another candidate to Shapley – Clinton B. Ford, the present Secretary of the AAVSO. He cited Ford's interest in variable stars since his high school days, his good work in observing, and his discussions on periods of LPVs. Ford, a Michigan alumnus, had studied for a Ph.D. before his war service. Campbell emphasized the need for someone who could retain the interest of the best observers: "I should hate to see the enthusiasm of members slacken because of lack of encouragement on the part of the Recorder."[3]

It is not clear how many candidates Shapley had in mind at this time, but by early March 1949, he set up a search committee at HCO consisting of HCO astronomer Cecilia Payne-Gaposchkin, and AAVSOers Charles H. Smiley (former president and Council member), Percy W. Witherell (current treasurer), and Charles W. Elmer (former secretary and president). By mid-March, the committee was considering at least two persons. AAVSO President Rosebrugh wrote to

AAVSO Secretary Ford that "because of the limited salary the job can pay it looks to me as if the selection will be between Dr. Joseph Ashbrook, now of Yale, and Mrs. Mayall."[4] Then on March 24th, reading between the lines of Rosebrugh's letters, Ford guessed that a third candidate might be in the running. That third candidate turned out to be Rosebrugh himself, as he later admitted to Ford.[5]

Ford had considered the Recorder's position about 1 year earlier, although it is not clear whether Shapley actually offered the position to him then or sometime later. In June 1948, Smiley wrote to Ford, "I have heard your name mentioned a number of times by people who wished that you might be available for that job." Ford responded that it would be impossible to take on the full-time duties of the recorder, citing his own work and residence commitments in Connecticut.[6]

An additional ingredient in this mix appeared at this time. In a letter to Rosebrugh, Ford mentioned that an amateur astronomer named Rolland R. LaPelle, an AAVSO member and active in the Astronomical League, had expressed an interest in the recorder's position after an earlier discussion with Rosebrugh. According to a July 11, 1949, memo from Campbell to Shapley, *Sky and Telescope* editor Charles Federer and his wife Helen had written Shapley early in July to suggest that LaPelle be considered.[7]

The order of contenders seems to have been Ford, Rosebrugh, Ashbrook, Mayall, and LaPelle, as well as several others who, as Rosebrugh stated, were not enticed by the limited salary. It is apparent that Mayall was not a prime contender early in the proceedings; she became the choice by default once it became clear to Shapley that additional funding for variable stars would not be forthcoming.[8]

In a memo from Shapley dated August 12, 1949, Mayall learned that the HCO Council voted that same day to offer her the position of Pickering Memorial Astronomer to begin on November 1, 1949. Shapley informed her that the HCO Council would recommend to the AAVSO President that she be appointed "Recorder for the AAVSO."[9] The language Shapley uses here is interesting: "Recorder *for* the AAVSO" not "AAVSO Recorder" – suggestive of a long-standing attitude that the AAVSO owed its existence to HCO and depended on the labor of HCO astronomers and staff.

Figure 10.1. The Walton general store and the home of Margaret Walton and her family, Iron Hill, Maryland.

MARGARET W. MAYALL

Margaret Mayall was born Margaret Lyle Walton on January 27, 1902, in Iron Hill, Maryland. Her father, William B. Walton, ran a general store in the early 1900s. Iron Hill was a country crossroad with a train station for the Philadelphia, Wilmington, and Baltimore Railroad; little else, except the general store, connected the countryside to the rest of the world. For its 45 residents, employment was primarily agricultural, and farm hands were paid with tokens from the Walton store that could be used only there for the purchase of groceries and other supplies. William's wife, Annie L. Walton, raised their two children, Margaret and her older sister Gladys, and assisted with the store. The family lived upstairs in the two-story building that housed the store on its ground floor.[10] In this setting Mayall recalled that she and her sister "led very sheltered lives" and did not have many opportunities to explore nature, although her father did take her out before dawn to view Halley's Comet in 1910.[11]

Education in the sciences

Mayall's exposure to the sciences was limited at her small-town high school, but in college she was convinced that her fields would be mathematics and astronomy. At the University of Delaware, her math professor gave her astronomical problems to solve, which raised her interest in astronomy, and encouraged her to apply to Swarthmore College, which was known for its excellent undergraduate astronomy program. At Swarthmore, Mayall studied with Dr. John A. Miller

(1859–1946) and Dr. Leslie J. Comrie (1893–1950). Mayall's first scientific publication as co-author was with Comrie, titled "Occultations During the Lunar Eclipse of 1924 August 14" in the *Journal of the British Astronomical Association*. Comrie, who studied math and astronomy at Cambridge University before joining Swarthmore's mathematics department in 1922, later became known for his innovations in scientific computing.[12] On the recommendation of her professors, Mayall spent the summer of her senior year in 1924 at Harvard Observatory. HCO Director Shapley assigned her to assist Annie Jump Cannon (1863–1941), who was working on the first extension of the *Henry Draper Catalogue (HD)* of stellar spectra, published in 1936 as Volume 100 of *Harvard Annals*. Mayall helped measure positions and determine magnitudes of stars from photographic plates, and she gained a working knowledge of the HD system of spectral classification.

Upon earning her B.A. at Swarthmore in 1925, Mayall immediately returned to work as Cannon's full-time assistant at HCO. Cannon gave Mayall extensive training in the classification of stellar spectra.[13] Mayall enjoyed working with Cannon, recalling that "the accidental discovery of variables from their peculiar spectra was an exciting occurrence." Such discoveries led Cannon to begin a spectrum patrol for novae in Sagittarius, which Mayall later expanded using spectrum plates taken in South Africa covering the Sagittarius-Scorpius region. For Mayall, "it never ceased to be exciting to find a nova spectrum and then to go to the regular patrol plates of the region and work up the light curve."[14]

During the 1925 and 1926 summers, Mayall was granted a leave-of-absence to work at the Maria Mitchell Observatory in Nantucket, Massachusetts, as an assistant to its director Margaret Harwood. It was here that she received her "first real introduction to the photographic discovery and observation of variable stars." Later, Mayall returned to her studies and received her M.A. in Astronomy from Radcliffe College in 1928.

Figure 10.2. Margaret Harwood, Margaret Walton, and Harlow Shapley, advertising an "Open Night" talk by Shapley at the Maria Mitchell Observatory, Nantucket, Massachusetts, in 1925.

Early exposure to the AAVSO

Mayall's early exposure to the AAVSO came in helping Campbell and other HCO staff to prepare exhibits for the AAVSO's annual meetings. Mayall recalled that "the AAVSO was an integral part of the Observatory, and Mr. Campbell's office was the clearing place for all kinds of variable star information." Over time she became more involved with AAVSO work, usually in helping Campbell with variable star identifications, selecting plates that might be used for new charts, checking reports of novae, and so on.[15]

It was at her first attendance at a 1924 AAVSO meeting that Mayall met her future husband, R. Newton Mayall – a landscape architect, sundial enthusiast and designer, amateur astronomer, and AAVSO member. They were married in 1927; they had no children.

Journeyman work at HCO

Mayall's first publications associated with her work at HCO appeared in 1925 and 1927. Of six papers, three were co-authored with Shapley, one on the period-spectrum relation for Cepheid variables.[16] Mayall also gave a paper at the annual meeting of the American Astronomical Society on observations of variable stars in the Messier 11 region made at the Maria Mitchell Observatory. From 1925 to the time she became AAVSO Recorder in 1949, Mayall's name appeared on 22 scientific papers and notes on photometry and spectral classes of variable stars, on 15 as the primary author.

Figure 10.3. Henrietta Swope, HCO staff astronomer, with Margaret and Newton Mayall at the Mayalls' home in Cambridge, Massachusetts, sometime in the early 1930s.

In 1936, Mayall prepared a paper about an unusual, long-period eclipsing variable star. She intended to submit it for publication in the *Harvard College Observatory Bulletin*, but Shapley thought the paper good enough to be included in the Harvard College Tercentenary Papers of the *Annals of the Astronomical Observatory of Harvard College*. Despite this implied confidence in her abilities, Shapley still sent her an itemized list of what he wanted to be included in the paper. Evidently Shapley thought her variable star work worthwhile but still viewed her as somewhat of a novice.[17] This paper, appearing with the title, "A New Eclipsing Star of Unusual Character," made use of her knowledge of spectral classifications, as did much of her other variable star work at HCO.

While at HCO, Mayall's work fell into a few broad categories: spectral classification; charts, sequences, and star identification problems; and other tasks such as public inquiries, which included communications having to deal with astrology. A good example of her research interest at this time was her 1940 article, "Spectral Curves for Thirty Cepheid Variables," which builds not only on her 1927 work co-authored with Shapley on the period-spectrum relation, but also her spectrographic work with Cannon and her own interest in variable stars.[18] The 1940 article was the result of work done for HCO's Milton Bureau for the study of the brighter variable stars, a program headed by Cecilia Payne-Gaposchkin and Sergei Gaposchkin.[19]

THE RESIDENT SPECTROSCOPIST

After Annie Jump Cannon's death on April 13, 1941, Mayall continued with Cannon's unfinished work compiling *The Henry Draper Extension*. It was published in 1949 as Volume 112 of the *Harvard Annals* and titled *The Annie J. Cannon Memorial Volume of The Henry Draper Extension*. Cannon made the spectral classifications in this volume, though Mayall contributed the supplementary classifications.

Mayall's HCO work was interrupted during World War II when she was granted leave to join the research staff of the Heat Research Lab of the Special Weapons Group at the Massachusetts Institute of Technology (MIT). Her work there from 1943 to 1946 involved thermocouples and artificial crystal technology and assembling and testing instruments for heat-seeking bombs. She spent most of one summer at the desert testing range at Tonopah, Nevada.[20]

Returning to HCO in 1946, Mayall resumed her routine work using HCO plates to classify the spectra of faint stars for other observatories. She also handled inquiries about star identifications from a broad cross-section of astronomers, many at the early stages of their careers, such as William P. Bidelman, Daniel M. Popper, Sidney W. McCuskey, and Balfour S. Whitney.[21]

In early December 1947, Dr. Otto Struve, chair of the International Astronomical Union (IAU) Commission 29 on Stellar Spectra, asked Mayall, a member of the commission then, to summarize her current work in stellar spectra for the Commission 29 report at the August 1948 IAU General Assembly.[22] Struve's query, and the list of projects Mayall mentions in her response, shows that there was interest in, and even a demand for, the continuing spectrographic work being done at HCO.[23] In 1948, Struve appointed Mayall to a Commission 29 sub-committee to correct and revise a list of standard stars for stellar spectra proposed by Dr. A. H. Joy of the Mount Wilson Observatory. Members of this committee were Joy, Struve, Dr. W. W. Morgan, Prof. J. J. Nassau, Dr. C. Schalén, and Prof. A. N. Vyssotsky.[24] Joy asked Mayall to join and work with this very distinguished group of astronomers with the intention of achieving IAU adoption of a set of standard stars for spectral classification "in order that the

Figure 10.4. Marjorie Williams, Margaret Mayall, and Leslie J. Comrie, in Zurich for the IAU meeting, August 1948.

Harvard system of stellar classification may be made more accessible to various observers."[25] In addition, Struve asked Mayall to prepare remarks on her work, future plans, "and the general outlook at Harvard in regard to the continuation of the spectral catalogs" for a discussion session of Commission 29 at the 1948 IAU General Assembly in Zurich.[26]

Mayall and Shapley, among other HCO astronomers, attended the Zurich meeting in 1948 – the first large meeting of astronomers since 1938. In the list of ambitious resolutions proposed by the IAU executive committee at the conclusion of the 1948 assembly, two must have greatly interested Mayall. The first stated that Commission 27 (Variable Stars) recognized the importance of spectrographic observations of variable stars in the southern hemisphere. The second recommended that a subcommission be established on variable star spectra that would include members of the Variable Star and Stellar Spectra commissions "in order to bring about prompt and effective collaboration between photometric and spectrographic observers."[27]

At the time of the 1948 meeting, Mayall was not a member of the IAU's Commission 27 on variable stars. Its membership did include some HCO astronomers (Campbell, Sergei Gaposchkin, and Shapley, plus his wife Martha, and AAVSO amateurs Reginald De Kock, Giovanni Lacchini, and Leslie Peltier). At HCO, Mayall only had a supporting role, whereas Campbell was still the Pickering Memorial Astronomer and Recorder, and the Gaposchkins were in charge of the Milton Bureau of variable stars. On the other hand, Mayall was the successor to Cannon as the resident expert in spectral classification and was recognized as such by her distinguished international colleagues. HCO astronomers on Commission 29 (Stellar Spectra) were Mayall, Armin Deutsch, Payne-Gaposchkin, and Dorrit Hoffleit.[28]

Soviet variable star catalog

When the IAU's Executive Committee met in 1946 in Copenhagen, it shifted responsibility from Germany to the Soviet Union for the naming and cataloging of variable stars. The USSR Academy of Sciences and the Sternberg Astronomical Institute volunteered to do this work.[29] The USSR variable star astronomers had taken on the imperative in an amazing flurry of productivity, which included a journal and series of circulars devoted to variable stars. Their most important accomplishment was the publication of the first *General Catalogue of Variable Stars* by Boris V. Kukarkin and Pavel P. Parenago in 1948. This catalog listed 10 912 stars and was distributed free of charge to any astronomers who requested it. This sudden productivity of the Soviet astronomers would soon unsettle HCO's century-long dominance in variable star research, which in turn would have a long-term effect on the existence of the AAVSO.[30]

Despite not being included among the HCO representatives in the IAU's Variable Star Commission, Mayall was always involved with variable star work in some way, both at HCO and in answering queries from outside the observatory. Her involvement with variable stars, to a large extent, stemmed from and was an outgrowth of her work with Cannon. Emblematic of this is her concern, in May 1947, that HCO's bibliographic card catalog of variable stars – which originated with Cannon – had not been kept up by the Observatory.[31] However, in April 1949, when Mayall responded to an internal HCO request for lists of staff research projects, she only listed her spectral classification work for the *Henry Draper Extension* and a search of spectrum plates for novae and other peculiar emission line objects.[32]

A new recorder

In offering Mayall the recorder position on August 12, 1949, Shapley explained: "It is our hope that after you get things under way we may be able to arrange the work in variable stars so that you can systematically have time to continue with the classification of faint stellar spectra." Was Shapley being truly accommodating, or was this a nicely worded expectation of double-duty without adequate compensation? Shapley was certainly making light of the nature and extent of the work involved as AAVSO Recorder so as to offer the enticement of continued involvement in spectral classification – a line of work that was really more meaningful and professionally satisfying to Mayall than the mostly administrative and managerial work of the AAVSO position. Interestingly, a memo Shapley wrote to Mayall 1 year later clearly showed his awareness of the burden of labor with the AAVSO Recorder's job, despite his earlier enticement that she probably could keep up her spectral classifications. The memo concerns a proposal by Ford to start a newsletter for the AAVSO. Shapley is in favor of the idea in principle, but is strongly against it for practical reasons – reasons that show how much he knew the Recorder's job entailed. "I don't see how you can possibly do anymore now than try to keep your head above water, and try at the same time to catch up," he wrote.[33]

Mayall accepted the recorder position in a memo to Shapley on August 13, 1949, the day after his offer.[34] Three days later Rosebrugh informed Ford of the HCO Council's selection.[35] As AAVSO Secretary, Ford immediately sent Mayall a letter of congratulations, which also included the latest problems involving preparations for the coming annual meeting and a quick update on the progress being made toward a final draft of the AAVSO Constitution Revision, which has been "in the works for some months now."[36]

Mayall responded a few days later: "I had thought over dozens of other people for the job, but had never considered myself! . . . I am delighted that you are my co-*it*! I sure have a lot to learn about my job, but luckily, Mr. Campbell seems very pleased at having me take over, and he is being extremely helpful and cooperative in getting me started." She also mentioned that the *Cannon Memorial Volume*, with classifications of about 87 000 stars was "out at last," which "makes a good breaking off place for me to go into another job." But she added, "I do plan to continue some spectral work, probably directed towards variables and novae, which after all, are the most interesting objects spectroscopically."[37]

According to Hoffleit, who was on the HCO staff at the time, the reason that Shapley offered the AAVSO Recorder's job to Mayall was simply because "support for large routine investigations" – such as Mayall's spectral classifications – had "waned after the war."[38] This reason is debatable, given the interest shown by Struve, the IAU, and the international astronomical community in Mayall's continuation of the HD classification work, and as shown by the IAU resolutions calling for cooperation between the Stellar Spectra and Variable Star commissions. It was, however, certainly a matter of financial and administrative convenience for Shapley and HCO to shift Mayall over to Campbell's Pickering Memorial Astronomer position, permitting her salary to be covered by the Harvard-controlled Pickering Memorial Endowment. She was already an HCO employee and had been assisting Campbell with a fair amount of AAVSO's day-to-day work, so little really changed on the side of HCO. All that HCO need give her was a prestigious title, little added compensation, and added responsibility.

Officially, the AAVSO Council was kept out of the loop in the selection of Campbell's successor. What little Secretary Ford knew was based on sporadic inside information delivered by Rosebrugh. It was about 5 months from the time that Rosebrugh and Ford first heard that a selection committee was set up by Shapley (March 7, 1949) until Shapley informed Rosebrugh (August 15, 1949) that Mayall was appointed Pickering Memorial Astronomer and was the "candidate for the Recordership" to be voted on by the AAVSO Council on October 15, 1949.

Most on the AAVSO Council and Shapley at HCO were acutely aware of the paltry salary available from Harvard for the position of recorder. Shapley made an effort to secure more funding for the recorder position but could only secure funding for areas of research other than variable stars. By 1949, the recorder's workload was considered to be full-time and more; it was recognized that a professional astronomer was required for the position from this point forward, and it was too late to suddenly attempt to bring the recorder's salary up to a full-time level. Moreover, by securing funding for solar and meteor work and failing to get money for the recorder position or variable star research and related staff, Shapley had in 1949 unwittingly set the stage for the AAVSO's ouster from HCO in 1954.[39]

THE CALM BEFORE THE STORM

Mayall, the new HCO Pickering Memorial Astronomer and AAVSO Recorder, moved from Building C (rooms 37 and 38) to Building A (room 12), the site of AAVSO's Headquarters. The October 1949 issue of the *HCO*

Figure 10.5. Margaret W. Mayall at the time she became AAVSO Recorder in 1949.

Gossip Sheet, an interoffice newsletter, noted that she replaced Campbell as Pickering Astronomer "and as the Observatory agent" of the AAVSO.[40] One of Mayall's first accomplishments as recorder was a thorough cleaning of Campbell's office, followed by a coat of paint. Her former rooms, once shared with Cannon, were, along with room C36, supposed to be converted to a "museum" of Cannon memorabilia, exhibited to honor her memory; but instead, by 1949, the rooms were used as graduate student offices.[41]

Once Mayall started in her new position, Ford wrote to long-time AAVSO member Lewis Boss to assure him that the Association was in good hands, also noting that his secretary's position was "nothing compared to what faces Margaret Mayall now. She writes that she even puts [her husband] Newt to work for a few hours every night, trying to get [observation report] ledgers up to date, L.C.'s files in some understandable order, correspondence up to date, etc . . . It is quite a mess to untangle, as you can well imagine, and the Association is very fortunate to have someone as thorough and conscientious as Margaret to do the job."[42]

Rosebrugh concurred that AAVSO Headquarters was really in "an awful mess when Mayall took over" – partly because of the war, partly because Shapley did not realize how much he was burdening Campbell, and partly because of Campbell's advancing age. He recalled one of the last times he was in Campbell's office: "I saw a note on his desk 'Mr. Campbell, There is no soap in the women's room.' This annoyed me, apparently he was being asked to act as supervisor of the janitor."[43]

The recorder's job

Mayall took up the recorder's job energetically, responding promptly and with confidence to internal problems and questions by Secretary Ford, the AAVSO Council, and other members. Publicly, too, Mayall presented herself as a knowledgeable and competent intermediary between the amateur and professional astronomy communities. An indication of her standing with professional astronomers outside HCO can be found in the relationship with Paul W. Merrill (1887–1961) of Mount Wilson Observatory. Merrill began his professional association with the AAVSO in 1919. He routinely made use of AAVSO light curves and photometric and visual observations, in connection with his spectral studies, especially for LPV stars having peculiar spectra. In 1938 he had published a book on *The Nature of Variable Stars*.[44]

As AAVSO Recorder in 1949, Mayall's knowledge of both variable stars and spectroscopy meshed perfectly with Merrill's research interests. In responding to his requests, Mayall sometimes wore the hat of the recorder, providing 10-day means of observations, tracings of light curves, dates of maximum and minimum magnitudes, and predictions. At other times she was the resident spectroscopist, returning to the plate stacks to confirm a star's spectral type or apply an analytical technique learned from Cannon.[45]

In the first letter Merrill wrote to Mayall on her role as the new AAVSO Recorder, he made it clear that he not only valued the work of the AAVSO, but that he also had confidence in Mayall's ability to do the job. "Some time ago," he wrote, "Dr. Shapley indicated that I might now apply to you for photometric data as I have in the past to Leon Campbell and then I received a Christmas card showing a little girl waving her feet around in the recorder's shoe. The feet did look a little small but I am sure the head is completely competent for this important work."[46]

Mayall's reply revealed that, although slightly overwhelmed by the recorder's job, she truly relished the work: "Thanks for the kind words! It makes me feel pretty helpless to suddenly get introduced to about 500 variables and 300 people, and feel that the personal history of each one should be remembered! Nevertheless, I find the new job fascinating, and am having great fun with it."[47]

Although she clearly enjoyed her new role as AAVSO Recorder, Mayall also learned that the position was not at all an autonomous one. A small, but telling example can be found in a memo she wrote to Shapley on January 9, 1950, asking if it "would be appropriate to ask the HOC [Harvard Observatory Council] to give consideration to a photographic program on variables for amateurs." In effect, Mayall was asking permission to ask permission: at first glance, indicating her lack of confidence as the new recorder. On closer reading, however, it shows how the hands of the recorder as AAVSO officer were tied by working concurrently as the HCO Pickering Memorial Astronomer and as a HCO staff member/employee. The AAVSO leader's exercise of independent thinking and initiative, even in matters pertaining to its own membership, was, at times, severely limited in this regard by its association with HCO.[48] But as the AAVSO's Recorder, Mayall was purposeful and equally at ease with professional and amateur astronomers (if not with her own Director and Observatory Council).

New goals

When Mayall gave her first paper as the AAVSO Recorder at the AAVSO's 1950 Spring Meeting, the future of the AAVSO as a real contributor to variable star research seemed certain as new avenues of investigation were opening up. Appropriately, she spoke on "The AAVSO Observing Program" in which "several lists of variables for special programs" would be available soon, among them a photographic program, stars to be observed with photoelectric cells, some bright stars that needed watching, and some special variables for those observers who wanted to keep watch on a small group of variables every night.[49] She and Campbell had already started on this latter group of stars with a list of 61 variables to be watched by experienced observers (semiregulars, irregulars, novae, Z Cam, U Gem and other types, RU Tau, and two flare stars). Peltier was one of the first observers to begin working from this list as part of his regular, monthly program.[50]

The pulse of HCO

Meanwhile, Shapley was taking the pulse of his HCO and finding it faint. On September 27, 1950, he wrote an unusually introspective and brooding memo to Sergei Gaposchkin and Cecilia Payne-Gaposchkin in which he conceded that HCO's dominance of the field of variable stars "has been taken away by the Russians. I suppose there is nothing to do about it." He referred to the several numbers of the Russian journal *Variable Stars* that came into the library "full of light curves, observations, – perhaps nothing very profound but very active."[51]

Shapley's rambling discussion of the state of variable star research at HCO did not mention the AAVSO; but someone, possibly Hoffleit or Mayall, had written in the margin of Mayall's copy, "And the valuable bibliography completely discontinued." The oversight suggests the degree to which the AAVSO had been taken for granted after some 30 years at HCO. Similarly, the December 1950 *HCO Gossip Sheet* described AAVSO observers as "these small but enthusiastic sub-stations of H.C.O." The marginal note also suggests that some of the blame for HCO's decline as the center of variable star studies was within, not outside its walls. Partly to blame was Campbell's years of emphasis on observation and study of LPV stars at the expense of other types of variables that Mayall would soon be emphasizing.

Shapley's musings about HCO losing its institutional prestige are in sharp contrast to Mayall's outgoing, cheerful, and publicly engaged depiction of the AAVSO's mission simply to aid anyone involved in variable star research. When measured by Shapley's yardstick, variable star work at HCO was indeed doomed, but by AAVSO's standard, as defined by its mission, it would prove to be full of promise and accomplishment.

FIRST YEAR AS RECORDER

When Mayall completed her first year as recorder on October 1, 1950, AAVSO Treasurer Percy Witherell reported at the AAVSO Council meeting that the Association's finances were "in excellent condition in spite of rising expenditures." Still, the Council looked for ways to save money, such as exchanging publications with the Astronomical Society of the Pacific (ASP) and a review of members with dues in arrears.[52] Mayall envisioned great savings with the purchase of a Multilith machine to publish the *Quarterly Reports*, observer report blanks, observer instructions, future issues of *Variable Comments*, and other publications. The purchase was made possible with $1000 raised (nearly one-quarter from benefactor Lawrence N. Upjohn), of which $817 paid for the Multilith.[53] In another money-saving motion, the Council empowered the recorder to consider "giving up" the AAVSO's Nova Search program to the Astronomical League. The Council discussed but took no action on several other proposals, including ways of increasing the general fund, an increase in annual dues, a campaign to increase sustaining membership, and a 40th anniversary fundraising campaign. The final item on the first day's agenda was a discussion of the probability that the National Bureau of Standards would no longer be able to support the Solar Division. The AAVSO would most likely have to assume the Division's operating expenses. HCO solar expert Donald Menzel, AAVSO's second vice president, urged that the Division's activities be continued "if necessary entirely at AAVSO expense," in spite of NBS cutbacks.[54]

On the second day of the Council meeting, Shapley, attending as past-president, suggested "raising a really large sum of money [$50 000], comparable perhaps to the Pickering Memorial Fund of 20 years ago, which would augment that Fund as a capital asset of the Association, and would provide, among other things, for a full-time salaried assistant to the Recorder."[55] That Shapley specifically recommended that funds raised be made an asset of the AAVSO certainly shows that he was supportive of the organization, but also demonstrates his awareness that any funds raised might be taken away by Harvard if not secured in this way. The Council adopted a motion presented by Shapley for the president to appoint a special committee to examine the needs and opportunities of the Association, to plan provisionally suitable new projects, and to outline methods of financially meeting the special and general requirements of the Association.

Along with his fundraising proposal, Shapley sent a three-page list to Mayall titled, "Enterprises that might be on the fortieth-year growth budget of AAVSO." The list covered many areas: library "overhauling and cataloguing"; adding new pictures to the slide collection; refurbishing AAVSO telescopes; providing special eyepieces for Solar Division observers; and "revival of the Harvard 12-inch polar telescope, and possibly a bi-prism photometer, for use on faint variable stars." He also suggested adding two part-time assistants ($300 each) to search Harvard plates for "early photographic maxima for long-period variables suspected of period variation"; special observation with photometers of quickly changing stars; special bibliographic studies; and typing Multilith masters "in order to catch up on the quarterly reports." Shapley included budget estimates for each item totaling $1650 and listed some sources of funds for the 40th anniversary project, including members of the Association, their friends, grants-in-aid from various scientific societies and academies, and observatories in need of special research.[56]

Worded more like a mandate than a suggestion, Shapley's Enterprise proposal showed his concern about how much weight the HCO could carry for the AAVSO in the future. Shapley recognized that the Pickering Memorial Endowment was insufficient and needed a fresh infusion of funding. Originally, it was intended that the income from the Endowment would pay the salary of the recorder and a part-time assistant. The AAVSO funds, mainly from dues and gifts, would cover all expenses of the recorder as an AAVSO officer, plus all operating expenses, postage, incidental expenses, and so forth. Eventually HCO took on the burden of added AAVSO expenses, including modest increases in the recorder's and assistant's salaries.

Shapley's proposal also reflected the anxiety he expressed in his memo to the Gaposchkins, in that he felt the need to urge the AAVSO to explore new opportunities and projects, presumably ones that would enhance the image of HCO as a leader in variable star astronomy. On the other hand, the proposal also suggested that, in 1950, there was a future ahead of the AAVSO. There was an important difference between Shapley's call for exploring new areas and the early initiatives taken by Mayall: Shapley directed his concern toward the image and reputation of HCO – he had no specific scientific goals in mind. Mayall, on the other hand, turned to current topics of interest in variable star research to enhance the value of the AAVSO's observations and services to astronomy.

The AAVSO began its general fund drive in conjunction with the AAVSO's 40th anniversary. President Neal J. Heines prepared a three-sheet fundraising letter in which he noted the successful Multilith campaign, quoted from Shapley's proposal to the AAVSO Council, and listed several of the budget items from Shapley's "Enterprises" outline, including the need for two part-time astronomical technical assistants. However, no mention was made of Shapley's desired $50 000 goal.

The 1951 drive raised a little over $400 by May 1951 and $754 by October 1951. After Campbell's death in May, the Council gave some thought to changing the Fortieth Anniversary Fund to a "Campbell Memorial Fund," similar to the Pickering Memorial Fund. The idea was dropped at the October Council meeting when it was clear that the fund "was not sufficient to establish an interest-bearing memorial fund, and that this money should probably be spent as needed to keep up with the increasing demands of the Recorder's office."[57] It was not until 1954, in the face of the AAVSO's survival crisis, that Ford and other AAVSO members began a strong, sustained, fundraising effort.

SHIFTING SANDS

In contrast to Shapley's hand-wringing over the state of variable star astronomy at HCO and his desire to increase the Pickering Memorial Endowment, Mayall focused on expanding the AAVSO's observing programs. Thus she believed the Association would be a useful tool to many more researchers, which in turn would generate professional interest in the AAVSO's services and, hopefully, lead to research grants. By the spring of 1951, perhaps swayed by her own interest in the spectrum novae searches she once conducted with Cannon, Mayall decided that the AAVSO ought to keep its Nova Search program, rather than turn it over to the Astronomical League.[58]

Mayall also considered expanding the observing program so that AAVSO observers would determine times of minima of a number of eclipsing binary (EB) stars. At her request, Professor Joseph Ashbrook at Yale sent her a five-page memorandum of practical suggestions for observing EBs, in which he noted that times of minimum were "of sufficient importance to warrant such a program."[59] Also, following the trend of current research interest that placed less emphasis on LPVs, Mayall encouraged several of her best observers to concentrate on irregular variables. Peltier in Ohio was one of the first observers to take up this program, along with Gunnar Darsenius of Sweden, Hermann Peter of Switzerland, Tom Cragg and Howard LaVaux in California, Ed Oravec in New York, and observers in Greece, including Stavras Plakidis, Chrysostomos Drakakis, Joannis Focas, Constantin Chassapis, and Demetrius Elias. "Taking it all in all," Mayall later noted, "I feel quite happy about most of the observers and their programs."[60]

Campbell died on May 10, 1951, less than 2 years after he retired from his work at HCO – he had been suffering from kidney problems earlier in the year. Mayall would also learn that Mrs. Clara Olcott – wife of William Tyler Olcott – also died on the same day.[61] The AAVSO's Spring Meeting began the next day at Georgetown College Observatory, Washington, DC. AAVSO President Heines and Mayall decided to carry on with the meeting, as they and other members present felt it would be a fitting honor to Campbell's memory. The Council unanimously presented the 10th Merit Award to Shapley and the 11th to Rosebrugh, both at the annual meeting. To the AAVSO membership in general, Shapley was not the cost-conscious and image-conscious HCO administrator, instead he was the champion of their cause – supportive of the AAVSO, its work, its members, and observers from his first day at HCO.

Eight days after Campbell's death, Shapley asked Mayall about the state of Campbell's study of LPVs. Mayall summarized the work in a short report: the first part of Campbell's volume consisting of the dates of observed maxima and minima of 400 LPV stars from 1920 to 1949 was completed, except for introductory and descriptive text; the second part comprising the mean light curves for the 400 variables was completed in table form, but not proofed; and what remained was the drafting of the mean light curves and completing the related descriptive text.[62] Mayall was optimistic about completing the Campbell volume quickly. Rosebrugh offered to prepare drawings of the mean light curves. Eventually, this task would have to be shared by other members as well. The Campbell volume work would hang like a millstone around Mayall's neck until 1955.[63]

The August 1951 *Popular Astronomy* contained two obituaries: one for Campbell, written by Mayall, and one for Curvin H. Gingrich (1880–1951), the *Popular Astronomy* editor, written by Frederick C. Leonard, the founder of the long-defunct Society for Practical Astronomy.[64] Just 1 year earlier, the AAVSO had made Gingrich an honorary member in recognition of his years of support. As the editor of *Popular Astronomy* since 1926, Gingrich had encouraged a whole generation of young amateur astronomers, including Ford, who – in writing an official letter of condolence to Mrs. Gingrich on behalf of the AAVSO – acknowledged her husband as "one of the cornerstones of the AAVSO's existence."[65]

Introducing PEP observing

The AAVSO Annual Meeting in October 1951 celebrated the organization's 40th anniversary. To mark the occasion, Rosebrugh wrote a "History of the A.A.V.S.O. Since 1936."[66] In addition, of the 22 papers given at the meeting, nine were on solar topics; five were on variable star topics; two each covered auroras, charts, and general topics; one was on planets; and one was on photoelectric photometry (PEP). Of all these, the most significant would prove to be the paper by John J. Ruiz (read by Rosebrugh in his absence).[67] Titled "The Gremlins and My Photometer," Ruiz's paper mostly described PEP pitfalls involving such issues as pier mounts, dust, dew, and mechanical problems. At the end of the paper, however, he showed a light curve for 12 Lac (DD Lac), a variable star with three overlapping periods, obtained from PEP observations. His plotted observations dramatically showed that PEP observations of a small amplitude, short-period variable star (0.08 magnitude and 4.5 hours) could be obtained by amateurs.

Ruiz – the only amateur PEP observer making quality observations of variable stars – called for more amateurs to contribute PEP observations. Ruiz further encouraged observer interest in a December 1951 *Sky and Telescope* article, the first by an amateur variable star observer in a general readership publication and significant because it presented an example of the type of valuable, accurate, work that could be accomplished by an amateur.[68]

In all, Ruiz had compiled 10 light curves of DD Lac. These greatly impressed Mayall who said they were "certainly beauties" and sent them to Dr. Frank Bradshaw Wood, a photometry expert and director of the Flower Observatory, Upper Darby, Pennsylvania. He, too was impressed with Ruiz's output, and wrote to Ruiz: "This certainly looks like an excellent piece of work and it is exceedingly encouraging to see work of such quality coming from amateur astronomers." The enthusiasm for Ruiz's work did not stop there. Mayall planned to publish the light curves in her "Variable Star Notes" in *Popular Astronomy*. She conveyed Wood's enthusiasm again to Ruiz, a few days after encountering Wood at a PEP conference in Philadelphia, noting: "He is very enthusiastic about your work. He wants to have your detailed light curves published sometime. I certainly agree with him that they should be printed."[69]

Figure 10.6. John J. Ruiz, with his daughter acting as recorder, demonstrating his photoelectric photometry equipment, 1953.

Publicizing variable star astronomy

Partly as a result of Gingrich's death, *Popular Astronomy* ceased publication after the December 1951 issue. Laurence M. Gould, president of Carleton College, Northfield, Minnesota, stated that the assets and liabilities of the magazine owned by Carleton College were to be taken over by Sky Publishing Corporation, Cambridge, Massachusetts, owned by Charles and Helen Federer, publishers of *Sky and Telescope* magazine.[70]

Before this agreement had been reached, several AAVSO members and others were concerned about the magazine's demise. Rensselaer Polytechnic Institute (RPI), Troy, New York, considered continuing the publication provided that financial support could be found to guarantee against a first year deficit. Charles Smiley, former AAVSO president, wrote to AAVSO President Martha Stahr Carpenter announcing that Brown University would underwrite $100, and possibly another

$200 would come from amateur astronomers in the Providence, Rhode Island area, and added that the AAVSO ought to consider doing likewise.[71] AAVSO Secretary Ford agreed that the AAVSO should underwrite the plan.[72] Rosebrugh, however, balked because the AAVSO had been supporting *Popular Astronomy* through its payments for printing the AAVSO's "Variable Star Notes" for "at least the last 15 years."[73] Ford went so far as to write G. Howard Carragan of the RPI Physics Department, telling him (without authorization) that the AAVSO was considering providing monetary assistance, pending a Council vote.[74] But there was more at stake than propping up a once-venerable publication. Ford recognized that the AAVSO would be at a great loss if its "Variable Star Notes" were not continued in some way. In a letter to Mayall, he described the "Variable Star Notes" as "the only medium through which observers are able to see their names and their hours of work appreciated by the astronomical world."[75]

In early January 1952, Mayall was hospitalized for surgery to correct a spinal disc disintegration condition. Throughout January, Mayall was not able to participate in the discussions about *Popular Astronomy*, although she did let Ford know that she agreed with him on the importance of publishing "Variable Star Notes" and lists of observer totals in a wide-circulation journal.[76]

Feeling uneasy about the quickly developing issues, the RPI Administration backed away from any further involvement in saving *Popular Astronomy*.[77] Furthermore, Sky Publishing – which held the assets of *PA* – eventually decided not to publish *Popular Astronomy* as an independent magazine. Ford learned that Charles and Helen Federer – from a purely business standpoint – had no interest in keeping *Popular Astronomy* alive. Thus the venue for monthly publication of the AAVSO's variable star notes was lost (the observations themselves were being published separately as *Quarterly Reports* since 1948).

It seems likely HCO chose not to get involved in finding an outlet for the AAVSO's "Variable Star Notes." Shapley, after all, was trying to find ways to economize, and the HCO had recently discontinued subsidizing the publication of AAVSO observations in the *Harvard Annals*. There was no good reason for HCO to support the publication of what it probably saw as "club" or membership notes, as opposed to hard scientific data. It is to the AAVSO's credit, and a sign of their developing maturity as an organization, that they sought possible solutions to the problem on their own and did not appeal to HCO for help.

If Mayall was trying to solve the publication problem during February, March, and April, she made no mention of it during this time. There were many distractions: she was still recovering from her operation, trying to catch up on the backlog of office work, and preparing for the AAVSO's Spring Meeting at Clarkson College, Potsdam, New York.[78] But apparently Mayall was working quietly behind the scenes with Canadian astronomer and past AAVSO President Helen Sawyer Hogg to secure arrangements for publishing "Variable Star Notes" in the *Journal of the Royal Astronomical Society of Canada* (*JRASC*). No word of this arrangement had surfaced until Mayall made an announcement at the AAVSO's Council and General meetings (May 23–24, 1952) "that the problem of finding a suitable journal in which to publish the AAVSO 'Variable Star Notes' . . . had been solved through the splendid cooperation of the Royal Astronomical Society of Canada . . . the fortunate arrangements, made largely through the efforts of past AAVSO President Mrs. Helen S. Hogg."[79]

The arrangement was indeed fortunate for both parties: the AAVSO would have its notes and observer totals published bi-monthly in a modest but respected international journal that reached a wide audience of professional and amateur astronomers, and the *JRASC* would see a healthy increase in the number of subscribers drawn from the ranks of variable star observers. The *JRASC* staff was so pleased that they were willing to accept US currency at face value for the $3 subscription – which was $1 less than the price of a *Popular Astronomy* subscription. The AAVSO agreed to pay the RASC $50 per year to cover publishing costs of the "Notes."[80]

Publication in *JRASC* was an important step forward for the AAVSO: it marked the first time that the AAVSO had presented itself on its own merit as an international scientific organization of amateurs and professionals that could appeal to a like audience. Mayall herself, in the first of the *JRASC* "Notes," pointed out that "the names of the observers may give some hint of the universality of interest in the visual observations of variable stars. During the fiscal year, 1950–1951, we received nearly 50 000 observations from 136 observers in 21 countries." The second *JRASC* installment of "Variable Star Notes" announced that July 1952 was

Figure 10.7. AAVSO Spring Meeting at Clarkson College, Potsdam, New York, May 1952. Front row, from left: Margaret Mayall, Helen Sawyer Hogg, Martha Stahr Carpenter, and Neal J. Heines. Middle row, third from left: Ruth J. Northcott, *JRASC* assistant editor; fifth from left: Peter M. Millman, Dominion Observatory, Ottawa; Back row, first from left: Clinton B. Ford; third from left: John J. Ruiz; second from right: Francis H. Reynolds of Clarkson College.

designated "AE Aquarii Month" by the AAVSO and also contained the important "Light Curves of 12 Lacertae" by John J. Ruiz. Mayall preceded Ruiz's article with an appeal for more PEP observers.[81]

Amateur-professional collaborations

Mayall soon undertook several international observing campaigns that were among the first amateur-professional collaborations of their kind. The first campaign noteworthy as an example of Mayall's first public outreach effort involving a worldwide call for observations, "AE Aquarii Month," began in collaboration with Yale astronomer Joseph Ashbrook. At the time, AE Aquarii was classified as a peculiar variable star, was known to be a white dwarf, and was "not a true flare star." Observers were asked to make an observation every 10 minutes over a period of at least an hour on every observing night during the month. In announcing the cooperation of observers in Europe, Africa, India, Australia, and New Zealand, as well as the Americas, Mayall noted it was "a chance for the visual observers to do something the professional astronomer would find impossible – to make observations of a peculiar variable star throughout a 24-hour period." The great success of the AE Aqr campaign that concluded in July 1952 encouraged Mayall and Ashbrook to plan observing campaigns for other variables. *Sky and Telescope* editor Leif J. Robinson later wrote: "Ashbrook clearly helped usher in the modern era of amateur-professional collaboration."[82]

An observing program on red dwarf flare stars began as a collaboration with Sarah Lee Lippincott of Sproul Observatory, Swarthmore College. Mayall agreed to publish Lippincott's paper, "Red Dwarf Flare Stars," as a part of the second installment of "Variable Star Notes" in the January–February 1953 *JRASC*.[83] The article listed known and suspected M-dwarf flare stars that Mayall urged observers to monitor. Mayall clearly saw the "Notes" in *JRASC* as an opportunity to reach beyond the ranks of the AAVSO membership and to increase the number of observations and observers.

With her mother's death in June 1952, the problem of the "Variable Star Notes" publication resolved, and her "forced vacations" of hospitalization and the spring meeting behind her, Mayall looked forward to sailing to Europe in mid-August with her husband Newton to attend the September 4–13 IAU meeting in Rome. Mayall was now a member of Commission 27 (Variable Stars) and Commission 29 (Stellar Spectra). Despite a very tight travel schedule, Mayall would also arrange a conference between the heads of various variable star groups present at the meeting and visit as many observers as possible in Italy and Switzerland.[84]

IAU report on LPVs

The "Observations" portion of the Variable Star Commission's report noted that an increasing number of observers were devoting their attention to the less-understood irregular or unique variables, rather than LPVs. The report cited AAVSO member Peltier as an example of someone who had ceased observing LPVs and added that they were now relatively well understood. This trend away from the observation and study of LPVs – and the fact that it was thought worth mentioning by the Variable Star Commission – is significant, given the discussions that would occur at HCO and the AAVSO in the months to come.[85]

As part of the Commission 27 proceedings, the Russian astronomer and editor of the *General Catalogue of Variable Stars*, Kukarkin, read a "Report on the Study of Variable Stars in the U.S.S.R. During the Period 1948–1951." In his energetic summary of recent projects and accomplishments in Russian variable star work, Kukarkin mentioned many types of variable stars that were being studied, but not LPVs. Also, given the fact that the Russian astronomers were now charged with compiling and updating catalogs of variable stars, suspected variable stars, and name lists, he emphasized the need for accurate reporting of star positions, identification charts, and the availability of data on which phase and period determinations were based. Kukarkin singled out the HCO "workers" for their failure "to present the data from which the phase of the mean light curves and the deviations of the epochs have been computed and graphs drawn, etc." In his recommendations to the Commission, he stated that "omissions of this kind preclude the possibility of utilizing new data for the verification of the characteristics of the variable star under investigation."[86]

The Commission's stand on the seemingly lessening importance of the study of LPVs, and Shapley's oddly absent encouragement of Campbell's decades-long LPV study, held implications for HCO's shift away from variable star work in the near future. Even so, these judgments were not firmly established. While the Variable Star Commission implied that LPVs were sufficiently well understood, Merrill – Mayall's colleague in the study of variable star spectra – disputed the notion in an article in *JRASC* in fall 1952 titled "Emission Lines in the Spectra of Long-Period Variable Stars." Merrill based this paper on a lecture he gave at a combined meeting of the American Astronomical Society and the Astronomical Society of the Pacific held June 1952 in Victoria, British Columbia. Later, Merrill would make an appeal for the continued, long-term study of these stars in his prefatory remarks to Leon Campbell's *Studies of Long Period Variables* (1955).[87]

EVALUATIONS

Since March 1952 an Ad Hoc Committee appointed by Harvard Corporation President James B. Conant began reviewing the Harvard astronomy programs. These studies would ultimately affect the relationship between HCO and the AAVSO. This review was part of a series of evaluations, retrenchments, redirections, and eliminations enacted by Conant and the Harvard Corporation after World War II.[88] Conant appointed Dr. J. Robert Oppenheimer, director of the Institute for Advanced Study in Princeton, as Chairman of the Ad Hoc Committee on astronomy. The other members of the committee were Drs. Robert F. Bacher (California Institute of Technology), Ira S. Bowen (director, Mt. Wilson and Palomar Observatories), C. Donald Shane (director, Lick Observatory), and Bengt G. D. Strömgren (Yerkes Observatory). It was Harvard's usual practice to draw Ad Hoc Committee members from outside of the University for such evaluations.

Conant had determined that the course of the Observatory's future and the choice of a new director should be determined by the Committee's report. Though he asked for suggestions for a new HCO Director, since Shapley was retiring, he made it known that the Corporation was "anxious to explore the whole future of the work in astronomy at Harvard" before a permanent HCO Director was appointed. Conant would explicitly reaffirm this policy in the coming weeks and months.[89]

The Ad Hoc Committee gave its preliminary report on June 18, 1952, summarized the issues, and listed possible courses of action. Specifically, the report recognized that the "great promise" of the four senior members of the HCO staff – Donald Menzel, Bart J. Bok, Cecilia Payne-Gaposchkin, and Fred L. Whipple – was no longer as great as when they first arrived at the observatory. Their research record, the Committee noted, was not as satisfactory as it was hoped it might be, yet each was doing important work. The Committee indicated that the senior staff alone "would not be able to create for Harvard an observational program and training of graduate students; they would not be able to lay the foundation for a really adequate and worthy program in purely theoretical astronomy. That is, Harvard hasn't got enough with these four people." None of the four, the report noted, had "any real feeling for contemporary instrumentation in astronomy, i.e., they are not advanced in contemporary experimental techniques." Other deficiencies in the Observatory were a lack of a junior teaching staff, telescopes and equipment inadequate for research, research interests of the senior staff having little in common, and the support of the Oak Ridge and Bloemfontein stations being pointless without changes in operating policy. If changes were to be implemented, the Committee noted that Shapley's replacement should have the full authority of a director, with the Observatory Council being relegated to an advisory, rather than a decision-making, role.[90]

The preliminary report indicated that the Committee's main interest was in finding ways to establish a strong graduate training program in theoretical astronomy. They agreed that this might be accomplished in one of three ways: (1) eliminate both observing stations to focus on establishing a small, but high-quality graduate program in conjunction with Harvard's strong physics and mathematics programs; (2) establish a larger graduate program, which would include upgrading the Oak Ridge observing facilities at the expense of restricting or eliminating operations at Bloemfontein; or (3) establish a large graduate program that would include an upgrade of both observing stations, along with the expenditure of several hundred thousand dollars of additional capital.[91]

The Committee suggested the names of three astronomers from outside of HCO as possible replacements for Shapley. The HCO Council later voiced its approval for two: Leo Goldberg and Lyman Spitzer.[92] At any rate, according to Conant's agenda, any consideration of Shapley's successor was greatly premature at this time.[93]

A few weeks after the Ad Hoc Committee report, Shapley wrote Conant, urging him not to delay in naming his successor. He explained: "The Observatory is suffering from fear and doubts and worry and uncertainty. During all this past year the tensions have been serious, and antipathies that formerly were trivial or absent have grown up and now are obstructive." This was in reference to the coming appointment of Menzel as chairman *pro tempore* of the HCO Council that Shapley had learned about from Provost Paul H. Buck. "If that acting chairmanship is of brief duration," Shapley stated, "the Observatory should soon recuperate and all be well." Shapley then elaborated: "Dr. Menzel is a first class scientist, and a first class personal friend. Experience has shown, however, that as an administrator, he is musty-minded, irregular and, because of his unswerving self-concern, at times ruthless (thoughtless)." Shapley then complained that Menzel had "never had an objective interest in the Harvard Observatory and its general problems except in terms of his own operations and advancement; that is perhaps the reason why his scientific work has been so extensive and in some respects so important."[94]

With his concern for the HCO staff, Shapley was also concerned about his legacy: "I have a life-time invested in the fame, happiness, and effectiveness of the Harvard Observatory and I feel obliged to do something to protect my investment." In other words, Menzel could not be allowed to remain in the chair once occupied by Shapley.

From Conant's point of view, the speedy appointment of a successor meant the continuation of the status quo at the Observatory, precisely what Conant did not want. "The one thing I am sure of," Conant wrote Provost Buck, "is that we will not do as Shapley wishes and make an appointment in a hurry." The other important point raised by Shapley – the title Menzel should be permitted as chairman *pro tempore* – was equally important to Conant. The title signifies the authority, and with the changes to Harvard's astronomy program envisioned by Conant before a permanent appointment would be made, it would be better if Menzel had a badge of authority.

Conant had earlier given oral approval to a suggestion from Buck that the Observatory Council "as a group be charged with the responsibility of defining policy and supervising the administration of the Observatory in the interim period and that the most senior member serve as presiding officer of the Council in which capacity he exercises under the Council supervisory administration of the Observatory." With this Buck recommended that Menzel be appointed to this interim position for 1 year beginning September 1, 1952. As for the title, after considering several alternatives, Buck suggested one that might satisfy everyone: "Acting Director."[95]

The matter of Menzel's title was important, but in a small way. Conant was more concerned with the primary issue of changes that would occur in the Harvard astronomy program. Conant stated his preferences in a memorandum to the Corporation summarizing the findings of the Ad Hoc Committee and sent a copy to Buck. In the memorandum Conant stated his preferences from the three possible actions suggested by the committee: "I would be inclined to go in for Item 1 [a small, quality graduate program] with no director and attempt to raise the money necessary to do Item 2 [upgrading of the Oak Ridge/Agassiz station and observing program, and closing of the Bloemfontein station], postponing the appointment of a director until such time as that money is raised."[96]

Menzel as acting director

On August 8, 1952, when Menzel wrote to President Conant to accept the appointment as acting director and chairman of the HCO Council, Menzel made a point of stating: "I recognize that, during such a transition period, the [HCO] Council should not initiate any radical changes of policy that should properly be the concern of the permanent director when he is appointed."[97] Despite this assertion, Menzel hit the ground running and immediately appealed directly to President Conant to increase salaries of the HCO staff. What he received, instead, was an admonishment from Harvard Corporation Secretary David W. Bailey, who said: "In regard to the budgetary problem you raised, I feel sure this is a matter which should be taken up through channels in the Provost's office before it is brought to the attention of the corporation." This is one of a number of instances to come in which Menzel sought to circumvent the chain of command to get what he wanted.[98]

It was a little over a week since Menzel had learned of his appointment, and Mayall was on her way by steamship to Europe for the IAU meeting, when her assistant Helen Stephansky wrote her a routine, unconcerned letter that Menzel was "having everybody in to his office for nice little chats, about their work." Stephansky listed some of Menzel's discussion points, such as the ways and means to make the AAVSO's work more efficient and whether it was necessary to follow the light curves of all the stars. On the maintenance of the light curves, Menzel had added that "if it is just to keep the amateurs happy, it doesn't seem worthwhile, but if it has a definite use to astronomy, is there some way we can make it of more use." Menzel had also complained about the storage problem and suggested having all the original records microfilmed and discarded. Stephansky also noted that Menzel was going to have each department presented to the HCO Council and wanted each department to discuss their suggestions for improvements with him.[99] Mayall seemed unconcerned and even confident about Menzel's efficiency drive. She had even considered asking him to give a talk about the IAU meeting at the AAVSO's annual meeting.[100]

On September 1, 1952, Shapley retired, and Menzel became the acting director of the Observatory and chairman of the Observatory Council. The HCO Council members were Bok, Payne-Gaposchkin, and Whipple. Not long afterwards, Menzel asked HCO's Business Manager, Sylvia L. Chubb, to tally the operating budgets of the HCO departments. Her summary showed that the AAVSO's percentage of the $134 289 total was precisely at the middle of the mid-range, or 5.2%. Others in the mid-range were Agassiz Station (6.2%), Bok's department (5.3%), Mrs. Gaposchkin's department (4.5%), the library (3.5%), and Sergei Gaposchkin (3.3%). The largest percentage was, of course, for HCO's basic operations (35.1%), followed by "Harlow Shapley Dept." (18.0%) and Boyden Station (13.7%). On the lower end of the scale, less than 2% each, were Hoffleit, the plate stacks, Whipple, and Menzel (Menzel written in as "0").[101] These budget items showed that the AAVSO was really an easy target. There were items that could not be eliminated (Cambridge operations, library, plate stacks), and Conant and the Ad Hoc Committee would soon mandate that the Agassiz Station be retained and that Boyden Station be eliminated. This left only the staff astronomers and the AAVSO.

Although Menzel had asserted to Conant that he would not take action on long-term policy changes as acting director, he nevertheless had to confront the immediate problem of HCO's short-term operating budget. The much-needed rehabilitation of HCO buildings and physical plant, Menzel learned, would create a large budget deficit. This, coupled with the need for salary and wage increases, and general operating costs, drove Menzel during the coming year.[102] Menzel and the HCO Council began to review all of these budget problems on October 1, 1952, in anticipation of the official release of the Ad Hoc Committee report, in response to President Conant's general call for retrenchment, and with a concern for the deficit that the rehabilitation program would generate.[103]

AAVSO re-evaluation

It may be that, at the beginning, Menzel had not singled out the AAVSO for elimination – he might have truly believed that all of the HCO departments only needed to be made more efficient, as his chat with Stephansky seemed to suggest. However, as the budget realities began to sink in over the next several weeks, Menzel and the HCO Council would have already been thinking about the AAVSO as a budget problem that had to be cleared away – perhaps easily so.

One of Menzel's first actions in this regard was to appoint HCO Council Member Payne-Gaposchkin to look into possibilities for relocation sites for the AAVSO. On October 16, one day before the AAVSO Council meeting, Payne-Gaposchkin sent a scribbled memo to Menzel with a few suggestions: Rensselaer Polytechnic Institute in Troy, New York; the Fels Planetarium, Philadelphia; Smith College, Northampton, Massachusetts; and the Boston Museum of Science. In her memo she specifically mentioned John Streeter, assistant director of the Fels Planetarium, and asked if Charles Federer would be a "good off-record adviser."[104]

On Mayall's return and just before the AAVSO's annual meeting in October, Menzel informed her that HCO "probably could not continue to support the AAVSO any longer, and that since office space was so crowded that AAVSO might have to give up its headquarters."[105] This news was quite a leap from the informal discussion points on improving AAVSO efficiency that Menzel had relayed to Stephansky in August.

The AAVSO's October 17–18 Council meeting began with a routine discussion of AAVSO operations followed by a discussion "on the usefulness of AAVSO data, and whether the general aims and purposes of the Association's program of work should be reviewed at this time." Ford noted that Menzel and Payne-Gaposchkin favored such a review. After "considerable discussion," the Council agreed to appoint a committee of five AAVSO members "to re-evaluate the AAVSO work program, and to recommend revisions of policy which . . . would result in the work of the Association being more useful to variable star astronomy."[106] Neither the possible loss of support, nor the specific topic of "re-evaluation," were on the official agenda for the Council meeting. If loss of funding, eviction from HCO, and relocation were mentioned during this meeting, they were not recorded by Ford in the minutes. Council Member Richard W. Hamilton, however, recalled that Menzel seemed to suggest that the AAVSO might be better off at some other location. If not explicitly mentioned during the discussion, the question might be asked whether Menzel and Mayall agreed to keep silent about the possible loss of support for the time being, on the hope that the situation might play itself out for the better, or had Menzel held out this false hope to Mayall knowing that the outcome was already decided?[107]

Although Menzel wanted to appoint professional astronomers for the AAVSO's Re-evaluation Committee, the AAVSO Council voted otherwise, preferring a committee of AAVSO members.[108] President Carpenter selected Dr. Smiley of Brown University's Ladd Observatory to head the committee, which he accepted in mid-November, with members Fernald, Heines, Mayall, and Witherell.[109] In the cover letter that Secretary Ford included with his draft copy of the meeting minutes for Mayall, he ruminated on what was already perceived as the impending HCO/AAVSO rift – indicating that Mayall had probably told him what Menzel had said about loss of support, eviction, and loss of the Endowment.[110] According to Mayall, it was at about this time – later in the fall – that Menzel told her that the AAVSO "could not consider the Pickering Memorial Endowment as belonging to the AAVSO. He had looked up the wording and it did not mention the AAVSO. I insisted there was a moral obligation there," Mayall stated, "since the AAVSO originated the fund, but I was told that the Harvard Corporation was a business organization, and did not recognize any moral obligations!"[111]

Change at HCO

Change now manifested itself at all levels at HCO. Gone were the informal, often humorous, director's memos to the HCO staff that Shapley hurriedly typed or jotted-off on small slips of paper. Menzel's memoranda were often spirit-duplicated, rather than personally typed or handwritten – their style was stiff and somewhat authoritarian, and under his facsimile signature was usually added "from the [HCO] Council."

The indirect address Menzel used was partly dictated by a change in University policy: in the interim President Conant and Provost Buck decided to do away with the position of Observatory director and replace it with a council. This gave Menzel a convenient way to deflect attention away from the fact that he was responsible for many unpopular and unfortunate decisions. He would frequently take advantage of this arrangement in the coming months.

The focus of Menzel's memos was nearly always on centralizing authority – that all requests go through Menzel's office, and so on. One of his first memos, dated October 24, 1952, bluntly referred to the "uneasiness" felt "throughout the Observatory" regarding many space-saving measures – reallocation of office space; moving of books, plates, and materials into dead storage or given to the main library or archives; and the imminent destruction of Building B. No mention was made of the future demolition of Building A or the status of its AAVSO occupants. At the end of this same memo, listed as item number 16, was the announcement of "Approval of the appointment of D. H. Menzel as Acting Director from 1 September, for one year pending appointment of Director."[112]

About 2 weeks after the AAVSO meeting of October 1952, Menzel received a letter from John Streeter, assistant director of the Fels Planetarium in Philadelphia, one of Payne-Gaposchkin's relocation prospects. Streeter indicated that he and Ashbrook were "ready to go into higher gear on AAVSO when you give the word."[113] Streeter's comments suggest that Menzel and the HCO Council were taking a series of pre-arranged steps. Clearly, they were approaching the AAVSO problem on two separate tracks: one public – HCO Chairman Menzel calling for a re-evaluation of all HCO departments – and one behind the scenes, as the Payne-Gaposchkin relocation prospects memo and the letter from Streeter suggest. Menzel replied to Streeter within a week suggesting that it would be better for the AAVSO "to prepare the lead. . . . I shall let you know anyway."[114] A day later, Menzel followed this up with a memo to the HCO Council: "If the idea gets out that Harvard can't support the A.A.V.S.O.," Menzel wrote, "then who in the world could? However, [Joseph Ashbrook] is planning to work up a memorandum during the next couple of weeks, covering all items of the subject, meeting specific recommendations for cutting down the costs of the A.A.V.S.O. operation in general. If the cost could be cut from say $6000 to $2000, then the idea easily could be sold to some other place." This memo shows that Menzel and the Council clearly intended to be rid of the AAVSO burden long before any re-evaluation report would be considered.[115]

Meanwhile, Ford was meeting with Dean Duncan Macdonald at the Boston University Graduate School and head of its Physical Research Laboratory on November 6. His visit may have been coincidental, but Macdonald was the person to whom Mayall later appealed for office space. It is also possible that, at this early date, the conscientious Ford sounded Macdonald on the availability of space, should the AAVSO need it in the near future. Mayall and Ford would have known Macdonald from his days as an instructor at HCO in the 1940s.[116]

With what seems to be a growing awareness of a coming crisis, Ford continued to spread the news among his confidants about what he saw developing. In a November 20 letter to former AAVSO Council member Claude B. Carpenter, Ford gave a blunt assessment of the emerging HCO/AAVSO controversy, especially his views on Menzel's role, and, in particular, Menzel's desire to appropriate the Pickering Memorial $100 000 by pushing the AAVSO out. Coupled with Menzel's apparent disdain for amateur astronomers, it all began to make sense to Ford, who surmised that Menzel was running short of funds to operate HCO and "is casting greedy eyes on the $100 000 Pickering Memorial endowment which is the life-blood of the AAVSO. He has always more or less looked down his nose at amateur activities, apparently just what he's trying to do."[117]

Conant's draft report

On November 11, 1952, President Conant prepared another "Draft Report on Astronomy at Harvard" for the Harvard Corporation. This memorandum, similar to one he had prepared in August, summarized the Ad Hoc Committee's findings, including the three courses of action they had suggested. To this, Conant added that Harvard had a continuing tradition in astronomy, and "any decision adverse to the profession would have a profound, negative effect on the support of astronomy in the United States."[118] Within 2 weeks of this report, President Conant and Provost Buck conferred, summarizing their meeting in a memorandum titled "Tentative Decisions Concerning Astronomy." They agreed to ask the Corporation to (1) appoint a Professor of Theoretical Astronomy; (2) authorize a $1 000 000 endowment for the Agassiz Station and up to $300 000 to equip and modernize it as a center for training observational astronomers; and (3) authorize a program of retrenchment and modernization of the Cambridge plant.[119]

This memorandum and the one preceding it show that (1) funding for astronomy at Harvard was not a critical issue – the money would be found one way or another; (2) astronomy was, according to Conant, as important as ever – although for new and different reasons – and so the emphasis on theoretical astronomy; and (3) graduate training in astronomy was an important part of the new direction being taken, and the upgrade of the Agassiz Station and HCO's physical facilities was a part of this training initiative.

In Conant's list of decisions regarding astronomy, only the item on retrenchment and modernization of HCO would have an effect on the AAVSO. This does not mean that Conant or the Ad Hoc Committee wanted to do away with the AAVSO; it was Menzel himself – with his newly found authority as Acting Director – who would seize on Conant's call for modernization in general as his reason for pursuing the eviction of the AAVSO in particular.

Preliminary inquiries

When Smiley was appointed chairman of the AAVSO Re-evaluation Committee, he wrote Menzel on November 14 to inform him of his appointment. He asked Menzel as acting director to state "what your attitude is toward AAVSO, in particular what changes, if any, you have in mind."[120]

Menzel did not formally acknowledge Smiley's committee appointment until December 10, noting that his "only official statement" at the AAVSO Council meeting was that after 40 years, it might be a good idea for the AAVSO to examine its programs and its relationship to HCO. Although not stated directly, this was probably in response to Smiley's question about what Menzel had in mind and also probably had to do with his earlier suggestion (which he now seemed to be denying) that the AAVSO might want to find another home. Menzel continued by saying that the University had requested that he review the operations at HCO, which would include the AAVSO, and that, "the Observatory Council, or members of the Council, would be asked by the University to express their opinions about the A.A.V.S.O. and its future relationship to the Observatory." Menzel further stated that because more than $6000 per year went for the operation of the AAVSO "from funds administered by Harvard University, our request for a report on the current status and future plans of the Association seemed both reasonable and necessary." Menzel acknowledged that "no member of the Observatory Council would wish to deny the usefulness of the observations made by the Association in the past. But," he added, "none of us feels that the future scientific returns from the program currently being pursued would be at all comparable to those of the past, in view of the changing emphasis in variable star astronomy."[121]

With this letter, Menzel enclosed a copy of the official description of the Pickering Memorial Endowment, from which he derived the last phrase in his letter regarding "changing emphasis." The phrase would become a key idea for Menzel in the coming months as he refined his arguments for the AAVSO's removal. Another interesting feature of this letter can be found in the previous paragraph, in which Menzel stated that the HCO council "will be asked by the University to express their opinions about the A.A.V.S.O. and its future relationship to the Observatory." The fact is that this never happened. Neither the Ad Hoc Committee nor Conant had ever singled out the AAVSO or any other department of HCO except for the Boyden and Agassiz Stations and the graduate program in general. Also, it is likely that Menzel at this time had already received some form of the Ad Hoc Committee report or Conant's synopsis, and he would have known what Conant and the Corporation had in mind. Indeed, on the next day, December 11, 1952, Menzel wrote to Conant to inform him that "as a part of [the HCO Council's] evaluation of current activities of the H.O.," he had requested the AAVSO to evaluate its present, and future, and its relationship to HCO, stating "I shall be pleased to discuss the matter with you if you wish."[122]

MAYALL TAKES ACTION

While Secretary Ford was pondering the developments and trying to make sense of it all, Mayall had taken action soon after Menzel's call for re-evaluation at the October 1952 Council meeting. Unfortunately, at about this time, Mayall also learned of the death of her brother-in-law – this came just 5 months after the death of her mother. Mayall had to return to Maryland to tend to family legal matters.[123] Before leaving, Mayall collected all of the historical and current documents she could find on the Pickering Memorial Endowment and the AAVSO's use of it. Then, on the day after Smiley accepted his nomination to chair the Re-evaluation Committee on November 18, 1952, Mayall, also on this committee, sent prepared letters to seven prominent professional astronomers: Harold L. Alden of Leander McCormick Observatory, Frank K. Edmondson of Indiana University, IAU President Otto Struve, Alfred H. Joy of Mount Wilson Observatory, Gerald E. Kron of Lick Observatory, and Jason J. Nassau of Warner and Swasey Observatory. Her letter briefly gave the history of the Pickering Endowment and a summary of the AAVSO's activities, along with the original Pickering Memorial Committee, the original 1921 Endowment proposal, and the final wording of the Endowment's establishment by the Harvard Corporation in 1931. She ended her letter by asking for each astronomer's "honest opinion of the work of the AAVSO as a service to astronomy."[124]

On November 19, Mayall sent a similar letter to her friend and colleague, Paul Merrill at Mount Wilson Observatory. In this somewhat more personal and frank letter, Mayall stated that the AAVSO situation was "really pretty desperate. The feeling of our new regime at HCO seems to be that the AAVSO has done a good job in the past, but it is not necessary to continue along the same lines. They tell me that the Observatory Ad Hoc Committee is very likely to recommend discontinuation of support to the AAVSO. . . . We are trying to keep things as quiet as we can at present, for I know it will cause a lot of hard feelings and resentment among our observers if they think Harvard is letting them down."[125]

Of course, the Ad Hoc Committee did not offer any recommendations so specific, and whether or not Menzel already knew this, it seems that he was purposely trying to convince Mayall that the AAVSO's days at HCO were numbered. In her letter, Mayall also mentioned that the AAVSO's Re-evaluation Committee planned to meet at Amherst College during the American Astronomical Society meeting December 28 through 31 in Western Massachusetts.

On the same day that she wrote to Merrill, Mayall also turned her thoughts to making arrangements for the 1953 Spring Meeting. This was not simply a case of optimism about the AAVSO's future: Mayall realized that holding the meeting in a centralized location would be an important part of her effort to keep the AAVSO's Council and membership informed and encouraged. On November 19 she wrote to Dr. Leo Goldberg at the McMath-Hulbert Observatory of the University of Michigan, Ann Arbor, to ask if a meeting could be arranged there. "A meeting at Ann Arbor," she said, "would certainly draw Wisconsin, Ohio, Indiana, and perhaps Minnesota members."[126] Ford noted that Mayall "grabbed the first date offered without further question, as this further emphasizes the continuing active interest of professionals in the AAVSO."[127]

Meanwhile, Menzel further formalized his call for re-evaluation, carefully worded to remove himself from any perception of a direct involvement, with a memo to AAVSO President Carpenter on December 1, 1952. He wrote: "The Council of Harvard Observatory has asked me to transmit to you a request that the A.A.V.S.O. submit its report on the long-range future and relations with Harvard Observatory not later than 20 January, 1953, in order that the Observatory Council can report back to the officials of Harvard University before 1 February 1953."[128]

Mayall and the Re-evaluation Committee members were taken by surprise by this notice. They had assumed over the past 2 months that there would yet be plenty of time to compose their report. Carpenter acknowledged Menzel's notice on December 16 with surprise and puzzlement: "This seems a rather early deadline for the type of work to be done."[129] Indeed, there was no obvious reason why Menzel suddenly presented a deadline for the AAVSO's report. It seems clear, however, that Menzel acted swiftly in order to move the AAVSO "re-evaluation" from an internal, informal process, to one that would be highly formalized at the corporate level, as a way to ensure that the process could not easily be revoked once it had been sent along that path.

Solar Division Chairman Heines reported to Ford on December 2, 1952, that Menzel had asked AAVSO Treasurer Witherell at the time of the 1952 annual meeting "to confer on the funds or fund (perhaps a particular fund) of the A.A.V.S.O." This observation seemed to confirm Ford's suspicion that Menzel already had his eye on the Pickering Memorial Endowment at the time he was calling for a Re-evaluation Committee to be established.[130] While Heines' letter was in the mail, Ford was once again meeting with Duncan Macdonald at Boston University on December 3, purpose unstated.[131]

Early confidence

One of Smiley's first steps as chairman of the Evaluation Committee (as he preferred to call it) was to have a talk with Shapley about the establishment of the Pickering Memorial Endowment and its relation to the AAVSO. Armed with this basic information from Shapley, he sent a letter to Harvard University President James B. Conant. In the brief letter, he mentioned having "very fine letters" from Struve "and other outstanding American astronomers" and then asked Conant to meet to discuss the Pickering Endowment and to learn "whether you plan any change in the Harvard policy with respect to [its] use." Smiley closed with, "I felt that the President is probably the only one who can speak with authority on University policy."[132]

Smiley's request to meet with President Conant was a move worth taking, but, not surprisingly, Conant countered with the suggestion that Smiley meet instead with University Provost Buck.[133] Provost Buck was the first to be named to the post that Conant created in 1944. When he created the post, Conant felt that the University President should not have to have direct oversight of the Arts and Sciences departments.[134] In a half-hour meeting on December 8, Smiley reported that he told Buck that the AAVSO was prepared to prove that it "has served astronomy well in the past and that many astronomers feel that it can reasonably be expected to continue to do so. In particular I mentioned Struve's generous supporting letter for the organization." Buck could give no opinion but, Smiley wrote: "He assured me that any suggested changes in the University's relations with AAVSO would be examined with care before

Figure 10.8. Helen M. Stephansky and Margaret Mayall at AAVSO Headquarters, Building A, HCO, 1952.

they were put into effect. I made the point that if the changes were drastic, an interval of time should be provided for the AAVSO to make the necessary adjustments."[135] In his actions, Smiley was fulfilling his responsibilities as the Evaluation Committee chairman, even though no progress had been made. If anything, the meeting with Buck seemed to confirm that, from the point of view on both sides, the fate of the AAVSO was already sealed.

In the meantime, Mayall left her file of documentary evidence and arriving testimonials in the hands of her assistant Stephansky for Ford to examine while she was away. When Ford arrived at Headquarters, Stephansky told him that Smiley was "hopping mad" about what was developing at HCO. Ford found the replies from the professional astronomers "glowing with praise" for the Association and uniformly expressing "great surprise that any move to curtail the AAVSO is in the wind." On her way to Maryland, Mayall stopped at the Ladd Observatory in Providence, Rhode Island, to meet with Smiley and discuss how their collected evidence could be presented in February to the HCO Visiting Committee (a group of professionals and academics from outside the University who regularly met to advise the Observatory director). The Visiting Committee, according to Ford, "was going to meet with Menzel at that time to hear his recommendation, to divert the Fund to other types of research (i.e. solar)."[136]

Buoyed with hope and inspired by Mayall's initial effort to collect documentary evidence to support the AAVSO's argument, Ford returned home and searched the historical meeting minutes in his AAVSO Secretary's file for all mentions of the Pickering Endowment from 1919 to 1936. He collected extracts pertaining to the Endowment from the Olcott correspondence in his possession, including letters between Olcott and Campbell, and distributed his typed notes to Mayall and others. "The net result," Ford wrote, "is a strong case in favor of the status quo and usefulness of AAVSO work."[137]

The much-anticipated meeting of Evaluation Committee members and other interested persons took place December 28 through 31, 1952, at Amherst College during the American Astronomical Society meeting. Those attending were Menzel, Ashbrook, Smiley, Witherell, Mayall, Jocelyn Gill, and Carpenter. Several other AAVSO members attending included former AAVSO President Dr. Alice H. Farnsworth, who had been a member of the AAVSO's Pickering Memorial Advisory Committee in 1930 through 1932.[138]

Upon her return to Cambridge, Mayall sent a second round of letters in December 1952 to at least 13 other astronomers, including Giorgio Abetti, Arcetri Observatory, Florence; Andre Danjon, president of IAU Commission 27; Carl W. Gartlein, Cornell University; Livio Gratton, Eva Peron Astronomical Observatory, Argentina; Guillermo Haro, Tonantzintla Observatory, Mexico; Philip C. Keenan, Perkins Observatory; Bertil Lindblad, Stockholm Observatory, Sweden; Daniel J. O'Connell, S.J., Vatican Observatory; Jan H. Oort, Leiden Observatory; Henry Norris Russell and Martin Schwarzchild, Princeton University; Otto Struve, U. California, Berkeley; and I. L. Thomsen, Carter Observatory, New Zealand.[139]

By early January, responses to the November mailing had arrived – they were unanimously positive about the value of the AAVSO to their research and in favor of seeing the AAVSO's operations continue.[140] Eleven specifically mentioned the value of the AAVSO's work on LPVs. Several responses have since proven to be prescient, such as one from Martin Schwarzchild, who wrote: "There seems to me very little doubt that when the theoretical work about the processes in the envelopes of long-period variables will be ripe for a more theoretical attack, the material accumulated by the AAVSO will provide a critical test for any theory."[141]

Shapley's opinions

On December 1, 1952, the same day that Menzel formally requested an evaluation report from the AAVSO, he sent a memo to Shapley saying that the

HCO Council had requested him to solicit Shapley's help on the AAVSO problem. He informed Shapley that the Evaluation Committee report was due in January, and the report from HCO would be sent to the Harvard Corporation in February. He then asked Shapley for "documentary evidence as may be available regarding the source and assignment of the $93 643.24 (the Pickering Memorial Endowment of $100 000) transferred in 1931 from HCO funds [i.e. $93 643.24 plus the original $6356.76 contribution from the AAVSO]." Menzel also requested Shapley's views on the future relation between the AAVSO and Harvard Observatory.[142]

Shapley responded on January 4, 1953, with a document titled "Statement About the A.A.V.S.O," in which Shapley defended the past value of the AAVSO to Harvard and science. However, in a pointed sentence, he wrote: "But here I want to interpolate that this general astronomical approval of what has been is no reason that changes should not now be considered." On the Pickering Memorial Endowment, he wrote that, "the University is confronted with an ethical rather than a legal situation. I have never doubted the sincerity of the Observatory or the University in this matter of providing through the Pickering Memorial for the continuity of the work of the AAVSO. . . . But the spirit of the statement implies of course the long-period and irregular variables . . . but there can hardly be any question about the continuing interest in and importance of variable stars in astronomical research."[143]

Shapley questioned the future value of the AAVSO based on this distinction between variable star research, in general, and the type of variable star work done by the AAVSO, in particular. His main reason was that the HCO Milton Bureau variable stars project was completed and "has perhaps reduced the urgency and significance of LPV star work with visual or with photographic telescopes." He did, however, see the value in continued study of changes of period, especially in irregular variables.

Interestingly, the strongest case that Shapley made for retaining the AAVSO was not the value of its scientific work, but as a magnet for benefactors to HCO and the University. In three-quarters of a page in section seven, he discussed donors and foundations influenced by the AAVSO's association with HCO. In the Harvard University archive copy of this document, this entire page is missing. Obviously someone had drastically elided Harvard's copy.

As a whole, Shapley's "Statement" is as unenthusiastic in its defense of the AAVSO as in its implied support of Menzel's/HCO's position. The greatest concern in the entire statement can be found in the concluding paragraph, in which Shapley stated: "I was sorry to find how widely known at Amherst was the affaire AAVSO, and apparently how completely misunderstood was the Observatory's current attitude. Amazing, incredible, mad, were three terms I heard from professional astronomers. I am afraid some damage has already been done to the Observatory's prestige, but a clear and full official statement in the near future, whatever the policy adopted, may easily rectify any harm that may have threatened."[144]

Mayall's search on the Pickering Endowment

Mayall, too, went on a search for information about the Pickering Endowment. On January 10 she found in AAVSO Treasurer Willard Fisher's files that the AAVSO's original contribution of $6356.76 was combined with $43 643.24 from the Wyeth Bequest to Harvard and matched by $50 000 from the Rockefeller Foundation to Harvard (see Chapter 7).[145] At Shapley's request, who may have been filling Menzel's request for information as well, Mayall then asked Arville Walker, Shapley's assistant, to find anything in Shapley's own files concerning the Rockefeller Fund that might help give the full picture. "I have talked to Dr. Shapley," Mayall wrote to Smiley, "and he is having Miss Walker look up old correspondence with reference to the Rockefeller grant of $50 000. If the AAVSO is mentioned in it, we should be all set [i.e., proven that the money was explicitly meant for the AAVSO] with the Pickering Fund."[146]

Because she found nothing new in Shapley's files, Mayall then contacted the HCO recording secretary 3 days later and only found that the Rockefeller Foundation gave a $500 000 matching grant toward the HCO development fund in 1931 and 1932 (see Chapter 7).[147] There was no more time for research: over the next 6 days, Mayall and other members of the Evaluation Committee frantically tried to pull together the pieces of their draft report to meet Menzel's January 20 deadline.[148] Smiley, sending a draft of the report to Shapley, tried to make the best of a declining optimism. "If an adverse decision is reached," he wrote, "we shall hope for a five year transition period in which we shall have the full income of the Pickering Fund."[149]

Upon receiving the final draft of the report, another Committee member, Cyrus Fernald, an accountant, responded to Smiley with one foreboding observation about the Pickering Endowment: "If this fund was largely obtained from the Harvard Observatory funds, by simple allocation, then I can't see where the AAVSO has even a moral claim on them."[150]

THE EVALUATION COMMITTEE REPORT

Smiley submitted the AAVSO Evaluation Committee report (its members dropped the "Re-" about a month earlier), "Report on the Value of Past and Future Activities," (dated January 15, 1953) on January 20th after only 7 weeks, in compliance with the HCO Council request.[151] The Evaluation Committee report on past and future activities was itself a little more than two concise, single-spaced pages; the remaining nine pages included Mayall's letter to astronomers and a selection of their replies. Despite an honorable record of past achievement and a list of scientifically useful current and future work, Mayall and the Evaluation Committee no doubt felt that the testimonials from professional astronomers spoke for themselves and were the AAVSO's best hope. "I hope their letters will have some effect on Harvard!" Mayall wrote to Fernald.[152] The same day, Mayall sent a postcard to Ford, saying: "The report got delivered today, and now we sit and wait." The two could now turn to the meeting arrangements. "Everything is OK for our Spring meeting at Ann Arbor May 22–23," she wrote. "I want to get the word spread around as soon as possible."[153]

In recounting the events to AAVSO Council member Hamilton, Ford wrote that the Evaluation Committee "met in a fighting mood during the AAS meetings at Amherst, and apparently Dr. Menzel was simply furious because everyone at the [AAS] meeting was asking him 'What's this about Harvard giving up the AAVSO?'" Ford continued that Menzel "up to that time" knew nothing about the letters of support Mayall had received. Ford noted that the Evaluation Committee report had since been submitted as requested by Menzel and included photostats of the letters. "Apparently," Ford continued, Mayall and Menzel "have all but come to blows, and now Menzel is requesting letters for his side of the argument, before submitting the whole thing to the higher powers at Harvard." Ford's letter updated a number of activities, including the fact that Menzel had "apparently cut off most of the funds allotted in the 1953 budget to the AAVSO," beginning July 1. In the meantime, Mayall "will have to raise additional money to keep the office going." Ford also referred to "an unfortunate situation" that Bok was the only member of the Observatory Council "friendly to the AAVSO." Ford concluded that, in spite of the letters Mayall collected, the report of the Observatory Council "will be biased against the AAVSO."[154]

WAITING FOR AN ANSWER

As the AAVSO's new recorder, Mayall set out to learn how to become the independent head of a growing organization, but the challenge from HCO, as unexpected as it was, had the dual effect of sharpening her focus on her work with the AAVSO and strengthening her sense of loyalty to the organization. At this point, Mayall had done everything she could, but she would soon have to do even more.

11 · Eviction from Harvard College Observatory

Certainly 100 years from now our observations are going to be of much more value than most of the theoretical work that is being done.

– Margaret W. Mayall, 1953[1]

J. B. Conant's return to the Harvard presidency after World War II set the stage for events that would play out over the next decade at the university and affect its relationship with the American Association of Variable Star Observers (AAVSO).

Conant was dedicated to modernizing the administration of Harvard University, streamlining the academic departments and upgrading the faculty while reducing costs. His initial focus in the sciences was on his home department of chemistry but included physics and classical biology. All three disciplines received massive infusions of capital and the addition of new faculty to shift their curriculums in response to new quantum mechanical and molecular knowledge and instrumental techniques.

Eventually Conant's attention reached the Harvard College Observatory (HCO). Instead of searching for a new director, Conant appointed an Observatory Council chaired by Donald Menzel to oversee Observatory administration and necessary restructuring of the deteriorated physical facilities and faculty.

Menzel's task was unenviable in many respects; he had been an AAVSO member since 1917 and served, at that very time, as elected vice-president of the organization. Menzel saw that to some small degree, he could solve some of the Observatory's space and financial problems by evicting AAVSO and appropriating the endowment that Shapley had arranged for support of the salary and expenses of the AAVSO Recorder.

Margaret Mayall (who had recently been appointed as the new AAVSO Recorder, a position supported by the Pickering Memorial Endowment) and the AAVSO Council committed themselves to resist both the eviction and the appropriation of the endowment in every way they could, but to no avail. This chapter unfolds the series of events that eventually led to the eviction of the AAVSO from HCO.

THE IDENTITY PROBLEM

Many amateur astronomers who were members of the AAVSO did not make a distinction between the Association and HCO. One example can be found in a 1935 Milwaukee Astronomical Society (MAS) article describing a public talk on Nova Herculis given by Shapley, which he followed with a chat with a few of the Society members afterwards. Although Shapley was speaking as the HCO Director at a non-AAVSO event sponsored by the Harvard Club, his presence was enough to suggest the article title, "An Impromptu Meeting of the A.A.V.S.O." Shapley *was* the AAVSO to many.[2]

Mayall later recalled that it wasn't until Campbell retired "that people outside the Observatory realized that the AAVSO really was an independent organization. Everyone thought that we [the AAVSO] were part of the Harvard Observatory." Although observers recognized the AAVSO, they thought they were sending their variable star observations to Harvard, not to the AAVSO, and they thought that Harvard, Mayall said, "was really dominating the whole AAVSO program."[3]

Of course, from the start, the AAVSO membership and officers felt grateful and privileged to be affiliated with Harvard and the HCO, no matter how official or unofficial the affiliation. The organization enjoyed and benefited from the prestige attached to its association with HCO.

The identification of the AAVSO with HCO even predated the 1931 Pickering Memorial Astronomer position and was encouraged by HCO. Shapley himself would later tell Mayall that she did "not fully appreciate" the fact that, "During my regime, and certainly during the last ten [sic] years of Pickering's directorship, the AAVSO was considered a responsibility and effectively a part of Harvard Observatory."[4]

After 1931, the recorder's position depended on the funding provided by the Pickering Memorial Astronomer appointment, and the scientific guidance of the AAVSO's observers depended on the recorder and other HCO staff. It was this arrangement that fixed the AAVSO's Harvard identity in the minds of all concerned.

Within and without the walls of HCO, the AAVSO Recorder position was conflated with that of the Harvard Pickering Memorial Astronomer appointment, the Harvard College Observatory, or with Harvard University in general.

Mayall's appointment in 1949 as the Harvard Corporation's Pickering Memorial Astronomer was distinct and apart from her election as an AAVSO officer. The search by Shapley and his HCO selection committee to fill both positions was a single process that the HCO Council then split into two recommendations: one for the Harvard Corporation (Pickering Memorial Astronomer), and the other for the AAVSO (Recorder). It was taken for granted on all sides that whomever HCO recommended for the Pickering Memorial Astronomer position would also be recommended for the AAVSO Recorder position.

From HCO's unofficial point of view, the AAVSO Recorder was seen as the "Observatory agent" for AAVSO-related work, rather than as an officer of the AAVSO that would be elected by the AAVSO Council. Even Leon Campbell had merged the two positions in his mind, finding it necessary to say, on recommending his successor to Shapley, "Perhaps the two posts, Pickering Memorial Astronomer and Recorder, should be, for the present at least, separated."[5]

This strong attachment of the AAVSO to HCO was a powerful influence on Mayall and the AAVSO Council in deciding to stand their ground in 1953. At first, it was simply a question of maintaining the status quo; this, in time, turned to fighting for the claim to the Pickering Memorial Endowment; and finally, to acceptance of the fact that the AAVSO could no longer enjoy the support of Harvard.

HARVARD OBSERVATORY FUTURE

No one would have expected any major changes to the Harvard College Observatory after Menzel became its acting director on the basis of his first statements. After all, he had sent a memo to HCO personnel saying that, "the future looks bright." On January 20, 1953, the same day that he had requested a report from the AAVSO Evaluation Committee, the HCO Council had received a report from President Conant regarding the University's policy relative to the future of the Observatory. Menzel then reported to HCO personnel that "the University looks forward to the appointment of a permanent Director at an early date . . . . Meanwhile no drastic changes are contemplated."[6]

Although HCO had a "bright" future, the reality of the distinction between HCO's future and that of the AAVSO was beginning to set in for the AAVSO's leaders. Mayall and the AAVSO Evaluation Committee knew that Menzel was contemplating drastic changes for the AAVSO and that he did not intend to delay them.

Just 2 weeks later, Menzel circulated another memo. In apparent contradiction to the promise of no drastic change, he announced the "liquidation" by Harvard University, of "its legal and financial obligations for maintaining an astronomical station in Bloemfontein, South Africa (the Boyden Station). This decision," he added, "is a mandate of the Harvard Corporation."[7] Menzel was entirely correct in stating that the Boyden Station decision was made at the corporate level, but he was mistaken to describe the Boyden decision as only a budgetary one, without taking into consideration the Corporation's broad plan to steer Harvard's astronomy into a new direction.

Menzel described the extent of the budgetary problem when he wrote to astronomer Walter Baade on February 10, 1953. He wrote about the "tough time" he had in explaining it all to the Observatory personnel who were blaming him, Cecilia Payne-Gaposchkin, and Fred Whipple for something that they "had absolutely nothing to do with. I had the extremely difficult job of calling together the entire Observatory and explaining to them the facts of life."

Menzel told Baade that "without saying one word of criticism," he had "pointed out the cold facts" contained in President Conant's recent report, which gave the figures comparing all departments of the University in 1932 and 1952. The annual income of the Observatory, he wrote, "had sunk from $178 000 in 1932 to $164 000 at that time. Over the same interval, almost every department in the University had an increase of 50 per cent and often more in the income."

Then, Menzel applied what he saw as the justification for the Corporation's decision about Boyden to his own problems at HCO, clearly expressing his determination to eliminate the AAVSO: "And still there are certain people in the Observatory who are fighting the Council (with special attention to me)," he complained, "when I point out that the only way we can save money is to cut out some of these extra curricular activities, like the A.A.V.S.O. and some others."[8]

Menzel was indeed preoccupied with his budget. HCO business manager Sylvia Chubb notified him the next day that Mayall was asking for a typewriter overhaul at a cost of $35. "Margaret does not know it," Chubb wrote helpfully, "but there is a small balance in the A.A.V.S.O. account, which could carry this charge.... They have requested practically nothing since the 1st of January. I await your decision." Menzel's reply, handwritten on this memo, was simply: "No money."[9]

OBSERVATORY PROBLEMS

On February 13, 1953, Menzel prepared an "Interim Report on Observatory Problems" for Provost Buck. The 11-page report listed 28 "problems" ranging from general comments about the budget deficit, salaries, and wages, to more specific items, such as renovations, publication storage, and even the HCO tennis court (recommending it be preserved). Most of these items were no longer than 100 words each; a recommendation that *Sky and Telescope* be allowed to maintain its offices at Harvard was about 180 words.[10] One item, "20. Dorrit Hoffleit" singles out Hoffleit for dismissal from HCO; it is unusual in negatively mentioning an individual.[11] Another individual, mentioned favorably, was optical expert James G. Baker. Hoffleit later noted that Menzel did not like to see her immersed in her spectroscopic work, suggesting she would be better off under one of his contracts at the Aberdeen Proving Ground. By 1956 Hoffleit could take no more; she left Harvard for Yale.[12]

Menzel listed other "problems" that are in some ways comparable to the AAVSO problem because of their connection to amateur astronomy. Menzel encouraged the Telescope Makers club since they had "more than paid their way" in providing HCO with specially made lenses and optical equipment. He also wrote favorably of the Bond Astronomical Club that held monthly meetings at HCO and had HCO staff and students as speakers. Menzel wrote: "They do represent a serious group of persons interested in astronomy and the Council feels that they should be allowed to continue. It should be mentioned that many observatories and astronomical departments, especially in large cities, have somewhat similar organizations directly or indirectly associated with them." Founded by Harlow Shapley in 1924, the Bond Club was a group of armchair astronomers. Several years later, Hoffleit, who had been president in 1952, recalled that within a few years of Menzel's directorship, the Bond Club was no longer permitted to hold its meetings at the Observatory and soon dissolved.[13]

By far the largest item in the Interim Report dealt with the AAVSO, which, according to Menzel, "presents a special problem." This item (number 13) covered nearly two single-spaced pages, approximately 600 words. It was really a rehearsal of the HCO Council report on the AAVSO that would be sent to Provost Buck in 3 days. The Interim Report is worth examining because it contains statements that were not carried over to the final report on the AAVSO. It begins with a very brief account of the AAVSO's beginnings and purpose, followed by a description of the Pickering Memorial Endowment. In quoting the official Harvard Corporation's wording of the Endowment's terms, Menzel makes what is perhaps his first use of the dubious tactic of inserting the word "such" in parentheses to "clarify what seems to be an obvious omission in the original phraseology." The insertion of "such" more definitely referenced the research being conducted by the AAVSO. The passage in question, which also appears in his final report to Provost Buck, reads as follows:

> The income of the Edward C. Pickering Memorial Endowment was to "be devoted to research in the study of variable stars with emphasis on those fields cultivated by amateur astronomers or for other astronomical research if, hereafter, the President

and Fellows, after consulting the Director of the Harvard College Observatory, shall be of the opinion that, owing to changed conditions, (such) research in variable stars is no longer of the same importance . . . ."[14]

Menzel continued his discussion of the AAVSO problem, using language he did not carry over to his final report, particularly his emphasis on the value of the Pickering Memorial Endowment to HCO. He stated that "amateur contributions to variable star research had reached the point of diminishing return and that the finances now devoted to the A.A.V.S.O., owing to these changed conditions, would be much more effective if devoted to other astronomical research." He also continued that "this activity does not fit with the trend of modern astronomy, since no other observatory maintains similar activities. The funds are of increasing importance to the Observatory." At the heart of his Interim Report was his statement that the Observatory Council would soon recommend to the Harvard Corporation that by July 1, 1954, "the income from the Edward C. Pickering Memorial Endowment no longer be used for support of the A.A.V.S.O." In the meantime, the report stated that the finances allocated to the AAVSO be limited to the Pickering Endowment beginning in July 1, 1953.[15]

PROBLEM OF AAVSO FINANCES

No one from the AAVSO expected to hear of any decision from the Harvard Corporation for at least a month after submitting their Evaluation Committee report to the HCO Council – in fact, they would receive no word for 2 1/2 months. In the meantime, the Association's work continued – President Carpenter had sent 15 new member certificates to Secretary Clinton Ford for his signature. Ford visited Mayall at Headquarters in late January; there Mayall informed him that there was an "immediate problem of raising more money to carry through the present activities in the Recorder's office," to which Ford responded: "This is evidently pretty serious."[16]

Ford at first did not think that a large fundraising effort would be possible "under the present circumstances," but his visit with Mayall changed this view and fortified his confidence in the survival of the Association. "Whatever happens," Ford declared, "the AAVSO is not going to go out of business – that's for sure."[17] Sharing this confidence was another AAVSO Council member, Richard Hamilton, who reminded Ford: "As she [Mayall] said she told you, she has learned how to fight from her [family] legal entanglements of last year, and she is with the AAVSO to the bitter end."[18]

RESEARCHING FUNDS

On February 11, 1953, Mayall composed a letter to Warren Weaver of the Rockefeller Foundation asking for clarification on the Rockefeller funds given toward the HCO development fund, which included the Pickering Memorial Endowment, "in particular those for the benefit of the American Association of Variable Star Observers." This letter, although signed, was never sent, and a note clipped to it in Mayall's handwriting read: "Do Not Mail."[19] Perhaps Mayall felt the letter was premature because she had not yet heard from HCO, or possibly, she believed she could not afford to wait for a reply. She may have been discouraged by Weaver's reply to a similar letter written by Bart Bok regarding the closing of Boyden Station: "Aid to any astronomical project now lies quite explicitly outside of the program and policy of the Rockefeller Foundation."[20] Mayall perhaps reasoned that if the Foundation was not now concerned about astronomical projects, they likewise would not be persuaded to research, or even speak for, their contribution to the 1931 Pickering Memorial Endowment. On the same day that she wrote the Weaver letter, Mayall sent Menzel what information she had about the Pickering Fund. She must have been hoping that any additional material about the fund would help the HCO Council make an informed decision. Along with the information, she also sent copies of letters of support from five more astronomers.[21]

The next day, February 12, 1953, Mayall obtained a copy of "Endowment Funds of Harvard University, June 30, 1947."[22] The description of the Pickering Memorial Endowment did not give her the detail she was seeking; she only learned that it was made up "from any unrestricted funds" of HCO (see Chapter 7).[23] Mayall had one other hope to go on: that the AAVSO would be allowed to remain at HCO if the AAVSO could come up with its own endowment fund in place of the Pickering money.

While she waited, Mayall wrote to AAVSO President Carpenter on February 19, expressing what she had heard through the underground: "When the Observatory Council finally took a vote on the AAVSO, it would be unanimous against us." She noted that HCO's 1953–54 budget was due to go to Harvard University that week, and the AAVSO was "an important item in the budget." Mayall noted that the HCO Council had met on February 16, and although she felt certain that the AAVSO question was voted on then, she had not heard a word.[24]

Unknown to Mayall, the HCO Council had already submitted their recommendations to University Provost Paul Buck on February 16. Despite the rumors, Mayall remained optimistic that the final outcome would be in the AAVSO's favor. "I think we have some very good fighting grounds," she wrote to Carpenter, "and it may be worth while to get hold of a good lawyer and let him go to town!"[25]

HCO "EXTREMELY UNPLEASANT"

Mayall then turned to conveying to Carpenter the immediate problems of life at HCO, where things were becoming "extremely unpleasant." Menzel, she said, had one of his students spreading the word to other students that the AAVSO's work was second-rate and not worth the university's support. She was also irritated by how he had accepted the AAVSO vice-presidency, knowing that he would be driving the AAVSO out of existence. She asked Carpenter whether she thought that he should be asked to resign.

On the other hand, Mayall was pleased to see that the Evaluation Committee report to the HCO Council and its letters of support were being read by the students. She was sure that they would not be fooled. Mayall noted that Menzel and the HCO Council "have been very much surprised at our fighting. I think they had no idea we would resist at all."

Aside from all this, Mayall was concerned about the growing number of expressions of concern from people who had been hearing nothing but rumors; the time was coming when she would have to inform the membership about what was happening. Lastly, it was now obvious that the AAVSO would have to establish an independent endowment to supplement, if not replace, the Pickering fund.[26]

While Mayall waited at Headquarters, Menzel – who was out west – expected that the Harvard Corporation would make their decision on accepting the HCO Council's recommendations during his absence. He wrote to Cecilia Payne-Gaposchkin to authorize her, when the decision was received, to release the HCO Council's report with an added statement beginning: "The Council has been asked to reply to a single question" regarding the current value of variable star research and continuing with, "in no case was the Council asked questions like 'would you like to keep the AAVSO at HCO – Is the AAVSO doing useful sci[ence] work?' " Menzel's use of the phrase "the Council has been asked" is intentionally obscure. Neither the Ad Hoc Committee nor President Conant or the Harvard Corporation had ever singled out variable star research. It was only Menzel who posed that question to the Observatory Council.[27]

MENZEL SEEKS SYMPATHY

Aware that there was little sympathy for him at HCO, Menzel explained his difficulties with the AAVSO to his professional colleagues. In response, Dr. Otto Struve, who had written a letter of support to Mayall, perhaps trying to distribute his sympathies more evenly, was diplomatic. He suggested that the AAVSO be encouraged to apply to several national societies, particularly the American Astronomical Society (AAS), and stated that he thought the AAVSO's work was important to astronomy. Then, he pointed out a contradiction in one of Menzel's AAVSO arguments: "Offhand, I do not see why the activities of the A.A.V.S.O. should be particularly expensive, and it is possible that, a more economical way can be found to continue these activities. If I understand you correctly, your own investigation has shown that the same results could be obtained at a small fraction of their present cost." In closing, Struve stated: "I should like to urge, as I have done with respect to other changes, that the transfer of the A.A.V.S.O. from Harvard be carried forward gradually and in as orderly a fashion as is possible . . . . I do believe that the continuation of the light curves without a sudden interruption would be desirable."[28]

Menzel also wrote to his mentor, Henry Norris Russell, who also had written a letter for Mayall. In declaring that all the pertinent information might not have been made available to Russell, Menzel then summarized the terms of the Pickering Memorial Endowment, including the "such" insertion and a discussion of the decrease in astronomy income. Here, Menzel added self-justifying, yet purposely vague language: "These financial considerations also made necessary the request for the report of the A.A.V.S.O. The Observatory Council, asked to answer the question concerning whether such research in variable stars is of the same importance now as it was in 1932, has answered unanimously 'No.'" Feeling confident that the Harvard Corporation would decide positively on the HCO Council's recommendations, Menzel asked Russell to provide a brief statement concerning the problem from the standpoint of HCO.[29]

Russell's reply was not as evenhanded as Struve's. He confessed: "My letter to Mrs. Mayall was, I think, less enthusiastic than the others, but I have nevertheless overstated the case and would express my maturer judgment here." Russell's reconsidered opinion was that amateur observers were not capable of providing precise magnitude estimates, particularly if photoelectric equipment was being used. It would be better, he felt, if existing institutions having large telescopes would undertake such work. He also felt that the funds of any institution could be put to better use than supporting an amateur activity.[30]

Menzel's confidence in his actions must have been bolstered when he received a long letter from Shapley 2 weeks earlier. Shapley was still as concerned about the unrest among the HCO staff as he had been 7 months previously. Now with the announcement of the Boyden Station closing, his legacy seemed to be crumbling away.

Nevertheless, in response to a letter of protest being circulated by a group of students, Shapley proclaimed that they needed "tolerance and consideration for the problems of others in this difficult year." "As you know," he wrote Menzel, "I am trying religiously to remain detached .... Also you have directly asked me to comment on the AAVSO. I have no contacts with students or personnel other than my own assistants. I hope that you and the council will appreciate (and advertise it if necessary) my hands-off policy, my pledge to Conant." Then Shapley expressed a more sympathetic stance in an added "personal note," unusual considering what he had written to Conant about Menzel in July and also remarkable in that he underscored his sympathy by denigrating Conant:

> I sympathize greatly with you in your strained and difficult assignment. This year is probably harder for you than any of my own thirty years were for me. I have worried that the strain may be too much, for you cannot be unaware of the wide-spread internal and external disturbance and dismay over some items in the re-direction of H.C.O .... But I fear (because of the universality of those reactions) that you and other members of the council may be personally penalized in the astronomical world. You must have considered this consequence. I sincerely hope there is no penalty. It would disturb me very much if members of the council, all of whom I value highly as scientists and individuals, lose in esteem because of Mr. Conant's inept and tardy handling of the observatory problem. On ethical grounds I cannot, of course, be happy about some of the decisions. Probably you share my uneasiness. But I make no unasked for protest or inquiry. That is the council's job.[31]

In reply, Menzel sent Shapley an unemotional memo thanking him for his comment and elaborating on his situation:

> The Council has noted with appreciation your detachment and I personally have been very grateful. The unrest that various persons feel is, I think, quite understandable. They simply do not and cannot know all the facts, and thus find it hard to adjust to the policy of retrenchment forced upon us. I accept as inevitable that I should receive the blame for decisions over which I have had little or no control. I am convinced, however, that both our science and our graduate school will, in the long run, be stronger, as hard as this retrenchment is to administer. I believe that we shall eventually regain, if we have ever lost, the leadership in astronomy that has become Harvard's tradition during your regime as Director.[32]

Shapley's surprising remark about Conant's handling of HCO stemmed from his mistaken belief that the retrenchment was due to Conant's negative view of the value of astronomy at Harvard. On the same day he wrote to Menzel, Shapley also wrote to Warren Weaver of the Rockefeller Foundation: "A new regime may fare less well because of the dim and definitely hard view of the astronomical future taken by the Harvard

Corporation under the direction and incitation of Mr. Conant. The financial situation, arising directly from the inflation, has afforded an excuse for the retraction. Astronomers the world over are mystified."[33]

Shapley's view contradicts Conant's own defense of the value of astronomy, which can be seen in his support of the Ad Hoc Committee's findings. The difference between the two is that Shapley (and the protesting staff and students) equated "Harvard Astronomy" with "the status quo," whereas the Corporation chose the course of revitalization through restructuring and retrenchment. At the same time, Menzel himself saw the inevitable restructuring of the astronomy program as the result of Shapley's overextending the observatory's resources – too much building, too many projects, too many staff.

RUMORS "THICK AND FAST"

While she waited for word from HCO, Mayall had to deal with another kind of problem: rumors, as she and Ford discovered, were "flying thick and fast." In early February, Mayall found it necessary to write to Council member Robert Greenley to explain what was happening. Her remarks show that the survival of the AAVSO was foremost in her mind, citing the "letters we have had from more than a dozen big-name professionals in this country and Europe, all enthusiastic about both the past and the future of the AAVSO! One thing I am sure of, and that is that the AAVSO is *not* going to fold up! Our work is much too important for that to happen." She emphasized that the greatest need was for a "good big endowment, so we can be financially independent, and I feel quite sure we will get that."[34]

Mayall considered sending an explanatory letter to the members, at least to acknowledge that the AAVSO was having financial difficulties. Some of the rumors "from all quarters," Ford remarked to Carpenter, "are wildly pessimistic," and he advised that "some kind of an acknowledgement should certainly be made."[35] It didn't help that on February 21, 1953, David Rosebrugh sent his unsolicited opinions to Ford, Cyrus Fernald, Roy Seely, and Neal Heines in two pages titled "Further notes on the AAVSO situation" and a similar more concise letter to Mayall. In a letter to Ford accompanying the "notes," Rosebrugh came across as an apologist for Menzel, saying, "I begin to think, as I have all along, that Don Menzel is merely acting upon instructions, so it might well show our confidence in him to elect him our president next fall."[36]

Mayall, however, was not so easily given to sympathy for Menzel. Within a week, she sent Rosebrugh a reply, thanked him for his letter, but then said: "Last night, while I was trying to compose a letter to you and some of the others, I decided the only way to let you know just what is going on would be through a good talk." After mentioning some ideas about how they could meet as a group, she continued – correctly noting who was to blame for what:

> I can't resist an answer to one of your suggestions – my idea would be to ask DHM to resign from the AAVSO – and I think both [Percy] Witherell and [Charles] Smiley would back me up on that! I know he has been a victim of the Ad Hoc Committee on other HCO matters, but he is responsible for AAVSO troubles.[37]

On the same day she responded to Rosebrugh, she also wrote to Fernald, second vice president and a member of the Evaluation Committee, that everything was uncertain around HCO about the AAVSO's future. She added: "Menzel is trying very hard to get rid of us, and has been quite surprised at our fighting. I still think if we can hold on until a permanent director is appointed, we will have a good chance to stay, or at least to keep the Pickering Mem[orial] Endowment. If we can get enough extra funds to be independent of the Univ[ersity], I am sure we would be very welcome around here.[38]

Mayall remained hopeful and in late February sent copies of the AAVSO Evaluation Committee report to all of the professional astronomers who submitted support letters. Further, in early March, she sent copies of the report to Joseph Ashbrook, Dirk Brouwer, and members of the AAVSO Council.[39]

ESTABLISHING AN ENDOWMENT

The AAVSO's first effort at establishing an endowment was a somewhat wishful stab at getting it quickly and easily. Acting on the authority of AAVSO President Carpenter and likely at the suggestion of Mayall, Endowment Committee Chairman Smiley wrote to Lawrence N. Upjohn, who had donated money for the Multilith campaign and the publication of Campbell's work on long-period variables. Smiley wrote circumspectly, noting that because the AAVSO had no expertise in finance, a committee member thought he "would be an excellent person to consult."[40] Responding, Upjohn felt that he did not qualify and easily read between the lines of Smiley's letter:

> I do not have access to philanthropic funds, other than some rather perfunctory association with certain modest and local affairs. The desirability of continuing the work of the AAVSO would appear to be unquestioned, if the amateur work is really of value.[41]

When she heard Upjohn was unable to help, Mayall noted: "It's too bad about Upjohn, but at least we made a good try." Then, referring to an item she had clipped from *The New York Times* – "$17,695,000 Grants Voted by Ford Fund" – Mayall looked to the Ford Foundation as "our best bet." She saw that they had recently given away nearly $18 000 000 and commented, "but perhaps they have a measly $250 000 left over! I think our project fits in very well with the sort of thing they seem to be supporting."[42]

If it could be called good news, Mayall received a memo on March 4 from Shapley that confirmed her impressions about the situation. A few days earlier, Smiley had written Shapley to ask his advice about using a letter of support from Polish astronomer Thaddeus Banachiewicz. Referring to the "cablegram from behind the iron curtain," Smiley discouraged the use of the letter because he felt there was "considerable rightwing feeling in the board of overseers at Harvard."[43] Shapley, however, addressing Mayall, discounted this idea, saying that it was not about international politics, but about very local politics, "because it looks to me as though the powers that are managing these affairs are completely indifferent to any opinions or suggestions that do not coincide with the decisions that were essentially made before any evidence was presented . . . The hopeful attitude in all this is the present inclination to find a successor within the next two or three months."[44]

Mayall's own view of the Pickering Memorial situation followed along the same lines as Shapley's: the University would likely act on the recommendation of the HCO director; hope for an outcome favorable to the AAVSO depended on appointment of a permanent director – which everyone seemed to assume would not be Menzel.

In the same week in early March when Mayall heard from Smiley and Shapley, she wrote Ford that she had received a letter from Rosebrugh concerning the AAVSO's problems and added: "He [Rosebrugh] has evidently been hearing lots of rumors and getting a wrong idea about things." Mayall was concerned enough about the rumors and misinformation that she sent a special delivery letter to Rosebrugh asking him to arrange a meeting of several "interested" AAVSO people at his home the following week.[45]

On March 3, 1953, Carpenter wrote to Mayall saying that she and Smiley agreed that fundraising should begin as soon as possible, but that, "we can't very well do this until we have had definite word from Harvard." Carpenter referred to some protocol after Mayall's suggestion: "The matter of asking Dr. Menzel to resign is a delicate one but one which I suppose we shall have to face if he doesn't do so of his own accord. Personally I have been expecting that he *would* resign."[46]

CHANGE INEVITABLE

The special meeting that Mayall had suggested to Rosebrugh was held on March 7, 1953, with Ford volunteering to take informal minutes. Present were Mayall; past-presidents Rosebrugh, Heines, Seely, and Brouwer, Yale Astronomy Department chairman; former Pickering Memorial committee member Farnsworth; Ashbrook; Chart Curator Hamilton; AAVSO Council members Emil Sill and Jocelyn Gill; Secretary Ford; Auroral Committee Chairman Donald S. Kimball; Grace C. Rademacher, co-editor of the *Astronomical League Bulletin*; and Newton Mayall. The purpose of the meeting, as Ford recorded in his minutes, was "to inform directly as many interested persons as possible concerning the critical financial status of the AAVSO."[47]

The first order of business was the wording of the Pickering Memorial Endowment and, Ford recorded: "The loop-hole therein pointed out, by means of which the Harvard Corporation and/or other interested parties intend in the near future to divert the Fund's income from its present use by the AAVSO." The party then discussed the Evaluation Committee report, including "the probable reactions of different members of the HCO Council to this report." The discussion then moved to the situation of Harvard University, in general, as other HCO changes were noted, "apparently reflecting drastic financial cut-backs and policy changes by Harvard University."[48]

It must have been apparent by this time that change was inevitable, whether or not the HCO Council gave a favorable recommendation about the AAVSO. The Harvard Corporation was intent on economizing resources and changing the direction of research programs throughout the University. Consequently, the question was not about the AAVSO's continuing at HCO, but whether or not the AAVSO would continue at all, anywhere.

Even so, the "most urgent AAVSO problem" addressed at the special AAVSO meeting was the immediate need for short-term operating funds "to replace the now unavailable HCO-budgeted monies which normally would have been used to run the Recorder's office between now and July 1, 1953." As Ford had noted in January, Menzel had cut off the AAVSO from these short-term, already budgeted funds, presumably in retaliation for the AAVSO's letter-writing campaign of the previous fall.[49]

Beyond this was the question of "finding a donor for an entirely new Fund" to replace the sure-to-be-lost Pickering fund. With this in mind, Mayall called for "specific purposes and recommendations" for a letter to foundations, which would be prepared by Smiley and AAVSO Treasurer Witherell.

Next came the question of finding a new headquarters location: everyone present preferred to see the AAVSO continue to work in affiliation with "a recognized professional observatory." Attention was then brought to the matter of "streamlining" the AAVSO's operations, including: "(1) allowing observers themselves to plot and determine maxima and minima of variables; (2) separating the present 400 or so variables into two groups, one an important 'representative group' to be followed in detail . . . and the other, the remaining variables, to be followed only casually; (3) 'farming out' the Recorder's duties to three or four non-paid regional directors."[50]

Someone, probably Mayall herself, pointed out the problems inherent in these suggestions: the "non-homogeneity of data which would result, and the question of how the 'important' stars should be selected." In a series of "what-if" questions posed by Rosebrugh, the most important were: "What is the worst possible situation that could develop?" and "Would Mrs. Mayall continue to carry on as AAVSO Recorder in that event?" The answers were, respectively: "Complete loss of the Pickering Fund income, and ejection of the AAVSO from HCO"; and "Yes, even at her home if temporarily necessary."[51]

A few days after the meeting, Mayall's assistant Helen Stephansky ended one of her routine business letters to Witherell to say that Mayall's meeting had been "rather unsatisfactory in many ways . . . . Apparently they [who attended the meeting] are all very discouraged and want her to CUT DOWN the work radically, including dropping many stars and observers. All along she has been planning to EXPAND . . . ."[52]

Indeed, as Hamilton wrote to Stephansky, Mayall's attitude about the problem was entirely positive. "We were pleasantly surprised by MWM's cheerfulness and optimism," he said, "I think everyone came away much more hopeful, and it was shown how much the AAVSO really means to a lot of people."[53]

Mayall in the meantime was seeking potential sources of funding. She wrote to AAS Treasurer Jason J. Nassau at the Warner and Swasey Observatory, asking him, "Do you have any suggestions as to where we might get a quarter of a million or so?"[54] Ford, too, hoped to keep the momentum going. On March 15, 1 week after the meeting, he sent Mayall a draft, "Proposed Letter to Foundations."[55] At the same time, Rosebrugh wanted another meeting in late March that would include Heines and others, but neither Mayall nor Ford saw the need. According to Ford, Rosebrugh seemed once again to have inside information telling them: "'My reading of the situation is that Harvard has definitely, already, ruled against us.'"[56] Unfortunately, Rosebrugh's inside information was once again not incorrect. On March 11, 1953, Harvard Provost Buck sent a memo to Menzel reporting that the Harvard Corporation had approved the recommendations of the HCO Council.

HCO CONFIRMATION

Ten weeks had passed – from January 20, 1953 to April 1, 1953 – from the time the AAVSO Evaluation Committee submitted its report to the HCO Council until Menzel gave them any word about what was happening. From the AAVSO's perspective, the time lag was large; but from Menzel's side, the AAVSO "problem" was just one of many issues. He was giving most of his attention to closing the Boyden station and upgrading the Agassiz Station.

On February 15, 1953, Ford learned from a collection of inside information being kept at Headquarters called the "secret file" that one of the HCO Council members, Bart Bok, had held up the HCO Council's report to the University Provost along with the AAVSO's report as late as the previous week. Although Ford did not know why it was delayed, he felt certain it had been sent on its way.[57] As AAVSO Headquarters learned 2 months later, the reports were held up by two of the HCO Council members. "Please don't give DHM [Menzel] credit for writing that [HCO] report," Stephansky wrote to Hamilton. "From what I hear it had to be rewritten by Dr. Bok and Whipple, to soften down some pretty hardboiled remarks."[58]

With AAVSO Headquarters still officially uninformed, Menzel sent the HCO Council's recommendations to University Provost Buck on February 16, 1953.[59] Buck had been delegated the responsibility of running Harvard University as a result of Conant's resignation in January. Signing the HCO report were Menzel, Bok, Payne-Gaposchkin, and Whipple. Mayall later wrote that "Dr. Bok announced at a public meeting that he had signed under protest and was very sorry he had to do it."[60]

Provost Buck acknowledged receipt of the HCO report 2 days later, stating: "I shall present its recommendations to the Corporation for action." In less than a month the Harvard Corporation declared its approval of the HCO Council's recommendations, and Provost Buck notified Menzel on March 11, 1953.[61] Menzel, who was traveling in the West at this time, received word of the Corporation's approval on March 12 from his assistant Mary Van Meter. Then on March 19, having not received any definite orders from Menzel, Van Meter wrote again to him, asking, "When and how is the AAVSO decision going to be released?"[62]

It was not until April 1 that Menzel wrote to AAVSO President Carpenter regarding the Corporation's decision.[63] He included copies of the HCO Council report (dated February 16, 1953) and Provost Buck's letter announcing the Corporation's decision (dated March 11, 1953). This was the first time that the HCO Council report on the AAVSO had been shown to anyone at the AAVSO. In his letter to Carpenter, Menzel remarked that she must have heard about the recent Boyden Station closing and said, "this current announcement relative to the A.A.V.S.O. can hardly come to you as a surprise. Even so we regret that budgetary considerations have forced us to such extremes." Mayall was also informed of the decision on April 1. Sets of mimeographed copies of these documents were sent to the AAVSO, with a blind cover letter by Menzel dated April 8.[64]

If Mayall still had any hope that it would be advantageous for the AAVSO to remain at HCO in some way, the HCO Council report made it clear that the AAVSO was no longer welcome. In little more than three single-spaced pages, the report made its case for revoking the income of the Pickering Memorial Endowment (allowing the AAVSO to have it for one more fiscal year until June 20, 1954), not allowing the AAVSO access during the year to any supplementary University funds, and asking that, at the end of the interim period, the Pickering Memorial Endowment be made available "for other astronomical research." The HCO Council did allow one small concession, recommending that AAVSO's original gift of $6356.76, or the annual income from it, be made available to the AAVSO. Mentioned, but deliberately left unsettled, was the question of where AAVSO Headquarters would be located in the future.[65]

Apart from the expected conclusions, the choice of wording throughout the document stung the most, revealing as Mayall already knew that the HCO Council's recommendations were not only driven by economic concerns, but also by Menzel's preexisting bias against the AAVSO. The first sentence spoke of the AAVSO "among other problems" at the observatory. In describing the AAVSO's membership, the report stated: "These amateurs use their small telescopes to estimate the brightness of certain types of variable stars visually, with special emphasis on the so-called long-period variables."

Then there was the deliberate alteration by Menzel (as tested by him in his earlier Interim Report) of the original wording of the Pickering Endowment. Menzel took it on himself to place the word "such" (which he enclosed in parenthesis) in the quoted language of the Pickering Memorial agreement to – as he stated – "clarify the phraseology":

> [the income from the Pickering Memorial Endowment was to] be devoted to research in the study of variable stars with emphasis on those fields

> cultivated by amateur astronomers or for other astronomical research if, hereafter, the President and Fellows, after consulting the Director of the Harvard College Observatory, shall be of the opinion that, owing to changed conditions, (such) research in variable stars is no longer of the same importance.[66]

The original Corporation language clearly states that the funds be "devoted to research in the study of variable stars" and be made available contingent on the value of variable star research in general. Certainly, in 1953 (as can be seen in the IAU Commission 27 reports, for example), variable star research was as important, if not more important, as in 1931. However, with the word "such" inserted, the sentence's meaning shifts to specify only "those fields cultivated by amateur astronomers."

The HCO report then turned to the matter of the numerous letters of support for the AAVSO, as well as the statement Menzel solicited from Shapley. Although the report acknowledges Shapley's remarks on the value of the AAVSO's scientific contributions and "the spirit of international cooperation engendered through the A.A.V.S.O.," it refrains from mentioning Shapley's main point that the AAVSO's association with HCO had attracted several large monetary gifts from donors and foundations. The report also chose its words carefully in mentioning the letters of support for the AAVSO, saying: "These persons generally express their own desire of seeing the status quo, i.e., the continued support by Harvard, maintained." And, although the report goes so far as to concur with the supporters' attesting "to the undisputed fact that the A.A.V.S.O. is performing a service of definite value to astronomy," it qualifies this as a service "especially useful for a small group of professional astronomers who study variable stars."[67]

In declaring that the study of variable stars was no longer as important as in 1931, the HCO specified four "reasons" that the variations in stars, especially long-period variables, is well-understood: that "modern observing techniques," such as photoelectric photometry, "cannot in general be effective in the hands of amateurs"; that "no major staff member at Harvard Observatory has had or is likely to have any special interest in the field"; and that the British Astronomical Association's Variable Star Section "has a long history," and AAVSO/BAA collaboration "should be explored."[68]

By the end of the report, the HCO Committee had reduced the AAVSO's function to "the recording and collating of the data [which] could no doubt be carried out as an amateur activity," whether or not a source of funding could be found.

MENZEL CALLS A MEETING

On Monday afternoon, April 6, 1953, an informal meeting of HCO personnel took place in the Observatory library. Taking "scattered notes" were Helen Stephansky, Virginia McKibben Nail, and Maribel Wetzel of the HCO staff.[69] The meeting was supposed to last only 10 minutes between scheduled classes, but, as the notes state: "Each time Menzel tried to close [the] meeting someone else asked a question, so it ran over until after four . . . . People seemed unsatisfied, would have liked to stay longer."

The meeting notes convey a certain tension among the staff concerning the AAVSO/HCO problem and include the note takers' parenthetical commentaries on the emotional state of some of the participants. A full record of those present was not made, but those mentioned were astronomers Bok, Payne-Gaposchkin, Sergei Gaposchkin, Mayall, and Menzel; observatory staff George H. Conant, Jr., Miles Davis (son of Watson Davis of Science Service), Edith Flather, Jeremy Knowles, Margaret Olmsted, and Campbell M. Wade; graduate students Owen Gingerich and Edward Upton; and undergraduate students Wayne Lowder, William H. Pinson, and William G. Tifft.

Menzel opened the meeting by remarking that it was his "honest hope" that the AAVSO would "still be able to maintain its headquarters for a long time to come." But he meant this only in a general sense, pointing out that supporting the AAVSO has "been a problem to us," just as its support of the Boyden Station had been. "Wherever there is a budgetary pinch," he continued, "it is obvious that some sort of retrenchment has to be made."

Menzel continued, stating as on earlier occasions, the misleading statement, "The Provost asked the HCO Council to study the AAVSO and its relation to the Observatory." He then read from the HCO report – "The entire picture is given here" – and explained how the Pickering Memorial Endowment had been funded partly from money raised by the AAVSO and substantially from funds arranged by HCO and Harvard.

Menzel stated that he regretted the action because he was a member of long standing since 1919. He said it was "the hope of the Council that with the additional year the AAVSO could find other support, reorganize its activities, strengthen the cause of amateur astronomy and plan a long range future." In ending his remarks, he said that the meeting had been called "at Mrs. Mayall's request," but the notes in parentheses indicate that the "meeting was already planned long before Mrs. Mayall even heard of it."

Mayall took her turn to explain how the Pickering Memorial was begun and "emphasized the part played by the AAVSO in securing [the] Rockefeller grant and the million dollar campaign in 1930–32" (see Chapter 7). When Mayall referred to the AAVSO gift as a moral and ethical obligation, Menzel responded: "The moral and ethical question was not what the Council was asked to investigate. Only the question as stated on Page 2 [of the HCO report]. The question raised by Mrs. Mayall was never raised by the Council."

Then HCO Council member Bok explained the Council's action in apologetic terms, saying that he "had no other choice but sign this letter," noting that the Rockefeller money was "a free gift to the University with no strings attached ... from here on the matter is between the AAVSO and the Corporation." He added that the Council went on record that it was "in no way unsympathetic at the present time. We shall do our very best to try and help the AAVSO in a positive way to solve its problems." Bok ended with, "One active thing is to call this problem to the Council of the American Astronomical Society and attempt to work this out on a national framework because if the AAVSO is worth saving, and I am convinced myself that it is, then we should expect other institutions to come in and help."

Menzel then listed the practical problems involved: Building B to be torn down, cost of rehabilitation of offices not economical, available space is limited, and "playing host to AAVSO is a consideration which must specifically be left to the future, and I hope it can be worked out."

Mayall then spoke about the future work of the AAVSO: flare stars, Air Force observations, "or other small jobs requiring skilled observers" and the introduction of irregular and semiregular stars to the observing programs. Sergei Gaposchkin asked about the AAVSO's immediate plans, to which Mayall replied: "Raise a larger endowment. It certainly would not be fair to ask for funds from HCO beyond income from [the] $100,000 Pickering Memorial."

At this point, Menzel tried to end the meeting, but more questions came. Wade asked: "Can the Observatory do anything to try and save the AAVSO?" Menzel responded by echoing Bok's statement that the AAS ought to be notified and that "other institutions should contribute to help the AAVSO," that it is "a national thing, not a private organization of Harvard Observatory." Flather asked about the escape clause in the HCO Council report. Menzel referred to page 3 of the report, but did not address the issue of the word "such" on page 2. Davis asked whether the HCO Council would be able "to act and campaign to help raise funds?" To which Menzel replied: "Fund raising must be done within the AAVSO, carefully planned and not sporadic attempts." A parenthetical note added states that Menzel "was extremely nervous at this stage."

The random give-and-take continued with a comment by Sergei Gaposchkin: "I have talked to many people out west, and they always say that the amateur should be protected by the AAVSO and not be protected by a big observatory. They should be independent." In reply, Mayall emphasized the international outlook of the AAVSO as a selling point to Rockefeller. (This fact was specifically mentioned by Shapley in his "Statement" to Menzel). To this, according to the notes, Menzel "muttered something about no facts of this sort were found in the records." Student Tifft chimed in: "Shouldn't the retaining of the $100,000 by the AAVSO be considered a *moral* obligation?" With this, the notes state, Cecilia Payne-Gaposchkin broke in, "very angry": "[the] Rockefeller gift was a free gift, they gave without stipulations and washed their hands of the responsibility of what is now being done with the funds. We should not pay any attention to *conditions* which cannot now be found out. It is impossible to find out now what selling points sold."

The conversation shifted again as Davis "pointed out the worthwhileness of the AAVSO observers, their high caliber and ingenuity, their flexibility and ability." Menzel responded by taking Davis' point on a different tack. "The most effective ways," Menzel said, "to cut down appreciably the work of the operational details do not concern us. With ingenuity amateurs could take care of their own recording." Olmsted answered: "It's a *full-time job* to process observations!" to which, according to the notes, Menzel replied angrily: "These are not details that should take our time, or concern us."

Bok once again interceded: "The training of people, the Council is aware, balanced with budgetary difficulties like the tragedy of Boyden Station, are not things that we do not feel, but we are forced to retrench our staff all along the line. There are half a dozen people in the room here now who are training to be professionals because of the AAVSO." Then according to the notes, Sergei Gaposchkin asked, "May I know who they are?" The notes indicated that Menzel "triumphantly" raised his hand, as well as Wade, Tifft, Gingerich, Davis, Lowder, Upton, and Knowles.

Then student Pinson stated helpfully: "Dr. Seyfert told me . . . that the National Science Foundation had funds they were just waiting to give away, and it seems to me that possibly the AAVSO might apply for money there, since it's apparently just waiting there for someone." Mayall had already done her homework on this question: "Those are year-by-year grants," she explained, "and perhaps might be used as stop-gap funds while the AAVSO is trying to get its own endowment. We need a large permanent endowment, not one on a yearly basis. And we now have an ONR [Office of Naval Research] Grant which is helping out."

Finally, as if on cue, Flather asked: "What is the Observatory planning to do with the $100 000 it's getting from the AAVSO?" The notes described Menzel's response: "Menzel (offhandedly as he closed the meeting with difficulty, walking toward the door): 'Oh, simply for other astronomical research, as far as this is concerned. It may help to remove some of the deficit in the budget in 1955.' "

IMPORTANT STEPS

With Menzel's report and the other documents in hand, Mayall now had a clear idea of what needed to be done once the AAVSO Council and Evaluation Committee were notified. There were important steps to be taken by the AAVSO leadership: a short-term effort to raise operating funds and a long-term endowment fundraising effort. The AAVSO membership would have to be notified and reassured of the continuity of the Association and of the value of its work.

Secondary to this was contemplation of what to do about the HCO report itself: make an appeal to the Harvard administration and obtain legal advice concerning the possibility of recovering part, or all, of the rescinded Pickering Memorial Endowment. At the same time, arrangements for the spring meeting in Ann Arbor, just 6 weeks away, were not yet final.

On the same day as the HCO personnel meeting, Mayall took her first independent action in response to the Corporation decision. She wrote to the president of the American Astronomical Society, Robert R. McMath. In her letter Mayall briefly outlined the HCO Council's recommendation and the Harvard Corporation's ruling and gave her understanding of the terms of the Pickering Memorial Endowment. She then itemized an estimated expense budget for the year totaling $20 000 and expressed the AAVSO's urgent need for operating funds and a long-term endowment. This letter was more a notification than an appeal for help; Mayall meant this letter to be only a prelude to a later direct appeal, saying: "Several of our advisors have suggested that we present our problem to the Council of the AAS and ask for their help and advice." Then, perhaps hoping to stir some sense of professional outrage in McMath, she added that the AAVSO felt very strongly that any decision should have been left to the new director when he was appointed and that the wording of the gift distinctly said "that the Harvard Corporation must consult with the Director of the Observatory before making any changes in the use of the fund."[70]

The next day Mayall wrote to Watson Davis, head of Science Service, noting that the meeting had been held the previous day to make the announcement, "and people were given the opportunity to ask embarrassing questions. I was delighted to have Miles here [his son], and you will no doubt get a full report from him," she wrote. With the letter, Mayall included the HCO report, the exchanges between the HCO Council and Provost Buck, and a copy of a news release she had just composed and planned to send out 2 days later on April 8.[71]

> FOR IMMEDIATE RELEASE
>
> Mrs. Margaret W. Mayall of Harvard College Observatory announced yesterday that the American Association of Variable Star Observers is about to embark upon a campaign for funds to support its important scientific work. With hearty endorsements from leading astronomers in many countries, it plans a broader program for observing variable stars, the sun, lunar occultations, and

aurorae, with a renewed emphasis on accuracy and the use of modern electronic equipment. The members of the AAVSO from forty states and thirty foreign countries have made more than 1,500,000 observations of the brightness of variable stars over the past forty years. Today this organization is the authority on the ups and downs of the brightness of long period and irregular variable stars. A book to contain some 400 light curves of variable stars – the work of the AAVSO – is due to come out this year.[72]

The news release is of course positive and does not even hint at any separation between the AAVSO and HCO. Indeed, at the outset Mayall boldly announces her HCO affiliation, which she was still entitled to do, quite aware of the cachet that the well-known and respected observatory still carried in the public's mind. The release seems almost inspired by Menzel's personnel meeting a few days earlier. The opening sentence suggests that it is HCO that is organizing the fund campaign on behalf of the AAVSO; also contrary to what Menzel had said, Mayall stressed that the AAVSO's work is "important," and "scientific"; its observations accurate; its observers capable of using the latest technology; and, most importantly, that the AAVSO "is the authority" on both long-period and irregular variables.

Mayall sent the news release to the Boston and Cambridge newspapers, the *Christian Science Monitor*, the *Scientific American,* several news services including Davis' Science Service, and at least 20 science writers. She also sent copies to several planetariums; to the Astronomical Society of the Pacific, which published it in the June 1953 *PASP*; as enclosures in correspondence to anyone she thought should be informed; and to Menzel himself.[73] The news release also appeared in the "Variable Star Notes" section of the *JRASC* May–June issue. Although the text of the release was unchanged, its presentation in *JRASC* was slightly different because the readers of the "Notes" would be mainly AAVSO observers and members. Mayall made an opening statement that the release had been given to the press in early April; then the release itself followed, and ended with a statement that the AAVSO was starting a drive with the hope of making its future more secure. "We should have a large enough endowment to make us self-supporting," she wrote, "so that we do not have to ask for special gifts or help from any institution to meet our operating and publication expenses."[74]

As of the April 15 "Notes" and the *JRASC* issue's publication date, neither the AAVSO's general membership, the general astronomical community, nor the public were yet aware of what had recently transpired at HCO. Apparently, Menzel was hoping to keep it that way when he sent a memo on April 14 to Mayall: "A check with the University indicates that they feel that no public announcement of the A.A.V.S.O. should be made at this time."[75]

Despite this warning, Mayall found another public relations opportunity. On May 7, 1953, *The Christian Science Monitor* printed an article about the Amateur Telescope Makers of Boston (ATMOB), the Bond Club, and the AAVSO. The article pointed out how the three amateur organizations were being hosted and encouraged by Harvard College Observatory. The article featured ATMOB members and mentioned briefly the Bond Club, but it discussed the AAVSO in the final one-quarter of the piece – in terms of its history and real value to science. The piece ended with a quote from Mayall in which she not only mentioned the fund drive, but perhaps for the first time publicly, although indirectly, spoke about the loss of HCO's support.[76]

On April 18, AAS President McMath replied to Mayall's letter of the previous week and acknowledged Mayall's news, "which is of course of very bad blow to the AAVSO." He suggested that Mayall present its problem to the AAS Council and ask for help and advice, but added, "you already know that the Society has no funds." McMath regretted the AAVSO's loss, but ended unhelpfully, "I must add, however, that it is always necessary to allow the other man his own point of view."[77] On the same day, McMath also wrote a much more supportive letter to Menzel acknowledging receipt of the HCO report and word of the Corporation's decision, noting: "your group is entitled to take whatever action you believe is best suited to the needs of astronomy at Harvard, and I can see no basis on which any valid criticism can be made."[78]

Mayall again wrote to her colleague J. J. Nassau of Warner and Swasey Observatory, who was AAS treasurer, partly to notify him of the situation and partly for a sympathetic ear: "I am sure that by this time you have heard the sad news of what Harvard is trying to do to the AAVSO. I hate to admit it, but the 'they' in my other letter means members of our own Harvard Observatory Council." She then went on to mention plans for making an appeal to the Harvard Corporation, finding short- and long-term funding, and her intention to ask the AAS Council for advice and support. "We have learned a bitter lesson!" she exclaimed. "Never again will we accept any funds unless we can have a part in the administration of them."[79]

Thinking about the AAVSO in more general terms, Mayall wrote to Fernald the next day that despite the HCO report, she was still very optimistic about the AAVSO's future. She intended to "go after a large grant from one of the big philanthropic organizations," and she added, "I am confident we can get one, with the wonderful backing we have from really big astronomers all over the world .... I am presenting our problem to the Council of the AAS [American Astronomical Society] and asking for their support."[80]

The root of the problem, Mayall declared, did not lie in the changes at the University or Corporation level but at the level of HCO itself: "The HCO Council has taken this action against us just to get an additional $4600 a year. (Harvard allows 4.6% on their investments.) They really expected us to fold up and move out without protest last fall!... I feel we owe too much to our members and to astronomers who depend on us for information. Certainly 100 years from now our observations are going to be of much more value than most of the theoretical work that is being done."[81]

Fernald responded with encouragement that he had confidence that most of their "worthwhile members would ride out the storm" with them, "whatever retrenchment may prove necessary to cut in way of present services to them. The astronomers who depend on us for information may prove to be of a great deal of help, if they are made to realize we need it shortly." He agreed with Mayall that the future value of the observations will be great, and, therefore, no stars should be cut from the program. In conclusion, he said, "Frankly I am rather disgusted by the lack of tact on the part of those in control on these matters. If the matter had been handled diplomatically it would have left good deal better memories than we are going to have, no matter what the outcome."[82]

Mayall asked Smiley to arrange a conference with Provost Buck at Harvard. She hoped that something could be settled before the May 22 Spring Meeting so that she would have something to report to the Council. She added that she and others at the informal AAVSO meeting in March felt that a meeting with Harvard Corporation was now "our only move . . . the Observatory is now out of the whole thing." Mayall felt that the meeting should include a member of the Corporation, their legal advisor, and Provost Buck. Then she speculated: "We all wonder if the members of the Corporation know anything at all of the problem, and we suspect they merely O.K.'d Buck's recommendations." Perhaps grasping at straws, Mayall added, "We have two good legal points in the wording of the Endowment: 1) the fact that the Director must be consulted, and 2) the Council's addition of the word 'such.'"

Mayall's letter ended with a set of dismal images, noting that "things were getting worse and worse around here [at HCO]." Aside from the closing of the Agassiz Station for a year, she wrote that there were plans to tear down the frame buildings in Cambridge, including the Director's residence and Building A, and that the old 15-inch telescope was being dismantled and would be taken to New Mexico. "I can see that HCO has nothing to offer the AAVSO for the future," she said. With nearly all hope shattered that the status quo might be preserved, she ended with an expression of a different kind of hope: "When we talk about our Pickering Fund, we must not forget the 'capital gains' over the last 22 years!"[83]

Over the next week, the rumors that Building A was marked for demolition were seemingly confirmed by their persistence and amplification. Stephansky described an incident: "Margaret went out with a trowel last Saturday morning to move some hollyhocks from one side of the 9" to a less trampled side, and his nibs came rushing downstairs to see what vandalism she was committing. Funny thing, his office is on the other side of the hall, but he has come over several times to see when something fishy is going on."[84]

Shapley comments

If she was going to send AAVSO representatives to meet with Harvard, Mayall felt she ought to gather as much background information as she could. She asked Shapley for his thoughts on how Menzel altered the wording of the Pickering agreement. Shapley responded on April 23 in a "non-public comment" that the insertion of the word "such" "seemed to be little less than a deliberate attempt to deceive." Shapley wrote "only in defense of my past association with the establishment of the Pickering Memorial Fund," saying, "I definitely owe it to the memory of my predecessor, and to my numerous friends in the AAVSO who were as startled and amazed by the ejection as were the astronomers all over the country." Regarding the wording for the establishment of the Pickering Memorial Endowment, and especially with respect to the escape clause, he recalled:

> Very intentionally I did not and would not put in the word "such." It is wrong to assume that the bequest makes no meaning without that word. With the word "such" in, any whimsical Director could eject the AAVSO, and repossess the funds on the slightest pretext. For instance, a mess of bad amateurish observations could be taken as a violation of the agreement. The most obvious reason for not putting the word "such" is that my experience had shown that the AAVSO is adaptable to changing conditions and therefore as new demands and standards come along the organization could modify its program. It has done so . . . . The word "such" had no business in the statement; and of course the procedure is highly questionable. By the introduction or deletion of one word "in order to make sense" all of the Ten Commandments could be negated, or any agreement that is solemnly entered into by any group anywhere at any time.

However, he doubted that this "gratuitous modification" of the condition would have had much effect on the decision. "I have a feeling that the decision was arrived at before much evidence was accumulated," he wrote. "Otherwise I feel that the Council's report to the Provost would have included actual quotations from the many letters from the country's top astronomers. And otherwise the ethics of the situation would have been given less hard-boiled consideration."

Shapley feared it would take a long time for the Harvard Observatory "to recover the prestige it has lost from the manner of handling the affair. It has gained greatly from the AAVSO in the past," he said, "not only through gifts but in the attraction to Harvard of some of our most effective students. I hope that the cause of amateur science could recover; and after the administration at the Observatory has been finally settled I might be able to make some useful suggestions to the officers of the AAVSO."[85]

Shapley told Mayall what she wanted to hear, but he was reluctant to have his statement used in his absence in any meeting with the Provost: "It must be used by you and Smiley only as background information – something that you know from me." In this statement, Shapley implied that the word "such" was at issue when the original Pickering Memorial wording was being deliberated. The implication is misleading. Deliberations over the wording of an escape clause in general did take place, as one would expect, but what Shapley neglected to mention – or perhaps merely forgot – was that the deliberations were over whether or not the AAVSO, as beneficiary of the endowment, should be specifically mentioned in the wording of the agreement. At that time, Professor Anne S. Young at Mt. Holyoke College and Harriet W. Bigelow at Smith College urged repeatedly that the AAVSO be mentioned specifically in any funding agreement with Harvard. Campbell and Shapley at the time strongly discouraged those demands and William Tyler Olcott went along with them, not wishing to rock the boat. They feared that the Harvard Corporation would withdraw their support if the deliberations were too drawn out or if the wording was not generalized enough to allow insertion by the Corporation of its escape clause pertaining to the future usefulness of variable star astronomy.

Writing to AAVSO President Carpenter on April 25, Mayall repeated most of the same information that she sent to AAVSO Council member Smiley 3 days earlier. Shifting to her other problems, Mayall thought they would be able to get a grant from the National Science Foundation (NSF) "to help out for the next year or possibly two" and was "writing to [Dr. Otto] Struve for help and advice with that."[86]

National Science Foundation

Mayall had good reason to place her short-term hopes in the NSF. In 1948, the National Research Council formed a Committee on Astronomy, Advisory to the Office of Naval Research (ONR). Its members were appointed on recommendations from the Council of the American Astronomical Society (AAS). In 1951, the Committee recommended that it be allowed to serve as an advisory body on astronomical matters to the NSF. With this, the Committee announced a proposed budget for astronomy for the first year of operation (1951–1952) of the newly formed NSF, with $35 000 allotted to publications support and the $160 000 budgeted for short-term research projects. Citing the success of short-term research projects being funded by the ONR, the Committee wished to see such a program continued under the NSF, but with the limit per project raised to $10, 000, twice that of the ONR's limit for similar contracts. "Besides the support of research at existing large observatories," the Committee wrote, "special weight has been and probably should continue to be given to projects which encourage research at smaller institutions, with a wide geographical distribution." The budget proposal appeared in the October 1951 *Astronomical Journal* and was summarized in the January 1952 *Sky and Telescope*. A second article appeared in the September 1953 *Sky and Telescope*, in which California Institute of Technology astronomer Jesse L. Greenstein further discussed the NSF's programs in support of astronomy. Greenstein was chairman and one of seven astronomers serving on a panel to advise the NSF on the merits of applications for the grants. The other advisors were Lawrence H. Aller, Dirk Brouwer, Gerald E. Kron, Gerard P. Kuiper, Martin Schwarzschild, and Fred Whipple.

The news of the NSF funding for small-scale astronomy research and publication grants – particularly in support of publishing monographs – must have been promising news for Mayall and the AAVSO. In fact, Greenstein emphasized that "Panel members are very conscious of the needs of the smaller institutions, and hope that they can assist the NSF in sponsoring a well-rounded development of astronomy in this country." Prophetically, Greenstein also predicted that, "Eventually, such government support of research will become more and more important. The NSF is still in an experimental stage, but we know that a satisfactory and efficient method of operation will be worked out." Mayall must have also felt encouraged by the fact that, with the exception of Whipple, who was on the HCO Council, all of the advisory panel members could be considered friends of the AAVSO.[87]

The legal pursuit

On April 28, Mayall met with two members of the law firm of Ely, Thompson, Bartlett, and Brown, who suggested meeting with Harvard's lawyers before any AAVSO–Harvard meeting. Mayall explained that she felt the AAVSO "should continue to have the use of the whole Pickering Fund, and failing that, we at least should have the $6000 plus the matching $6000, plus the Capital gains over the 20 years."[88] The lawyers could only promise that they would see what could be done to get the Harvard Corporation to postpone the crisis until there was a permanent director and they could consult him. "I am sure the new director," Mayall tried to predict, "whoever he may be, will certainly not be any less friendly than our acting director." Mayall decided that the Evaluation Committee would have to postpone writing to Harvard Corporation Secretary David W. Bailey until the lawyers had made their try.

A RETROSPECTIVE REBUTTAL

In concluding this chapter, a retrospective rebuttal of the HCO Council's four "reasons" why the study of variable stars was no longer of the same importance as it was in 1931 seems especially appropriate.

(1) "That the nature of variations in stars, especially of long period variables, is well-understood." Well-known professional astronomers such as, but not limited to, Merrill of Mount Wilson Observatory had already strongly asserted the contrary position that variable stars, even the LPVs, were not well understood, nor was there a theoretical basis for understanding, and until that theoretical basis was established and tested, the accumulation of additional data was essential to the development and testing of such theories.
(2) "Modern observing techniques . . . cannot in general be effective in the hands of amateurs." There was already ample evidence in the work of John Ruiz, which had been lauded by professionals, that given time, amateurs would make

some contributions in photoelectric photometry. This point, moreover, clearly ignored the long-standing justification that redundant observations by a number of visual observers were more than adequate for theoretical understanding, especially LPVs.

(3) "No major staff member at Harvard Observatory has had or is likely to have any special interest in the field." The agendas of the HCO Council's meetings clearly put the lie to this point, as Bok and Payne-Gaposchkin of the council, as well as Sergei Gaposchkin, were all deeply involved in variable star astronomy in many ways.

(4) An AAVSO-BAA collaboration "should be explored." The BAA-VSS involved fewer than 20 observers observing about 30 stars, as compared with more than 100 AAVSO observers recording observations all over the world for around 400 variable stars. Clearly, the HCO Council was badly uninformed and/or was misrepresenting these points about the AAVSO for the sake of justifying their decision.

12 · Actions and reactions

It made me nearly sick to think our beloved organization is endangered.

– Richard W. Hamilton, 1953[1]

With the information that Harvard College Observatory would not be supporting AAVSO operations in the future, Margaret Mayall now had a clear idea of what needed to be done once the AAVSO Council and Evaluation Committee were notified.

There were critical steps to be taken by the AAVSO leadership: a short-term effort to raise operating funds and a long-term endowment fundraising effort. The AAVSO membership would have to be notified and reassured of the continuity of the Association and of the value of its work.

Secondary to this was contemplation of what to do about the HCO report itself: Mayall would make an appeal to the Harvard administration and obtain legal advice concerning the possibility of recovering part, or all, of the rescinded Pickering Memorial Endowment. At the same time, arrangements for the spring meeting in Ann Arbor, just 4 weeks away, were not yet final.

These were just a few of the steps that Mayall, the second Pickering Memorial Astronomer, would take to refute Harvard's actions and attempt a recovery of the Pickering Memorial Endowment. She and the AAVSO would fight back, an action that was totally unexpected by the HCO and Donald Menzel, its newly appointed director. These events are covered in this chapter.

MEETINGS

Life in the local astronomical community went on while Mayall was wrestling with these issues. On April 16, HCO astronomer Dorrit Hoffleit, president of the Bond Astronomical Club, invited AAVSO members to the picnic of the Bond Club and the Amateur Telescope Makers of Boston honoring Shapley, who had "probably done more than any other professional astronomer to encourage the amateur." Hoffleit invited "any amateur astronomers" to attend the May 9 picnic at the HCO's Agassiz Station in Harvard, Massachusetts.[2]

The picnic offered a respite to the AAVSOers who knew about the coming eviction, with part of the entertainment being the opportunity to observe Menzel. "The picnic is the main topic this morning," Mayall's assistant Helen Stephansky wrote from Headquarters to AAVSO member Richard Hamilton the following Monday, "(mostly the DHM reaction, you might know!) . . . . Hurray for Shapley – we shure [sic] showed whose side we all are on. They think over 400 came."[3] A memo from headquarters, probably written by Stephansky to AAVSO Treasurer Percy Witherell, described the "wonderful picnic" and the ceremonies honoring Shapley.[4] At the ceremony, David Rosebrugh, as AAVSO's representative, thanked the Shapleys for all that they had done in the past. Hamilton described Menzel's behavior: "He gave about three little claps of applause after Dave R[osebrugh's] speech! Most of the time he stood there, with a bored, deadpan expression."[5]

The gathering boosted everyone's spirits.[6] Even Rosebrugh's chronic pessimism seemed to lift. He had learned there was a further reason to be hopeful because Harvard had backed down from its action 3 years ago taking away some $750 000 of endowment funds from Arnold Arboretum. Hamilton, too, thought their conversations at the picnic took some of the sting out of their troubles: "It reassures us that we've got what it takes to fight the demons who would cast us out."[7]

Mayall informed the small group of AAVSOers at the picnic that the Association would be supported by HCO until June 30, 1954; that the $2000 per year received from the US Naval Observatory for information on certain stars would probably continue; that she had retained a law firm to represent the AAVSO's interest in the Pickering Endowment; and that she was in the process of applying for a NSF grant to cover 2 years of operating expenses and "seems optimistic of getting it." Above all, everyone's greatest hope was that all problems would dissolve once a new, permanent, HCO director arrived to make order out of the chaos.[8]

Perhaps less optimistically, Ford, responding to a letter about the picnic from Rosebrugh, summarized his thinking about the AAVSO's situation up to the moment: support from the NSF would be temporary, and year-to-year at best; a long legal battle with Harvard held little promise, and there would be lawyer's fees. "The only healthy solution," he concluded, "would be to find a professional astronomer of high stature to state the AAVSO's case to one of the large charitable foundations, and also to influence potential donors to the fundraising campaign." Lastly, "if it becomes necessary to move from Harvard, that would be a major struggle for MWM [Mayall], but it would have to be done." He hoped that they would be able to work out something definite at the AAVSO Spring Meeting and before the fateful June 30, 1954, when the Pickering Memorial Fund would end.[9]

Figure 12.1. Percy Witherell, AAVSO Treasurer 1931–1960, at the AAVSO Spring Meeting, Ann Arbor, Michigan, May 1953.

The AAVSO 1953 Spring Meeting

The AAVSO held its spring meeting on May 22 and 23 at the University of Michigan in Ann Arbor. The council meeting agenda of routine topics included one weighty item: "General discussion of the present situation regarding the discontinuation of Harvard's Pickering Memorial Fund support to the AAVSO, and proposals for future action to offset this impending loss of income to the Association."[10]

The good news was that the AAVSO would not lose its HCO funding until June 30, 1954. The hopeful news was that final action on the status of the Association at Harvard College Observatory (HCO) was still pending appointment of a new director. Plans for any legal action rested on the HCO Council's insertion of the word "such" in the Pickering agreement and on the fact that, according to the minutes, the HCO Council "apparently took action in the name of the Director whereas legally there was no Director of the Observatory at the time that report was submitted." Regarding financial problems, the Council agreed that a new fund should be established, whether or not the Pickering Fund was taken away, and a figure of $250 000 was set as a preliminary goal.[11]

At the general membership meeting the next day, Mayall described the situation for the first time to an audience that had only heard rumors or nothing at all. She covered the history of the Pickering Memorial Endowment, described the Harvard Corporation's intention to "divert the use of the Fund's income from the AAVSO to 'other astronomical research' as of June 30, 1954," and summarized plans for raising a new endowment to provide "adequately for the maintenance of the expanding work of the Association." Mayall commended the work of Charles Smiley and the Evaluation Committee and summarized the letters of support from professional astronomers.

Secretary Clint Ford did not describe the members' reactions, but only that several members asked questions, which led to a further discussion of the Association's financial outlook.[12] A statement was made that "the value of the AAVSO as a stimulant to education and international cooperation made it an organization, however small, whose activities are worthy of financial support from large Foundations."

Technical papers were presented as usual: Dr. Dean B. McLaughlin, a stellar spectroscopist and featured speaker, spoke on "The Vastness of Celestial Distances"; Newton Mayall reported progress on Leon

Figure 12.2. Dean McLaughlin, Margaret Mayall, Helen S. Hogg, and Leo Goldberg, at the AAVSO Spring Meeting, Ann Arbor, Michigan, May 1953.

Campbell's long-period variables volume; and Margaret Mayall ended the session with "Recent AAVSO Light Curves of Special Interest."[13]

Overall, the spring meeting shifted everyone's focus away from HCO to the problem of keeping the AAVSO running smoothly and independent of any further influence of the HCO Council.

However, there was an undercurrent of tension and mistrust growing between the HCO and the AAVSO. About 1 week later, Ford wrote to Mayall about some comments made by AAVSO President Martha Carpenter that if Menzel came to the fall Council meeting he should be ignored, and especially if Mrs. Gaposchkin came, "we should be firm in not giving her a voice in any policy making, as she is not even an AAVSO member, let alone a Council member." Furthermore, Ford thought an effort should be made to inform the other Council members of developments. "If we don't do this," he warned, "these people will arrive at the last minute and very possibly be railroaded by whatever tactics DHM [Menzel] may resort to."[14]

Mayall had just completed a study that was funded by the Office of Naval Research (ONR). Back in November 1951, at the urging of Shapley, Mayall prepared a list of 750 semiregular and irregular red variables for the ONR study. The AAVSO observations of these stars, Shapley had noted, had never been discussed, and most of them had not been published. Mayall discussed her results for two of the stars studied, R Pictoris and W Persei, in the July–August 1953 *JRASC*.[15]

On her mid-June visit home in Maryland, Mayall met with Smiley in Providence, Rhode Island, and with several individuals, including the NSF Director, to discuss the possibilities of obtaining a grant.[16] Before making the trip, Mayall had written no fewer than 11 letters to friends and officials asking for their opinions and if they could make preliminary inquiries at NSF on whether or not she should apply for a grant. In some letters she included a summary of her proposed projects; in others she included budget outlines and even a sample application. All urged her to go ahead and apply to the NSF.[17]

Later on July 28, Mayall sought Ford's opinion about her NSF application and a draft letter for AAVSO members. She described both letters as "the result of several weeks hard labor and consultation with a number of people" on her way to the NSF in Washington, DC. Her meetings with others indicated the different directions she was pursuing: Watson Davis of Science Service; Father Francis J. Heyden at Georgetown College Observatory, a graduate student and post-doctoral fellow at HCO during the 1940s; Herbert B. Nichols, *The Christian Science Monitor* science editor; Smiley; and others closer to home, including Shapley, Witherell, and the AAVSO's lawyers.

Mayall sent her letter of application to the NSF on August 4 "for a grant to help the AAVSO through a current emergency" while a drive for a permanent endowment was being launched. Mayall titled the proposal: "The light variations of more than 500 variable stars" and planned "to continue and expand the scientific program of the AAVSO." The plan included about 400 LPVs and more than 100 irregular and semiregular types. She included an annual budget and requested a total of $23 000 to cover the AAVSO's operation expenses and salaries for 2 years. Her budget included the $4600 income from the Pickering Memorial Endowment for the coming year.[18]

LETTER TO AAVSO MEMBERS

Despite having put pen to paper, Mayall was unsure about sending her letter to the members, especially after a phone conversation with AAVSO President Carpenter, who had felt it would be better not to send it at all.[19] Ford encouraged her to go through with it, commenting that he did not see how she could "avoid putting matters directly on the line pretty much in the form that you have presented them."[20] Although the letter might be "disturbing to foreign members," he felt certain, from the view of the US and Canadian "regulars," that the letter "would be sympathetically understood and beneficial to the cause." Noting that there was "a lot of pessimism among members," Ford "heartily" endorsed the letter to members: "I really believe your letter would be a valuable shot in the arm to clear the air and let everybody know where we stand and what we see ahead."

On August 5, 1953, Mayall sent out a mimeographed letter that began: "Rumors have been flying around the world of all sorts of dire happenings in the AAVSO." The four-page letter summarized the AAVSO's financial situation and the Pickering Memorial Endowment and presented a chronology of the events leading to the Evaluation Committee report and the decision of the Harvard Corporation. Mayall also critiqued the HCO Council's rewording of the Pickering Memorial document and commented on other semantic elements of the HCO Council's report to Provost Buck.[21]

In presenting the basic facts, Mayall did not want to diminish the respect that she knew all amateur observers felt for Harvard College Observatory. In describing the AAVSO's financial situation, Mayall began: "I am sure none of us realized just how much Harvard Observatory has done for us in the past"; and further down, she wrote: "In recent years, the annual income from the [Pickering] Endowment has amounted to approximately $4600. This was about half of our annual budget, and Harvard Observatory took care of the rest, including office space and 'overhead' from various unrestricted funds and stocks of supplies."

However, regarding Acting Director Menzel's actions, she did not try to hide the unpleasant details. She made it clear that Menzel had informed her before the 1952 Annual Meeting that the AAVSO could expect both the loss of HCO support and its office space and that soon afterwards he also informed her that the AAVSO "could not consider the Pickering Memorial Endowment as belonging to the AAVSO," and followed with his assertion that "the Harvard Corporation was a business organization and did not recognize any moral obligations!" Mayall described the HCO Council's alteration of the original wording of the Pickering Memorial terms, even borrowing Shapley's line from his personal statement to her how "the meaning of *any* agreement or law (including the Ten Commandments) could be changed if words could be inserted indiscriminately!"

Borrowing again from Shapley's statement and adding her own emphasis, Mayall questioned the motivation for Menzel's action against the AAVSO. "It is regrettable," Mayall stated, "that an interim administration of the Observatory should risk international criticism by taking action against the organization which has received, and is receiving, so much acclaim . . . . I hope that the members of the AAVSO will not feel discouraged over this attitude taken by the present interim administration of the Observatory."

Mayall was still convinced that the problem was "budgetary and temporary." Because Menzel was the interim acting director, Mayall was saying, it was not his problem: he could have held the entire situation, including HCO's own budget crisis, in abeyance until the appointment of a permanent director. Instead, she implied, Menzel *chose* to take action against the AAVSO.

The last part of her letter discussed the AAVSO's future, the need for funds, changes in observing programs, changes in publications, and so on. "The work of the AAVSO continues to be just as important as ever," she wrote. Confidently, she concluded, "Let's all look forward to the day when we have all our troubles straightened out, and everything goes along smoothly once more. In the meantime, don't let any of the variables put on an unobserved show. It is your observations, more than anything else, that invariably prove the worth of the AAVSO."

Perhaps as an afterthought that something might yet be salvaged from the crisis, Mayall attached a "Special Note" on bright yellow paper asking the members to write to the Harvard Corporation Secretary David W. Bailey if they felt "that the decision of the Harvard Corporation was unjust in allowing the Observatory to use the Pickering Memorial Endowment for other than AAVSO purposes." In addition to AAVSO members and friends, she sent the letter to AAS President Robert R. McMath, Corporation Secretary Bailey, at least two of the Harvard Corporation members and one of the Board of Overseers, some of the HCO Visiting Committee, and to Menzel himself.[22]

Now, once again, Mayall could only sit and wait. Although she wrote her friend, University of Toronto astronomer Helen Sawyer Hogg, on August 15, 1953, that things were "rather quiet" with Menzel away for several weeks, she fully expected a "big blowup" over her letter to the AAVSO members. In the meantime, she had a "very nice letter" from David W. Bailey, Secretary of the Harvard Corporation, saying that he would present their side to the Corporation at their next meeting. She also told Hogg that AAVSO member Cyrus Fernald had written a "wonderful" letter to the Corporation stating that he was changing a trust he had set up for the Harvard Observatory some years previously.[23]

Members "Shocked"

"Shocked" was the most frequent comment by members who wrote to Mayall on hearing the news from her August 5 circular letter to the membership. A number of them sent Mayall copies of their protest letters to Bailey. "Things look hopeful around here," her assistant Helen Stephansky wrote at the end of the month, "or perhaps it is just that we are looking at them hopefully." She reported that the response to Mayall's letter was growing and that the Harvard Corporation was receiving "a good many. The carbons we get are very persuasive."[24] Shapley had been among the first to respond, noting that Mayall did not emphasize "and perhaps did not fully appreciate" that "During my regime, and certainly during the last ten [sic] years of Pickering's directorship, the AAVSO was considered a responsibility and effectively a part of Harvard Observatory."[25]

A few AAVSO members were more realistic than hopeful. Regarding the letter campaign, amateur observer Frank J. Kelly commented, "it can hardly be expected to raise a ripple if letters from those 'persons' [as Menzel dismissively called the supportive professional astronomers] didn't – but we can do no less than try."[26] Walter P. Reeves, vice president of the Maine Central Railroad, stated matter-of-factly: "I am afraid that in view of the provision you quote, the Corporation has the legal right to use income from the Pickering Memorial Fund for other purposes than support of the A.A.V.S.O., if this is the judgment of the Director of the Observatory."[27] Lewis J. Boss, an executive for the Philco Corporation, remembered that when the Pickering Fund was started by the AAVSO in 1921 and was later increased and taken over by HCO, there was "no intention, expressed or implied, that this Fund's earnings would, or could, be used for anything except the AAVSO." Boss felt that if Menzel did not want the AAVSO around any longer, "the feeling was mutual and . . . the Association would be happy to remove itself from his immediate vicinity."[28]

Each letter Mayall received in response to the August 5 mailing provided a unique point of view; different parts of her letter resonated with different readers. Charles P. Olivier, director of the Flower Observatory in Philadelphia, commented that, "If universities forget morals, then they had better close up!"[29] His letter to Secretary Bailey, though more formal, praised the AAVSO as "the most valuable amateur society in Astronomy, in this or any country. Its work has been of the greatest assistance to professional astronomers in the study of variables, and the amount of work done by its members is immense . . . . Numerous young people have been attracted to a scientific career through membership. I dare say some of your best Harvard graduates got their first inspiration in this way."[30]

Long-time member Walter Scott Houston could not see how the AAVSO could flourish under an antagonistic director. "I met Menzel back in the thirties," he recalled in a letter to Mayall. "[H]e was most impolite then and shocked me by stating he didn't think much of amateurs. I suspect his attitude is of long standing."[31] He scolded Bailey, haranguing him for Menzel's "open disregard of the obvious intent of the Pickering fund, his manipulation of the clause inserted only to save the income if the AAVSO should become senile, and his alteration of the text of the original grant – all these make one suspect all his other statements. That he should establish some new top in insolence by referring to the professional opinions of some of America's most eminent astronomers as if [they] were only private individuals outside of Cambridge is downright shocking."[32]

Perhaps the most eloquent appeal to Secretary Bailey came from amateur Theodore R. Treadwell, a member and observer since 1935. Instead of discussing the endowment or the parties involved, he explained why the amateur variable star observer contributes his work to the AAVSO:

> We amateurs have had a feeling of intense satisfaction in knowing that our humble efforts were of use to the professionals, that our observations, systematized and tabulated by our Recorder, would be filed in all the important observatories of the world where they might be put to future use in solving some fundamental problems. In short, we amateurs have had a feeling

> that we were making a small, but perhaps essential contribution to the sum total of astronomical measurements.[33]

By the end of August, Mayall still could not know what the future held for the AAVSO, but the letters must have strengthened her resolve. "The future of the AAVSO is still very uncertain," Mayall wrote to Paul Merrill on September 1, "but at least the Harvard Corporation knows that we exist and have plenty of enthusiastic and loyal members. They are getting lots of very good letters, and Mr. Bailey tells me he will present our problem at the September 21st meeting."[34]

By mid-September, Mayall had received enough encouraging words from members to know that her circular letter was the right thing to do. Professor Alice Farnsworth of Mt. Holyoke College, a member of the Pickering Fund Committee in 1931, told Mayall, "I want to congratulate you on the letter which was distributed to AAVSO members. It seems to me it is a masterpiece! and I am sure it was not easy to compose – to say enough yet not too much."[35] On the same day, Mayall also received a memo from Menzel informing her that Building A would be demolished "sometime this fall." He made no mention of Mayall's circular letter.[36]

The "AAVSO problem" was also generating interest at the Astronomical League (AL), probably in a large part due to its contacts with Rosebrugh. Earlier in the summer, Charles H. LeRoy of the AL wrote to Mayall stating that there was word that the AAVSO would present their plight to the AL Council. Most likely, the "word" came from Rosebrugh; neither Mayall nor the AAVSO Council had made such plans. Rosebrugh suggested that the AAVSO send a speaker to the AL convention in September to present the history and current status of the AAVSO problem. Instead, Mayall asked Ford to prepare a report in late August, which was then published in the AL's September *Proceedings*. It was hardly a plea for AL support; Ford stated confidently that the AAVSO's work was continuing more now than ever and that the AAVSO planned to raise an endowment of its own. He only briefly mentioned the loss of the "22 year-old" HCO Pickering Fund.[37]

AAVSO member Claude Carpenter provided his own impressions about AL's interest in the AAVSO. He wrote that at the Western Amateur Astronomers convention, AL President Rolland R. LaPelle, once a contender for the AAVSO Recorder position, "threw out his chest . . . and said 'the trouble with the AAVSO is that there is no more use for the damn observations anymore.'"[38]

Ironically, HCO astronomer Sergei Gaposchkin, who had just published "How to Estimate the Brightness of a Star" in the September 1953 *Sky and Telescope*, began his discussion by stating, "Let us limit ourselves to the specific case of estimating the brightness of a variable star, since variables yield results of the greatest importance to astronomy and will always remain on the front line of research."[39]

Despite the negative attitudes on the part of some outsiders, and the doubts sometimes expressed by the insiders, the AAVSO was fortunate to have three important things going for it: a grand tradition, its scientific mission, and the loyalty of its members, officers, and staff. Chart curator Richard Hamilton paused in the middle of one of his usually jocular and irreverent letters he exchanged with Helen Stephansky each week to reflect on what the AAVSO was all about:

> T'other nite was looking thru some AAVSO stuff, and it made me nearly sick to think our beloved organization is endangered. Letters from Olcott, Pickering; photos of groups at meetings; V.C.s [*Variable Comments*]; souvenirs; Annals; reprints from P.A.; on and on they went. And the same day a splendid letter from S[imon] Archer in South Africa, who went over and found errors on a couple of our southern variables charts. Now nobody can tell me the AAVSO has outlived its usefulness, is second-rate, etc., etc."[40]

Menzel responds

Menzel was again traveling in the West when his secretary, Velma Adams, wrote him August 18 with her opinion of Mayall's circular letter:

> Let the AAVSO hang themselves. They will do it soon enough if they continue to give out with such snide remarks as the latest report. More and more people will see the light, particularly if they continue to discuss Harvard policy instead of the stars. Just sit tight, and give them the rope. They'll do the rest.[41]

Because Mayall had broadcast her response to the "problem," Menzel drafted a two-page, paragraph-by-paragraph rebuttal to Mayall's circular letter.[42] However, many of his comments were quibbling and trivial. He was especially sensitive to Mayall's account of what he had told her concerning the impending loss of HCO support, and loss of office space, saying: "It was not a shortage of space that prompted his suggestion that the AAVSO should take stock of itself, but a desire for the welfare of an Association of which he was at the time a Vice-President."

Another sensitive point for Menzel was Mayall's emphasis on his tenure as acting director. He once again deflected the blame, noting that the implication was that the HCO Council "had acted arbitrarily and summarily." He responded by saying: "In fact we believe that our recommendation in this matter conformed to the general pattern set for us by the University. The *action* was taken by the University (which, therefore, should be blamed rather than the 'interim administration'). Had the University considered our recommendation out of line with its plans, it would not have taken the action."

The draft rebuttal was so poorly conceived that on August 21, his secretary Adams felt it necessary to send a telegram to Menzel: "Comments true but are you sure want to circulate in present form? Consider type campaign unworthy you. Adversary negligible and self-destructive."[43] On the same day that Adams sent her telegram, the new Harvard Corporation President, Nathan Marsh Pusey, informed Menzel that the Corporation had voted to appoint him Acting Director for another year. Pusey also informed him that the Corporation authorized the expenditure of $150 000 for the rehabilitation of the HCO physical plant – to be charged against the total appropriation of $300 000 that was made available to the Agassiz Station by Corporation vote of January 5, 1953.[44]

HARVARD'S LEGAL RESPONSES

The AAVSO problem, including the statements in Mayall's August 5 circular, were discussed in late August 1953 by Harvard Corporation officers Secretary Bailey, Provost Buck, and Dean McGeorge Bundy. They were concerned about the legal implications of any response that Menzel might make to Mayall's circular. Bailey reported that the AAVSO lawyer had met with Harvard's lawyer, O. M. Shaw. Shaw cautioned that any communication from Menzel that might be construed as official Harvard University policy should be reviewed by counsel. These warnings from the Corporation checked Menzel's impulse to strike back on his own. He did not issue a public rebuttal.[45]

Still concerned about the effect of Mayall's circular, Menzel warned Secretary Bailey on September 3 to expect letters seeking reversal of the Corporation decision, adding uncertainly, "I do not think that they are representative from the standpoint of the astronomical world at large."[46] Bailey wrote back a week later, also sounding concerned that the Mayall letter campaign was actually underway. "Something over twenty [letters] are now at hand," he wrote. "[P]ractically all of them are letters of protest."[47]

Bundy and Bailey decided on September 10 that a rebuttal might still be in order, as long as it was carefully prepared. However, Bailey warned Menzel that if he did prepare a rebuttal, he must send an advance copy to the University counsel for approval.[48] To a certain extent, the Corporation apparently did not have confidence that Menzel would say the right things.

Menzel prepared a more polished version of his rebuttal, without the minor issues and biased language, and more evenly discussed the major points of budgetary problems, the Ad Hoc Committee mandate, and the Pickering Memorial wording. It does not seem that a final version was ever sent to Mayall, but the revised draft was sent to Bailey on September 15 for his consideration.[49]

In the same September 10 letter, Bailey made a point of cautioning Menzel on some other issues. Contradicting Menzel's repeated assertions that it was the HCO Council that made the decisions about the AAVSO problem, Bailey noted that Harvard attorney Shaw emphasized that Menzel, "as the occupant of the Director's chair at the observatory" was the individual responsible "under the terms of our agreement with the AAVSO for determining when it may be advisable to take advantage of the escape clause in the vote establishing the Pickering Endowment."

To make it possible for the Corporation to review the policy decision reached some months ago by the Observatory Council, Bailey advised Menzel that it would be advisable for him to submit his own recommendation in writing, transmitting it to the Corporation through the Dean of the Faculty, who would add his own comment and recommendation. "It will, of course," he said, "be perfectly proper for you to explain that the decision which you have reached is concurred in by the other members of the Observatory Council. But the Council itself was not in existence at the time the Pickering Fund was set up and is not recognized in the agreement with the Association."[50]

On September 18, Bailey acknowledged receiving Menzel's draft rebuttal and sent copies to Bundy and Shaw, who would confer with Bundy. Bundy would then write an interim report to the Corporation on the subject, and "further action will follow after he has had an opportunity to explore this matter direct with you [Menzel]."[51]

One of the remaining sticky points for the Corporation officers was whether or not Menzel had given the AAVSO any notice to vacate their office. On September 3, 1953, Menzel's secretary Adams informed him that it would be "a matter of months" before demolition of Building A, adding that she was "doing nothing about informing Mrs. Mayall." "She was asked months ago to look for other quarters," she recalled, "and if this evacuation sneaks up on her now, it will be her own fault. I assume that you are going to take some clear and definite action soon, and we are just standing by. Mr. [Charles C.] Pyne [of Harvard Buildings and Grounds] did make the quip that Harvard had evicted people before this."[52]

On September 9, Adams telephoned and wrote to Menzel, who was in New Mexico, that Harvard's Administrative Vice President, Edward Reynolds, wanted to know what Menzel had said to Mayall about the loss of their office space and if any written notice was given to the AAVSO. Adams noted that Mayall saw contractors examining the HCO grounds and that Mayall then called Reynolds to say that she [Mayall] had not been told about the coming demolition. In response to Mayall's protest, Reynolds asked Adams and Menzel for a full copy of any letter relevant to the AAVSO which indicated that they could not expect permanent occupation of Building A.[53]

Menzel promptly wrote to Reynolds to explain that he had never given written notice to the AAVSO. He noted that loss of support mentioned in the HCO Council report might have implied loss of office space, but he denied telling Mayall in October 1952 that the AAVSO might have to vacate their headquarters. "Later on," he said, when it was determined that Building A could not be saved, "I told Mrs. Mayall that building A would go and with it the AAVSO Headquarters. I believe this definite statement, verbal with Mrs. Gaposchkin present, was in early April."

Menzel ended his letter – which was neutral in tone up to this point – with a touch of sarcasm, saying that he felt it "would be better if they went to another town." He said he had been told that "several observatories from smaller institutions might offer them a home." "On the other hand," he continued, "if say a dozen or so larger observatories directed by some of the astronomers who wrote those nice letters, were to offer adequate financial support, we should at least consider giving them a headquarters. (I admit that Mrs. Mayall's letter somewhat nullifies whatever charitable feelings I might have had toward the organization, of which I have been a member for more that thirty years and am now a V.P.)." He then added, "Mrs. Mayall's principal reaction to all this has been the statement 'What a pity to tear down so beautiful a building,' meanwhile complaining about the inadequate toilet facilities therein."[54] Five days later, on September 14, Menzel typed a memo to Mayall, stating: "Although we do not know the exact date of demolition, it will probably be done some time this Fall."[55]

The flap over whether or not Menzel had given notice to Mayall was, apparently, taken very seriously by the Harvard administration. Reynolds was neither mollified nor amused by Menzel's letter of explanation. On September 28, Reynolds admonished Menzel for not having gone through the proper channels with regard to the AAVSO. He mentioned that Mayall's lawyer had been asking the Corporation "lawyer questions," one having to do with the supposed demolition of Building A – Mayall said that she was not notified of the demolition until she saw contractors on site, taking measurements, and so on. The Corporation, Reynolds said, insists that "of course" she had been notified.[56]

NEARING A CORPORATION VOTE

As the AAVSO's annual meeting drew near, Harvard's Dean Bundy was tying-up the loose ends in preparation for the Corporation's vote to redirect the Pickering Endowment. He received a note from Shapley, which included a copy of his postwar development plans. It is not clear whether this is something Bundy had asked Shapley for or whether Shapley sent it on his own, perhaps in a last effort to speak in support of variable star work at Harvard. There is no mention in Shapley's postwar plan of cutting AAVSO work, and it does include re-instatement of the Milton Bureau work.[57]

A few days later, Bundy received a letter from legal counsel Shaw with a draft ballot for the Corporation on the Pickering Endowment. Shaw wanted to be sure that the wording of the document and the opinions of the principals involved were in agreement between Menzel, the Council, and the Harvard Corporation. In his draft, he had left unstated "the other astronomical research" to which the income of the major part of the fund was to be devoted. "Since the other astronomical research will undoubtedly be a far cry from research in variable stars, it seems to me to be desirable from a public relations point of view not to pin it down in this vote," he explained. Then, referring to the wording of the AAVSO's original gift to the Corporation, that the income could be diverted (if variable star work is "no longer an important field of research") – Shaw ended with: "I should think that you and the others involved might have some difficulty in reaching their conclusion."[58] This last sentence acknowledges, in effect, that there was a great deal of variable star work going on at HCO in spite of the assertion it was no longer of the same importance. Variable star astronomy was, if anything, of even more importance than at the time of the original resolution.

Perhaps seeing the end of a troublesome time, Menzel wrote to Board of Overseers member Lawrence Terry on September 23, 1953, noting that the AAVSO was "one major problem," along with Boyden, Agassiz, and the need for new buildings. "Things are looking up, in the broadest respects," he proclaimed; "I think that next year may be more interesting than the last, since a lot of the ground-breaking work is over."[59]

HCO COUNCIL REVISITS ITS DECISION

In accordance with Bundy's request, the HCO Council revisited the AAVSO problem, and Menzel reported to Bundy that "they see no reason to change their previous stand." Menzel quoted from the wording of the Pickering Endowment – keeping strictly to the text that would agree with the Corporation vote, no longer using his device of the inserted word "such."[60]

A few days later, Menzel wrote an unusual letter to Bundy. He reported that the National Bureau of Standards had cancelled its $4000 grant to the AAVSO's Solar Division. He quoted Rosebrugh, who told him that the Solar Division would continue on an amateur basis. Menzel cited this as a sort of proof that he was right to withdraw support from the AAVSO: "The foregoing is interesting in view of the cessation of the Harvard Support of the stellar side. The author [i.e., Rosebrugh], however, shows the right spirit by wishing to continue on an amateur basis. I have felt that the stellar observers could do the same."[61]

Menzel also felt it necessary to write to Harvard Secretary Bailey, sending him three copies of the letter of support he received from Henry Norris Russell in March. One of the copies was meant for legal counsel Shaw. Apparently, Menzel felt that Russell's letter would perhaps bolster the University's contention that variable star research as conducted by amateurs had changed since 1931. Menzel stressed the importance of Russell's opinions, saying, "I have regarded this evaluation very seriously." It is possible that, with the recent change in presidents from Conant to Pusey, and with the Corporation about to review last spring's AAVSO decision, Menzel was worried that the decision might yet be reversed.[62]

HARVARD MEETING WITH THE AAVSO COUNCIL

Even though the membership letter was out of the way, Mayall did not have time to rest. Not only was the annual AAVSO meeting approaching, but suddenly a number of issues were developing – or not developing – at once. On September 28, Mayall learned indirectly, through Menzel, that Dean Bundy had offered to meet with the AAVSO Council "to explain to them the current situation between Harvard University and the Association." Mayall wrote to Bundy thanking him for his kindness and scheduled the meeting for the afternoon session of the AAVSO Council on October 9.[63] Mayall excitedly reported to her friend Margaret Beardsley on the same day that "the new Dean of the Faculty of Arts and Sciences is coming up to talk to us and explain the University action. I feel it is quite a triumph for us, for the Dean to consider us of such importance. I am quite sure it is the result of the many letters the members have written to the Corp[oration] and also somewhat due to the action of our lawyers."[64]

Mayall did not learn until later that AAVSO Council member Domingo Taboada of Puebla, Mexico, had

Figure 12.3. Dedication of the Leon Campbell Observatory at the home of Domingo Taboada, Puebla, Mexico, October 1951. From left: the Mayor of Puebla, Harlow Shapley, Domingo Taboada, Watson Davis, and Luis Munch.

played a part. Taboada, according to Mayall, "asked Harvard to send a representative to the AAVSO Council Meeting." Taboada, a wealthy businessman and considered the leading amateur astronomer of Mexico, met Shapley, Menzel, Bok, and other American astronomers at the dedication of the Mexican National Observatory in 1942 in Tonanzintla. As a founding member of the Astronomical Society of Mexico, with AAVSO members Augusto S. Maupome and Luis Enrique Erro, he had joined the AAVSO in 1945.

In 1951 when Shapley attended the 400th anniversary celebration of the National University of Mexico to receive an honorary degree, Taboada invited him to give the inaugural address at the dedication of his newly built "Leon Campbell Observatory." Shapley gladly accepted the invitation. This was at a time when Shapley was encouraging HCO's involvement with the international astronomy community. Taboada had that kind of influence at Harvard, Mayall recalled, and "so they sent Dean McGeorge Bundy to represent Harvard and explain why we were being put out of HCO."[65]

But there was disappointing news as well. Mayall had also learned that the NSF would not consider her application for emergency support until January. "That means I may not have enough money to keep my one assistant (Helen Stephansky) through this winter," she wrote to Beardsley, "unless we can scrape up some funds somewhere." She had counted on NSF to carry them over during the emergency period until the AAVSO could get its endowment drive going.

Looking ahead to the inevitable day when the AAVSO would have to leave HCO, Mayall was worrying about the storage of the AAVSO's library, telescopes, and equipment, including the 6-inch Post refractor in the dome of Building A. She thought that Charles Smiley could probably take the telescopes to Brown University.[66]

Earlier, on August 18, Mayall had written to the AAS Secretary C. M. Huffer to report on the AAVSO's plight. On September 29, Huffer replied with a curiously disappointing letter that referred to the "delicate situation."[67] Huffer reported that the AAS Council "discussed your problems, and instructed the secretary to inform you that it is deeply sympathetic with all the problems of the AAVSO, and hopes that everything will be settled to your satisfaction. However, the Council, because of the delicate situation involving the Corporation of Harvard University, decided that it can take no official action at present." Somewhat guiltily, and with a bureaucratic spirit of passing the buck, Huffer added a personal note that every member of the AAS Council felt that the AAVSO should be continued. "I hope that the National Science Foundation will help you out of your difficult situation, and I feel that the Astronomical League should back you to a greater extent than the American Astronomical Society can do," he wrote. "We all realize that from your point of view the situation is quite critical, and under slightly different circumstances I am sure would be glad to back the organization. Perhaps in the future we can still do something." For a number of reasons (as will be seen), which included wanting to wait until the annual meeting had passed, Mayall did not reply to this letter until 2 months later.

Her next most important step was to inform the AAVSO's officers, Council members, committee chairs, and past presidents of the special meeting with Bundy, which she did September 30. She asked all the Council members to help in resolving the current crisis in the AAVSO by attending the "extremely important gathering" on October 9 in Building A. Mayall wrote that the University was "very much disturbed about the AAVSO problem" and was sending "Dean of the Faculty of Arts and Sciences, Dr. McGeorge Bundy (ranking official next to the President of Harvard University)."[68] Mayall also wrote to Carpenter, saying that Menzel had told her he would not be around for the meeting and she added that "My private opinion is that he has probably been told it is best to keep out of the affair as much as possible from now on. Any discussion should be between the AAVSO and the Corporation."[69]

It is not known how much truth there was in Mayall's opinion, but what she did not mention to Carpenter was that Menzel had told her that he would be attending a conference on radio astronomy in Ottawa while the AAVSO's Annual Meeting was taking place. Mayall did include this information in a letter to astronomer Peter Millman written the previous day. Mayall also added, "Menzel is certainly getting rid of all tradition. The latest thing to go is Pickering's Director's Round Table. Many of us felt like weeping when we saw it pulled apart on the ground outside. It's a sad world!"[70]

Mayall hoped that having a substantial number of past and present officers at the AAVSO Council meeting would ensure that Dean Bundy would be subjected to many hard questions. She had also "heard" that Bundy would help find office space for the AAVSO somewhere at the University until the time that their Pickering support would run out on July 1, 1954. On another Council matter, Mayall was dismayed that Smiley's Fundraising Committee had not yet taken any action and would have no recommendations to make at the Council meeting as she had hoped.[71]

To help prepare Bundy for the meeting, Menzel sent him a letter of facts and advice. In very even and controlled language, he gave the general background of the AAVSO's history, then added what he thought Bundy should say to the Council: highest regard for its work; hope for a successful fund drive; ready to give advice; and, although official ties are dissolved, hope that the AAVSO will "hold meetings at Harvard observatory from time to time." Menzel then added, incongruously, "one wishes to appear cordial, without implying that there is to be an open house on any occasion."[72]

Although keeping to the sidelines and never eager to give his own opinions about the AAVSO problem, Shapley nonetheless remained interested in what was happening. Always the overseeing director, he even gave Mayall a little tip on the eve of the special Council meeting: "For that Friday," he wrote, "you should have *handy* a statement of how long you have been at H.C.O – your 'war' work, *and* a bibliography." It is not known whether or not Mayall took this particular advice, but filed with this letter was a copy of Shapley's HCO Director's report for the year ending September 1952 (touting the AAVSO as "the most widespread and active international operation at the Observatory") and a typed list of past AAVSO presidents, with a penciled notation by Mayall, "11 out of 20 are professional astronomers." This seems to indicate that Mayall was not so much interested in defending herself before Bundy as she was in defending the value and qualifications of the AAVSO.[73]

As yet unknown to Mayall and the AAVSO Council, the redirection of the bulk of the Pickering Memorial Endowment to other astronomical research had at last been made formal "at a meeting of the President and Fellows of Harvard College in Cambridge, October 5, 1953." The resolution read as follows:

> After consideration and discussion of the recommendation and of the questions involved, upon motion duly made and seconded it was unanimously
>
> Resolved: That after consulting, through the President, the Acting Director of the Harvard College Observatory and the Observatory Council, this board is of the opinion that, owing to changed conditions, research in variable stars is no longer of the same importance that it was at the time of the establishment by this board of the Edward C. Pickering Memorial Endowment (1931).
>
> Voted, pursuant to the foregoing, that commencing July 1, 1954, and until otherwise ordered by this board, the income from $93,643.24 of the principal of the Endowment (the amount contributed by this board from University funds) be devoted to such other astronomical research as this board may from time to time direct, and the income from the remaining $6356.76 of the principal of the Endowment (the amount of the gift of the American Association of Variable Star Observers) be devoted to research in the study of variable stars under the supervision of, and in accordance with the terms of the gift by, said Association.[74]

AAVSO members could take little comfort in noting that the wording of the resolution matched that of the original 1931 agreement: "Research in variable stars is no longer of the same importance." It does not specify, as Menzel had, "(such) research as done by amateurs."

An interesting sidelight to this meeting was that Taboada of Mexico attended with his nephew acting as translator, signifying the importance he placed on the meeting and the interest he had in what was happening. Mayall later commented: "Well, Don Domingo dominated that meeting, and poor Dean Bundy was really taken aback. I don't think he's ever been quite so speechless as he was at that meeting.... I almost felt sorry for him. We gave him such a bad time." Unfortunately, Taboada's remarks were not recorded in the transcript of the meeting, and there is no record of what was discussed in a planned meeting between Mayall and Taboada beforehand.[75]

Bundy began the October 9 meeting with a lengthy account of the why and how the Corporation's decision was made.[76] He assured everyone that everything had been done legally, "that we have been at some pains to be quite sure that the words are quite legal words, and that is always the intent of the University." Next, he outlined that the real problem was a lack of money and that many departments throughout the University were going through the same thing. In the AAVSO's case, he stated that it had become clear "from the scientific standpoint and the size of [the University's] resources and its commitments that we must reorganize and get ready to bring it into fundamental first-rate quality, and President Conant instructed the Observatory Council that it should examine into the affairs of the Observatory and make recommendations as to those parts . . . which must now be regarded as less important, and which, therefore, could be given up in order that we could continue to do what we do in the best possible way." He referred to the recommendation made by Dr. Menzel "that the University could no longer properly and wisely continue to extend to the work of the AAVSO the support which it had been given over a very long period of time." He further stated:

> I should like to give my assurance that after talking to the [HCO] Council about the AAVSO work this is not the kind of decision made hastily and without careful consideration of what the AAVSO had done, and without a sense of the very great accomplishment that had been brought about. What we are dealing with is the fact that in 1953 and in a situation in which we have to ask, what things are most important, given resources and opportunities that we have. And it was impossible for the Council to believe that this kind of variable star work was of the same importance. That recommendation therefore came to the Administrative Head of Harvard Observatory and was transferred to the Administrative Committee and the Committee approved the recommendation.

Bundy went on to explain that with the arrival of a new University administration, after the recommendations had been approved, "it seemed to me important . . . that we should reconsider and re-examine." However, the re-examination, according to Bundy, was only at the level of the HCO Council itself. The new administration had asked for the HCO Council's opinions along with those of "other scientists to consider the whole question of the Observatory." The University asked the HCO Council to make a formal recommendation, which with Bundy's covering recommendation, had been forwarded to the University. "Dr. Menzel and I had a conference on the subject, and got the best kind of advice to the Council on this matter," Bundy said, then reported that the President and Fellows had met 4 days ago and formally voted.

At this point, Bundy began reading from the Harvard College resolution – presumably this was the first time that the AAVSO had learned of the formal declaration. He stopped to point out that it was at first resolved to return the original gift of $6356.76 to the AAVSO, but that he had had a better idea. "The corporations' view at the moment is to stay with the law and it occurred likely to me important, regardless of the AAVSO, to have the canny Harvard Corporation pay a rate of interest at a rate of almost 5%. This is the substance of the decision of the Corporation."

When Bundy agreed to answer any questions, AAVSO Treasurer Witherell asked for the names of those who were on the University's Ad Hoc Committee that had originally studied the problems of the Observatory. Bundy only stated that the committee chairman was J. Robert Oppenheimer. Bundy correctly added that the Ad Hoc Committee "made no specific recommendations in regard to the Association," and that the [HCO] Visiting Committee did not make any definite recommendation.

Margaret Harwood of the AAVSO Council made the point that the HCO Council had ignored some of the letters of support, to which Bundy only replied: "We don't like to wash this kind of business in public." Harwood then went into the issue of the original intent of the Pickering agreement, saying that she remembered that the money Shapley used to bring the AAVSO's contribution up to $100 000 was paid for "primarily AAVSO or for amateurs," but that the wording was changed to more generalized terms by the University. Bundy replied that "Dr. Menzel's position has been very helpful" on this point, that it was a "matter of opinion that [variable star research] is or is not of same importance."

The questions that followed covered much of the same ground from the past 6 months. Newton Mayall made the point that the agreement specified that the Director be consulted: "I personally do not recognize that there is a director of Harvard Observatory," he said. After pressing this point a few times, Bundy gave a somewhat exasperated opinion: "We have a recommendation from the HCO Council, a personal recommendation from the Director, and the Corporation has consulted with everybody it can consult, and has come to this conclusion."

AAVSO Council member Robert M. Greenley made a point about the monetary value of the observers' labor in making observations. Mayall stated that variable star work was always considered a part of HCO research. Ford remarked that the board's decision "clearly is opposed to [the] majority of astronomers," and he also mentioned that the HCO Council inserted the word "such" in their interpretation of the agreement. Witherell raised the issue of Harvard's moral obligation, to which Bundy replied, "That is really not the way that the Corporation does business."

With each point raised, Bundy repeated the terms of the agreement as stated in the University's final resolution, all revolving on the question of whether or not variable star research is of the same importance. From the AAVSO's viewpoint, and that of many professional astronomers worldwide, it was clear that variable star research was still of the same, if not greater, importance to astronomy. But from the University's point of view, it was never a question of whether variable star research was of the same importance to astronomy, but only a question of whether variable star research was of the same importance to Harvard University and the University's best interest.

Bundy's performance did not fit his position as Harvard University's number-two man. He avoided speaking with any authority, stating at the outset: "I am in a position of being an unconfirmed dean." He proclaimed his ignorance about the questions at issue – "You've got me," he responded to repeated questions about the importance of variable star research, "I don't know what kind of research is of the same importance," and he always deferred to the decisions and advice of the HCO Council, Menzel, and the University authorities while lamely appealing to the AAVSO's sense of fair play. "I'm bound to say that some of the questions put to me here indicate a lack of good faith," Bundy responded at the halfway point in the meeting. "We are dealing with the simple fact that we're not rich." After further points were made by Ford and Newton Mayall, Bundy simply turned off the discussion. "If you want me to I shall make it clear," he said: "The form of the vote cannot be changed, and [we] will not discuss it publicly. I'm happy to say that although Harvard University and the Harvard Observatory Council no longer can assign the Pickering Memorial fund . . . does not mean that we do not regard the work as important. . . . With this encouragement the Association can go on from here."

Despite this final declaration, the attendees did not let the meeting come to a close without a few follow-up questions. In response to Witherell's questions about razing the building, Bundy said, "We will house the recorder until the end of June." When Harwood asked if it would include the library, Bundy said, "We are not proposing to throw anything away." When Witherell asked about temporary quarters at the Observatory, Bundy said: "The association between Harvard University and the AAVSO is coming to an end. The intent has been clear for quite a while. I am in the position of trying to expel fog. Even if we cannot agree, we can understand each other." The meeting was over.

"But it didn't do any good," Mayall recalled, "except for the satisfaction, as far as we were concerned. We still had to leave HCO."[77]

RESISTANCE ENDED

The decision of Mayall and the AAVSO Council to resist the loss of the Pickering Endowment in 1953 was influenced by the strong attachment of the AAVSO to HCO. Mayall's view of Menzel as "interim director" is what informed her decision to wait out the crisis – even if the Pickering Endowment would be lost – and stay at HCO until a new director was appointed.[78] Mayall and the AAVSO Council had committed themselves to resist both the eviction and the appropriation of the endowment in every way they could, but to no avail. Their next steps would determine the survival of the AAVSO.

13 · In search of a home

Don't let the good work of the AAVSO cease! Nothing, no nothing, can take the place of a lot of well-established facts.

– Paul W. Merrill, 1953[1]

The 1953 Annual meeting of the AAVSO brought heavy demands on Margaret Mayall, the Association's director, and other officers for making arrangements for the usual Council business, the reports from committees and her recorder's reports, and the technical sessions with their scientific papers. Perhaps even more stressful for the Council were events looming in the Association's future – moving from its headquarters at Harvard College Observatory and finding the necessary funds for its continuance.

The Council meeting with Harvard University's Dean McGeorge Bundy brought no other hope that the Harvard Corporation's decision would be rescinded. Mayall's position would only be supported for 1 year, up to June 1, 1954, from the Pickering Memorial Endowment. After that date, Harvard College Observatory would divert the money to other programs. Even more prominent was the date June 1, 1953, when all other assistance from Harvard would be discontinued.

Not only did Mayall and the officers face these dates, but there was urgency in obtaining funds from other grant agencies, particularly the National Science Foundation. They approached fundraising with trepidation, not knowing if their disagreements with Harvard University would bring a cloud over their heads. The prospects of moving their offices to Boston University seemed hopeful, but they were still waiting a final approval from the Boston University officers. In the midst of these difficulties, an emergency with the Solar Division meant Mayall had to consider a whole new procedure to handle the reports from that group. Thus multiple crises fell upon her shoulders in 1953, difficulties that tested and proved her resilience.

LOOKING TO THE FUTURE

The special meeting of the AAVSO Council and other officers with Harvard University's second-in-command, Dean McGeorge Bundy of the Faculty of Arts and Sciences, was a significant but small part of the weekend's activities. This was, after all, the Annual Meeting of the AAVSO for 1953, which meant that the usual Council business would be conducted, committee and recorder's reports would be read, scientific papers would be presented, and – true to AAVSO tradition – acquaintanceships would be established and renewed.

The invited speaker for a pre-meeting evening lecture was Dr. Albert P. Linnell, assistant professor of Astronomy at Amherst College in Massachusetts, who spoke on "Electronics and Variable Stars." Dr. Linnell not only described photoelectric photometry, but he also demonstrated how a compact photometer and amplifier worked and stressed the importance of making photoelectric observations of variable stars. The informative talk was timely and, as AAVSO Council member Jocelyn Gill reported, "in line with the note of 'looking to the future.' "[2]

Amateur interest in photometry was beginning to grow – due in part to the postwar availability of electronic tubes and components. Linnell's talk generated much interest at the meeting and inspired several to consider photoelectric observing.

Mayall certainly appreciated Linnell's encouragement of the amateur's ability to use photometric equipment for variable star observing. His talk provided a much-needed antidote to discouraging remarks by Donald Menzel, Henry Norris Russell, and others who claimed it was beyond the reach of the amateur. Also, Linnell's talk echoed Paul Merrill's repeated calls for more amateur photoelectric observers, as well as Merrill's remarks about the greater degree of accuracy of variable star observations gained by this method.

In her *JRASC* "Variable Star Notes" for that fall, Mayall publicly endorsed Linnell's lecture, writing: "Electronic equipment will make possible a valuable expansion of A.A.V.S.O. programs, and will enable us to add eclipsing stars and other variables with small rapid changes." Yet she also hastened to add a word of support for the AAVSO's primary observing programs, saying, "but it can never replace the long, continuous visual observations needed for the long period variables."[3]

There was also cause for celebration: Curtis E. Anderson of Minneapolis, Minnesota, brought the Association's total observations to 1.5 million on June 10 with an observation of Nova Her 1934 (DQ Her); Newton Mayall reported that the final layout of 53 plates containing 387 mean light curves plotted by David Rosebrugh for the Campbell memorial volume was complete and ready for the printer; and Dr. Martha Stahr Carpenter of Cornell University became the only president in AAVSO history to be elected for a third consecutive term. Writing in *Variable Comments*, Jocelyn Gill recognized her re-election as "a tribute to her inspired leadership and devotion to the interests and work of the Association through this difficult period."[4]

Continuation of AAVSO work

In her 1952–1953 Annual Report, Mayall gave no direct mention of the HCO problem at all; her introductory paragraph saluted the past and positively marked the continuation of the Association's work.[5] Given the uncertainty and stress of the previous year, Mayall's report was, in contrast, succinct and purposeful, placing some emphasis on the donations and gifts to the general fund. But her main topics were the business of the Association, the observers, and the observing programs. The only direct mention of what had transpired at Harvard came when President Carpenter asked Secretary Ford to summarize the current financial situation of the AAVSO, particularly the outcome of the Council's special meeting with Dean Bundy. Concerning the special meeting, Secretary Ford simply reported that it "resulted in no developments not already known to the Council, and also to the general AAVSO membership following the Recorder's circulated letter."[6]

After her report, Mayall spoke briefly about the AAVSO's future. Among the items covered were the possibility of having Boston University as the AAVSO's new headquarters, NSF funding for the coming year that would probably be approved, and the appointment of a new Fund Committee for the long-term project of raising a new AAVSO Endowment Fund to replace the Pickering Memorial income. Ford noted: "These announcements precipitated favorable comments and discussion by Dr. Smiley and others."[7]

Value of long-period variables

There was one other important item on Mayall's meeting agenda: she wanted to clear the air about any doubts on the value of continued observations of long-period variable (LPV) stars. To this end, Mayall invited Dr. Paul Merrill of Mount Wilson Observatory and Dr. Dorrit Hoffleit of HCO to present papers on the value of mean light curves of LPVs, from both the astronomical and statistical perspectives. Writing on behalf of Mayall, Hoffleit had asked Merrill to emphasize the value of amateur observations and to discuss future needs of professional astronomers. Hoffleit also asked him to speak on the value of mean light curves, specifically mentioning "one very active member" (David Rosebrugh) who had been "quibbling loudly" that mean light curves were worthless. "He obviously understands nothing about probable errors," she wrote. "Perhaps you would care to include a few remarks bearing on this situation."[8]

Merrill, who was unable to attend the meeting, asked Hoffleit to read his paper in his absence. In his regrets to Mayall, he wrote, "I certainly hope the future of the AAVSO will be decided wisely. Its future possibilities as well as its past accomplishments deserve careful and friendly consideration."[9] In his paper, Merrill acknowledged the valuable contribution that AAVSO observers had made toward the knowledge of LPVs and stressed the value of, and need for, mean light curves, predicted curves for planning observing programs, and complete light curves. He addressed a number of questions concerning LPVs: the stability of light cycles, changes in period length, differences between types of variables, correlation between spectral type and light curve characteristics, the relationship of mean curve deviations to the star's physical properties, and many others. The tone and even some of the language of Merrill's paper recalled the sense of urgency in Friedrich Argelander's nineteenth-century article, "Variable Stars." Merrill concluded his paper with this Argelander-like plea:

> To learn about long period variables by detailed and prolonged studies of their light curves and of their spectra is certainly a hard way, but any other way will be longer and harder and far, far less certain. Don't let the good work of the AAVSO cease! Nothing, no nothing, can take the place of a lot of well-established facts. Theories without sufficient facts are sophomoric exercises that keep the tools of analysis free from rust and ready for use. Theories based on essential facts, capable of coordinating existing knowledge and of predicting quickly and correctly many results from a minimum of data are monuments of science and guideposts of advancing civilization.[10]

After the meeting, Hoffleit wrote to Merrill to tell him that his paper "was received with great enthusiasm . . . ten hands went up at once to ask where to obtain copies."[11] Merrill's paper, "Some Questions Concerning Long Period Variables," would later be included as a foreword to Leon Campbell's *Studies of Long Period Variable Stars* prepared by Mayall. Hoffleit followed with her presentation on the value of LPV observations, "On Errors of Mean Light-Curves." She responded directly to questions about the value of LPV visual observations raised by Rosebrugh in 1949 and in 1952. Similar questions had been raised at the 1952 IAU conference and in those main arguments by Menzel when he was arguing against supporting the AAVSO because amateurs were incapable of contributing to any other form of variable star knowledge. Rosebrugh was even expressing his doubts about the irregular appearance of many of the mean LPV light curves he had been plotting for Mayall, although she had told him – correctly – that some of the irregularities are "true reflections of the stars' variations."[12]

With such questions continually arising, Mayall asked Hoffleit to feature in her talk the analysis of light curves. Hoffleit mentioned this in her letter to Merrill: "The differences from the mean from cycle to cycle are apt to be so different that the joy of observing is greatly enhanced precisely because of the unpredictable differences from the means. If this were not so, too many might, like our present acting administration [i.e., Menzel], argue effectively on the futility of observing any long-period variable that has already been observed more than a few cycles."[13]

Hoffleit also wanted to dispel the growing common perception that visual observations of LPVs were of little use owing to their inaccuracy and scatter. She showed how the value of such observations in preparing mean light curves depended on "the degree of accuracy intrinsic in the observations themselves" and how the curves were to be used. She concluded that the systematic differences between individual observers are not of sufficient magnitude (rarely exceeding $\pm$ 0.1 mag.) to justify the reduction of those differences to one standard system. Hoffleit also stated that "the application of spurious systematic corrections is both time-consuming and can cause more eventual harm than good." She noted that the procedure adopted by Campbell for his *Studies of Long Period Variables* seems the "most desirable from the standpoint of attainable accuracy of the accumulated observations, reasonable economy and efficiency in compromising time expended with accuracy and usefulness to astrophysicists."[14]

The stimulating meeting concluded the next evening with the usual banquet, followed by Harlow Shapley's traditional "Highlights of Astronomy" speech. "Our distinguished past-president and enthusiastic supporter," Ford reported, "delivered a Highlights speech second to none in the long list of his AAVSO dinner speeches, and was greeted with a spontaneous rising applause reminiscent of the ovations accorded Mr. Leon Campbell on his retirement in 1949."[15]

Proposal for relocation

In the following week, Mayall and her husband prepared a four-page appeal to Dean Duncan Macdonald at Boston University regarding a prospect for a new headquarters there. They had not yet received any definite word from either Macdonald or Boston University. Their appeal included a brief history of the AAVSO and its observing program, a short account of its present problems, its organizational structure and officers, and its services and value to amateurs and professionals.

The authors framed the primary problem – a need for office space – in the language of an investment broker: "The support of the AAVSO is not unlike the acquisition of assets of a small but growing concern by another; they both may be enhanced by the resulting closer association. Therefore, what assets has the AAVSO to offer that might be attractive to Boston University?" Along with a general description of the AAVSO's make-up and activities, they stressed that "the AAVSO is not a loosely knit group of amateurs, but rather a legally responsible and well organized body whose activities are directed jointly by professionals of known repute and amateurs of ability; and its data is correlated, studied, and made available by its own recorder, – a professional astronomer. This is an asset." The paper went on to list other "assets": having the AAVSO on campus; having young people involved in the AAVSO who would follow through and take astronomy courses and perhaps enter science careers, as others had done (Menzel is listed among the examples); community involvement that would focus attention on the university; AAVSO's connections with other national and international programs; and its physical assets, including its 40 years of observations, its publications, library, and equipment.

On the funding side, the Mayalls made it clear that the AAVSO would be launching its own fundraising effort with a goal of $500 000 in not more than 5 years. "This however," the authors continued, somewhat hedging their bet and adding a dash of caution learned from their recent experience, "does not preclude the possibility of the AAVSO fund raising in close cooperation with its associated university. In fact, due to its present position, it will be advantageous to all concerned to work together on such a program and take whatever legal steps are necessary to insure the advantages each is to derive and to provide proper allocation of funds if dissolution were ever found to be required." The appeal, which was scheduled for delivery October 16, concluded with a glowing promise:

> The AAVSO is eager to associate itself with Boston University not only because of the opportunities offered by B.U. but also because of the contribution it can make and the active part it can take in revitalizing the department of astronomy, and helping in its future growth. It is our sincere wish that Boston University can house the organization and work out a mutually beneficent association.[16]

With that task completed, Mayall responded to the September 29th letter from C. M. Huffer, secretary of the American Astronomical Society, who had written that the AAS was unable to help the AAVSO due to the "delicate situation" involving the Harvard Corporation. Now Mayall painted a rosier picture of the AAVSO's situation, perhaps hoping the AAS would recognize that there was no longer a "delicate situation" between the AAVSO and Harvard and therefore would be more willing to help. With cheerfulness she elaborated about the AAVSO's successful meeting, noting that "the situation with Harvard was cleared up" when Dean Bundy came to the AAVSO Council meeting and reported on the Corporation action. She noted that "all official connection" between the AAVSO and the University would end on June 30, 1954, and that they had offered them housing until that time, "even though the building should be torn down."

In turning to the AAVSO's future plans, she noted they were exploring the possibility of a future association with Boston University. "It sounds extremely attractive for both sides," she wrote. "Of course we will still have a financial problem, but that will be much simpler to solve if we are connected with an institution that shows such remarkable growth over the last few years, and has such plans for a brilliant future as Boston University."[17]

A few days later Mayall received a letter from astronomer Peter M. Millman of Canada's Dominion Observatory, who sent a copy of a letter he had written to the Harvard Corporation in support of the AAVSO. He realized that it was too late, having been away on an extended trip, so his eloquent letter could have no effect on decisions already made. Mayall, nonetheless, kept it on file with the other letters of support. Millman aptly stated the role of the AAVSO:

> . . . throughout its existence, performed a signal and unique service to astronomy and, in a broader sense, to the whole field of natural science. It has brought together professional and amateur scientists and united them in a common organized effort based on the idea of learning more about the universe within which we live. The indirect results of this association have been wide spread and I could quote specific cases where we are now reaping the benefits of this in some of our most recent technological advances. The importance of interesting an influential group of laymen in worthwhile scientific endeavour is all too often forgotten. . . . The huge number of observations contributed by this Association is unique and could

> never have been duplicated by the professional astronomers. The variable stars probably hold the key to the secret of stellar evolution and, hence, to the general evolutionary process of our whole universe. The American Association of Variable Star Observers has provided the main body of observational data for one great group of these variable stars.[18]

FINAL WORDS

McGeorge Bundy apparently easily recovered from the contentious meeting with the AAVSO Council and other interested individuals earlier that fall. Bundy wrote to Menzel on October 13 that "our main battles with the AAVSO are ended . . . and it was understood at the end of the meeting that this was a question which is not going to be reopened." He also reported that Mayall had indicated some possibility of finding a home at Boston University.

In a cooperative tone, Bundy said he told her that such a solution would be "wholly satisfactory" to them, and that "should it work out that she is able to move her office at the time she has to leave Building A, we shall certainly be prepared to cooperate by maintaining our commitment to pay her salary through June 30 even if she is no longer on the Harvard premises." He further stated: "My basic feeling in this case is that since it is now clearly understood that the AAVSO must be off on its own by the first of July next, we should not hesitate to cooperate in any way that may help them to get settled in some new location."[19]

Bundy's statements suggest none of the animosity toward Mayall or the AAVSO that was so often expressed in comments made by Menzel, his office staff, and by members of the HCO Council, and that was so often perceived by Mayall and the AAVSO's supporters. Menzel was well aware that some in the astronomical community might question the wisdom or motivation behind his recommendations and Harvard's decision to accept them. He proposed to Bundy an idea for damage control:

> If worse comes to the worst, it seems to me that one of the lawyers could provide a one-paragraph opinion of the legal side, to be followed by a paragraph from you of the moral side. This sent to the President of the American Astronomical Society, for use in any way that he deems desirable, would probably have a quieting effect. But it may not be necessary.[20]

On October 26, 1953, Mayall received a letter from the AAVSO's legal representative, Edward Boit of the law firm Ely, Bartlett, Thompson, and Brown. He enclosed the original letter from the Harvard Corporation's lawyer, Shaw, of Ropes, Gray, Best, Coolidge, and Rugg. Boit had earlier written to Ropes-Gray with questions about the Pickering Endowment and the AAVSO. Shaw's response to Boit's inquiry was to send a copy of the "formal resolution and vote adopted October 5 by the President and Fellows" of Harvard College, which stated that the income from the Pickering Endowment would be devoted to other astronomical research, and that only the annual income from the AAVSO's original gift would be paid to the AAVSO. The letter from Shaw, dated October 9, stated that the Harvard College resolution, "was taken upon the recommendation not only of the Acting Director of the Harvard College Observatory and of the Observatory Council, but also of the new Dean of the Faculty of Arts and Sciences, who made a thorough study of the entire problem. Before it acted, the Corporation had seen and considered the views expressed by many members of the AAVSO as to the continued importance of the search in variable stars."[21]

"That," Shaw seemed to be saying, "was that."

SOLAR DIVISION CRISIS

During this most stressful time for Mayall – the summer and fall of 1953 – a second crisis developed that she also had to confront, control, and resolve: the near-collapse of the AAVSO Solar Division.

Throughout its brief history, there was never any doubt that the AAVSO Solar Division did useful work and deserved whatever monetary support it could find. In the June 1952 AAVSO *Solar Bulletin*, Chairman Neal Heines proudly quoted from a letter of commendation written by Alan H. Shapley of the Central Radio Propagation Laboratory, which noted that the Division "does play a role in ionospheric physics and radio communications which is highly significant."[22] However, the Solar Division was not so much a part of the AAVSO as it was a part of Neal Heines.

Figure 13.1. Neal J. Heines, Chairman of the AAVSO Solar Division 1944–1955.

Dependent as it was on yearly government contracts, and never having a physical presence at AAVSO Headquarters, the Solar Division under the enigmatic Heines nevertheless led a productive, if hand-to-mouth, existence. From 1946, when Heines declared that more than half of his National Bureau of Standards (NBS) contract would be his "chairman's salary," until mid-1953, Heines' full-time and only occupation was as chairman of the Solar Division. The problems that arose during the summer and fall of 1953 were triggered by the loss of funding, but the root cause had to do with how Heines viewed himself in his role as the Division chairman.

Beginning in April 1950, Alan Shapley of the NBS and others made repeated hints and predictions that funding for the Solar Division would be lost. In June 1953, Harlow Shapley dashed off a quick note to Mayall that Alan Shapley had told him by telephone that as of July 1, Heines would not be on the payroll. Alan Shapley ascribed this to government cutbacks in general, and NBS cuts in particular, all apparently a result of the continuing Korean war crisis. Harlow Shapley darkly concluded: "It looks as though the solar work will have to be settled along with variables – perhaps perish."[23] Mayall received word from Heines on July 7 that NBS "has terminated our contract," adding, "I have arranged a temporary fund drive for the Solar Division."[24]

Heines, in fact, had already drafted a letter to observers concerning what he called his "continuance fund" for the Solar Division. He sent a copy to Alan Shapley for his comments but did not send it to Mayall, nor ask her approval to write a fundraising letter. Alan Shapley immediately discouraged Heines from starting such a fund drive, advising him that the solar program should be suspended until sufficient funds could be organized. He added that the NBS position had worsened during the last 2 months and reminded Heines that the sunspot program could not count on the continuance of the pre–Korean war plans of the NBS. "We have the impossible job of getting the most and the most important work done while hurting as little as possible," Alan Shapley concluded, "Nevertheless, a lot of people are being hurt; I know of no escape."[25]

Despite this advice, Heines sent out his fund appeal with the July 1953 *Solar Bulletin*, announcing the loss of the 2-year NBS contract and asking for pledges. He also indicated that virtually all work of the Division would cease.[26] The July 1953 *Solar Bulletin* was the last to be published under Heines' editorship, and it would not reappear until January 1954.

It was becoming obvious that Heines had fallen into despair over the loss of "his" NBS contract. "Unlike yourself," Rosebrugh wrote to Mayall, "he has had a sinecure, and cannot lose face by admitting it." However, seen from another perspective, it is likely that Heines' despair had less to do with the loss of money and more to do with the loss of professional recognition and esteem which that contract implied. No one seemed to know how Heines was making a living now. "Neal has a job in some sort of work," Rosebrugh wrote, "variously reported as 'Industrial Air Conditioning Sales' and 'shift work.' He is a hard man to pin down."[27]

Alan Shapley wrote to President Carpenter (via Mayall) on October 2 stating "on record" that the NBS hoped that ways could be found "to continue the Solar Division as an active, interested and effective group." He was clearly worried, however, that the backlog of unreduced observations would increase in the meantime. He stressed that it was important to resume the reductions and the publication of data tables and the *Bulletin*. "The assistance we have given to this work in the past indicates our interest," he concluded, "Withdrawal of this assistance was forced on us by extraneous considerations which I am sure you can appreciate."[28]

On October 10, 1953, Mayall responded to a letter from Martin Schwarzschild, who had written 2 weeks earlier on behalf of the IAU. Schwarzchild had sent a form letter asking IAU commission members to evaluate their qualifications for serving on a commission and asking them to limit their membership to three separate commissions. After confirming her preference to remain with the Variable Star and Stellar Spectra Commissions, she noted that Neal Heines was no longer active in any astronomical pursuits and should no longer represent the AAVSO on the IAU's Commission 10 (Photospheric Phenomena). Mayall added that a new chairman would be named soon, saying, "I believe the work of the AAVSO in producing the American relative sunspot numbers is important."[29]

Faced once again with a rising storm of uncertainties, Mayall took steps to reach a solution. She and her husband Newton visited Heines on November 29 in New Jersey. They told Heines a special Solar Committee created by the AAVSO Council agreed to have him remain as chairman while some of them took over the work "for this emergency." Heines approved of the idea and agreed to send the papers. Back in Cambridge on December 1, Margaret Mayall wrote to the special Solar Committee members to coordinate the times and places that they would meet and perform the computations.[30] Then on December 7, Mayall sent a letter to all solar observers, not mentioning that the backlog in the reductions was the result of Heines' refusal to work without a salary. She simply noted that with the loss of NBS funding, Heines no longer had time to devote to Solar Division work and asked that all reports and correspondence be sent directly to AAVSO Headquarters where they would be forwarded to the appropriate persons.[31] In taking this step, Mayall recognized that an interruption in the flow of information such as Heines caused could not be tolerated and repeated; the procedure for handling reports had to be centralized. Given the division of labor for the reorganized Solar Division, centralization was now necessary. Lastly, Mayall's letter also had the effect of recording the fact that Heines had promised to release the unreduced reports to the Solar Committee.

In a flurry of letters over the next several days, Mayall and the Solar Committee members coordinated their schedules to meet for the reduction session, which would be held December 12 and 13 at Rosebrugh's home in Meriden, Connecticut. Those attending the 2-day session were Margaret and Newton Mayall, Harry L. Bondy, Bill Reid, Heines, Alan Shapley, and Rosebrugh. In four pages of notes, Mayall jotted down ideas for revising the sunspot report form; a note about how dues payments by solar observers might help to cover postage costs; a proposed editorial board for the *Bulletin*; a question about the role of Heines' supplementary solar projects – Gleissberg Foreshortening, Granulation, and Migratory Birds – in future Solar Division work; an idea for providing solar observers with their own "sunspot number reductions chart," and an outline of the reorganized system for handling reports and correspondence.[32]

As soon as she returned to AAVSO Headquarters, Mayall typed a short notice for the February *Sky and Telescope* that the AAVSO Solar Division had been reorganized and the reduction of the American sunspot numbers had been brought up to date by a committee composed of Heines, Alan Shapley, Rosebrugh, Bondy, Newton Mayall, William Reid, and Margaret Mayall. She specifically added, "All correspondence and observations should be sent to the Recorder of the AAVSO, until further notice."[33]

By the end of the month, several solar observers answered Mayall's December 7th letter that they were relieved that the solar program would continue. Mayall responded to one writer, Ralph N. Buckstaff, giving a short summary of how the Solar Division would be operating from that point on. All reports and correspondence would come to her, and she would distribute them to the proper persons. Rosebrugh would do the reductions of the standard observers, and Reid would take care of the observers who were in training. Bondy was preparing a bulletin for the solar observers and would be assisted by an editorial committee including Alan Shapley, Rosebrugh, and Mayall. Heines remained as chairman in name only since he was doing nothing except acting as an advisor. All financial matters would go through the AAVSO Treasurer Witherell, who was setting up a special fund for solar expenses. Mayall noted that they had already collected $15 in donations for the *Bulletin*. The weekend at Rosebrugh's house had been very profitable since they completed the reductions for July through October – all that Heines brought along.

Mayall could not resist a comparison of the minor, finished Solar Division crisis to her major, unfinished AAVSO crisis. She confessed, "Right this minute, I feel very low, for nothing seems to be clicking. The B[oston] U[niversity] move has not been cleared, and our funds have not come through from Washington, as yet. And we have to move out of our building by Jan. 4 or 5. Something has got to work fast, before long."[34]

IN SEARCH OF STABILITY

As uncertain as were the prospects of short-term NSF funding, a home office at Boston University, and the survival of the Solar Division, the most important problem now facing the AAVSO was whether or not the funds for an endowment could be raised.

At the October 1953 Annual Meeting, the AAVSO Council appointed Secretary Ford and Dr. Charles Smiley of Brown University to be the Fundraising Committee's co-chairmen. Other appointments to the Committee were Margaret Beardsley (publicity), Percy Witherell (finance), Mayall (scientific), Cyrus Fernald (business), Francis H. Reynolds (academic), and Martha Carpenter (ex-officio).[35]

Ford began working on funding as soon as he returned home from the Annual meeting. "My biggest worry," he wrote Carpenter, "concerning the Fund drive and literature directed toward raising money is the following: How are we going to explain why Harvard kicked us out? After all, everybody knows (or at least thinks) that Harvard has more money than any other school in the world, so – if Harvard thinks we're useless, why should anyone else support us? Of course we can say it's an economy move on Harvard's part, but the trouble is that it actually is not an economy move – the money is simply going to [be] used for something else, and that is pretty common information. Don't think that I am discouraged, tho[ugh]. We will make it somehow even if it takes years."[36]

Ford now had time to rewrite his March 1953 draft letter for the foundations. Over the next few months, successive revisions were sent out for comments from the members of the Fundraising Committee. He wanted to pursue one large source of funds rather than a "piecemeal approach" favored by Newton Mayall to gather small amounts from many sources, which Ford thought would take an inordinate amount of time. He wanted to concentrate the effort on an appeal to the Ford Foundation.[37]

An important part of the fundraising plan was building a list of important scientists, educators, business people, and dignitaries who would be willing to lend their names to the fundraising literature as sponsors. Dr. Paul Merrill of Mount Wilson Observatory was the first such person that Ford, Mayall, and Carpenter agreed should be approached. Ford asked Merrill on November 5 if he would serve as the honorary chairman of the Fund Committee and noted the prestige connected with Merrill's name and his "position as the leading professional astronomer making regular use of AAVSO data."[38] Merrill graciously accepted the appointment, although he flatly stated he was unwilling to participate in any of the fundraising committee's work.[39]

By the last week of November, Ford had completed his first revision of the letter for the Ford Foundation. Accordingly, Ford emphasized the humanistic value of the AAVSO, "encouraging world-wide international cooperation in a project of mutual endeavor on a high educational plane"; the value of the AAVSO's work to "the pure science of astronomy"; the modest funding required to continue and expand the AAVSO's operations; and the fact that the AAVSO is "in no sense an applied science project capable of appealing for permanent government-sponsored support." The main justification for the AAVSO's work, according to Ford, was "international cooperation."[40]

Mayall, Smiley, and Carpenter thought that the letter was generally a good one, although they agreed that it should be kept brief and positive and that the mention of AAVSO's difficulties with Harvard should be kept to an absolute minimum. Carpenter began working on a short, informal fundraising letter that would be sent to individuals. This letter stated a goal of $250 000. Mayall drew upon some of her local acquaintanceships. She thought it would be fairly easy to find a contact at Ford through two trustees in Cambridge, Judge Charles Edward Wyzanski, Jr., and Donald David, dean of the Harvard Business School.[41]

Dealing with uncertainties

Time now was running short. Mayall was concerned that she had not heard about the NSF grant. Despite having been told that the NSF would not consider her proposal until January, she nonetheless wrote to Watson Davis in Washington on November 25 and asked him to look into the status of the proposal. She noted that, "in the last week or so I have been asked rather pointed questions by several people."[42]

On the same day, Mayall also wrote to Martin Schwarzschild at Princeton University, asking if he knew anything about the grant.[43] She recounted that Dr. McMillen at NSF had called her about a month ago "to ask if I would be willing to accept $10 000 for

1 year instead of the $23 000 for 2 years that I had requested. My answer, of course," she wrote, "was yes!" She wrote that he told her "no final word could be had until after the meeting of the committee in the middle of December."

She further explained her predicament: "I felt from the conversation with him that it was fairly certain we would get the $10 000. But in the last few days I have been asked questions by several people that make me wonder if anything has gone wrong with our request. It is all very vague, and I hate to ask McMillen about it." She had also found out that McMillen had talked with Menzel before calling her. Menzel had then told her that "McMillen had asked him if he would consider it embarrassing to him or to the Harvard Observatory if the NSF gave us a grant. Menzel told me his answer was *no*."

Mayall wrote that she also told McMillen she expected the AAVSO would be moving its headquarters to Boston University. Duncan Macdonald, she wrote, "offered us space in his Physical Research Laboratory, and is very anxious to have us there. We do not have the final OK from Boston University officials, but I hope to get it soon."[44]

Waiting for BU response

Mayall was still hopeful that the new AAVSO Headquarters would be in the Physical Research Laboratory at Boston University (BU), but she did not yet have any confirmation.[45] Ford wrote to AAVSO member Claude Carpenter on December 3 that the arrangement had been discussed through a personal contact with Macdonald. "The whole thing has possibilities, since B.U. is a growing institution," Ford wrote. Mayall had a talk with BU President Harold C. Case, who, Ford noted, "wanted to make something of their astronomy dept. and felt the prestige of the AAVSO would be a step in that direction." So, Ford speculated, "possibly there will be some money from B.U. too, as time goes on."[46]

The Macdonald contact is intriguing. Ford seemed to suggest that there was a personal connection between Macdonald and either Mayall or himself. When the HCO crisis first became apparent, Ford made two trips to Boston to speak with Macdonald sometime in the period from October to December 1952. At any rate, Macdonald must have been acquainted with Mayall, and possibly even Ford, when he was part of the HCO staff.

Macdonald's Physical Research Laboratory began in 1946 as the BU Optical Laboratory. BU formed the optical lab around a nucleus of members of the Amateur Telescope Makers of Boston (ATMoB) and others led by HCO optical expert Dr. James G. Baker, who had contributed to optical research during the war. When HCO discontinued the program after the war, the Army asked BU to begin its own research program using the surplus HCO equipment and HCO researchers. Macdonald, who had been one of Baker's students, was among the researchers moving to the BU Optical Laboratory, renamed the Physical Research Laboratory (PRL), with a staff of more than 100 persons. Their fundamental purpose was to conduct secret research for the US government relating to aerial reconnaissance. The PRL also had a public face: it publicized its work on an aerial panoramic camera, for example, and BU's alumni magazine published stories about the PRL's work on a "spherical shell" aerial camera.[47]

It is also intriguing that BU President Case met with Mayall and expressed his desire to develop the school's astronomy department, citing the prestige that the AAVSO would bring to that program.[48] This would have been an entirely plausible development: Case was in the second year of his presidency and had begun a postwar expansion of the university, which included establishing a School of Fine and Applied Arts, a College of Engineering, and an African Studies Program.[49]

The imminent move

The long-expected "eviction notice" was almost anticlimactic when it arrived on December 8, 1953. Donald Menzel sent a memorandum addressed "AAVSO (Att. Mrs. Mayall) / Subject: Moving." Menzel wrote: "We hope to start tearing down Building A as soon after the first of January 1954, as the contractor can undertake it. We should appreciate it, therefore, if you would plan to make the move by that date."[50]

This memorandum took its place behind Mayall's other concerns. "I don't know yet whether we will move Christmas week or wait until Jan. 4," she wrote to Ford on December 9. "Of course," she continued, "I don't know where we are going, either!" She had just talked with Macdonald and found out that BU had not yet made a decision. "They evidently do not want any financial responsibilities," she noted. She would have to wait a week for a final answer, she told Ford, and added, "We are not worrying (we don't have time for that). . . . " She also said she had an offer of optical expert Jim Baker's office in Harvard Square and storage space at Brown University in Providence. Still confident, she remarked, "I am quite sure we will get around $10 000 from the N.S.F. and hope to hear from them next week."[51]

In October, Mayall had felt fairly sure that the offer from Macdonald would be realized, so she visited Bundy at that time to tell him so. Bundy reiterated Harvard's commitment to paying her salary through June 30, 1954, even though she might no longer be on campus. At no time did Bundy withdraw his offer to house the AAVSO up to July 1, 1954, but now that the prospect of the AAVSO actually moving away was on the table, both Bundy and Mayall had, in effect, stepped back from that housing commitment. Although Mayall was now uncertain about the BU offer, her mention of temporary office space from Baker and storage space in Providence seemed to indicate that she no longer wished to stay at Harvard under any circumstances – and perhaps it was now also a matter of pride that kept Mayall from asking for Bundy's help.[52]

To further complicate matters, Harvard Observatory would cease paying a salary to Helen Stephansky, Mayall's assistant, in 1 week. Mayall quickly made arrangements with a local business to pay Stephansky a full-time salary for 1 month, presumably augmented with funds from the AAVSO, or possibly out of Mayall's own salary, allowing Stephansky to work 3 days a week for the business and 2 days for the AAVSO.

On December 14, less than a week after speaking with Macdonald, Mayall wrote to assure him that the AAVSO "does not ask or expect any financial help from Boston University." She added, "The only thing we ask is space for our headquarters. Of course, we would be glad to have it heated and lighted, but if necessary, we could probably pay something toward that." She even offered to have their lawyers "draw up papers, agreeable to the University, to the effect that we are not asking financial help." She ended by stating that "the AAVSO should be financially independent of any institution," adding that "this does not affect our willingness to cooperate with the University in every way possible."[53]

A CHRISTMAS EVE CONFIRMATION

Ten days later on Christmas Eve, Mayall received unofficial word that the NSF had approved her grant application, allowing her $10 000 for 1 year. Then, on Monday night, December 28, just 2 days before they had to vacate the HCO offices, Mayall learned there would be no BU headquarters. She wrote to AAVSO Second Vice-President Fernald that somehow this did not surprise her, given the past 18 months of bad news and general nastiness: "The Boston University deal fell through at the last minute (we suspect some dirty work back of that) . . . although I suspected it was coming."[54]

The next day, December 29, Mayall rented Room 212 at 4 Brattle Street in Harvard Square, 200 yards from the Harvard University main campus and 0.6 mile east of the Observatory. She planned to move out of Building A during the weekend of January 1–3, 1954.[55] On December 31, the AAVSO's last working day on the Harvard College Observatory grounds, Mayall convened a last-minute meeting with Carpenter, Smiley, Witherell, Harwood, Hoffleit, and her husband Newton to compose the final version of the fundraising letter to members. The AAVSO's new 4 Brattle Street address would be at the top of the page.[56]

As she had always done, Mayall took this latest development in stride – keeping an optimistic attitude, and being somewhat philosophical – as can be seen in her letter to Fernald:

> We have had to work fast! It really is going to be much better for us to have our own place for this year, while we are working on our endowment. This way we will be independent and on our own. We have a good address, in a very nice office building, and can be proud that the National Science Foundation think us worthy of support."[57]

14 · Survival on Brattle Street

One thing I can promise you, and that is if we are penniless I will not close the door and walk out!
– Margaret W. Mayall, 1954[1]

For the next decade, the American Association of Variable Star Observers, now without its Harvard Observatory support, found itself facing deficits in its operations and struggling to raise its own Endowment Fund with which to fund its operations.

Survival in this period must be credited to the continued tenacity and optimism of Director Margaret Mayall and many other officers and friends. Although there were "low" spots, she was also being encouraged by the responses from grant associations who were taking an interest in her initiatives in observing programs and publications. Yet, at the same time she was trying to make the Association's operations more efficient.

This chapter traces some of the ups and downs for the Association during the next 9 critical years of operations.

MOVE TO BRATTLE STREET

A small group of AAVSO members and friends gathered on a Saturday, January 2, 1954, to assist with the move to the new headquarters on Brattle Street. Among the helpers were Jackie Sweeney Kloss, Jean McCroskey, Virginia McKibben Nail, and Dorrit Hoffleit, all from the HCO scientific staff; and Franklin Marsh, Percy Witherell, Helen Stephansky, Newton Mayall, and Margaret Mayall of the AAVSO.[2]

There was not enough time or space to move everything into the new office. Over the next few weeks, the Mayalls, Stephansky, Clint Ford, and Charles Smiley moved books, publications, records, and equipment into temporary storage, and a considerable number of old files were discarded. Newton Mayall got busy building sets of bookshelves for the new office.[3]

Symbolic of the loss of an earlier dream, the AAVSO removed its 6-inch Post refractor from the dome of HCO's Building A on Monday, January 4. Smiley and members of Skyscrapers, Inc., the Providence, Rhode Island, astronomy club, took the dismantled telescope to Brown University for storage.[4] The final logbook entry by Mayall that day was surprisingly unemotional: "6″ Post dismantled for temporary storage. Building A, HCO to be demolished to make way for new building. Work on 6″ Post done by Walter Locke & Margaret Harwood / Margaret W. Mayall, Recorder."[5]

The first public announcement about the new Brattle Street headquarters appeared in the March–April 1954 *JRASC*.[6] Mayall painted a pleasant and confident picture as she emphasized stability, physical comfort, and operational efficiency in a professional setting, while acknowledging an increasing workload.

In describing the new situation, Mayall said, "It was fortunate that ample space comparable to that previously occupied could be obtained in a modern building, and within the means of the association. In many ways the present location makes for a more orderly arrangement of the working material. The office is a comfortable place in which to work, with ample light, and conveniently located. Everything is being done to streamline the activities of the A.A.V.S.O., thereby making it possible to carry on its work more efficiently, and serve better the ever-increasing demands upon its facilities."

Moving to a new location was exciting, yet there was also an element of culture clash. The AAVSO had left its academic cocoon and thrust itself into the commercial and business center of Cambridge. Headquarters was a second floor, one-room office (405 square feet) near busy Harvard Square. Brattle Street was only a short stroll from Harvard's main campus, but it was an altogether different world. No longer working amidst benevolent and understanding astronomy colleagues, the AAVSO was now just another business occupying one room in a nondescript office building on

Figure 14.1. A composite image showing the AAVSO's one-room headquarters at 4 Brattle Street, Cambridge, Massachusetts, about 1961. AAVSO Director Margaret Mayall is seated at her desk in the right background.

a noisy street. No one in the neighboring offices had any idea what "AAVSO" was – passersby only saw a room crammed full of books and papers. Stephansky described their new space to AAVSO Chart Curator Richard Hamilton: "Our office here is quite a place. We managed to get all the books in, having bookcases built from floor to ceiling all along one wall. The other wall is packed from door at one end to window at other with eleven filing cabinets! I've been having a grand time throwing out old correspondence."[7]

The task of cleaning out files was grueling work but still involved a few lighter moments, according to Stephansky: "It's really loads of fun reading Mr. Campbell's old correspondence; he had such a sedate way of writing the funniest episodes that I am in chuckles most of the time. A lot of it we can't bear to part with, yet it must go. All day long we are sorting, sorting, sorting, and evenings Newton and Margaret do the same thing."[8] Stephansky had "bales of it at home" and had discovered "another three drawers full of 1931–34 stuff."

SOME DISAPPOINTMENTS

In December 1953, while still waiting to hear about offices at Boston University, Mayall had an unpleasant encounter with John Streeter, assistant director of Fels Planetarium and AAVSO member (Menzel's secret ally in his earlier push to relocate the AAVSO). Referring to the prospective arrangement with BU, Streeter told her that, "she had absolutely no right to accept a salary from another university while still employed at Harvard." This upset Mayall so much that she called Harvard Dean McGeorge Bundy during the Christmas holiday. Bundy, Stephansky recalled, "told her to come over and he would unlock the door and let her in his office (he was the only one in the whole building) – and they had a nice frank chat, during which he said in no uncertain terms that Streeter had no right nor reason to say untrue things." Bundy must have conveyed his displeasure to Menzel because, according to Stephansky, "nothing further was ever said."[9]

On January 6, 1954, Mayall received a letter from Bundy, who was "delighted to hear" about the AAVSO receiving an NSF grant for the coming year. Apparently, Mayall had not yet received any official word of the grant. In a note that seemed to express his sincere sentiments, he wrote: "This seems to me a very satisfactory outcome, and I am very glad that you have been able to find quarters at No. 4 Brattle Street."[10]

The day after receiving Bundy's letter, Mayall received a letter from Duncan Macdonald at BU rescinding their offer. Somewhat distant and bureaucratic in tone, he began: "In spite of the inherent value in the work of AAVSO and the many positive potentials that such an organization would furnish, it is now clear that Boston University is not the place to locate the headquarters of such a group." Macdonald cited several reasons: space problems, commitments to existing programs, the minor academic importance of astronomy at BU, the university being unable to undertake additional fiscal obligations, its physical plant "strained to capacity," priority placed on existing student needs and academic programs, and lastly, because of "the low enrollment in Astronomy and the non-existence of Astronomy majors, there is no present student function to be served by the organization." He concluded: "Although I have regrets that this union cannot be worked out, may I express my sincere hopes that you may find a suitable location to enable you to carry on in pleasant surroundings and with plentiful and technically profitable productivity."[11]

The sudden change in attitude at Boston University – from enthusiasm to coolness – did not seem to surprise anyone at AAVSO Headquarters. Stephansky related the news to Hamilton that Mayall "had even gone down to look at office space and could take her pick." Stephansky suspected that "their officials, consulting with the head of HCO as per orders, were told some nasty tales and vetoed the whole thing. Not content with kicking us out of HCO, Menzel had to justify himself by blackballing us at B.U., and probably at any other University that showed a little interest." Stephansky then quoted Smiley who said, "it was a rotten thing to stick a knife in us in the first place, but then twisting it afterwards showed the real character of the beast."[12]

But being on their own had its advantages, she reflected, "as I say, that is the end of bureaucracy, dictatorship, et al. Business will go on as usual, now that I have uncovered my typewriter at long last."

NATIONAL SCIENCE FOUNDATION

By January 15, 1954, Mayall had heard that the National Science Foundation (NSF) had sent a letter approving her grant application, but it had not yet arrived. She asked AAVSO President Martha Carpenter if she had heard anything, then conveyed the latest news. "We have been here until after midnight nearly every night," she wrote, adding the latest from HCO: "DHM [Menzel] is now Director!" She also reported that the Solar Division was "coming along well," with Rosebrugh finished with the November reductions, and Bondy having a *Solar Bulletin* ready for publication.[13]

At last, AAVSO President Carpenter received the NSF's 1-year grant check for $10 000 (about $78 500 in 2009) and forwarded it to Headquarters, where it arrived on January 29, 1954. Mayall happily conveyed the news of the check's arrival to astronomer Peter van de Kamp, director of Sproul Observatory, Swarthmore College: "We are getting all settled in our new headquarters, and like it very much. We find we get a lot more work done in a nice, calm, peaceful atmosphere!"[14]

THE ENDOWMENT FUND

Early in February 1954, Mayall wrote to editor Harry Bondy to congratulate him on his first *Solar Bulletin*. She mentioned the plan to send a fundraising letter to members, including Solar Division members, and promised a paragraph for the next bulletin announcing the NSF grant award. She felt it necessary to add a cautionary note: "Off the record – the less said about the B.U. affair the better. All I can say is that at the last minute, plans suddenly fell through. Dean Macdonald did everything he could and felt very bad about the whole deal. But actually I think we are much better off as we are. If we had gone to B.U. it would have been much more difficult to raise our Endowment Fund, for we would have had to consider them and their fundraising problems. Now we can do as we please."[15]

During the week of February 8, 1954, the AAVSO Endowment Fund Committee sent an informal fundraising letter to AAVSO members and friends. The Committee saw this as an expeditious first step in the appeal for funds, which freed them to spend more time composing a formal letter to foundations and institutions. Raising an endowment would not be easy, Ford admitted to fellow Committee member, Treasurer Percy Witherell. "The magnitude of the task is pretty frightening."[16]

Release of this first, informal letter to members helped ease the Committee's stress. Ford's next step was to send out letters to potential sponsoring Committee members. Mayall especially wanted the list of sponsors to include "prominent non-astronomical names." Eventually, the list of nominees grew to 18 "National Sponsors."[17]

Among the first to be nominated for the sponsoring committee were Watson Davis, director, Science Service; Harlow Shapley, former director, Harvard College Observatory; Otto Struve, president, International Astronomical Union; and Lawrence N. Upjohn, chairman of the board, The Upjohn Company. Of the initial group of nine nominees, all but one had agreed to be on the sponsoring committee. Paul Merrill, Mt. Wilson and Palomar observatories, had earlier agreed to serve as Honorary Chairman of the fundraising effort. C. E. Kenneth Mees, vice president and director, Eastman Kodak Company, would later join the list of sponsors. Those who accepted all felt pleased and honored to be asked, though most kept to the sidelines.[18]

The Endowment Fund Committee held another meeting on February 27 at the Brattle Street headquarters in conjunction with an interim meeting of the AAVSO Council. Fund Committee co-chairman Ford brought copies of the third revision of a proposed letter to foundations, a five-page document with a list of eight enclosures.[19] Because the Committee wanted to address a broad range of philanthropic organizations, the letter stressed the AAVSO's value to international cooperation and education first, its value to "the pure science of Astronomy" second, and emphasized that "the AAVSO is in no sense an applied science project capable of appealing for permanent or continuing government-sponsored support." In this version, Ford kept much of what he had first drafted in November.

The Committee also drafted two short fundraising letters: one for individual professional astronomers, and the other for individual AAVSO members. By early March, Mayall sent edited versions of these to Ford, Smiley, and Carpenter.[20] In the meantime, a number of astronomical societies sent their membership lists to Headquarters in response to the informal appeal letter. "It certainly looks like the societies are going to support our drive," Mayall hopefully remarked.[21]

Fund Committee member Fernald was encouraged by the amounts of the early individual contributions. Five had contributed $100 apiece, but he was discouraged to see that only 27 had responded so far, although "probably it was too soon to expect too many." Still, he concluded that they could not count on individual contributions. Furthermore, he felt that the worst outcome should be anticipated, and plans made for the AAVSO to continue on a limited footing.[22] Mayall agreed, but remained cautiously optimistic about reaching their $250 000 goal. In case they did not reach it, she wrote they would be prepared for other action. She had already been seeking volunteer help for the many routine office, computation, and publication tasks. Then showing her mettle, Mayall declared:

> One thing I can promise you, and that is if we are penniless I will not close the door and walk out! As long as I am physically able, I will always find plenty of evening and weekend hours for the AAVSO.[23]

Fernald responded that her letter "improved my outlook on the situation no end. The volunteer work you have had done and the leads you have for future development would seem to assure a good deal." Fernald, an accountant, estimated the AAVSO's possible income: a minimum income of $1000 per year, with $300 from the Pickering Memorial, $600 from dues, $100 from the life member fund and other income, and any income from the Endowment Fund, if enough was raised.[24]

In the March–April 1954 *JRASC,* Mayall briefly mentioned the reorganization of the Solar Division, saying that it is "again active in all of its programs." She then discussed the coming fundraising campaign, noting the NSF grant but emphasizing the need to raise a $500 000 endowment and frankly acknowledging the difficulty of even raising half of that in the coming year.[25]

In general, the outlook for grants was improving nationally. In its third annual report, NSF recommended its budget request be expanded from $8 million in 1954 to $14 million in 1955. The report emphasized that "a high proportion of [new scientific knowledge produced] may prove useful in ways unforeseeable today."[26]

Demolition began on HCO's Building A during the last week of March. On April 2, Stephansky wrote, "today Margaret Olmsted told me Bldg. A is gone – the last wall fell this morning."[27]

FUNDRAISING RESPONSES

In mid-April, Ford was surprised by a letter from the executive secretary for T. J. Watson, Jr., president of IBM Corporation. The letter confirmed that Watson would not serve as a sponsoring member of the Fund Committee but, amazingly, contained a $250 check and note: "we feel that your Association is extremely important and we are glad to lend our monetary support." Ford immediately forwarded the check to AAVSO Treasurer Witherell, saying, "This is a significant gift."[28]

By May 8, 1954, first day of the spring meeting, morale at AAVSO Headquarters had risen quite a bit. The AAVSO held its meeting at Columbia University, New York. Columbia Professor Jan Schilt hosted the meeting, which was held in conjunction with Columbia's bicentennial celebration.[29] "Perhaps," Mayall impishly observed, "in the year 2111 we may be able to return the favor." Many members of the Amateur Astronomers Association of New York were present, as well as AAVSO founding member Helen M. Swartz, 76 years old.

During the 1954 spring Council meeting, Treasurer Witherell declared the Association's finances in good order. Chart orders totaled 2200, plus 25 finder charts and 26 copies of the *AAVSO Atlas*. Headquarters received 23 applications for membership, the largest number in several years for a spring meeting. In response to the Fund Drive, Ford contributed 10 shares of IBM stock. "A very generous donation," according to the meeting chronicler Roy Seely.[30]

The only ominous note was a discussion of the legal status of the Endowment Fund; it would be necessary, Ford wrote, "to assign a final benefactor for the Fund to cover the remote possibility of an eventual disbanding of the AAVSO and cessation of its corporate activities." After a long discussion, the Boston Museum of Science, Boston, Massachusetts, was appointed as a final benefactor. No binding vote was actually taken, but reflecting 3 months later, Mayall told Ford that, if it came to that, she would rather see any AAVSO legacy go to the American Astronomical Society for variable star research.[31]

At the 1954 spring general membership meeting, there was little hint of the crisis of the past 6 months. Mayall reported that she and her assistant were feeling very much "at home" at the new headquarters, a $10 000 NSF grant was in hand, and professional astronomers were requesting that observations be made available by individual observers, rather than by 10-day means. In announcing the NSF grant, Mayall noted: "Astronomers all over the world approve our work." Her outlook for the fund drive was "very optimistic." One wag at the meeting, Edgar Paulton, suggested that AAVSO now stood for "Accept Any Variety of Sincere Offers."[32]

In mid-May, Margaret Mayall received "a nice surprise" when C. E. Kenneth Mees, vice president and director of Eastman Kodak Co., agreed to be a national sponsor and contributed $100 to the fund.[33] In thanking him, Mayall said: "The Harvard-AAVSO relationship this last year has been a very sad affair, but I think that in the end it is going to be a good thing to be out on our own. We can really appreciate now what Dr. Shapley and Harvard did for us in the past." Mees' sponsorship was just the sort of non-astronomical support she had been hoping to find.[34]

On July 1, 1954, the sustaining income from the $100 000 Pickering Memorial Endowment ceased.

Figure 14.2. Margaret Mayall while with the Copper Harbor, Michigan, eclipse expedition, June 1954.

HCO had cut all other support 1 year earlier. From then on, the AAVSO would receive only the income from its original $6356.76 contribution. Though Mayall was optimistic, Ford was worried enough that he sent a cheerleading letter to rally more AAVSO contributors to the fund. "Come on," Ford urged, "let's get going!" By August, he reported only 25 percent of the members had "fed the kitty," which now totaled $2871, plus $3000 in stock shares. Hoping to further inspire his readers, Ford ended by saying: "Very soon the Fund Committee's official brochure will be sent out to the corners of the world, astronomical and otherwise . . . . Come on, let's go – here's to more dollars for the AAVSO!"[35]

The Fund Committee hoped that their recently completed brochure – professionally prepared – would show results. "In any case," Ford wrote to Professor Francis Reynolds of the Chemical Engineering faculty at Clarkson College in Potsdam, New York, "the fund raising will probably go on for years; however, in the end we will obtain a permanent endowment wholly in the name of the Association and sufficient to carry on its activities indefinitely."[36]

CHANGES AND ADJUSTMENTS

The AAVSO's first year on its own was rapidly coming to a close. It was again time to plan the 1954 Annual Meeting, to be held at Brown University, Providence, Rhode Island. This would be the first annual meeting held away from Cambridge, Massachusetts. Mayall queried Ford on finding an able successor to AAVSO President Carpenter. She hoped that Fernald would be willing to take that post before becoming treasurer. Because she and her husband Newton would be traveling to Maine to visit the Fernalds, she wrote, "I will try to sound him out."[37] Although Mayall was thinking of someone with financial experience as president, Ford was searching for someone who would bring academic prestige to the position. He thought that Council member Reynolds of Clarkson College might be a good candidate.[38]

About 3 weeks later, Mayall wrote to Ford recommending Cyrus Fernald for president, mentioning she was "quite sure he is willing to accept." She thought Charles Smiley brought academic prestige as Co-Chairman of the Endowment Fund and thought he should be kept as first vice president. She recommended Reynolds for second vice president to follow Fernald as president. In recommending Fernald, she thought they needed a good businessman and organizer and stated, "Fernald can do that. He also has the time to devote to it, which I doubt Reynolds would." Ford responded quickly to Mayall's letter to assure her that he thought Fernald was an excellent choice for president.[39]

Mayall then turned to the mechanics of the fundraising effort. "Contributions have been coming in again as a result of the brochure," she wrote. "I have been stamping and stuffing [contribution solicitation envelopes] as a midnight job." Dorrit Hoffleit and Arville Walker were also helping with some of the folding during their lunch hours. AAVSO member Lewis Boss had mimeographed the Solar Instructions and had volunteered to do more. She also noted that Boss would make his quota of $600 toward the fund. Another Headquarters friend, Mrs. Robert Dunn, helped with typing every night between 4:30 p.m. and 1 a.m. By the end of September, Headquarters had mailed more than 2000 fund brochures to amateur and professional astronomers.[40]

Mayall expressed nothing but optimism in her 1953–1954 Annual Report. "The past 40 years have been good ones," she wrote. "The future holds even more for the organization, with its almost unlimited opportunities." With a small degree of stability attained, Mayall now needed to restore the integrity of the AAVSO's operations. Some modest reshuffling of the Nominating and Chart committees needed to be done.

The Endowment was the biggest challenge: it stood at about $15 000, and serious thought had to be given to investing the fund "as soon as possible." Mayall hoped that Fernald would bring his financial acumen to bear as either president or treasurer. With enough members to form a quorum, the Council held a special business meeting on October 7, 1954, at AAVSO headquarters to satisfy the legal requirement that the annual meeting be held in Massachusetts. There, the Council voted to accept a rewording of an article in the constitution pertaining to the Finance Committee: the Committee would, in effect, have "all the powers of the Council with relation to the investment and disposition of funds of the Association." This was an important step in the development of the endowment.

The next day, the Council held its regular meeting at Brown University in Providence, Rhode Island. Officers elected were Cyrus Fernald, president; Margaret Harwood, first vice president; Richard Hamilton, second vice president; Margaret Mayall, recorder; Percy Witherell, treasurer; and Clinton Ford, secretary. Members elected to the Finance Committee were Fernald, president, ex-officio; Witherell, treasurer; and Walter P. Reeves, Ralph N. Buckstaff, and Clint Ford.

In early December 1954, a note from Donald Menzel arrived at AAVSO headquarters asking Mayall to allow Cecilia Payne-Gaposchkin to examine light curves for two of her projects. Repeating a request he made in November, he wrote, "Mrs. Gaposchkin is quite anxious to know whether she will be able to have access to the AAVSO light curves." "How do you like that!" Stephansky huffed; "Tell the AAVSO their work is second-rate, kick them out, and then come running for help . . . . By the way, they [the Gaposchkins] still owe a couple years dues and are suspended!"[41]

Ending the week on a happier note, the Astronomical Society of the Pacific with some help from Paul Merrill sent a $500 check for the Endowment Fund. More contributions came in from members and friends: James Baker, $100 as a first installment of a pledge (total unknown); founding member Helen M. Swartz, $100; and IBM stock dividend checks began arriving regularly.[42]

Headquarters operations

President Fernald and his wife Emily paid a visit to the Brattle Street headquarters on December 16, 1954.

"While there is nothing finished," he wrote to former president Roy Seely, "there is progress on several of the vital things that need to be done. It is too bad that Margaret has to spread herself so thin, but I can't see how it can be helped."[43]

On December 31, Smiley wrote a letter to President Fernald resigning as the fund drive co-chair effective January 1, 1955. Smiley felt that the appeals for year-to-year funding would undercut the larger Endowment Fund appeal. He also cited "internal friction" within the AAVSO, but did not elaborate. As a third problem, Smiley merely added there were two fund drive chairmen. Oddly, neither Mayall nor Ford, it seems, expressed any immediate reaction to Smiley's resignation.[44]

Fernald reluctantly accepted Smiley's resignation. He did not respond directly to any of the points Smiley raised, feeling that he had not been involved enough to form an opinion. He did offer some remarks based on his recent observation of Headquarters' operations, noting that "the drive for funds is taking headquarters time that should be put on other things, but here again: What else can we do? The necessary routine work is far behind, and where is the time that should be put on constructive analysis of our results?"

In referring to Mayall's articles in the last two issues in the *Journal of the RASC* on S Ori and Z And, he noted that although Mayall and Stephansky were "doing a remarkable job to keep things up as well as they do," he wondered, "just which will come first, the running out of our funds or the breakdown of our present recording system."[45]

The search for funds had to take precedence. On January 14, 1955, Mayall applied for a grant from the Research Corporation, New York. Mayall's main selling points were her plans to research 800 regular and irregular variables over the next few years using observations made by volunteers as well as published and unpublished observations on record, to publish 4 years of unpublished AAVSO observations and make them available to researchers, to work on a 25-year light curve of Mira Ceti for Leo Goldberg, and to train young scientists. She noted that the Endowment Fund drive was underway, but that the need for interim funding was urgent.[46]

In February 1955, Ford received the sobering news that the NSF money was used up, and "the financial situation is now very critical . . . and there is not enough in the new Endowment Fund to pay Margaret much of a salary after all the running expenses are taken care of."[47] In fact, Mayall had been working without a salary since the first of the year. Knowing that money available for operating expenses and salaries would soon be scarce, she deposited her paychecks into a salary fund that she opened for her assistant, Stephansky. If the AAVSO was presently struggling just to accomplish its routine tasks, Mayall knew that it would be impossible to accomplish anything at all without the help of Stephansky, who had been at her job since the time of Leon Campbell. Mayall decided that she would rather lose her salary than lose her right arm.[48]

Fund drive progress

On March 23, 1955, Secretary Ford sent a progress report on the fund drive to the national sponsors. At the end of fiscal year 1954, fund contributions totaled $3314, and stock contributions totaled $3000. This, plus the life membership fund of $2393, brought the fund total to $9200. The Pickering Memorial remnant of $6356.76 was not added to the total, as only the income from that amount from Harvard was available to the AAVSO.

The anticipated 1955 income from all sources was $2000. Interim help was needed badly. Ford asked the sponsors for their suggestions, adding, "a number of foundations and individuals have been contacted, with negative results, but the possibilities are far from exhausted."[49]

Just 4 days later, Lawrence N. Upjohn responded to the report, regretting that he could not come up with any ideas about who the AAVSO might contact for support. Instead, he enclosed a check for $200 toward current operating expenses and promised a contribution to the Endowment Fund in the future.[50] Ford thanked Upjohn for his generous contribution: "This gift comes at a time when it is sorely needed"; and in response to Upjohn's inquiry, Ford added that the AAVSO "is now legally incorporated as an educational non-profit institution, so that all contributions to it for any purpose are deductible for income tax purposes." Ford also expressed his apprehensions to Upjohn regarding a sign of the times:

> At present there is an unfortunate deep-seated public distrust in science, resulting probably from the implications of the H-bomb and the much

> distorted past connections of a few famous scientists with communism. Also there is the inevitable feeling that if science of any sort needs money, the government should furnish it. It is just no good to point out that astronomy (not to mention the AAVSO) has nothing to do with all this, and that the "increase of knowledge" in the Smithsonian sense is the important thing in the long run. It is rather discouraging, but we will somehow survive.[51]

In addition to Upjohn's $200 contribution, Walter P. Reeves, Vice-President of the Maine Central Railroad and an AAVSO member since 1941, sent $100.[52]

Impatience with headquarters

A small group of officers was now becoming impatient with what they saw as a lack of progress at headquarters. Mayall was particularly upset that Witherell was "very dissatisfied with everything we are doing . . . . He seems to feel that everything we do is very inefficient. Of course, the trouble is, he has no idea what we really do here in the office."[53]

At least Mayall could find occasional satisfaction. She had just returned from the American Astronomical Society meeting held in Princeton, New Jersey. Stephansky noted that Mayall was "full of enthusiasm. She talked AAVSO to everybody she met!" Donald Menzel, who was AAS President that year, responded to Paul Merrill's "inspiring talk" on red variables. Noting they were "so very important," Stephansky wrote, "he hoped they would have many many more papers on them in coming meetings! Apparently this set up a good deal of snickering, with much comment afterwards to Margaret on the old refrain 'no longer of the same importance' [a reference to the wording of the Pickering Endowment terms]."[54]

On April 18, 1955, the Research Corporation wrote to say they would not support organizations – particularly not for what was essentially an appeal for operating expenses; they could only support individual science research projects.[55]

The AAVSO reached what was probably the low point of its financial situation in May. Operating funds were diminishing. Ford estimated available funds at $1450 ($300 from the Pickering Memorial Endowment remnant; $350 from the endowment now on hand; $300 income from possible gifts; and $500 from the annual dues and operating gifts. There was enough to cover operating expenses, but not salaries.[56]

Expressions of discontent with the slow rate of progress at headquarters continued to come from Witherell, Fernald, Seeley, and Rosebrugh. Knowing that there would not be enough money to cover salaries, they took it for granted that Stephansky's position would be dropped, and so she became an easy target for their criticism.

Witherell and Fernald, especially, began raising questions not only about Stephansky's productivity, but her "attitude" as well. They did not elaborate, but they probably mistook her upbeat and often irreverent outlook to mean that she was not serious about her work. Furthermore, they felt that Mayall was too easily distracted from seeing important tasks to completion, and she lacked the will to give more of her laborious routine work over to Stephansky.

On fundraising, Fernald began to see that one of the reasons Smiley resigned as fund co-chairman was because he did not agree with Newton Mayall's ideas about how to raise money. Newton Mayall was, in part, chasing after dubious contacts obtained through mutual friends. Vice-President Harwood added her voice to this disconsolate group, though she at least offered her criticisms in the form of constructive suggestions to all, including Mayall: make the Campbell volume a priority, have the Finance Committee draw up an annual budget, shift more of Mayall's tasks to Stephansky, cease individual acknowledgment of observations received, cease the informal newsletter compilations or shift the work to volunteers not involving headquarters expense or time, and fundraising should not be a function of the recorder, her husband, or her assistant.[57]

At the May 27–28, 1955 Spring Meeting in Pittsburgh, Pennsylvania, Ford reported little progress in the endowment campaign – "all requests for large endowments so far having been unsuccessful." Mayall reported that many reacted favorably to her suggestion at the April AAS meeting that professional astronomers would be asked to pay for information provided by the AAVSO.[58]

At the 1955 spring technical paper session, Dr. Nicholas E. Wagman, director of the Allegheny Observatory, spoke about "Flare Stars." Later, some 25 members visited the J. W. Fecker plant, where they were shown "the famous optical shop of Dr. John Brashear,"

Figure 14.3. Clifford Seauvageau, Claude Carpenter, Jim Breckinridge, Leo Scanlon, George Diedrich, and Clinton Ford examining the latest photoelectric photometry equipment at the J. W. Fecker Plant, Pittsburgh, May 1955.

as well as the latest in PEP equipment.[59] Remarkably, Pittsburgh television station KDKA gave "special feature" coverage (1.5 minutes) of the AAVSO meeting.[60] Also at the meeting, the Finance Committee voted to pay the salary of Mayall's assistant, Stephansky, up to November 1, 1955.[61]

Probably reacting to what had been said during the Council meeting, Fernald later told Mayall that he never had much hope that the appeal to foundations and corporations would succeed, pointing out that tax loopholes favorable to large corporate donations had been closed after 1952. This was contrary to the view he held in March when Smiley wanted to put the most effort in the foundation appeals. Fernald's position now was that he wanted to see the fundraising effort concentrated on observatories, individuals, and professional astronomers. "If our work has the value that MWM tells us some of the professionals tell her it has, then they should support it."[62]

In summarizing his spring 1955 post-meeting news to Mayall, Ford noted the regrettable professional–amateur relations emerging from HCO: "Have you read [Payne-Gaposchkin's] newly published volume *Variable Stars and Galactic Structure*?" Ford saw that "numerous light-curves had been reproduced from AAVSO data without any acknowledgement for same, while she goes to great ends to give complete bibliographies for all sources that she considers professional, including several light curves produced in full from unpublished studies by Sergei Gaposchkin."[63]

The AAVSO's reputation among professional astronomers seemed to hinge on completion of Campbell's *Studies of Long Period Variables*. Rosebrugh told Fernald that he had finished his plotting for the book nearly 2 years previously, and to him, this meant there was nothing more to be done, so he couldn't understand the delay. In reality, Stephansky explained that Mayall, in preparing the Campbell volume, had "absolutely no notes to follow, for of course L.C. had it all in his head. We have several drawers full of materials which all must be checked to see what is applicable in the volume and what is outdated, before she can work on the introduction."[64]

Seemingly in response to Harwood's letter of suggestions – but in reality it was the culmination of 4 years of steadfast, if not uninterrupted, work – Mayall sent her final text for the Campbell volume to Harwood on June 28, asking her to proofread it. Mayall, in the meantime, obtained price quotes from printers and hoped to have the book out in time for the annual meeting.[65] About 2 weeks later, Mayall forwarded the text to Ford for proofing: "Margaret Harwood has had it," she said, "and, for a wonder, made very few changes!"[66]

While Mayall was hard at work, Fernald was expressing his displeasure to Ford about what he and others perceived as the lack of progress at headquarters, and worried that the Campbell book "will be permanently lost" if it was not printed by that fall.[67] Fernald, Witherell, and the other critics of the headquarters staff did not realize just how much work was getting done, despite the large amount there was to do. Their accusations of inefficiency at AAVSO Headquarters were not justified.

In early August 1955, AAVSO Headquarters sent out "special pre-publication" order forms for the Campbell volume. A week later, the Mayalls departed for Dublin, Ireland, and the IAU meeting. They would tour the Irish countryside, attend the meeting, visit Iceland (including a sightseeing flight over the arctic circle), and return to Boston in early October as passengers on an unscheduled freighter. "This trip," Stephansky noted, was "partly AAVSO business and partly to help Newton recover from his operation . . . . Poor Newton had cancer of the throat and had one vocal chord removed . . . . He was in the hospital and had the operation before they hardly had time to realize what was happening. Dreadful! You can imagine how Margaret felt, what with the AAVSO difficulties and their own personal problems as well."[68]

Question of salaries

At the 1955 Annual Council Meeting, the question of salary for Mayall's assistant was partly alleviated when Council member Walter P. Reeves "generously contributed $300 to the General Fund and moved that Miss Stephansky be retained until May 1, 1956." Because Mayall was willing to keep working without salary, the Council agreed to the proposal and allowed for the possibility of an extension for Stephansky after that date.

The Association's financial health was the foremost topic. Ford reported that the Endowment Committee still had no success in securing any large gifts. Mayall mentioned the possibility of an ONR grant and believed that the ASP's recent generous donations would increase in the future.

Mayall also read a letter she had sent to professional observatories informing them that AAVSO's services and publications would no longer be provided for free. Nearly all those responding, she noted, were willing to pay any fixed charges. "The organization is in considerably better condition than I expected," President Fernald later wrote Mayall; "As near as we can estimate 1955–6 income and operating expenses, we will have about $1000 leftover." This, he noted, did not include salaries for Mayall and Stephansky.[69]

In her report at the 1955 Annual Meeting in Springfield, Massachusetts, Mayall admitted that the past year was financially one of the most trying years, yet scientifically, and in terms of the loyalty and enthusiasm of members and friends, it was one of the most satisfying. She regretted that she had to spend far too much time during the year engaged in raising funds: "In the AAVSO files there is much fine material for research programs, but work on those can be postponed better than our immediate need for money to keep our office open." To this she added – publicly for the first time – "Since January 1955 the Recorder has volunteered her services in order to keep the work going and to stretch our meager funds as far as possible."

The Yankee accountant in Fernald, too, must have realized the value in keeping Mayall's assistant employed and Mayall herself unsalaried. In a November 1955 letter to Rosebrugh, he wrote: "we have little to offer, so must be careful to get the most for it. Think the present set-up does that, so let's keep it going as long as [we] can." He also reiterated a backup plan he and Rosebrugh had discussed 3 years earlier, by which observations would be recorded and light curves generated and published by amateur work alone, in the event that Mayall decided to find employment elsewhere. He and Rosebrugh agreed that "the AAVSO income would get a lot of work out of a dozen amateurs." Fernald went so far as to suggest that "unless the financial picture improves . . . a strong committee should be appointed to develop such a system." What Fernald and Rosebrugh failed to appreciate was what it might mean in the eyes of the astronomical community if there was no longer a professional astronomer at the AAVSO's helm.[70]

Despite her irritation with applying for grants, Mayall had no choice but to spend a substantial amount of time preparing material and filling forms just so the AAVSO could survive another year. In her NSF proposal to study and prepare for publication of AAVSO observations, she noted that some 60 000 observations were now being received each year, she quoted from Merrill's article on the importance of obtaining observations of LPVs, and she included excerpts from her collection of endorsements from professional astronomers. She asked NSF for $16 000 over 2 years.[71] Her application to the ONR requested $4945 to support the publication of observations and for special studies over a 2-year period. Here, too, she quoted from her collection of letters of support.[72]

The next few months of 1955–1956 brought good news on their funding efforts. The NSF approved a $12 000 grant for 2 years with an additional $400 toward the purchase of an electric typewriter. And as Bart Bok reported to Mayall, the ONR approved a $4370 grant, saying Mayall's proposal received the "full and enthusiastic support of the whole astronomy panel."[73] Reporting this news to treasurer Witherell, Mayall wrote: "It is a wonderful feeling to know the next two years will be OK for us . . . . I certainly will be glad to get back on the payroll . . . . I am also delighted that Helen will be back again on full time starting April 2. We are going to have a busy year, with all the work I am planning to accomplish."[74]

Stimulated by the turn of events, Mayall's assistant Stephansky thought it worth writing to President Fernald, not only to tell him how hard Mayall had been working to get grants, but also to point out how little the AAVSO Council seemed to appreciate Mayall's effort. "She is so delighted to get the grants," wrote Stephansky, "for of course she has worried and worked tremendous hours to go after them; nobody except me (and her husband!) knows just how much work has gone into it. Every night she is here until 8 o'clock, and some nights until midnight! This is *every* night, not just once in a while as some people may think . . . . Oddly enough, she has such a wonderful standing in the professional world . . . (you should hear what they say about her), that I sometimes wonder why our Council is so apathetic in saying that they appreciate her."[75]

Director – a new title

On receiving the official word from the ONR and NSF grants, Mayall wrote a follow-up letter to Treasurer Witherell quoting from the NSF's letter, which said: "We hope that this grant will enable you to continue and expand the important work you are doing." To this, Mayall added: "Needless to say, I am extremely happy over all this. It makes our struggles over the last year or so very worthwhile." Here, Mayall also noted her wish to make a change to her title: "In order to make my duties and responsibilities clearer to outsiders, particularly in handling government funds and applying for grants, I would like to propose that my title be changed from Recorder to Director." She hoped that the change could be voted on at the spring meeting in May.[76]

Whether enlightened by the letter from Stephansky, or simply moved by the turn of events, Fernald – writing to Hamilton, Buckstaff, and Ford to report the news of the grants – proclaimed "this simply goes to show that we have as our leader a worthy successor to Leon Campbell, one who is facing and solving even tougher problems than LC had on his hands." He thought it was unfortunate that Mayall had to spend so much of her time and energy on applying for grants when "that should be done by other officers of the AAVSO." Admitting that the AAVSO was "not organized so that could be done," he said "some thought should be given to the problem for the future."[77]

At the 1956 AAVSO Spring Meeting at Cornell University in Ithaca, New York, the Council voted to continue Stephansky's employment for at least the next 2 years and spent some time on the question of what new title to give to Mayall; "Executive Officer" did not seem right. Mayall pointed out that, not only would "Director" make her duties clearer to outsiders, but there was also a precedent; the BAA had a "President" and a "Director" of each of its Sections. They finally agreed to change the title of "Recorder" to "Director" in the next revision of the Constitution and bylaws in the fall.[78]

The year 1956 ended with guarded optimism. Mayall applied for a renewal of her ONR grant, asking for $6670. The grant would be approved at the committee level in February, and final approval would be made in March 1957 for the full amount requested. Professor Edwin B. Wilson at the Boston Branch of the ONR suggested that Mayall apply for other funds, including the Bache Fund of the National Academy of Sciences. She asked for $700 to print the results of her 1952–1953 observations studies. Writing to Paul Merrill, she said that she was hoping "to get enough money to get the next 2 years observations (1954 & 1955) published, for I hate to be so far behind."[79]

Some Council members contributed financially as often as they were able to: Reeves, Witherell, and an anonymous donor (likely Clint Ford) each made contributions of stock to the Endowment Fund. Mayall was delighted, saying, "This gets our endowment up over $30 000 and our income nearly $1000 – pretty good for 3 years, don't you agree?"[80]

THE AAVSO WOULD SURVIVE

"The year is starting out very well for us," Mayall wrote to the AAVSO's officers in March 1957, bringing them up to date on the financial situation: the second payment of the NSF grant would carry the AAVSO through March 1958; the ONR grant would be renewed for another year; a Bache Fund grant would cover printing costs for publication of 1952–1954 observations; and the Bureau of Standards provided $50 toward the Solar Division work.

"With this help from various grants," Mayall wrote, "I hope we can turn most of our current income from dues, etc., into the Endowment Fund, and help build it up. It is really growing, as the Finance Committee report shows, thanks to Mr. Ford for his work in getting us an anonymous gift of 10 shares of IBM stock, and to Mr. Reeves for his generous gift of five shares of International Paper. We are getting wonderful cooperation everywhere."[81]

Slowly, in 1957 it seemed, the Council and membership began to realize that the AAVSO would survive. "Mrs. Mayall has done a remarkable job of keeping the Assn going," Ford wrote Francis Reynolds, "and it now looks as though the 'Dark Ages' which began in 1952 are drawing to a close. . . . The AAVSO is still treading on ice, but the ice is a lot thicker than it was a few years ago."[82]

The Council meeting that spring 1957 was once again devoid of any urgency or controversy. The most pressing concern was the Finance Committee's handling of the Endowment Fund: what types of stocks to purchase or sell, the immediate cash needs, and ensuring the future growth of the fund as well. And importantly, Mayall had to spend as much time as possible catching up on her preparations of the *Quarterly Reports* to meet the terms of her government contracts.[83]

A DARING PREDICTION

At the AAVSO's October 1958 Council meeting, Mayall reported that the Endowment Fund growth was slow but steady. Making a prediction that would have been unimaginable even 1 year ago, Mayall declared that the fund was "continually growing and may be expected to eventually take over complete financial support of the AAVSO."[84]

Self-sufficiency had its price. In December the NSF awarded the AAVSO a grant for $16 000 over 2 years. However, Dwight E. Gray, program director for NSF Publications and Information Services, wrote a follow-up letter to remind Mayall that, as had been earlier discussed with NSF officials, the following "understandings" had been decided: "First, the Foundation is not committed to granting further support to the AAVSO in the future; and Second, the Foundation hopes that the AAVSO will be able to make greater progress in becoming self-supporting during the period of this grant." Yet, Gray ended his letter by saying:

> You will be glad to know that a number of those who reviewed the AAVSO proposal praised the program highly. One suggestion which you may be interested in, came from an astronomer who said the *Bulletin* would be even more useful "if it were used as a means of notifying astronomers quickly whenever particular variable stars start to show peculiarities in their current light cycles."[85]

In her reply, Mayall assured Gray that everything possible would be done to fulfill the contracts. She made a point of mentioning the Endowment Fund's growth, thanks to large and small gifts, but especially 20 more shares of IBM stock. The Endowment Fund value as of November 12, 1958, Mayall noted, was $52 000, not including the recent IBM gift. She also noted that, "We feel that we are progressing, because a little over four years ago our total assets were less than $7000." She acknowledged the NSF suggestion about providing information for quick notification of star activity, saying, "I will certainly look into the possibility of [publishing current information]. I realize that the AAVSO is the only possible source of up-to-the-minute news on current activities of the variable stars."[86]

Figure 14.4. Margaret Mayall, member/observer George Ripley, and Florence Campbell Bibber at AAVSO's Brattle Street headquarters, 1963.

The Endowment Fund passed the $100 000 mark in 1960. Stephansky resigned to work as a secretary for a Boston insurance company. In the interim, Florence Campbell Bibber, who had been working at headquarters part-time, increased her hours.[87] Following a unanimous vote by the members at the 1960 Annual meeting in October, the Council elected Stephansky to Life Membership in gratitude for her loyalty and service to the AAVSO.[88]

By the end of fiscal year 1960, the Endowment Fund stood at $154 529, and produced more than $3300 annual income. The Finance Committee considered turning the administration of the fund portfolio over to a trust company.

The NSF "understandings" did not dissuade Mayall from applying for another grant, over 5 years this time. "We have made considerable progress in becoming self-supporting," she wrote, "but have not been able to reach the final stage as yet." The NSF would eventually meet her halfway, granting a 2-year contract for $20 000. In her proposal, Mayall mentioned the importance of obtaining good comparison star magnitudes if more stars were to be added to the AAVSO's observing programs. "Thanks to Thomas Cragg and his colleagues at the Mt. Wilson and Griffith Observatories," she wrote, "we are getting sequences based on photoelectric and photometric measures."[89]

With the AAVSO's growing financial independence, the Council voted to transfer the assets of the Endowment Fund to the New England Merchants Bank for investment management with a formal letter of transmittal sent on August 7, 1962.[90]

ON ITS OWN WINGS

In the first spring meeting held on the West Coast in 1963, Treasurer Richard H. Davis announced that, for the first time in 10 years, the AAVSO would not be receiving any direct support from contracts with government agencies. Although the Council anticipated a budget deficit of $6500 during the coming fiscal year, the deficit could be mitigated by diverting income from the Endowment Fund to the General Fund. This would leave a deficit of about $3500. With this in mind, the Treasurer proposed that the Finance Committee authorize him "to transfer to the General Fund so much of the current and accumulated income of the Endowment Fund as may from time to time become necessary to meet operating deficits of the Association: but that no portion of the principal of the Endowment Fund may be expended for any purpose whatsoever without specific authorization therefore by vote at an annual meeting of a majority of those members then present." The motion was unanimously approved and would become part of a proposed amendment to the Constitution.[91]

At the 1963 Annual Meeting in Worcester, Massachusetts, that October, Mayall reported that she had secured another ONR contract, making $8000 available for 1 year with the possibility of renewal. Her planned work for this contract was to extend the data in Campbell's *Studies of Long Period Variables* on the dates of maxima and minima of long-period stars. With this news, the Treasurer stated that the General Fund "would show a credit balance of about $5000 instead of the anticipated deficit of $3200." The Council unanimously approved an increase the Director's salary by $500 per year, and the wages of the Director's assistant from $1.65 to $1.75 per hour, effective October 14, 1963. The Endowment Fund Chairman Ford reported that the value of all the Fund's assets totaled $173 187, with income from all investments totaling $3871.[92]

The survival of the AAVSO during its first years without the support of HCO depended on Margaret Mayall's determination and hard work. She found the short-term grants needed to carry the Association through its most difficult years.

Long-term survival depended on the establishment and growth of an endowment. Clint Ford led this effort with the help of the Council.

But the survival of the Association also meant maintaining the continuity of fundamental office operations. With the very capable and dedicated help of her assistant Helen Stephansky and, later, Florence Bibber, Mayall kept up with recording monthly reports, publishing observations, and fulfilling requests from astronomers.

With the Association on a stable footing, it could once more devote its effort to improving its services to astronomy and exploring new areas of variable star research.

15 · AAVSO achievements

Our observations do not yet tell the full story. Nor may they for many decades or perhaps centuries to come.

– Cyrus Fernald, 1954[1]

Even while she was struggling with the funding proposals, Margaret Mayall was encouraged by the Association's observation reports and activities. Several developments marked a turning point for the Association, especially when Mayall and the Association's achievements were recognized by the International Astronomical Union.

In spite of the upheaval from HCO, Mayall had no reservations about expanding the scientific programs of the AAVSO: she had already been encouraging observers to shift their efforts from long-period to irregular variable stars, and, through her association with Paul Merrill and having seen the work of John Ruiz in photoelectric photometry (PEP), she was convinced of the value of PEP observations in the study of variable stars.

The Association had drawn together many outstanding observers who were willing to initiate new observing programs and try new technologies, sometimes on their own initiative. This chapter follows their successes and some failures.

NEW OBSERVING PROGRAMS

In keeping with her eagerness to move the AAVSO observing program in new directions, Mayall's first "Variable Star Notes" column in *JRASC* for 1954 featured several faint-magnitude stars. She had, in prior months, been requesting that observations be made, if possible, as faint as 16th magnitude. Mayall's column listed some results, mostly faint observations of 25 stars classified by 5 types: U Gem, R CrB, Z Cam, flare stars and other peculiar variables, and semiregular variables.[2] Cyrus Fernald, AAVSO member and amateur observer, responded to her "Notes" on RU And, a long-period variable showing changes in period and amplitude. He contended that although HCO might be right in stating that "our 40 year collection of observations appears to establish the broad principles of variable star research on sufficiently firm ground," there was "reasonable evidence that our observations do not yet tell the full story. . . . Changes in the length of the period or in the amplitude of the light fluctuation would furnish additional information for those who speculate as to the physics and chemistry that makes these stars act as they do."[3]

A lot was happening in astronomy, as mentioned in Mayall's 1953–1954 Annual Report, which "promised an ever-increasing demand for the AAVSO's services." The value of visual observations of the long-period and semiregular variables, Mayall noted, had become "even greater in those times when observatories such as Harvard found it necessary to curtail their photographic patrol programs." She declared that the AAVSO Nova Search program would be even more important with fewer patrol plates available, stating, "a negative report from the Nova Search observers can narrow down the time in which the nova might have blazed up."

Mayall also noted in her 1953–1954 Annual Report that Boris Kukarkin, International Astronomical Union (IAU) Commission 27 (Variable Stars) president, stressed the value of the AAVSO's work on long-period, U Gem, and Z Cam variable stars. His request for observations of LPVs with maxima of at least 11th magnitude meant "adding about 200 more of that class to our program." With that request came the enormous job of preparing visual sequences of comparison stars. Kukarkin also pointed out the need for observations of southern stars and suggested that "closest possible contact should be maintained with the variable star section of the New Zealand Astronomical Society." Kukarkin gave the AAVSO a vote of confidence by writing that such future work "can certainly be accomplished by such a powerful organization . . . known throughout the world by its variable star observations."[4]

At the end of the 1953–1954 fiscal year, 36 AAVSO observers abroad had made 17 404 observations, and 100 observers in 25 states of the United States made 34 624 observations. "On the whole," Mayall noted, "the variables on our program are being covered very well, and we can usually furnish light curves and dates of observed maxima and minima when they are needed." She cited the many requests from professional astronomers for current or past information on variables. In thanking the observers, she concluded: "The spirit of cooperation throughout the astronomical world is very heartwarming. From every quarter come fine expressions of appreciation of the work of the AAVSO."[5]

PUBLICITY EFFORTS

The AAVSO's early publicity efforts, some long in the making, were now appearing at an opportune time. Two articles about the AAVSO appeared in the Boston papers, and the United Press International wire service carried one of the articles to newspapers nationwide.[6] A "much-heralded" *Saturday Evening Post* article on amateur astronomers featured amateur observers in a good light, but as Clint Ford noted, its author also felt it necessary to go into the topics of "peeping toms" and youngsters "necking" at star parties.[7] A well-written *Mechanix Illustrated* article by AAVSO Headquarters assistant Helen Stephansky featured variable star observing – including photoelectric photometry – and the AAVSO.[8] An article by C. L. Stong in the *Scientific American* described photoelectric photometry and how amateur astronomers could become involved, particularly by making photometric observations of variable stars. Much of the article quoted AAVSO photometrist Ruiz.[9] The publicity efforts were paying off in popular interest, if not in contributions: "As a result," Stephansky noted, "our mail has been quadrupled!"[10] A five-page pictorial in the weekly *Life* magazine, "U.S. youngsters invade the grown-up world to spur a hobby boom in astronomy," was a lightweight piece, but surprisingly had a brief inset titled "Stargazing of genuine value," which featured an R Scuti light curve, a photo of 17-year-old variable star observer Charles Aronowitz of New Jersey, and a very brief mention of the AAVSO and its function.[11]

VINDICATION

The Ninth General Assembly of the IAU, August 29–September 5, 1955, in Dublin, Ireland, was a turning point for the beleaguered AAVSO and the harried Mayall. In his report on the activities of Commission 27 (Variable Stars), Commission President Kukarkin made it clear that the astronomical community recognized the value of both the past and future visual observations collected by "the variable star associations," which, mentioned in order, were the AAVSO, Association Française des Observateurs d'Étoiles Variables (AFOEV), Nordisk Astronomisk Selskab (Scandinavia), Royal Astronomical Society of New Zealand–Variable Star Section (RASNZ-VSS), Astronomical Society of South Africa (ASSA), and unspecified groups in Japan and Italy.

The Commission specifically called for closer coordination between the AAVSO and the RASNZ-VSS. In his remarks concerning the AAVSO, Kukarkin called for an extension of its LPV program, "on account of all the newly discovered bright Mira Ceti type stars to which no attention has been paid for many years." He momentarily departed from the usual business-like tone of the IAU reports to say, "One cannot refrain from expressing regret that in the course of the period under report the American Association was separated from the Harvard College Observatory, where it flourished during more than forty years of its existence."[12]

The affirmation by her colleagues of the AAVSO's work, and that of the other amateur observing groups worldwide, was a much-needed boost of morale for Mayall and the AAVSO membership. At one point, Mayall "made an impromptu speech on behalf of amateur work, and was greatly applauded."[13]

Margaret Harwood, who also attended the Dublin meeting as a Commission 27 member, wrote, "Everywhere, I could see appreciation of the work of the A.A.V.S.O. and disgust and amazement at the H.C.O. attitude and treatment of Variable Stars."[14] The Commission's recognition of the importance of variable star studies, and of the value of amateur visual observations, helped to set a new tone for professional–amateur cooperation. "Everyone who attended," Mayall reported, "came away filled with enthusiasm for future work on variables."[15]

The Commission planned for cooperation between observatories and visual observers throughout the world. "Astronomers present," Mayall noted, "continually emphasized their need for complete light curves of the long-period, semi-regular, and irregular variables. All were in agreement that the only way they can be sure of getting these curves is through the assiduous work of the many groups of amateur observers." Mayall agreed to add stars to the AAVSO observing program "as rapidly as we can get charts and suitable sequences of comparison stars. This will mean extra work for all of us," she noted, "but it is very necessary work. Eventually it will add several hundred stars to our program."

The Commission also acknowledged the importance of photoelectric photometers "for observations of rapidly varying stars and other variables with small ranges." Several professional astronomers made very specific appeals to the amateur community for the types of stars that they needed to have observed. A group of astronomers who were interested in the programs of the amateur variable star associations held an informal meeting at Mayall's request. All agreed on the importance of using accepted comparison star magnitudes and to adopt any changes on a prearranged date.[16]

The IAU meeting must have put the Mayalls in a mood of triumphant accomplishment. Not long after they returned to AAVSO Headquarters, a news item appeared in the *Cambridge* [Massachusetts] *Chronicle*, titled "Famous woman astronomer back from Dublin." The details were most likely provided to the newspaper by Mayall herself, who was not only aware of the article's publicity value for the fund drive (which the article mentioned), but who certainly knew that it would be read by Donald Menzel, his HCO staff, and the Harvard community. Perhaps with that in mind, the opening sentence began with a flourish: "Cambridge's famous woman astronomer, Mrs. Margaret W. Mayall, has just returned from the meetings of the International Astronomical Union in Dublin, where she was entertained by [Prime Minister and Irish Nationalist movement leader] Eamon de Valera, Ambassador W. H. Taft of America, and many other notables interested in astronomy."[17]

MILESTONES

Marking another milestone: for the first time the AAVSO held both its annual business and general meetings on October 22, 1955, outside of Cambridge in Springfield, Massachusetts. Harlow Shapley was present, and Carl L. Stearns, director of the Van Vleck Observatory, Wesleyan University, gave a talk on trigonometric parallaxes.

Reporting on the AAVSO's services to observatories and astronomers in her 1955 Annual Report, Mayall mentioned fulfilling many requests for information and light curves. She also determined dates of maxima for 420 variables for use in the second edition of Kukarkin's *General Catalogue of Variable Stars*. "It is not a one-way service," she added, "for the observatories have been very generous in their help to the AAVSO." She cited the staffs of the Warner and Swasey, Lick, and Mount Wilson observatories who had indicated their willingness to help in identification problems and to supply photovisual plates for new blueprint charts. Mayall also reported on the IAU General Assembly meeting and how the astronomers there "were enthusiastic in their appreciation of the work of the AAVSO, and of the other amateur observing groups."

Mayall enumerated their many publication efforts: Leon Campbell's *Studies of Long Period Variables* was completed and would be published in December; *Quarterly Report #18*, the first to list 10-day means, was in progress; five bulletins containing bimonthly predictions of variable star brightness and the annual max-min predictions were published; "Variable Star Notes" appeared in each of the six *JRASC* issues for the year; and she and Newton Mayall completed their revision of Olcott's *Field Book of the Skies*.

Headquarters had received 64 990 observations made by 35 observers in 19 countries. She ended her report by acknowledging the able assistance of Helen Stephansky. In addition to her routine office tasks and many "urgent" data request interruptions from astronomers, Stephansky successfully and almost single-handedly completed much of the year's publication production work and virtually all of the annual meeting preparation and mailings without supervision while the Mayalls were abroad. Clint Ford was equally appreciative: "Helen is a cornerstone of the present set-up," Ford wisely noted, "and to lose her would mean drastic cuts in the Assn's activities."[18]

Up to this time, the end of 1955, there were only four scientific committees in the AAVSO: Solar, chaired by Neal J. Heines; Occultations, chaired by Margaret H. Beardsley; Auroral, chaired by its sole committee member, Donald S. Kimball; and Nova Search, chaired by Mayall. In contemplating Committee chair nominations for the coming year (1956), AAVSO President Cyrus Fernald raised a significant point, saying that the AAVSO Council's role in such selections would be much more crucial than in the past. At HCO it had only been Campbell and a few others who controlled AAVSO operations.[19]

PUBLICATION SUCCESS

With the end of the year 1955 came the publication of Campbell's *Studies of Long Period Variables*, which had been started in early 1940, interrupted by the war, and halted when Campbell died in 1951. Mayall, who continued the work at Shapley's urging, wrote in the preface: "It has been my duty as the present Recorder, to complete the descriptions and supervise the drawing of the mean light curves." She noted the difficulty she had in compiling the book's unfinished second part, working from only a few notes that Campbell had left. "We are indebted to Dr. Harlow Shapley," she concluded, "whose interest in variable stars and enthusiastic encouragement of the amateur observer helped make it possible for Mr. Campbell to collect the material for this volume." Mayall acknowledged the work of AAVSO member David Rosebrugh, who plotted the mean light curves; her husband Newton, who prepared them for the press; and Mrs. Florence Campbell Bibber, Leon Campbell's daughter, who typed the tables that made up most of the book. Its publication was made possible by an anonymous gift from Lawrence N. Upjohn of $1500 through his "Civic Fund."[20] A small amount from Mayall's NSF grant went toward final proofing and preparation. Important in this volume was the introductory article by Paul W. Merrill of Mount Wilson and Palomar Observatories. Based on his 1953 annual meeting paper, Merrill praised the value of the observations made by amateur astronomers and compiled by the AAVSO. Mayall sent a copy of the book to Campbell's wife. Her daughter replied that she had never seen her mother "so delighted with anything in her life." Mayall also sent copies to each of the Campbell children who "were all very happy about it." Sadly, Frederica Campbell died not long afterward on December 30.[21]

The publication of the Campbell volume, the affirmation of the professional community worldwide, and the continued survival of the AAVSO were perhaps small indicators of progress in December 1955, but they had seemed unimaginable less than 3 years earlier. As a token of its appreciation and gratitude, the AAVSO Council voted unanimously to present the 14th Merit Award to Margaret and Newton Mayall, an event scheduled for the spring meeting.[22]

SATELLITE COMMITTEE

In March 1956, the AAVSO formed a "Satellite Committee" to work with scientists at the Smithsonian Astrophysical Observatory (SAO). SAO had recently been relocated from Washington, DC, to the Harvard College Observatory grounds. Mayall formed the AAVSO committee after receiving a request from SAO's Armand Spitz to serve on a National Committee for visual observers of satellites. Professor J. Allen Hynek was associate director of the SAO Satellite Tracking Program; Spitz was Coordinator of Visual Observations, assisted by Leon Campbell, Jr., of the HCO staff as supervisor of station operations. The AAVSO committee consisted of Newton Mayall, chairman, with Margaret Mayall and Cyrus Fernald, members.[23]

The idea behind this program, as Mayall pointed out, was that observers would serve as visual spotters when a satellite was first launched. Their tracking would be especially important if radio signals should fail. "As soon as [the satellite] has been located and its orbit determined," she reported, "the big telescopes will take over, but there will be another time at the end of its life when it begins to spiral in to the earth when only visual observers will be able to follow it."[24]

Visual tracking of satellites would also provide scientists with information about the density of the upper atmosphere. The SAO's satellite tracking program was the national headquarters for Project Moonwatch – the nationwide coordination of amateur observers enlisted in satellite tracking. AAVSO members were a part of its National Advisory Committee from the start and included Thomas Cragg, Edward Halbach, Russell Maag, and Mayall. The SAO tracking program was begun as part of the International Geophysical Year (IGY), in anticipation of the launching of artificial satellites that were planned as a part of the IGY.[25]

SAO Director Fred Whipple and the tracking program coordinators knew the value of having a network of amateur visual observers and eagerly sought the assistance and advice of the AAVSO, the Astronomical League, the Milwaukee Astronomical Society, and other such groups. Whipple knew he would be able to draw from an already existing base of highly experienced amateur observers. The leaders among those experienced amateurs would form the basis of regional observing stations – AAVSO member Walter Scott Houston in Manhattan, Kansas, was one particularly successful regional leader.[26]

The regional leaders would recruit and train observers for their stations – many of whom would be raw novices – who would eventually come up with amazingly accurate and useful observations over a 5-year period. Although in the past it had taken Edward C. Pickering decades to realize his goal of "co-operation" (see Chapter 2), Project Moonwatch was off and running in 1 year and grew to include scores of international stations, with an especially strong showing in Japan, thanks to an already existing nucleus of experienced and dedicated amateur astronomers around the world.[27]

HEADQUARTERS COMMUNICATIONS

At the start of 1956, Mayall took a moment to compose a short letter to the AAVSO's members. She apologized for not being able to answer all letters promptly or at all, but added, "please don't stop sending them." When observers heard nothing from the recorder, they naturally assumed that nothing was being done with their observing reports other than being filed away. Mayall wanted to dispel this notion right away. "One of our most important jobs," she wrote, "is to keep all the plots of light curves up to date. Your reports will not be filed away for 'future use'... they are plotted [by hand] immediately and are available for reference at any time! And the plotting is no small job, when you realize that last year we received nearly 65 000 observations on 700 reports."

Her next important announcement concerned the temporary discontinuance of the bimonthly predictions of brightness in the interest of saving staff time and publication and mailing costs, but Mayall did not give these as her reasons. Instead, she stated that the bulletins were no longer necessary since the Campbell volume was published – observers could refer to the published mean light curves to make their own predictions. Headquarters would publish the predicted dates of maxima, and observers would use a paper scale to measure time intervals on the published mean light curves. Mayall even gave instructions on how to do this.[28]

PROCESSING OBSERVATIONS

In late March 1956, Mayall attended the American Astronomical Society (AAS) meeting in Columbus, Ohio. She gave a paper on predicted and observed light curves of nine LPVs, based on 5-year curves she drew for Merrill in 1952 with 3 years of additional observations.[29] Of course, the preparation of such light curves depended on the careful and systematic processing of observations received at AAVSO Headquarters.

Of the papers presented at the 1956 AAVSO Spring Meeting held in May, Newton Mayall gave the membership an unusual, but much-needed, lecture entitled "What goes on!"[30] He emphasized the improvements in administrative efficiency at AAVSO Headquarters since the Association's reorganization and showed how this allowed the expansion of observing programs, which in turn led to improved service to astronomy. Almost sounding like an angry schoolmaster scolding his pupils, he began by saying, "Now examine these figures," then explained in detail the many steps involved in receiving, recording, and processing observations each month – performed by a staff of only two – but made it clear that this was only a glimpse into the AAVSO's daily operations. "It is obvious that the Recorder and her assistant do not watch the clock," he said. This lecture must have been a real eye-opener for many of the members and officers, particularly those on the Council who had been offering nothing but criticism about staff inefficiency, delays, and lack of communication at headquarters.

The high point of the 1956 Spring Meeting was the presentation of the fourteenth Merit Award to Margaret and Newton Mayall, "for their unfailing devotion to the ideals and interests of the Association during its recent years of great trial, and for their combined efforts in completing the Leon Campbell Memorial Volume." The joint award surprised them both; it not only pleased Margaret Mayall very much, but, as Stephansky noted, was also "quite a boost for her morale, which had been kinda low lately."[31]

The AAVSO held its annual meeting on October 19–21, 1956, at the Springfield Museum of

Natural History, Springfield, Massachusetts. For the first time since 1953, there was no sense of urgency at the Council meeting about the AAVSO's existence. The NSF and ONR grants went a long way to help to stabilize operations. Ford still could not report any success in finding a large contributor to the endowment.[32] Shapley was present at the banquet to give his annual highlights of astronomy. The publicity-conscious Mayall sent out copies of Shapley's highlights in a press release.[33]

INTERNATIONAL COOPERATION

In the summer of 1957, Frank M. Bateson, director of the Variable Star Section, Royal Astronomical Society of New Zealand (RASNZ-VSS), and his wife Doris visited several observatories in Canada and the United States to discuss how RASNZ-VSS could be of service to professional astronomers. Frank Bradshaw Wood of the University of Pennsylvania arranged for Bateson to give a series of lectures at observatories across North America to help pay his way.[34] An important part of the trip – in keeping with the 1956 IAU Commission 27 directive – would be a meeting with Mayall at AAVSO Headquarters that fall. Bateson and Mayall held a number of meetings "to arrange even better co-ordination of their respective programs than had existed in the past." Bateson had many questions concerning observing programs, charts (including chart making methods), sequences, methods of reduction and analysis of observations, and general needs and requirements. He also asked Mayall to provide him with as many copies of charts or photographic prints of southern stars from Harvard's collection as possible.

Their "numerous meetings" in late September and early October led to a number of important decisions: the VSS would drop all variable stars north of –20 degrees declination, with exception of U Gem, R CrB, and other special types (to help fill the "longitude gap") from its program; data would be "freely exchanged" between the two organizations; the VSS would provide the AAVSO with information on new sequences whenever available; and revised southern sequences would (with the cooperation of the other international observing groups) be adopted universally on a given date whenever possible. If necessary, Mayall would propose to the IAU that an 18-inch refractor with photoelectric equipment be set up at Rarotonga or New Zealand. Mayall and Bateson formally signed and dated a two-page summary of these and other points of agreement.[35]

The cooperation between Bateson and Mayall in 1957 was nothing new: the two had been exchanging letters on matters of sequences and observing programs since Mayall's days at HCO. Astronomer I. L. Thomsen of the Carter Observatory, Wellington, New Zealand, was greatly impressed with the zeal and generosity extended by Mayall when, in 1952, Bateson needed sets of IAU charts of southern variables, *Harvard Bulletins*, and other publications.

Bateson established working relationships with other professional astronomers as well, particularly Alan W. J. Cousins and Reginald de Kock at the Cape Observatory of South Africa, to obtain observations of southern comparison stars. By 1959, Bateson was confident he would be able to generate 500 charts for southern variables. Over the next 50 years, he would produce more than 1300 charts.[36]

SOME AAVSO SUCCESSES

The AAVSO's public side was now beginning to show itself as something more than mere publicity articles and press releases. Mayall's new title of "Director" made its first appearance in the February 1957 *JRASC*.[37] A John Ruiz article, "A Photoelectric Light Curve of u Her," appearing in the June 1957 *PASP* was based on observations he made during 1952–1955.[38] Both the AAVSO's auroral program and the Solar Division's Sudden Enhancement of Atmospherics (SEA) program were involved in the International Geophysical Year's (IGY) efforts. The IGY loaned AAVSO several recording devices for the SEA work, which was being done in cooperation with Dr. Walter Orr Roberts of the Climax High-Altitude Observatory.[39]

In 1957 and 1958, Mayall began to receive some recognition for her professional contributions outside the AAVSO. She was the first woman to receive the G. Bruce Blair Gold Medal from the Western Amateur Astronomers for contributions to Amateur Astronomy, awarded to her in August 1957.[40] Most notably, in December 1957, the AAS nominated Mayall for the Annie J. Cannon Award for Outstanding Contributions to Astronomy. On hearing of the nomination, Menzel sent a brief but cordial letter to Mayall in which he managed to deflect some of the glory his way, writing: "It

Figure 15.1. Joint meeting of the AAVSO and the RASC, Montreal, June 1957. From left: Isabel K. Williamson; Charles M. Good, RASC Montreal Centre President; Margaret Mayall; Dorothy Yant; Allie Vibert Douglas; Charles Fox; Richard W. Hamilton, AAVSO President; Katherina Zorgo.

is especially a pleasure for me to offer these congratulations since much of the work for which you received the honor was done at the Harvard Observatory and hence Harvard also shares, in a small minor way, in the honor awarded you."[41] The AAS awarded the Cannon prize to Mayall in the Spring of 1958.

In May 1958 Mayall notified the ONR that she had satisfied the terms of her contract, having completed all work for *Quarterly Report 19* (1951–1953 observations), the *Annual Reports* for 1956 and 1957, and *JRASC* "Variable Star Notes" on five stars.[42] Then, by June, with *QR 20* and *21* also completed and *QR 22* nearly finished, Mayall announced that "the gap between observations received and published" was finally closed as far as it could be. Coincident with this accomplishment, Mayall noted Peltier's 40th year of observing, Ford's 30th year of observing, Fernald's 90 000th observation, and Hamilton's 10th year as Chart Curator.[43]

Mayall attended the IAU's Tenth General Assembly, held August 12–20, 1958, in Moscow. In the Variable Stars Commission meeting, Commission President Boris Kukarkin acknowledged the AAVSO's "continued intense observations," the publication of observations (only as 10-day means), and the publication of "the important" Campbell Memorial volume. However, Kukarkin also noted that the AAVSO could not further expand its observing program to include 9th and 10th magnitude maxima of more than 200 bright Mira-type stars due to a lack of comparison star sequences.[44]

NEW DIRECTIONS

With a certain degree of stability attained by 1958, Mayall now had the luxury of considering ways of improving and expanding the AAVSO's operations. She considered "the possibility of processing all variable star data at headquarters by means of IBM punched cards." The Council recommended that she conduct further investigations "on this long-range project." The estimated conversion cost was $5000. Mayall was also now able to enjoy the regular occurrence of milestones as the AAVSO approached its 50th year. She noted Peltier's "500th consecutive monthly variable star report" and De Kock's 100 000th observation – the first such achievement in the AAVSO.[45]

Mayall established a number of special observing programs during this time, accommodating observers' wishes as much as possible. There were several projects undertaken by observers: Leif Robinson with the help of other California observers began a program of simultaneous observations of Orion Nebula variable stars; Donald Engelkemeir made PEP observations of flare and suspected flare stars; Harvard Observatory graduate student Andrew Young, Tom Cragg, and others monitored CE Cas a and b – a close double star, both components being Cepheid variables – and another Cepheid, CF Cas. Mayall noted that all three of these stars, members of galactic cluster NGC 7790, were "of great interest in the study of stellar evolution." In addition, Robert Adams and others made simultaneous observations of the peculiar variable V Sge. Observers were encouraged to continue observing faint LPVs at minimum phase.[46]

"ALL IS FORGIVEN"

On November 22, 1960, Mayall wrote to Ford: "We had the Menzels down for tea recently, and Don seemed very enthusiastic about our [AAVSO 50th anniversary] meeting, planned at the Harvard College Observatory. Of course it is a feather in his cap to be able to point out to [the AAS] that 'all is forgiven.' "[47]

The AAVSO celebrated its 50th anniversary at the 1961 October meeting held at the Harvard College Observatory with HCO Director Menzel and his wife as the meeting hosts. Menzel attended a buffet supper on the first day and welcomed the attendees, but he regretted not being able to stay for the remaining sessions; his wife, however, stayed through the meeting's

Figure 15.2. R. Newton Mayall, Thomas Cave, Tom Cragg, Claude Carpenter, and Margaret Mayall, at Griffith Observatory, August 1961.

conclusion. Dorrit Hoffleit, now director of the Maria Mitchell Observatory, Nantucket, Massachusetts, conducted a special Symposium on Variable Stars. Participants in the symposium were Valfrids Osvalds, Leander McCormick Observatory; Harlan J. Smith, Yale University Observatory; Helen Sawyer Hogg, David Dunlap Observatory; and Cecilia Payne-Gaposchkin, Harvard College Observatory.

During the end-of-meeting festivities, Ford read a congratulatory telegram that was sent to Lacchini celebrating "his 50 years of continuous observing for the AAVSO." Ford also read greetings and telegrams received from around the world. Shapley was once again present for the first time since 1956 to deliver the "Highlights of Astronomy" over the past 50 years – number one being the "Founding and growth of the AAVSO." Reginald P. de Kock of South Africa received the AAVSO's 15th Merit Award for contributing more than 100 000 observations of southern stars, and Ford received the 16th Merit Award "in grateful appreciation of his moral and material support during trying times, his long observing record, and his service as Secretary and President." The evening concluded with film clips (1928–1958) titled "Through the years with the AAVSO" compiled and edited by Lewis Boss, Claude Carpenter, and Ford.[48]

The successful meeting at HCO did much to help put the past in its place; Mayall, writing to Ford, said, "it went off very well.... I think we can have future meetings at HCO when we want to... now that the barrier is broken down."[49]

Guests who attended the 50th anniversary banquet found at their seat a special copy of *The Review of Popular Astronomy* (*RPA*), which featured "The Story of the AAVSO" by Newton Mayall.[50] The article (the first of several parts) summarized the 50-year history of the organization. Long time AAVSO member and observer Donald D. Zahner, editor and publisher of *RPA*, proposed to Margaret Mayall that variable star observing be made a continuing feature of his magazine. Mayall and Ford both thought that this was an excellent idea and promised to help as much as possible. In addition, Zahner wanted to collect observations from his readers, print the observers' names and totals, and forward the collected observations to Mayall each month. Mayall agreed to this plan with enthusiasm.[51]

In early 1962 he decided to have Margaret Mayall begin a regular series of "Director Reports," which included monthly observer totals and featured a variable star topic with a chart. This column was similar to what was being done for *JRASC* but with more emphasis on amateur observing rather than variable star research. The regular appearance of Mayall's column in *RPA* was an important development, since, as Zahner pointed out, her limited-circulation *JRASC* column was "the only source of information for non-members on the activities of the AAVSO." By the fall of 1963, Mayall was able to confirm that, judging by the response she was getting, her *RPA* column was certainly helping to attract more amateur astronomers to variable star work.[52]

COMMUNICATIONS AGAIN

With the Association now on "thicker ice," some observers and members must have felt justified in raising their voices to complain during 1962 through 1964 about what they perceived to be a lack of communication from headquarters. Ford relayed a complaint from member Claude Carpenter to Mayall, who felt that observers should receive an acknowledgement card for each monthly report submitted, something Campbell had routinely done. Even for the old hands, Carpenter said it would be a morale booster. The observer totals published in the bimonthly *JRASC* were not enough of an acknowledgment. Ford noted that he had been hearing many similar complaints from other AAVSO members, and he quoted from letters and cards he had received from Claude Carpenter, Fernald, Anderson,

Robert Adams, and Carolyn Hurless over the past year or more. Ford believed that the reinstitution of acknowledgment cards would go far in quelling the disquiet. He also felt that a revival of some form of Stephansky's chatty news bulletins of 10 years previously would be a big morale booster, but acknowledged that Mayall's staff was already overworked.

Ford estimated that Mayall and her assistant must now do four times the work as Campbell and his assistants did – and they also had their research contract work to fulfill.[53] There was really little that Mayall could do or say to remedy the situation; she was, as usual, too preoccupied with keeping the organization alive. Even so, during these years she did her best to respond – and respond cheerfully – to every letter that she received, if not to every monthly report.

Figure 15.3. Lewis J. Boss, Chairman, AAVSO Photoelectric Photometry Committee 1954–1967.

VISIBILITY IN PROFESSIONAL COMMUNITY

Mayall attended the second IAU colloquium on variable stars held September 1962 in Bamberg, West Germany. As a result of the discussion in Commission 27 at the General Assembly of the IAU, she announced that the AAVSO was setting up a program for all variable star groups throughout the world to send their observations to AAVSO Headquarters so all observations could be combined into mean light curves. "This will make all observations available to the astronomer in one place, and the mean light curves should be more complete," she reported.[54] In the years to come, many important amateur variable star organizations around the world did voluntarily contribute their members' observations to the AAVSO.

One interesting indicator of the AAVSO's growing visibility in the professional community could be found in what was possibly the earliest example of correlation of AAVSO visual observations with infrared spectroscopic observations. After successfully observing o Cet, R Leo (with the help of AAVSO finder charts), and several other stars during the Stratoscope II Balloon flight on November 26, 1963, Princeton astronomer Martin Schwarzschild wrote to Mayall to ask her if she would have her good observers obtain estimates of Mira and R Leo for the date of that flight and also include the entire light curve for a couple of months surrounding the flight date.[55]

A PHOTOELECTRIC PHOTOMETRY COMMITTEE

Early in 1953, Mayall had contacted HCO's James Baker, saying that she wished to encourage amateur astronomers to take up PEP work, and seeking his suggestions for a simple photoelectric photometer for amateurs. Baker passed her query to John S. Hall at the US Naval Observatory (co-author of the 1949 *Sky and Telescope* PEP article). Hall outlined the basic requirements for such equipment but noted that the cost and the time needed to construct a PEP set was beyond the reach of many amateurs. One solution, he said, was to find a grant to cover costs and enlist a team of competent amateurs to mass-produce a simple photometer.[56]

Besides John Ruiz, Mayall found another AAVSO member who expressed an interest in photoelectric photometry. Lewis J. Boss was an executive engineer with the Philco Corporation in San Francisco and the nephew of the former Dudley Observatory director, also named Lewis Boss. Offering to help with the production of a PEP unit that could be used by amateurs, Lewis J. Boss wrote to Mayall several times wanting to know what equipment designs might be available. Mayall, unfortunately, was too overwhelmed with her other AAVSO work to give him the material he needed.[57]

Mayall did not forget Boss' interest. At the October 1954 Council meeting, she proposed "that the Council authorize the formation of a 'P.E.P. Committee' to encourage the interest of observers in photoelectric photometry." She also recommended that Boss be appointed the PEP committee chairman.[58] With Council's approval, President Fernald wrote to formally notify him of his appointment as PEP committee chairman, thanking him "for accepting this important post in our organization."[59]

In his position at the Philco Corporation, Boss was in an ideal position to know the latest equipment developments. He wrote to Fernald in January 1955 that he could probably "assist in procurement of equipment and possibly to guide design trends. If PEP is far enough along by the Spring Meeting," he added, "perhaps an invitation can be extended to those interested to volunteer their ideas and services to initiate a small observing program." Boss also noted that photometry expert Gerald Kron was at Lick Observatory, only 50 miles from him in San Francisco. He hoped to "enlist his interest" in the AAVSO PEP program.[60]

One of Boss' first official actions was – at the suggestion of Mayall – to write to Ruiz and Claude Carpenter asking them to serve on the PEP committee. In his letter to Ruiz, Boss proudly mentioned that he [Boss] pioneered PEP work over 30 years earlier by "constructing and using a selenium cell photometer" with which he observed using Frank E. Seagrave's 6-inch Clark refractor in Rhode Island.[61]

Ford, who was enthusiastic about Boss and the PEP Committee, suggested that Claude Carpenter and Ruiz would make good choices for the committee. He also mentioned three others with an interest in PEP: Francis Reynolds of Clarkson College; Edward Halbach of Milwaukee, Wisconsin; and John E. Welch of West Springfield, Massachusetts.[62]

During the early 1950s, Ruiz was the only amateur astronomer and AAVSO member who was experienced in developing and using PEP equipment for variable star observing.[63] Now interest in photoelectric photometry among other AAVSO observers was beginning to stir. Ford and amateur Domingo Taboada in Mexico were bitten by the electronic bug, and Ruiz would be guiding them through letters and visits in the coming months.[64]

Despite these ambitious beginnings, the nascent PEP program was still trying to center itself. Perhaps

Figure 15.4. Clinton B. Ford at his Wilton, Connecticut, observatory in 1955.

the most interesting item to be brought out at the 1956 Spring meeting was a collection of mimeographed papers by Ruiz on the construction and use of a photometer, which included circuit diagrams, observing instructions, sample observations, and examples of data reduction. Copies of this "handbook" were distributed to anyone who was interested.[65]

Ruiz made a series of evaluations on the use of transistors in place of vacuum tubes in the amplifier section of PEP equipment. He was encountering input impedance problems with the transistors he tested, and he intended to make further investigations. Boss was convinced that the newer types of transistors could be used for PEP amplifiers, if the right circuits could be built for them, but Ruiz was skeptical, although admitting that transistors "came after my time." Boss nevertheless wrote up a supplement for the PEP handbook, on the "types, design, and application" of transistors.[66]

Over the next few years, the PEP program continued to make progress, including a second, more formal edition of the handbook. *The A.A.V.S.O. Photo-Electric Photometer Handbook* appeared in 1958, and Mayall reported very good photoelectric observations of AD Leonis from Donald W. Engelkemeir of Hinsdale, Illinois. He observed a flare of this star and monitored it throughout its cycle. PEP Chairman Boss noted that Engelkemeir "also has the distinction of being the first amateur to obtain a photo-electric curve of the light from a recurring nova – RS Ophiucus."[67] Engelkemeir and Boss considered possible PEP programs, including one suitable for beginners. They also suggested to

Figure 15.5. Margaret Mayall with Lewis and Gertrude Boss, Long Beach, California, May 1963.

Mayall that the timing of minima of eclipsing binary stars might be worthwhile for the experienced observers. Mayall liked the idea, but cautiously made no commitment to it – she deferred instead to the expertise of Frank B. Wood at the University of Pennsylvania regarding the selection of stars and reduction of observations.[68]

Despite the interest shown by its small band of observers, the PEP program did not quite catch on at the AAVSO. In the spring of 1960, Boss wrote to Mayall saying that he wished to resign as PEP chairman, citing a lack of time in his busy schedule. He was satisfied with the program's beginnings but felt that it was time for an active observer to take over.[69] Surprisingly, Boss continued as chairman for another 7 years – always agreeing to stay on just 1 more year – finally giving up his post when he retired in 1967, and Arthur Stokes took over.[70]

SOLAR DIVISION RECOVERS

Harry Bondy, who replaced Neal Heines as editor of the *Solar Bulletin* in December 1953, reported in his first editorial statement of the January 1954 issue that the Solar Division "is again functioning." He noted the loss of NBS funding, but gave it a positive spin by emphasizing the fact that the Division was now reorganized, thanks to Mayall and the special Solar Committee members. He stated that Heines would remain chairman of the Division, but cryptically added that Heines would "be able to devote very little time to the work." Bondy understandably reported nothing else on the Heines problem, but he made it clear that "All work in the near future will be done strictly on a voluntary basis." Bondy's emphasis on volunteerism, of course, stemmed from Heines' willingness to do the Solar Division work only on salary, and his refusal to do any astronomy work whatsoever unless he was being paid. Heines – now chairman in name only – added no more than a muted postscript of gratitude to the editor's page.[71]

Although the Solar Division crisis was past, there was still the question of replacing Heines as chairman. "I do not think we should keep Neal in any longer," Mayall told Ford, "because he has shown absolutely no sign of interest in the AAVSO since he lost his contract." She felt that Bondy would like to take on the position, in addition to his "fine job" of editing the *Solar Bulletin*; but Mayall was not convinced that he was the right person for the job: "We should have someone who can command respect from the professionals as well as amateurs." Mayall had two possibilities, Ralph Buckstaff or Tom Cragg. She thought Cragg would be very good and would carry the prestige of Mt. Wilson with him, but doubted that he would accept the job. She also questioned whether it would be wise to have the chairman on the West Coast. Considering her problem, Mayall wrote to Ford, "It worries me dreadfully, for we must do something."[72]

After "much discussion" at the October 1954 Council meeting, the Council decided to replace Heines as Solar Division chairman "inasmuch as he had taken no active part in its work for over a year." The Council hoped, however, that Heines would agree to remain on the Solar Committee and voted to send him an official note thanking him for his service. The Council discussed appointing Cragg as Heines' replacement, but reached no conclusion.[73]

When the Council met the next day, they read a letter from Cragg stating that he did not wish to be appointed chairman of the Solar Division. Most on the Council thought that Cragg "should be pressed further to accept the appointment."[74] Mayall, who wanted Cragg to take the place of Heines, wrote to Fernald, who was head of the Nominating Committee, to express her hope that he would officially ask Cragg to be a candidate for the Solar Division chairmanship. "Another thing," Mayall said, "I don't think Neal should be given an 'Honorary' title. He has really disgraced himself and the amateur astronomer by his unwillingness to cooperate in the work on a volunteer basis. I fought for him last fall, but I think we now have leaned over too far in our consideration of him. If he had any interest at all in science, he would at least send in his observations when he is one of the Standard observers."[75]

On January 25, 1955, Cragg again wrote with his reasons for not accepting the Solar Division chair nomination. He stated that he had been active in the Solar Division since 1944, and wanted to be of service, but his circumstances worked against it. Cragg felt that his work in the solar department at Mt. Wilson Observatory demanded that he devote all of his research energies there – he could not be distracted with taking on the AAVSO chairmanship.[76]

After a 4-month delay caused by a series of miscommunications between Cragg and the AAVSO, the AAVSO Council offered the position to *Solar Bulletin* editor Harry Bondy. This news made Bondy "very happy."[77] Bondy would serve as Solar Division chairman until October 1963, when he resigned and was succeeded by Casper H. Hossfield.[78]

On August 21, 1955, Neal Heines died of a cerebral hemorrhage. Helen Stephansky conveyed the news to the Mayalls, who were abroad.[79] According to Bondy, Heines had suffered a stroke in February 1955, but did not seek medical help. His wife was then hospitalized in early Summer 1955.[80] As a result of these unfortunate circumstances, Bondy was unable to obtain the Solar Division archives up to 1954; the heirs were not familiar with the nature of the work and thus were not inclined to release any of Heines' papers.[81]

GUARANTEE OF SURVIVAL

The AAVSO operated in survival mode for nearly 10 years after separating from the Harvard College Observatory. This meant maintaining a physical presence and continuity of operations, finding year-to-year financial support with which to accomplish that, reorganizing the Association to adjust for its release from HCO control and also bringing its observing programs in line with emerging research interests, reaffirming and strengthening the amateur–professional relationship that was the nexus of the AAVSO, streamlining AAVSO procedures and operations, and lastly, establishing an endowment and developing it to the point where a comfortable and dependable operating income might be attained.

Although confidence wavered at times, Director Mayall and Secretary Ford were determined to not let the AAVSO fail – their willfulness led the Council through the worst moments. Yet that would not have meant very much in itself. What really kept the AAVSO moving forward was awareness of its primary mission – to provide astronomers with useful data. Thus it was the observers themselves who guaranteed the AAVSO's survival and who would serve as a strong foundation for the AAVSO's future.

16 · Breathing room on Concord Avenue

The AAVSO that we see today, a thriving and financially independent organization still devoted to its original aims and purposes, would delight Leon Campbell. We are now assured that the Association will indeed "grow, and grow, and grow" thanks above all to the determined efforts and unfailing devotion of its retiring Director Margaret W. Mayall.

– Clinton B. Ford, 1973[1]

Margaret Mayall and the AAVSO went through the remainder of the 1960s and early 1970s with few distractions and accomplishing many of the Association's goals.

During this time, Mayall expanded the number of observing program stars, issued new series of charts, formed new observing committees, began using modern computer data processing methods to compile and plot observations, and approved the start of new publications. Above all, she maintained a plodding attention to the fundamentals of the organization: collecting and plotting observations; disseminating them in the form of publications and request fulfillment; and chart production, revision, and comparison star sequence evaluation.

By these steps the work of the AAVSO continued to grow – and that itself could be called Mayall's greatest accomplishment as Director. This chapter follows the history of the Association in a new headquarters location, the expansion of observer programs and publications efforts by its members, and how it finally reached financial stability.

SOME GROWTH

The returns from many years of hard work were finally accumulating for the AAVSO. Accomplishments in publication during 1964 included *Quarterly Report 26*; "The Story of the AAVSO, Part Four," which listed observer totals for the AAVSO's first 50 years as compiled by Curtis Anderson; and Part Three, previously published in 1961, which listed officers, meeting locations, and awards.[2] In August 1954, Mayall was "delighted and highly honored" to learn the Astronomical League had selected her as the 1964 recipient of its annual award.[3]

At the 1964 Annual Meeting, Mayall reported to the membership that headquarters had been providing data to seven professional astronomers who were doing spectrographic work; that there was a continued increase in the Endowment, as well as growth in membership; Leslie Peltier had won a sixth AAVSO Nova Award; and an all-time high of 83 599 observations were received from 303 observers in 18 countries.[4]

The Endowment continued to grow by small increments – its value in May 1965 was nearly $240 000 – but as yet there never materialized the hoped-for great infusion of cash from any one benefactor or philanthropic organization. AAVSO Treasurer Richard Davis noted that unless new research contracts or other income sources were forthcoming, the General Fund (covering operations) would be "in the red" starting about February 1966. He anticipated a deficit of $5500 during the 1966–1967 fiscal year. He further informed the Council that the Endowment Fund would have to be doubled "in order for its income to provide the major portion of the Association's operating budget, now estimated at between $15 000 and $20 000 annually."[5]

Despite the treasurer's dour prediction, the Finance Committee ended its fall 1964 meeting with a discussion of the "long-range possibility of instituting a new fund to be called the Building Fund, devoted to eventual purchase or construction of a permanent building or office for the Association's headquarters." Mayall had pointed out that "the present office space in Cambridge had been inadequate for some time, and that rental rates there would probably be increased again in the immediate future." The Director and Treasurer

Figure 16.1. Edward Oravec, Roger Kolman, Leslie Peltier, Carolyn Hurless, Thomas Cragg, Curtis Anderson, and Clinton Ford, in front of Peltier's comet seeker telescope and merry-go-round observatory, Delphos, Ohio. This was an informal gathering of AAVSO members organized by Carolyn Hurless in August 1964.

Figure 16.2. AAVSO Headquarters at 187 Concord Avenue, Cambridge, Massachusetts.

agreed to draw up a proposal for a Building Fund. Architect Newton Mayall agreed "to sketch tentative plans for a new building or office space facilities to provide for present and foreseeable future needs."[6]

Possibly inspired by the opening of the Schoonover Observatory in Lima, Ohio, constructed and operated by amateurs through a philanthropic gift of $38 000; by the recent progress being made in construction of the Ford Observatory in Southern California; and by the dedication of a telescope at the Stamford Museum in Connecticut, the AAVSO Council asked architect Newton Mayall to develop a set of preliminary plans for a headquarters building. At the October 1965 meeting, Newton Mayall presented plans for an elaborate three-story office building plus basement, including an observatory dome on the roof to house the Post refractor. Surprisingly, he drew a second set of only slightly less elaborate "Alternate Plans" in May 1966. These plans and the "building and fund endowment" were apparently never discussed again, and this grand vision of AAVSO Headquarters never materialized.[7]

Although the AAVSO Council might be commended for its ambitious wish to see the Association in an expanded, permanent home, this was just not the right time. The AAVSO, like any small business, was not isolated from the general economy or culture. The Kennedy assassination and escalation of the war in Vietnam led many to realize that the economy would suffer from inflation. General prospects for the immediate future were not good. Up until now, an optimistic outlook at headquarters helped the AAVSO survive, but an unchecked optimism could have only led to disaster.

In the meantime, necessity once again reared its head: on July 1, 1965, Margaret Mayall wrote to Clint Ford to announce that the AAVSO would be moving from 4 Brattle to 187 Concord Avenue, a first floor office with apartments on the upper floors. She found the place "after a dreadfully time-consuming couple of weeks" and decided to take it. The offices consisted of a large room approximately 18 × 28 and a private office, 11 × 11, and alcove space. Rent, she reported, was $160 per month, "much better than anything else of comparable size, location, and type of building." Though much more than what they were currently paying, she said she had consulted with Council member Richard Hamilton and he had approved.[8]

Only 600 feet west of Harvard College Observatory, the AAVSO's third official headquarters brought Mayall and her staff conveniently close to the Harvard astronomers, the astronomy library, and the Sky Publishing Corporation offices, then located in HCO Building C. In time the storefront adjoining 187 Concord became available, and the AAVSO rented that, too.

Her note to Ford gave enough other detail to indicate that things were hectic: she felt she "ought to stick to Cambridge" over the next 2 months, but she had meetings arranged over the weekend with the Fernalds, the Fords, and the Fairfield County Astronomical Society. She had a "rush job" of a data request she "promised to do for Keenan [Philip C., of Perkins Observatory] months ago!" She would also have to move the office before sailing for Bamberg on July 28 to attend a conference.[9]

ECLIPSING BINARY STARS

In the foregoing chapters it was shown how HCO's years of neglect of AAVSO's observing program gave the HCO Council the convenient "reasons" to evict the Association and retain the Pickering Endowment income. Eclipsing binary (EB) star systems offer a telling example of that neglect. It is unclear, knowing the importance of EBs for the advance of fundamental astronomical knowledge, why HCO Director Harlow Shapley or AAVSO Recorder Leon Campbell, who himself had observed EBs from the HCO southern station in Arequipa, did not vigorously pursue an EB program for AAVSO observers.[10] It is not surprising that EBs would be a high priority for Margaret Mayall as she considered useful ways to enlarge the AAVSO observing program and return it to the mainstream of variable star astronomy.

EB star systems had been a vibrant field of investigation for both amateur and professional astronomers well before the founding of the AAVSO in 1911. The importance of understanding these systems was twofold: EB systems offered the sole means at that time of estimating the diameters of stars from the surface of the Earth, and in conjunction with spectroscopic analysis, some EB systems yielded useful stellar masses.[11]

Little knowledge of stellar masses existed at the turn of the century. Using his own observations of southern hemisphere binary systems made at Lovedale, South Africa, amateur astronomer Alexander William Roberts exploited the light curves he developed during binary star eclipses to estimate the densities and diameters of the individual stars in a few such systems. From that information, Roberts surmised that the stars involved were extremely large and tenuous.

Interestingly, Shapely was very knowledgeable about EB systems. He had worked for Henry Norris Russell at Princeton University as a graduate student. Russell estimated the combined densities of the two stars in a larger number of binary star systems, but came to the same conclusion as Roberts; the stars in these systems were what are now known as giant stars, very large and tenuous.[12] Russell developed analytical procedures for reducing EB light curves to provide the shapes, sizes and masses of these systems more rigorously. Shapley applied Russell's methodology for more than 80 systems. His thesis on the subject earned him a Ph.D. degree in astronomy from Princeton, and a later publication with Russell documented an analytical process that remained standard in the field for many years.[13]

The AAVSO EB program

The roots of the AAVSO eclipsing binary program that finally emerged in 1965 can be traced to a couple of articles published in *Sky and Telescope* magazine in the early 1960s. The first was a discussion by Alan H. Batten of the University of Manchester, England, entitled "Why Observe Stellar Eclipses?" Five months later *Sky and Telescope* editor Joseph Ashbrook published his own article on RZ Cas with a light curve and its star field chart including comparison stars. Ashbrook called for interested readers to send their RZ Cas observations to him "for analysis and evaluation of the times of minimum." *Sky and Telescope* followed its RZ Cas campaign with publication of sequences and charts for other EB stars.[14]

Two AAVSO members became interested in EBs. David B. Williams was one of the observers who responded with enthusiasm to the *Sky and Telescope* campaigns.[15] Marvin E. Baldwin was another amateur observer who became the keystone of the AAVSO's EB program. When he joined the AAVSO in 1962, Baldwin was, according to Williams, still practicing his observing, making estimates of the Cepheid variable RT Aur with binoculars. "He was annoyed when one of his comp stars faded suddenly, so he began watching that star as well." Eventually, he produced a light curve and sent it to Mayall at AAVSO Headquarters. She responded: "Congratulations, you just discovered an eclipsing variable star." Baldwin was, said Williams, "apparently so turned-on" by discovering the eclipsing binary WW Aur, "that he started observing eclipsing binary stars full time."[16]

At about the same time, another young amateur, Thom Gandet, a high school student in Pittsburgh, started an organization for stellar observers. Modeled on the AAVSO and the Association of Lunar and Planetary Observers (ALPO), Gandet intended to achieve useful scientific results observing stars and named his organization the National Organization of Stellar Observers (NOSO). EBs were an important part of Gandet's plan for the NOSO's scientific program.[17]

The announcement of the formation of NOSO may have alarmed AAVSO Headquarters, for such an organization would represent a challenge to the hegemony of AAVSO in amateur variable star observing. Whether or not that is true, shortly afterwards Mayall wrote Williams asking him to organize an AAVSO EB committee, which Williams agreed to do. This seemed natural because he was already an AAVSO member and his group was using some of the AAVSO charts, including EBs in the field of AAVSO program stars. Most of those who had been observing these stars for *Sky and Telescope* comprised the initial participants in the AAVSO's EB program.[18]

At the AAVSO's Annual Meeting, held October 1965 in Cambridge, Massachusetts, the Council voted to establish an Eclipsing Binary Committee with Williams as its chairman. There were at this time 12 observers of EBs who were contributing observations to the AAVSO.[19] The EB program's mission was to obtain accurate times of mid-eclipse from which orbital period changes due to mass transfer and other phenomena could be detected. Dr. Ashbrook at *Sky and Telescope* served as professional advisor to the EB program. From his extensive knowledge of the literature, he compiled a list of almost 100 EB stars that were suitable for visual observation and had shown some indication of period changes. Chairman Williams prepared finding charts with comparison sequences for all the stars previously featured in *Sky and Telescope* and began creating charts for additional stars using the University of Illinois' observatory library. He also used an adding machine to prepare monthly predictions of minima for a growing number of stars, which were distributed to interested observers. Observations were collected and forwarded to Ashbrook for reduction and publication. As the number of observations increased, Leif Robinson, *Sky and Telescope* assistant editor and soon Ashbrook's successor as editor, assumed those responsibilities. When the lists of minima timings grew too long for publication in *Sky and Telescope*, Robinson began publishing the lists in IAU Commission 27's new *Information Bulletin on Variable Stars* (*IBVS*). His papers in this publication, widely read by variable star astronomers, were among the first three or four on the topic of eclipsing binary stars to appear there.[20]

Although enthusiasm was high, and the number of observers grew, the collection and forwarding of observer reports lagged more than usual in 1968 when

Figure 16.3. At the AAVSO Annual Meeting, Nantucket, October 1969. Marvin Baldwin is on the right.

Williams began working for Donald Zahner's *Review of Popular Astronomy* as assistant editor (a misleading title, because he actually performed all editorial and production functions for the magazine). The delay was enough to concern Mayall, who wrote to Williams to encourage him to catch up with the work. She was, in fact, not only afraid that observers would begin losing interest, but she also worried that she would lose her EB program to *Sky and Telescope*. She suspected they wanted "to get the whole program out of the hands of the AAVSO" and back to *Sky and Telescope*. Statements that the AAVSO had taken it over and made a failure of it concerned her. "That really hurts. I am fighting to keep it an AAVSO Committee, but it remains to be seen what the future holds," she wrote.[21]

Acknowledging the difficult situation, Williams resigned as chairman in May 1969, with Baldwin named as his successor. Not long afterward, Robinson assumed the chief editor position at *Sky and Telescope* and ended his participation in the program. The new EB Chairman Baldwin also assumed responsibility for reducing and publishing the times of minima.[22]

With the impetus of amateur interest, the AAVSO's EB program reorganized, expanded, and improved. Donald C. Livingston, who had access to a computer, began preparing ephemerides, which resulted in the enlargement of the program because the computer could effortlessly calculate predicted times of minima for any number of additional stars. Baldwin had a scientific calculator and programmed it to calculate the heliocentric corrections for observations. He later simplified the whole process by plotting a graph of each star's heliocentric corrections from which a value could be read at a glance for any day of the year.[23]

Another AAVSO member important to the progress of eclipsing binary observing was Edward Halbach, who, with Gerard Samolyk and other members of the Milwaukee Astronomical Society (MAS), made enlarged tracings from the Vehrenberg *Atlas Stellarum* to produce accurate charts in AAVSO format for more than 300 EB stars. This monumental project led to many more stars being observed because observers could now locate them.[24]

CATACLYSMIC VARIABLES

Although the AAVSO's development of an EB program came quite late, the Association's interest in cataclysmic variables (CV) was earlier, even preceding that of many professionals. Mayall's call for observations of faint U Gem-type stars (including the prototype dwarf nova) in the January 1954 *JRASC* was followed immediately by the IAU's acknowledgment of the importance of the AAVSO's observations of these stars and led the emerging professional interest into CVs and their role in understanding stellar evolution.[25]

Two articles by professional astronomers in the popular press called the attention of amateur observers to this class of stars. The first was by George S. Mumford, III, of Tufts University, Medford, Massachusetts, who published a series of articles on dwarf novae in the February and March 1962 issues of *Sky and Telescope*. He began his article with a discussion of U Gem, using data from Mayall's table of 100 years of observed maxima of U Gem, which she had published in *JRASC* 4 years earlier.[26] The second article was by Robert P. Kraft of Mount Wilson Observatory, one of the first astronomers to explore CVs. Kraft discussed his understanding of dwarf novae and novae as close binary stars in the April 1962 *Scientific American*. He published a formal scientific study of U Gem stars as cataclysmic variables in the same year.[27]

Not long after these articles appeared, AAVSO members Larry C. Bornhurst and Thomas Cragg, both in the Los Angeles, California, area, decided to observe the U Gem eclipse. They wanted to find out if astronomers using moderate-sized telescopes and AAVSO standard observing methods could make a contribution to the study of the dwarf nova. Although Cragg worked at Mount Wilson Observatory, he and Bornhurst had free use of the recently constructed Ford Observatory at Wrightwood, California, and made observations there.[28] The difficulty of observing an eclipse is in knowing precisely when it will happen. Using a time of minimum based on W. Krzeminski's recently published ephemeris, Bornhurst observed the magnitude of U Gem dropping rapidly just a few minutes before the predicted time.[29] He noted some fluctuation at minimum, then observed a rise in magnitude after about 10 minutes. This was, in effect, a "practice run" to confirm Mumford's published period for the star. With Bornhurst's observed time of minimum as a zero point, Cragg computed an ephemeris for the coming month. He sent this to several observers who had instruments capable of seeing the faintest comparison stars of the U Gem field. On February 10, 1965, Cragg and Bornhurst observed a "lack of an eclipse . . . when U Gem was continuously fainter than normal." They conferred with Dr. Olin J. Eggen, who was working at Mount Wilson Observatory where Cragg was employed. Eggen "said somewhat facetiously," Cragg wrote to Mayall, "it's as if one of the binary components just disappears for a while." Cragg continued:

> Dr. [Robert P.] Kraft [who was also at Mt. Wilson] was most impressed with the data and is very interested to see it continued. His contention of course is so few astronomers ever have a chance to follow it closely in any kind of continuous fashion that such coverage can't help but bring much new evidence to bear as to what really goes on in the system. Our present simple models are simple because we don't know what complications occur and on what sort of time scale. He was practically begging me to continue those runs as long as possible as he knew of no other similar coverage.[30]

Cragg followed this with a discussion of the observations and also informed Mayall that Robinson, *Sky and Telescope* editor, was anxious to get them. Cragg felt, however, "it would be much better from Headquarters where you MAY have additional observations from others, AND to make it additionally evident that this is an AAVSO effort rather than an individual." Cragg also cautioned that it was "just a pilot set of observations... merely to get an idea how to set up the real program next season" on the Ford Observatory 18-inch telescope.

Cragg and Bornhurst continued making observations of U Gem through February and March. In April, they observed an eclipse using Bornhurst's 10-inch telescope on Mt. Peltier. While observing the star at minimum during this eclipse, Cragg noted: "There were several distinct very rapid drops of nearly a full magnitude and recovery in what appeared to be less than one minute.... Each of the very deep drops were substantiated by the other observer [i.e., Bornhurst]. This is a new effect as far as I am aware, never having seen or heard it mentioned by anyone working on these eclipses." Cragg also noted observing an "obvious double minimum" during a following U Gem eclipse.[31]

Ford was very interested in what Cragg and Bornhurst were doing, and he soon began his own U Gem observing program. Ford felt that Cragg and Bornhurst should present a paper on U Gem at the 1965 Spring Meeting in Toronto. Cragg, however, felt that a paper would be premature until more observations could be collected. "The situation is so irregular," he wrote Mayall, "it really is rather difficult to say what is normal." Aside from this, Cragg did understand Ford's reasoning: "Clint's contention however, is to demonstrate rather forcefully that there exists an opening for visual work to be of great help in a current astrophysical problem; with which I must agree that he has a good point."

Cragg then discussed his and Bornhurst's consultations with Kraft at Mt. Wilson Observatory, who thought the material should be in the literature since the already published work of Krzeminski and Mumford was all the astrophysicists had to go on. "Clearly," Cragg continued, "we have material which is of considerable astrophysical interest since they don't have this kind of time coverage (a two month stretch). My contention is that I should like to hold onto the material at least until another maximum occurs so that we have some idea as to what went on during one whole 'quiescent' period." Instead of a paper, Cragg asked Mayall to "comment on the fact that a concerted effort is currently going on in the AAVSO on this problem, and that preliminary and sketchy returns so far indicate there's a lot more to this system than meets the eye."[32]

Despite Cragg's reluctance to publish, Bornhurst wrote a short paper on "Dwarf Novae as Binary Systems," which Ford read at the joint meeting of the AAVSO and RASC on May 22, 1965, in Toronto.[33] Bornhurst began by pointing out that "the theory of an association of dwarf novae with binary stellar systems is of recent origin," citing the early papers on this subject by M. F. Walker, A. H. Joy, and R. P. Kraft.[34] Then, he briefly explained how and why he began making observations of U Gem and presented a composite phase plot showing about one-quarter of the star's cycle, near primary eclipse. The plot was based on 767 observations of 29 different eclipses made by Cragg, Ford, Carolyn Hurless, Leslie Peltier, and Bornhurst. The wide scattering of the points in the plot was, Bornhurst noted, not due to errors of observation – the observers were careful to all use the same comparison stars – but to real differences between the light curves of individual eclipses. Bornhurst ended with a call for more observations: "Further observations are required to check the suggestion of a reflection effect observed rarely before primary eclipse and the indication of a double minimum often suspected during the eclipse."

Mayall included a brief account of the Bornhurst-Cragg U Gem observing series in a paper she gave on visual observations of 13 U Gem stars at the IAU Colloquium on variable stars held August 1965 in Bamberg, Germany. "This is another case," she reported, "where the amateur should be able to make a real contribution to our knowledge. Not many observatories can afford to tie up their photometers and large instruments for coverage of a single star over a long period of time. From the minima so far observed, it is very evident that we cannot consider any minimum a typical one."[35]

From spring 1965 through December 1966, Ford led a program of U Gem eclipse observations. He prepared tables of observable predicted eclipse minima for U Gem, distributing them to 15 AAVSO members for eclipses of October 1965–December 1965, 22 observers for eclipses of January 1966–May 1966, and 23 observers for eclipses of August 1966–December 1966.[36]

TOWARD THE FUTURE

Once again, in the absence of any government contracts, Treasurer Davis anticipated a deficit for 1965–1966. He again sought to divert Endowment income to the General Fund: this, he noted, "would permit the Director, for the first time since the inception of the Endowment Fund in 1953, to devote 100% of her time to work for the AAVSO." The Council favored this move, noting that the Fund had a current value of more than $260 000 and "was on a firm footing and could forego the investment of its income for at least one year." The highlight of Mayall's Annual Report was that for the first time, more than 100 000 observations were received in a single year (102 026 observations from 320 observers).[37]

One disastrous event occurred in 1965 when a Boston blueprint company hired to mass-produce charts for the AAVSO lost 277 original chart tracings. Newton Mayall arranged for the personnel at his architectural consulting firm, Planning & Research Associates, to make new replacement tracings inked on Imperial Cloth from copies on hand at a cost of $15 each. Newton Mayall not only supervised this operation, but made most, if not all, of the tracings – an enormous task.[38]

NEW CHART PROGRAMS

The AAVSO answered the IAU's call for adding more variables to amateur observing programs in earnest late in 1965 when Ford began a project making pencil-traced preliminary AAVSO charts for many previously unobserved variables. The basis of this work was the photometric observations of faint variable stars made by Charles P. Olivier of the University of Pennsylvania's Flower Observatory between 1928 and 1941, and two papers Olivier published in *Astronomische Nachrichten* (*AN*) in 1958 and 1959. Mayall sent a reprint from one of the later papers to Ford, suggesting that he contact Olivier to ask about borrowing some of the charts for long-period variables on the list. In early 1960, Olivier agreed to loan "a dozen or two of the more interesting charts" from the Flower Observatory files. Ford's original intention was to reproduce the charts for himself and other observers, like Cragg, who were interested in observing variable stars that were not yet part of the AAVSO's program.[39]

Early in 1961, Mayall wrote to Olivier asking about variable star charts used at the Flower and Cook Observatories near Philadelphia, Pennsylvania. Olivier told her about the set he loaned to Ford that were mentioned in Olivier's two latest articles and offered to send her any additional charts for reproduction from the stars listed. He also gave Mayall permission to reproduce any of the charts that he had loaned to Ford.[40] Perhaps having not seen any progress from Ford in 1960, Mayall might have been following up on her original suggestion for Ford to contact Olivier; at the time, however, Ford had been distracted with an illness in the family. In any case, it is clear that Mayall, acting on her own initiative after seeing the *AN* articles, was interested in expanding the AAVSO's list of LPV program stars by making charts available for a new series of stars.[41] In 1961, Olivier released his original photos and field sketches for 18 of the variables to the AAVSO, and Ford began using this material. Olivier sent the remainder of his usable photographs and field sketches to Ford over the next years.[42]

Ford's preliminary chart program is significant because it allowed hundreds of new variable stars to be added to the AAVSO observing program. What began as a short-term project that took about 5 years to complete led to a greatly expanded new charts program that continued for many years. Charles E. Scovil at the Stamford Observatory in Connecticut and the Rev. Ronald E. Royer at the Ford Observatory in California both contributed to the effort by making photographic surveys of variable star fields. Others contributing information about star fields and magnitudes during the early stages of the project were Cragg, Wayne M. Lowder, John E. Bortle, Larry Hazel, Howard Landis, and Dr. Arlo Landolt of Louisiana State University.

EXPANDING PROJECTS

The rapid broadening of the AAVSO observing program was evident on several fronts at the May 1966 meeting held in Chicago, Illinois. The Photoelectric Photometry (PEP) Committee reported on the installation of PEP equipment at the new Ford Observatory in California and on its collaboration with the newly formed AAVSO EB Committee. A special symposium on visual observation of EB stars was held at which Chairman Williams introduced his new committee. Marv Baldwin and Robert Monske also participated as panel members. Cragg led a symposium on eclipses of U Geminorum with Hurless and Ford participating.[43]

The AAVSO now had a total of nine committees or divisions: Occultations, Solar, Eclipsing Binary, Photoelectric Photometry, Cepheid, Nova Search, Chart, Photographic, and Telescope. A tenth committee on RR Lyrae stars would soon be added.[44]

Figure 16.4. Owen Gingerich, Barbara Welther, and Dorrit Hoffleit, Maria Mitchell Observatory, Nantucket, Massachusetts, October 1966.

Figure 16.5. Leif Robinson and Walter Scott Houston in Leslie Peltier's transit room, Delphos, Ohio, 1968.

In October 1966, AAVSO member Owen Gingerich of the Smithsonian Astrophysical Observatory (SAO) led a Council discussion on the possibility of processing AAVSO observations using punch card techniques. The Council and director had been seeking ways to improve data handling at headquarters, but now an excellent opportunity to make a significant change was at hand. SAO, Gingerich noted, would soon own its own computer, and he believed that the AAVSO would be allowed to have some free access time. The only outlay at headquarters would be to purchase or rent a keypunch machine so that individual observations could be recorded on cards suitable for computer reading.[45] Gingerich volunteered to help with programming the necessary compilation and plotting routines. He eventually would be of great help in obtaining funding and computer time, and AAVSO member Barbara Welther, an SAO researcher, would help with the development and implementation of many important computer programs. The AAVSO Council decided to rent a machine, and at the May 1967 meeting, Mayall announced "the invasion of the grey giant . . . a beautiful IBM Key Punch, that is supposed to solve all our problems, except lunch."[46]

At the October 1967 Annual Meeting, the Council voted to raise the director's annual salary from $5000 to $7500 and the assistant's wages from $70 to $80 per week. Secretary Ford noted that "they were the first major increases in pay scale for the two paid employees of the AAVSO since the inception of the Endowment Fund in 1953." With the Endowment Fund now at more than $316 000, the treasurer felt confident that the Association was able "to absorb limited operating deficits."[47]

It turned out that the hoped-for free computer time did not materialize due to budget cuts at SAO. The AAVSO would have to find grant support for this work over the years, which included the cost of renting equipment, additional office space, and hiring of additional employees. In the meantime, the Council approved short-term funding of the computerization program, using money from the Endowment's accumulated income account. The Council decided it was important "to get the program under way immediately."[48]

During the next few years, Mayall applied the new computer plotting techniques developed by Welther toward the publication of *Report 28* containing the light curves of 464 variable stars. With this publication, headquarters established its basic data processing format and procedures, which would continue with numerous modifications over the next 36 years.[49]

The AAVSO held its 1968 Spring Meeting in Lima, Ohio, where it honored member Leslie Peltier for 50 years of continuous observing and reporting for the AAVSO. At the banquet, the AAVSO presented a scroll and citation to Peltier. Russell Maag, on behalf of the Astronomical League, gave him a painting depicting him at the eyepiece of his comet seeker telescope.[50]

Despite a sudden surge of political unrest in Hungary, Mayall traveled to Budapest in September 1968 to attend the Fourth Colloquium on Variable Stars. She presented a paper, "Eclipses of U Geminorum," in which she discussed the results of observations of

U Gem eclipses during four observing seasons, 97 of which were observed well enough so that the width of the eclipse curve could be determined. Of the AAVSO observers who contributed data on the 97 eclipses, Peltier observed 75 percent with his 12-inch Clark refractor – one-quarter of those in collaboration with Hurless, the two making alternate observations 1 minute apart or less, and the other 25 percent by Ford, Hurless, and Cragg. Other contributions to Mayall's study were made by Baldwin and Bornhurst, and Vicki Schmitz and Diane Lucas who lived in the same Ohio region as Peltier and Hurless. Mayall concluded: "I have great confidence in these visual light curves, and feel sure that a large number of minimum observations give good evidence that U Geminorum eclipses are irregular in shape and there is no difference in the shape before or after the outburst."[51]

LOSS OF PUBLICATION OUTLET

The AAVSO once again lost its outlet for publication of observer totals and feature articles about variable stars when *The Review of Popular Astronomy (RPA)* suddenly stopped after the August 1969 issue. Mayall wrote to Donald Zahner seeking an explanation, saying: "I have not seen the last four articles I sent you."[52] *RPA* was a fine magazine and there was substantial regret among amateur astronomers when it folded. After several years of *RPA's* success, Zahner lost interest in astronomy and switched to fly fishing, then his main avocational interest. *RPA* had never gained a large circulation or a full-time editor. Then the firm that was printing *RPA* went bankrupt; the addressograph cards for the *RPA* mailing list were lost to one of the creditors and there were no funds available to restore the situation. Zahner simply gave up and stopped publishing *RPA* with no notice to subscribers or in any other publication. The loss felt by some amateur astronomers when *RPA* stopped publishing was acute; *RPA* filled a niche for active observers that *Sky and Telescope* was not filling at the time.[53]

NEW PUBLICATIONS

With operations more stable than ever, and in keeping with the expansion of its observing programs and a general renewal of member interest, it was an appropriate time to enlarge the AAVSO's catalog of publications.

The interest in making acquaintances and sharing experiences with other members at the annual and spring meetings was shared through the pages of *Variable Comments* during the time of Olcott and Campbell. *Variable Comments* served that function well with its accounts of meetings and publication of the Director's Annual Report. *Variable Comments* began in 1924 and lasted until September 1954. When the *AAVSO Abstracts* appeared in 1951, brief meeting descriptions began to be added to the start of each *Abstracts* issue – these were expanded over time; also, the annual reports were being distributed to members separately, and so *Variable Comments* no longer served a useful purpose and was discontinued. There was always, however, interest in having some kind of membership newsletter that might be less formal and more wide-ranging than reports of meetings.

Ford attempted to start such a newsletter in November 1950, even created a mockup issue to show Mayall what was possible. Mayall liked the idea, noting, "KEEP. A good idea for some time," at the top of the page, but she was simply too busy to consider its development.[54] In 1954, Administrative Assistant Helen Stephansky attempted to fill the need for membership "news" when she added pages called *Bi-Monthly News* to the AAVSO's mailings of bimonthly predictions of maxima and minima of LPVs. Stephansky's pages contained brief items of interest and humor gleaned mostly from correspondence received at headquarters. These pages ran for only 1 year. Headquarters tried again 10 years later with *AAVSO News Notes* (1965–1974). These notes had a more regular format, with more emphasis on news and less on humor. Member Carolyn Hurless edited this publication using material she collected herself, as well as items collected by Mayall at headquarters. Like its predecessor, Hurless' *Notes* was an official publication of the AAVSO. It was printed at headquarters and included in mailings to members, usually sent together with a copy of the *AAVSO Abstracts*.[55]

Hurless also launched her own newsletter called *Variable Views*, which she published from 1964 to 1979. Although not an AAVSO publication and not officially sanctioned by the AAVSO, Mayall did support its publication. As Hurless often said, Mayall gave *Variable Views* "her blessing," and at least once, in 1968, Mayall even sent Hurless a personal check to help cover its costs.[56] With Hurless as editor and publisher, the aim

Figure 16.6. Carolyn Hurless with Margaret Mayall, AAVSO Spring Meeting, Louisville, Kentucky, May 1969.

of *Variable Views* was to serve as an informal meeting place where observers could describe themselves and their observing through short notes or news items. Hurless' original intention was for *Variable Views* to serve as a forum for "inner sanctum" observers, those who regularly made variable star observations fainter than 13.8 and lived mainly in the Midwest. Although many *Variable Views* participants were in this group, the only "requirement" for participants was a love of variable star observing. Each issue became progressively larger in size and scope with Hurless' boundless enthusiasm for variable star observing, for the dissemination of her *Variable Views*, and with the positive and growing response from its readers. Readers' contributions grew in length, and Hurless even began adding clippings from published sources, many times duplicating entire articles. Admitting that she was never a properly disciplined editor, she ceased publishing the newsletter. Suffering from painful neurological problems, Hurless never mentioned her illness in any detail in print. Instead she described her decision to give up as simply because the issues were too large, the "subscribers" too many, and the cost of postage, materials, and time too great.

The AAVSO discontinued its *Bi-Monthly Bulletin* of maxima-minima predictions in 1958 because of the increased workload at headquarters. In 1959, Mayall replaced that publication with an annual *AAVSO Bulletin*. At the May 1965 AAVSO meeting, Ford described taking the predictions from the *Bulletin*, together with the time intervals of maximum and minimum from Campbell's *Studies of Long Period Variables*, and combining them into a graphical table. As Ford pointed out, Cragg had already been combining these data in tabular form for a number of observers who were interested in making a special effort to trace out the minimums in the light curves of LPVs that fell below magnitude 14. These observations took on special significance as so-called "inner sanctum" observations and were one of the early regular features of Hurless' *Variable Views*.

Ford's graphical form of the tabulations proved to be even more convenient for observers. This was a calendar-like depiction using solid horizontal bars with various symbols of the times when a star was expected to be at minimum magnitude. Bortle, a dedicated binocular observer, then adopted this format and began preparing a similar graphical prediction of the times when long-period variables would be brighter than magnitude 9. This format made it easy for an observer to know at a glance when any of the listed variables could be expected to be visible during any given month; the old *Bi-Monthly* format, by contrast, merely listed each star and the month and day of maxima or minima.

Eventually, both Ford and Bortle tired of preparing the predictions and the work was turned over to headquarters. Mayall adopted the graphical form in 1966 as a supplement to her annual predictions bulletins. This graphical method became the standard format for future *AAVSO Bulletins* for the next 40 years.[57]

The *AAVSO Circular*

November 1970 saw the inauguration of the *AAVSO Circular*, which was initiated and edited by John Bortle. The *Circular* filled the need for a timely monthly source of information about any unusual behavior in some of the variable stars in the AAVSO's observing program. It would regularly provide observations of maxima of U Gem– and Z Cam–type stars; report on

Figure 16.7. John Bortle during the AAVSO Annual Meeting held at the Maria Mitchell Observatory, October 1969.

Figure 16.8. Edward Oravec, Charles Scovil, Leslie Peltier, Clinton Ford, and John Bortle, Delphos, Ohio, 1968.

activity of novae, old novae, recurrent novae, and irregular and R CrB–type stars; print director's comments on the activity of Mira variables; and list stars in need of observation. Observers mailed their observational data directly to Bortle at his home for the stars that were part of this program. They sent their reports to Bortle at the same time that they prepared and submitted their full monthly reports to headquarters. Assisting Bortle in editing, preparation, and distribution of the *Circular* were Lowder, Scovil, and Ford, eventually with Mayall advising. The *Circular* was a popular innovation among observers who hungered for information about some of these stars on a regular basis. The idea for the circular originated with several AAVSO members who were also members of the Fairfield County Astronomical Society in Connecticut, including Bortle, Scovil, Lowder, and Ford, and also Ed Oravec for a time. Their intention was to publish information with as little interference with AAVSO Headquarters routine as possible. Mayall was agreeable to the idea, provided that she be allowed final approval of what was published in each issue.[58]

Bortle, in particular, felt that with *The Review of Popular Astronomy* no longer publishing, "it was very important to the observer membership that some sort of publication illustrating their compiled results continue." Observers, he knew, needed to see that their effort was being used in some way but received little, if any, acknowledgment from the overburdened headquarters staff.

Bortle and the other Fairfield amateurs agreed that it was time for "a limited variable star newsletter independent of AAVSO." Actually, the *Circular* was semi-independent from headquarters, as Bortle pointed out, "benefiting from their advice and assistance but not obligating them in any significant manner." The *Circular's* independence from the direct control of the Association was stated in a *Circular* editorial somewhere in the first half-dozen issues. After a time, Bortle recalled that Mayall's oversight "became less and less." Eventually, Bortle came to know more about the behavior of the cataclysmic variable stars listed in the *Circular* than anyone at headquarters. He recognized that he had actually become "the primary source of specific CV outburst predictions for a number of real-time research requests from HQ for astronomers."[59]

The *AAVSO Variable Star Atlas*

The same Fairfield group saw the need for a variable star atlas, according to Scovil, who noted that their only motivation was that they regretted the lack of an atlas that showed where all their favorite variables could be found. Early in 1972 Scovil proposed the production of a new variable star atlas, provided that a suitable source could be found for the foundation stars. The first choice for a starting document was *Webb's Atlas of the Stars*, on which Harold B. Webb, a long-time member, had plotted many of the older AAVSO variable stars with their comparison star sequences.[60] Bortle, Lowder, and Oravec were all still using the Webb atlas while observing with binoculars. When contacted by Bortle, Webb, who lived on Long Island, was at first agreeable to that appropriation and gave Bortle an unopened copy to use for that purpose. However, when a potential publisher contacted Webb for a release of copyrights, he balked; an agreement could not be reached with the AAVSO on using his historical atlas.[61]

After considering, then rejecting, *Webb's Atlas* as a foundation, Scovil then conceived of updating the Smithsonian Astrophysical Observatory (SAO) *Star Atlas of Reference Stars.*[62] He found that the *Smithsonian Astrophysical Observatory Atlas*, a newer and more accurate atlas, would make an excellent foundation on a larger scale, say that of a typical AAVSO (b) chart, $60'' = 1$ mm. The prepared AAVSO Atlas charts would include photoelectric magnitudes where needed. A trained draftsman, Scovil prepared a number of sample pages and discussed the idea with Yale astronomers Dorrit Hoffleit and Adriaan Wesselink. Finding them supportive, he then pursued the matter with Margaret Mayall and Katharine Haramundanis, the director of the SAO's star atlas project.

Mayall wrote that they "certainly met all our specifications and expectations. It is our hope that the ready availability of this information in one volume will stimulate more amateur astronomers to contribute to the science of astronomy through the observation of Variable Stars. . . . You have our full approval to go ahead with this project."[63] Haramundanis was also enthusiastic and agreed to loan AAVSO the original working drawings from which the SAO *Atlas* had been reduced to its final size and printed; she quickly sent those to the AAVSO.[64] Using those samples, Scovil made crude examples of chart pages to show what was intended. Mayall then sought funds for the project from The Research Corporation. Unfortunately, they rejected the project as "too far from their area of interest."[65]

Scovil then presented the idea to the AAVSO Council at its 1972 Annual Meeting.[66] The Council authorized the expenditure of up to $250 to support further development of the project. By early 1974, contacts with several other foundations, including the National Science Foundation (NSF), had identified no financial support for the project. Ford finally agreed to provide support with income from the remainder of his father's estate to the extent of $25 000 spread over a period of 2 years.[67]

The *Journal of the AAVSO*

At the end of the May 1972 Council meeting, Scovil read a letter from Council member Bortle calling for the publication of a journal to be issued twice each year. This idea, like that of the *Circular*, originated with Bortle after discussions with Scovil, Lowder, and Ford. Lowder told Bortle about visiting the HCO library and noticing that the *AAVSO Abstracts*, "basically a stapled bunch of separate pages but the organization's only real annual publication of papers, Council minutes, and information – was not being filed like other astronomical publications received by the library." As he described the situation, he noted that they were "simply left out on a table . . . until the table was periodically cleaned off and the *Abstracts* consigned to the waste basket!" Lowder noted that the same thing was probably happening elsewhere.

"Admittedly," Bortle wrote, "the *Abstracts* did look very amateurish and probably not worthy of any filing or cataloging," but Lowder's observation prompted Bortle to visit headquarters and informally discuss the idea of a more formal publication with Mayall. At first, he only considered giving the *Abstracts* a more professional look, with proper covers and binding. Instead, he found that "this didn't seem of much importance to her." The idea of a formal journal took hold in further discussions with the Fairfield group. Bortle wrote a proposal, which Scovil read to the Council that spring. Preoccupied with editing the *Circular*, Bortle then withdrew from any further involvement with the journal.[68]

As with the *Circular*, Mayall declared her approval of the proposal for the formal journal with the understanding that she would have final approval of what was published. The Council authorized the director to compile and edit material for the first issue in collaboration with Bortle and Scovil, associate editors, and William H. Glenn, editor. With the recent loss of *The Review of Popular Astronomy* (and perhaps the earlier loss of *Popular Astronomy* still fresh in her memory), an AAVSO journal must have seemed like a good idea to Mayall. This quick approval, however, came with little thought regarding how the journal would be produced. Less than

1 month later, Mayall wrote to Scovil with many production questions, such as manuscript typing and final editing. "I definitely will want to see the copy before it goes to press, since it will be an official AAVSO publication," Mayall wrote. She also wanted to see Ford's Secretary's minutes in each issue adding, "There are so many things to be decided about the Journal, and this is just a starter!"[69]

The *Journal of the American Association of Variable Star Observers* (*JAAVSO*) began publication with its Spring 1972 issue appearing in September 1972. *JAAVSO*, Mayall wrote in an introductory note, "will be a place where professional and non-professional astronomers can publish papers on research of interest to the observer."[70] By the October meeting, *JAAVSO* editor Glenn reported that the reception was excellent from both amateurs and professionals.[71] By the next annual meeting, however, the Council, director, and editorial staff all expressed differing opinions about *JAAVSO*'s purpose. For Mayall, the key idea behind *JAAVSO* was, in her words, "papers on research of interest to the observer" rather than "papers of research interest to astronomers." She further stated that her understanding was that *JAAVSO* was to include "all miscellaneous publications previously mimeographed at AAVSO Headquarters such as meeting minutes, reports of the Director, and AAVSO Abstracts, etc." She questioned "the acceptance of papers from professional astronomers outside the AAVSO, lest *JAAVSO* become a dumping ground for rejects from the [*Astrophysical Journal, Publications of the Astronomical Society of the Pacific*] and other professional journals."[72]

In response, Lowder re-stated the policies of the *JAAVSO* founders (Bortle, Scovil, Ford, Lowder, Glenn, and other active observers), namely that *JAAVSO* "should be a publication outlet for valid work in variable star astronomy done primarily by amateurs, that its papers should be of sufficient quality to attract professional readers and . . . be referenced in standard catalogs and libraries, and that it should contain the business reports of the AAVSO, including abstracts and/or complete texts of qualified papers presented at AAVSO meetings." Cragg suggested that only papers given at an AAVSO meeting should be published. Mayall wanted "all AAVSO members, as well as outside amateurs and professionals" to be encouraged to use AAVSO data, "thus stimulating observers to contribute more papers." Davis, Stokes, Hossfield, and Hurless thought that two journals, "one technical and the other exclusively for business and news items" should be published.[73]

Since the initial impulse to produce an AAVSO journal was a somewhat offhand one, and that *JAAVSO*'s purpose at the start was only vaguely defined as a vehicle for material of interest to observers, it is not surprising that no one gave thought to an editorial policy. It was not until after its first issue that the Council realized how significant – or potentially significant – the publication of the AAVSO's own journal was. With this realization, a sense of seriousness about *JAAVSO*'s purpose ensued. Scovil reported to the Council that, since the *JAAVSO's* publication, "the prestige and recognition of the AAVSO has markedly increased as evidenced by many unsolicited letters from professionals as well as amateurs." Following this, Baldwin moved that *JAAVSO*'s editors and Editorial Board "be required to draft an editorial policy statement to be submitted to all Council members at the next meeting." This motion passed.[74]

WALTER B. FORD BEQUEST

By the end of 1970, the AAVSO's Endowment Fund had a value of more than $335 000. Treasurer Davis stated that the coming year's operating deficit would be met by withdrawals from the accumulated income account, as done in recent years. Although tenuous, this degree of financial stability allowed Mayall to relieve the workload at headquarters by hiring temporary student help. The students, who were put to work operating card punch machines, were available through work-study programs at Harvard, Radcliffe, and MIT.[75]

On February 24, 1971, Ford's father died at the age of 96. A few days later, Mayall received word from the New York State Court that the AAVSO would receive $500 000 from the estate of Walter B. Ford.[76] A week later, Mayall sent a letter to AAVSO friends and members announcing the bequest, saying, "I am sure that all members of the AAVSO will join with me in my appreciation of this wonderful bequest, the income from which will go far to alleviate this Association's recurrent financial problems and which will make possible many things which till now have been little more than dreams."[77]

Dr. Walter B. Ford was a retired mathematics professor who taught at the University of Michigan from

1900 to 1940. He died with an estate valued at more than $10 million. In 1912 his father had left him less than $1 million, and Walter Ford invested it in International Business Machines (IBM) stock. A *New York Times* article announcing the bequest quoted one of Ford's teaching colleagues: "Walter was an astute, careful investor, and his money increased noticeably, but I remember that before the twenties he lived a quiet, frugal life on a teacher's salary." The *Times* article reported that Ford bequeathed more than $5.2 million of the estate to several colleges, churches, and other philanthropic and charitable organizations. Harvard University, where Walter Ford received two degrees, was given $200 000. Other colleges receiving $200 000 each were Amherst, Hobart, William Smith, Eisenhower, Ithaca, and Hartwick State. The largest gift to an organization, $500 000, went to the AAVSO. The remainder of the estate, some $4.8 million, went to his son, Clinton B. Ford.[78]

Always short-staffed, the AAVSO was at last in a position to remedy the problem. At the 1971 Spring Meeting, Margaret Mayall stated that, in addition to her current full-time assistant, Florence Campbell Bibber, she planned to hire a second full-time person, one new part-timer, and five work-study students with at least one full time that summer. With the additional staff, Mayall intended to bring the publication of observations up to date and to publish computer-plotted light curves. The Council also authorized Mayall and Treasurer Davis to negotiate a lease of the office space adjoining the current headquarters.[79]

Of course, all who worried with the AAVSO through the previous 17 years were immensely pleased to hear about the bequest. Mayall received a number of letters from well-wishers; among them, Alan Shapley who wrote "My father [Harlow Shapley] and I said 'whooppee' when we read your circular about the . . . bequest." Yale astronomer Ida Barney wrote, "you and your staff certainly deserve this good fortune." Member John Ruiz said, "Donald Menzel, eat your heart out."[80] The bequest would be paid in time for the AAVSO's 60th anniversary meeting at HCO. "Our relations with HCO are fine now," Mayall gleefully wrote to Bart Bok, "especially since the announcement of the Ford bequest! . . . Pretty good 17 years after being stranded with total resources of $6356.76 plus many friends, eh what??"[81]

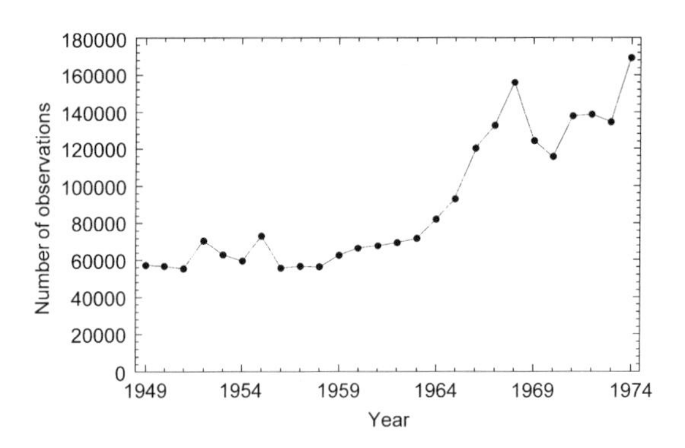

Figure 16.9. Annual totals of observations received at AAVSO Headquarters, 1949–1974.

The Ford Bequest helped the AAVSO reach its goal of financial independence. By May 1972, the total endowment fund market value was more than $946 000. Treasurer Davis felt confident enough to predict that fiscal year 1971–1972 would "probably be the last year in which operating budget deficits would have to be met by further withdrawals from the accumulated income account of the Endowment Fund, thus indicating that by October 1973 the AAVSO would possibly reach its long-time goal of full self-sufficiency within the income provided directly by the Endowment Fund plus dues and other non-recurring income sources."[82]

By the October 1972 meeting, Davis modified the prediction; he expected a $3000 deficit the next fiscal year. A $5000 pledge of a cash gift, most likely from Ford, put the organization's finances "in the black" until October 1973.[83]

READY TO RETIRE

At the 1971 Annual Meeting, Mayall told the membership that she was ready to retire and asked the Council to form a search committee to find her successor. She also announced that the Council had voted to increase the director's annual salary from $7500 to $10 000. (The search for a new director – which began shortly after the annual meeting – will be examined in detail in the next chapter.) Among the year's accomplishments, Mayall noted that more than 1 million IBM cards had been punched and verified, more than 125 000 observations were received, and there were more than 600 requests from professional astronomers and others (see Figure 16.9).[84]

Figure 16.10. Variable star conference at Stamford, Connecticut, 1973. From left: AAVSO Director Margaret Mayall, Joseph Ashbrook (behind Mayall), Charles Scovil (seated), Dorrit Hoffleit, and Clinton B. Ford.

Although busy as ever, Mayall was able to relax a little and reflect on what she had accomplished in 23 years as director. "It is a wonderful feeling," she wrote to Bok in January 1972, "to have an endowment fund and know the future of the AAVSO is secure . . . we have been able to expand our office space and hire more people. . . . You know we had built our fund to about $250 000 ourselves, which I think was pretty good after our total of $6000 in 1954."[85]

While the search for a new director proceeded behind the scenes, Mayall turned her attention to completing as many projects as she could. Scovil produced a new catalog of blueprint charts, and Ford made an updated catalog of preliminary charts. Mayall acquired a complete set of southern variable charts from Frank Bateson, and she and Cragg would plan on how to distribute the charts as they were needed.

In her 1971–1972 Annual Report (her penultimate report and the first to be published in the new *JAAVSO*), Mayall was pleased to describe "increased activity both at Headquarters and among the members." She began by pointing out that the card punch training program for student help at headquarters had produced "some very efficient operators," who nearly completed keypunching observations as recent as spring 1971. Her goal of publishing these observations, she reported to the Council, was "finally in sight." Her report for the next year, her last before retirement, also acknowledged the diligent and efficient work of the student punch machine operators who were up to early 1972. The year ended with a record number of observers (373) from 19 countries and 43 states who contributed a total of 121 889 observations. Margaret Mayall had done her best to bring the work of the AAVSO up to date.[86]

Part V

ANALYSIS AND SCIENCE – THE JANET MATTEI ERA

17 • The growth of a director

We have found a real treasure in Janet Akyüz, she knows a lot about variables.

– Margaret Mayall, 1972[1]

Margaret Mayall had many reasons to be optimistic about the future of the AAVSO in 1971. The Association had considerable financial assets in the Endowment Fund, more than enough to guarantee the salaries for a director and the staff, and the membership and observer programs had grown to all-time highs. It was time to pass the reins to the next person. She had served as the director for 24 years, and at 69 years of age, she was ready to retire.

It did not take her long to realize that a new assistant on her staff would fit the job description very well. She told her friend Bart Bok, who had left Harvard College Observatory, about several new hires in the AAVSO office. Mayall spoke enthusiastically about a young Turkish woman named Janet Akyüz who was helping her make some progress with her backlog of work.[2] Akyüz initially came to the United States to study for a medical career, but once here she discovered astronomy. It offered intellectual challenges she relished. Her early years in Turkey, education, and the events leading her to this transition and employment by the AAVSO provide necessary understanding for much that follows in AAVSO history.

As its fourth leader, Janet Akyüz would have a profound influence on the American Association of Variable Star Observers as she began to exert the director's authority over the many directions that the Association had taken.

EARLY YEARS IN TURKEY

Born January 2, 1943, in Bodrum, Turkey, the girl who would become AAVSO Director was the oldest child of Bulisa neé Notrika-Ishbir and Baruh Akyüz. Her father, a native of Bodrum and educated at the University of Ege in Izmir, participated in the operation and management of his family's businesses in Bodrum. The family owned several shops selling fabrics, clothing, and general merchandise. Janet's mother, Bulisa, was raised in Milas, a small city not far from Bodrum noted for its carpet manufacture. Together, Baruh and Bulisa raised a family of five children that included, in addition to Janet, two other girls, Beki and Kadem, and two boys, Yusef and Hayim.

According to her brother, Yusef, the children have fond memories of their home in Bodrum, which was located in Charshi, a village that was part of Bodrum on the shore of the Aegean Sea. The roof of their home offered a convenient spot for the children to sleep on hot summer evenings; they could hear waves breaking on the beach in front of their home while watching the stars parade overhead in the dark night skies. The cistern on their roof also provided fresh water for the neighbors around their home.

Janet's reputation as a bright student was first established in a one-room elementary school. The teacher in this limited school, impressed by Janet's intellect, made a special point of this to Janet's parents. They, in turn, recognized that there would be limited opportunity for any further education in Bodrum, which had no middle school or high school for any of their children. As Janet ended her elementary school years, her parents sold all their assets in Bodrum and moved the family to Izmir. Once settled there, they registered their children as students in the American Collegiate Institute (ACI).[3]

At ACI, Janet received the remainder of her primary and secondary education, continuing to exhibit the brilliance that caused her to stand out from the bulk of the student body. She excelled particularly in mathematics and science. The ACI teachers were all Americans; the language of instruction was English. More than half of the teachers were women; this stood in vivid contrast to the prevailing Islamic culture in which a Turkish woman's place was in the home and women were still not widely employed. These female teachers provided a model for Turkish women who had not achieved the full measure of equality American women enjoyed during college and afterward. They no doubt encouraged their best students – Janet included – to continue their education outside Turkey.[4]

Akyüz applied to four universities in the United States, winning a 4-year Wien International Scholarship at Brandeis University. Originally intent on a career in medicine, she earned a B.A. in Physics at Brandeis, graduating with honors in 1965. After her graduation from Brandeis, she accepted employment as a supervisor in a cardio-pulmonary laboratory, thinking that might be a useful prelude to medical school and her intended career. However, the work in that laboratory did not fully capture her interest.

Akyüz returned to Turkey, entering graduate school in physics at the University of Ege, located in Bornova, a suburb of Izmir. While studying there, she taught high school physics and mathematics in the nearby ACI, where she was very popular with students. After her first year at the University of Ege, an astronomer, Paris Pişmiş, who was visiting her hometown in 1969 suggested that Akyüz apply for a summer internship at the Maria Mitchell Observatory (MMO) on Nantucket Island back in the United States. Akyüz applied for and was invited to spend a summer as a research intern to MMO director Dorrit Hoffleit.[5]

Hoffleit's customary program for summer interns at MMO always involved variable star research, using the photographic refractor to make new plates and the observatory's archival plates that preserved images of star fields from earlier summers. Akyüz's project for that summer involved determining light curves and periods for RR Lyrae variable stars, a project that introduced her to both telescopic observing and observing with historical plates, then tying those results together with analysis of the combined data set.

That autumn, MMO served as the host for the 1969 AAVSO Annual Meeting. Because of that scheduled meeting, Akyüz remained at the observatory to help with the meeting and to complete her research before the meeting began. When poor weather delayed Hoffleit's return to the island from a meeting at Woods Hole on the Massachusetts coast, Hoffleit delegated the handling of the meeting to Akyüz. With the assistance of fellow student Nancy Gregg, Akyüz took charge of the observatory's role as host of the meeting.

Figure 17.1. Janet Akyüz as a summer research student at Maria Mitchell Observatory, presenting her first paper at an AAVSO meeting, October 1969.

Mayall and the AAVSO officers exhibited the flexibility required to conduct their usual meeting without any difficulty.[6] At that meeting, Akyüz joined the AAVSO and, as was customary for Maria Mitchell summer assistants, presented a paper titled "RR Lyrae Type Variables with Two Periods," describing the results of her work at MMO.[7] As young as she was at the time – just 26 years old – Janet Akyüz made a distinct impression on the meeting attendees. Also, at that fateful meeting, Akyüz met Harvard College Observatory assistant and AAVSO member Michael Mattei, who would later become her husband.

When Akyüz returned to Turkey at the end of the AAVSO meeting, she was determined she would follow a career in astronomy. She had learned to love variable star astronomy while working for Hoffleit. After completing her M.S. in Physics at the University of Ege in 1970, she returned to the United States and entered graduate school at the Leander McCormick Observatory of the University of Virginia. During the next year, she worked for the summer as a teaching assistant at the Hayden Planetarium in New York City. Akyüz earned her second M.S. degree, this time in astronomy in 1972, presenting a thesis entitled *The T Tauri Phenomenon* at the University of Virginia.

Hoffleit encouraged the 29-year-old Akyüz to apply for a position at the AAVSO as Mayall's scientific assistant. With Hoffleit's recommendation, Akyüz was hired in June 1972.[8] Akyüz and Mike Mattei were married later that same year.

Mayall, who had been thinking about possible successors "for several years," concluded that Akyüz was the right choice by September 1972.[9] "We have found a real treasure in Janet Akyüz," Mayall wrote enthusiastically to Clint Ford. "She knows a lot about variables," she continued, "works well with Barbara [Welther] on the computing programs and is very good on our many information requests. Things are really moving in the office now!"[10]

Although a Search Committee for a new director[11] had been formed in October 1971 on Mayall's request, Ford, the committee chairman, did not submit a list of names to the Council until spring 1972. He also listed for the Council a few minimum qualifications: the candidate should be a professional astronomer with specialization in variable stars, preferably with a Ph.D.; have an "understanding and appreciation for the work of serious amateur astronomers, and an ability to inspire their confidence and enthusiasm"; and have a "willingness and ability to cope with the large amount of office routine work demanded by the job."[12] The basic qualifications were not much different from those listed by Leon Campbell in 1949.

At the same Council meeting, Mayall suggested a letter be sent to the AAS job center. Ford did not do this until 3 months later, after it became apparent that personal inquiries by the Search Committee members to observatories, schools, and colleagues – although yielding a few additional candidates – were not the most efficient way to go about the search.[13]

Akyüz, now Mrs. Mattei, submitted her résumé as a candidate for the director's job in January 1973.[14] After two rounds of balloting during the spring of 1973, the Search Committee narrowed the field to seven finalists. The Search Committee members ranked the finalists in order of their individual preference. Only three of the five committee members selected Janet Mattei as their first choice. The other two were unanimous in their first choice, but their ranking of all the other candidates varied widely. The seven finalists were sent a list of general and specific conditions: the position had no fringe benefits such as vacation or travel pay, and the office of the director had to remain in Cambridge, Massachusetts. The detailed list of duties – a full page – was grouped by responsibilities for various functions: organization and office operation, observations, publications, and IAU responsibility.

In his August 1973 memo to the Search Committee members, Ford reported that all the candidates continued to express strong interest in the AAVSO directorship, "as does Mrs. Akyüz-Mattei who has been Mrs. Mayall's assistant for over a year and is currently in complete charge of headquarters operations during the Director's absence in Europe for the 1973 I.A.U. meetings."[15] In his October 1973 Council meeting minutes, Ford noted that Mayall "strongly recommended" that Mattei be selected as the new AAVSO director. In response to questions, Mayall reported she had interviewed four candidates. At the request of Council member Martha Carpenter, the qualifications of Mattei "were discussed at length." Ford did not record the names of the other candidates or details of the discussion. The Council decided to vote the next day.

The next morning, Mayall read a list of the director's duties and responsibilities as originally compiled for the Search Committee. She detailed the experience of Mattei in meeting each of the job requirements, "renewing her recommendation that Mrs. Mattei be elected." After a secret ballot, Ford noted, the tellers reported a unanimous vote in favor of Mattei, who would be the new director "effective immediately as to title [and] as of November 1, 1973 as regards to salary."[16] Later that day, after the general membership business, the committee reports, and the Director's Annual Report, Secretary Ford announced the election of officers for the next fiscal year. The last name read was "Director: Mrs. Janet Akyüz Mattei." She "received a standing ovation, and gave a short speech of thanks and acceptance to the general meeting." Ford also announced that the Council had elected the retiring Director Mrs. Mayall "to the special office of Consultant to the Director for a period of at least one year."[17]

MAYALL'S RETIREMENT

Because Mayall was still the Director through October 31, 1973, there was no formal retirement ceremony for her at that year's annual meeting. As consultant to the director, Mayall continued to go to the office every day. One of Mattei's first acts as director was to initiate a "secret" appeal to all members, asking

Figure 17.2. The new AAVSO Director, Janet A. Mattei, with Director Emeritus Margaret Mayall, in a 1973 publicity photograph.

Figure 17.3. AAVSO member/observer Diane Lucas of Elyria, Ohio, with Janet Mattei in 1974.

them to send her their photos, reminiscences, and letters of appreciation for Mayall. Mattei collected these in an album that she presented to Mayall during a special program of celebration at the next annual meeting.[18] In a farewell note in *JAAVSO*, Mayall wrote: "After spending 24 years as Director of the AAVSO, I decided to retire. It was with very mixed emotions that I made the decision, for my work with the Association has always been most satisfying, with never a dull moment."[19]

With only 1 year as an assistant in the office, Mattei faced enormous challenges. Mattei would need all the advice and counsel that Mayall could provide her. For example, shortly after assuming her duties on November 1, 1973, she received an angry letter from a member who believed that headquarters resented his sending observations to John Bortle for the *AAVSO Circular* (how he got that impression is unclear). He stated bluntly that he planned to continue doing so because he received acknowledgments from Bortle but never from headquarters. Mattei's way of dealing with such problems – the perennial need for care and feeding of members – would go from bad to worse before she mastered the necessary interpersonal and managerial skills to finesse such anger.[20]

Mattei was quick to assume control of the routine functions of headquarters, such as planning and execution of two meetings per year. Indeed, she launched into those tasks vigorously in her first Council meeting as director at the spring 1974 assembly in Winnipeg, Saskatchewan, Canada.[21]

Assuming the leadership of an ongoing Association, Mattei was forced to begin dealing with a multitude of problems that had been developing during the difficult years after the eviction from Harvard. The problems that had plagued Mayall since the eviction were problems that never went away, and new problems emerged to complicate the life of the new AAVSO Director. Nevertheless, beyond those administrative issues, Mattei was responding to increased requests for the Association's observations. Interest in variable stars grew among professional astronomers as theoretical understanding improved, more variable star types were recognized, and stars were added to the AAVSO program. More professional astronomers took up the study of variable stars and sought access to the AAVSO database for information. Membership increased annually, as well as the number of nonmember observers from outside North America. Although the growth in nonmember observers indicated the extent to which the AAVSO had advanced in importance, it also increased the daily data processing load – a problem the Council only slowly recognized. Thus the problems Mattei faced were not static problems; her targets moved constantly.

MEMBERSHIP GROWTH AND ADJUSTMENTS

First, and in retrospect, foremost of the problems that Mattei faced was the fact that the Association's members had never adjusted to the new environment for the Association's expanding operations. The need for member communication was a necessary, but heavy, commitment of time and energy that had limited Mayall's ability to handle other necessary work, though she tried. Mattei had neither the easy facility for such communication, nor the time to maintain it.

Furthermore, several external factors contributed to a substantial growth in the public's interest in astronomy and, perhaps, the growth in membership. The lunar landings and the space program accelerated interest in science education in schools. Other factors were the discoveries by astronomers at Mount Palomar, Kitt Peak, and many other observatories, which appeared regularly in the news. Teachers looking for attractive science projects occasionally realized the potential of variable star observing. The interest in science in general and specifically in astronomy was coupled with rapid population growth as the Baby Boom generation passed through the school systems and into full and productive employment. All combined to increase public awareness of astronomy as an avocational opportunity.

OBSERVATION GROWTH AND DATA PROCESSING

As Mattei struggled to manage the office routine, the director's work load increased continuously even as it evolved in new directions. As a consequence of both changes, Mattei experienced an increase in complaints from both observers and professional astronomers. Observers complained because they were no longer receiving a monthly acknowledgment from headquarters that their observation reports had been received, were appreciated, and were being added to the AAVSO's database of observations. Such monthly feedback had been a first priority for Leon Campbell, and for Mayall, briefly after she replaced him. However, all that ended with the eviction from HCO and the loss of the Pickering Memorial Endowment income. As described in Part 4, those events pitched the AAVSO into darkness.

At about the same time, during the 1950s and 1960s, professional astronomers began to take more interest in variable star astronomy as they studied stellar evolution. With high-speed computing capability during the 1950s, theoreticians like D. E. Osterbrock, C. Hayashi, and L. G. Henyey in the 1950s, and later P. Ledoux, S. A. Zhevakin, and J. P. Cox with Charles A. Whitney in the 1960s enjoyed some success in explaining stellar pulsation with models that showed promise.[22] Furthermore, astrophysical surveys produced catalogs that included variable stars, and relationships between those data and empirical observations of variable star behavior allowed not only refined categorization of variable stars, but also indicated the existence of inter-category relationships and continuities that were previously obscured.

Advances in galactic and stellar observational astronomy also exposed the importance of variable stars as distance markers, population tracers, age indicators, or for stellar mass measurements, and so on. Thus the number of professional astronomers who wanted access to AAVSO observational data continued to rise. Their complaints rose rapidly when they could not have instant access to AAVSO records. Such complaints were unreasonable under the circumstances, but the charges were nevertheless leveled at Mattei too frequently for her comfort.

The commitment to modern data processing technology had been made, but in the face of growing volume, headquarters could not keep up with the incoming data because of staff turnover. Mayall established routines for manually plotting 5- and 10-day means during her last years, and those graphs – though tedious to maintain – proved essential in answering questions that arose with increased frequency. However, nothing had been done to expedite conversion of recent observations recorded on keypunched cards into the plotted light curves that were needed to answer the frequent and increasing questions from the professional community. The hand-plotting of light curves and arrangements for semiannual meetings constituted an overwhelming work load for Mattei and her staff and left her too little time for membership communication and development of professional relations.[23]

Mattei recognized that the one way out of this dilemma was to capitalize on the promise of data processing. If incoming data could be converted to machine-readable cards quickly each month, then the

Figure 17.4. Janet Mattei and Owen Gingerich, 1981.

observations could be added to the growing inventory of data points for each star that already existed on computer tapes at the Smithsonian Astrophysical Observatory (SAO). Then, the observations could be plotted by SAO computers on a monthly basis. There would be no need to spend hours plotting the observations by hand as Campbell and Mayall had done for decades before her.

Although she had a solution, Mattei's hours during the day were, nevertheless, spent in the office answering telephone requests for data from professional astronomers, making arrangements for meetings, and handling other routine aspects of the director's job. But at night, assisted by her husband Mike Mattei, she loaded cards in their car and carried them to the SAO computing center, where the latest data could be read onto tapes and appended to the existing database. The observational database was then processed through a system developed by SAO astronomer and AAVSO member Barbara Welther and other SAO staff. Arrangements for this data processing were made by another SAO astronomer, long-time AAVSO member Owen Gingerich. Mattei hoped that by using modern data processing methods she could remain better aware of what was going on in the night sky and more expeditiously and accurately respond to the ever-increasing volume of questions from professional astronomers.

In an ideal world, this process might have matured and provided relief from the tedious pencil plotting of results. In reality, nothing worked quite right with those earliest computer routines; the plotting technology was limited, as well as the funding needed to refine those procedures. Thus the daily routine of answering questions continued to call for hand-plotting of the latest observations to maintain and extend the existing graphical records. In addition, a series of other major developments distracted Mattei as she attempted to master this fundamentally important organizational need.

FINANCIAL PROBLEMS – AGAIN

After the AAVSO received the Walter B. Ford bequest in 1971, the AAVSO Council settled into a comfortable mode of operation in which annual financial reports were considered and no action was required. After the move to 187 Concord Avenue, rent increases began to strain the Association's finances. In previous years, there had been sufficient accumulated income in the endowment account to cover the budget deficits without encroaching on the principal. Inevitably though, with rising costs, even that source of financial flexibility dried up.

The financial honeymoon that the Council enjoyed for a few years came to an end at a time when pressures grew from a number of other areas. Pressure to stay current on converting incoming observation reports to punched cards increased as both the number of incoming observations and the requests by professional astronomers increased. The obvious solution – enter data on cards more rapidly – was limited by the number of keypunch operators who could work on the machine at any time. With only one machine, that machine had to be used for both punching cards and for verifying them. The obvious solution – rent a second machine – required a parallel increase in staff to operate the machine. The Council finally agreed to that increase in monthly expense.

With inflation, salaries in the United States had risen to keep pace, but those at the AAVSO remained unattractively low. Another type of increase – occurring in a time of constrained budgets and rising costs of living – involved the Association's first experience of untimely staff turnover. The static salaries imposed by the strained financial conditions limited Mattei's ability to hire and retain needed support staff. The AAVSO budgetary constraints severely complicated the problem of maintaining a trained and competent staff in the AAVSO office. She could not offer benefits to

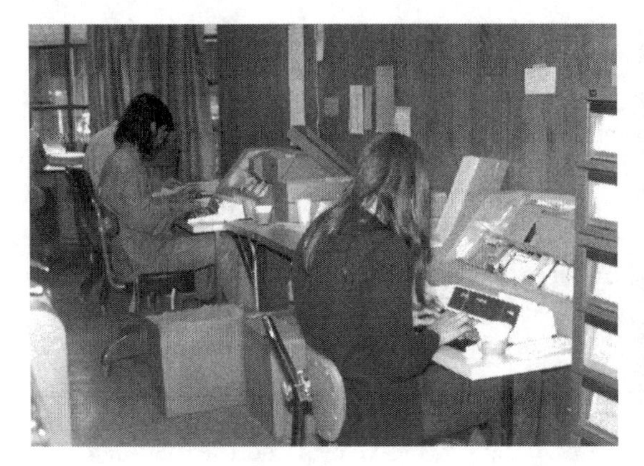

Figure 17.5. Work-study students keypunching observations at AAVSO's Concord Avenue Headquarters, about 1978.

Figure 17.6. AAVSO Headquarters staff, December 1981. Standing, from left: Jean Souza, Margarita Vargas, Katherine Hazen, Shelly Pope, Barbara Silva, Elizabeth O. Waagen, Dorothy Haviland, Mary Collins; Director Janet A. Mattei, seated.

employees in spite of the increasing competitive pressure by other employers.

Mattei addressed some of the issues by hiring part-time employees under a government sponsored work-study program, where much lower pay and no benefits were expected. She then employed a few better-qualified college graduates to handle some of the more technical work associated with processing and plotting observations. Elizabeth Waagen recalls that when she was first employed as a new graduate from Smith College in 1979, the staff consisted of four individuals who, like herself, had degrees in astronomy or some relevant technical field, and six individuals who performed data entry work (card punching and verification) for observations and handled other routine clerical work.[24]

Indeed, Mattei's own salary from the AAVSO in this period was well below the market for a professional person of her education and experience. As AAVSO Director, her salary was not very different from what she had received as the director's assistant and remained at a low level for 5 or more years after she accepted full responsibility. In retrospect, the AAVSO could consider itself quite fortunate that she loved the work so much, for only her husband's income and his interest in the AAVSO allowed her to remain with the AAVSO.

The *AAVSO Variable Star Atlas*

One of Mattei's earliest challenges as AAVSO Director emerged soon after she assumed that role. In 1972 Mayall had, perhaps too hastily, agreed that a suitably scaled variable star atlas with AAVSO comparison star sequences proposed by Charles Scovil and Clint Ford would serve AAVSO binocular observers well and consented to their development of such a project (see Chapter 16).

In early 1974, Scovil and Ford returned to Cambridge to brief Mattei for the first time on the atlas project. In preparation for the meeting, Mattei discussed the matter with several professional astronomers. Her notes in preparation for the meeting with Ford and Scovil indicate a fairly strong resistance to the project on her part, but her reservations did not dissuade Ford and Scovil.[25] Mattei and Mayall concluded that a committee should be formed to advise and assist with the project. Scovil and Ford, on that same trip to Cambridge, picked up the SAO drawings and arranged to produce the necessary working copies of those originals on a larger scale at an industrial copying firm. Scovil then proceeded with the drafting work, preparing 12 final sheets by April 8, 1974, and indicated that "work [was] proceeding at a more rapid rate."[26]

Ford initiated the project formally on March 6, 1974, by sending a check for the first monthly payment of $1000 to AAVSO Treasurer Richard Davis.[27] In his letter to Davis, Ford detailed the way that his donation was to be allocated by the AAVSO to include a salary for Scovil, drafting materials, necessary reproduction work, and an allocated portion applied to AAVSO overhead. As attachments to his letter, Ford included a proposed one-page agreement between the AAVSO and himself and a prospectus for the project that included a committee to oversee the project and advise Scovil. Ford highlighted several problems not specifically covered in his one-page agreement, including what would happen if Scovil had not finished the project in 2 years and what would happen if the actual publication costs exceeded his donation. On the first problem, he proposed, "we'll cross that bridge if and when we come to it," with a similar comment for the second problem.

The four-page prospectus included the appointment of an Advisory Committee by AAVSO President Scovil which, in addition to Mattei, Mayall, and Ford, included seven observers representing the user community for the atlas: Ed Oravec, John Bortle, Wayne Lowder, William and Florence Glenn, Lawrence Hazel (who regularly assisted Ford with chart work), and Henry E. M. Specht, a leading photographic observer. The prospectus outlined the background, technical specifications, preparation process, financial arrangements, and the expected progress reporting.

In her penciled notes on a copy of Ford's letter, Mattei expressed her concerns more explicitly. She wanted a more definite project duration limit to the 2 years and a definite statement that the atlas would become the property of the AAVSO at the end of the project and would be published by the AAVSO; she also questioned whether royalties would accrue to the AAVSO. Her notes also reflect a concern for Scovil's allocation of time to the project as opposed to his other duties as Curator of the Stamford Observatory.

Davis, a practicing attorney as well as AAVSO Treasurer, had even more objections, initially expressed as concerns about whether or not the AAVSO could justify the project in terms of receiving "full and fair value" for the expenditures.[28] Davis' objections to the project, expressed in a series of nearly daily letters over the next month, devolved into a more explicit concern about protecting the AAVSO's status as a tax-exempt organization under the provisions of the federal government's tax codes.[29] In a series of increasingly bitter letters exchanged between various parties to the project over the next several weeks, Davis confronted Ford directly on the "pass-through" nature of his donation to the AAVSO and expressed the view that Ford's intention was to receive a tax deduction for his support of Scovil by using the AAVSO's name and exempt status inappropriately. In addition, Davis pointed out that Ford's support for Scovil amounted to a salary nearly equal to what the AAVSO paid its director, and he questioned whether other members who volunteered their services to the organization might not also take exception to the arrangement, which was clearly not in keeping with the "true spirit of the AAVSO." To Davis, Scovil's involvement provided a particularly uninspiring example for the membership who "unstintingly and uncomplainingly" donated their services to the AAVSO.[30]

The discussion spiraled out of control as Davis shot from the hip in letter after letter, in effect establishing a legal trail that might, in itself, possibly have provided a damaging appearance of inappropriate illegal intent on the part of Ford and Scovil. Ford, on the other hand, took a more cautious and reasoned approach. After receiving a second letter from Davis that he regarded as insulting, Ford shifted his focus to communicating with Mattei.[31] Instead of replying to Davis, Ford sent an annotated copy of the letter to Mattei with a note inquiring: "Would you please advise me to what extent you or Mrs. Mayall were consulted before Dick wrote this letter? Also I would be interested, of course, in your reactions and comments – especially to what extent, if any, you agree with his statements."[32]

After reading Davis' two-page letter attached to Ford's note (if she did not already have a copy, which seems more likely), Mattei must have realized that the situation was worsening. Her concerns, undoubtedly expressed during telephone conversations with Davis and Mayall, explicitly included the implications of the project in terms of demands on her own time and are pretty clearly expressed in Davis' letter. Ford did not miss her implied involvement; by asking Mattei to clarify her relationship with Davis, Ford rendered an obvious warning against any such future behind-the-scenes maneuvering by either her or Mayall.

Davis continued to pepper Ford and Scovil with missives, which they apparently chose to ignore. On April 14, Ford wrote another note to Mattei in response to her note that is not in the files and outlined plans for another trip to Cambridge for a meeting at which he and Scovil hoped to resolve the problems regarding the atlas. In that note, Ford makes clear what was at stake in the matter as far as Mattei and the AAVSO were concerned: "Both Chuck and I hope that some agreement may be reached. However," Ford wrote, "the continued invective from Davis is more insulting than ever, and if he continues in that vein, I can see no solution. The main purpose of Chuck's and my trip to Cambridge will be to determine exactly where HQ stands in the matter. In any case, the Atlas project will continue – with or without the AAVSO."[33]

On April 16, Davis finally provided a draft contract to Mattei for her consideration.[34] The contract, now 11 pages in length compared with Ford's single-page proposal, provided what Davis characterized as the "minimum" legal protections for the AAVSO that allowed the work to proceed. In a key difference in approach, however, Davis insisted that Scovil not be paid a monthly salary, but instead a fee for each chart completed and approved. Scovil objected and countered with editorial changes designed "to bring the contract in line with the prospectus on the project written by Clinton Ford." Among other things, Scovil had to correct Davis on the fact that the new atlas could not be copyrighted, though Davis had required that in his contract, since the underlying document (the SAO atlas) had been underwritten by a government grant.[35] Davis, however, refused to negotiate, claiming the "doubly-favorable tax status of the AAVSO" would be jeopardized by Scovil's counterproposal.[36]

Frustrated and angry with the situation by May 15, Ford again directed his communication to Mattei, who had not participated formally but whose presence is visible behind the scenes in the preceding correspondence from Davis. Disappointed, Ford stated that the project would proceed without AAVSO participation. Furthermore, he declared "I anticipate making no further gifts of any kind to the AAVSO. I deeply regret having to make this decision, but it is irrevocable."[37] Ford then departed for Wrightwood, California, and an observing run at his observatory on Mount Peltier.

Scovil, however, was unwilling to leave the matter alone. In telephone conversations with Mattei and Davis, both the director and the treasurer appeared to have climbed down from their precarious positions. Negotiating from a position of strength, Scovil offered to meet with Mattei and Davis to discuss the matter further. Scovil pointed out, in a letter dated May 28, 1974, that Davis had rejected Scovil's counter-proposal without offering any alternatives and apparently without consulting with Mattei in the matter. Scovil continued to insist that both he and Ford had the AAVSO's best interests at heart. As officers of the Association, they both recognized the need to protect it from a legal perspective, but felt that Davis' wordy legalistic contract was unnecessarily restrictive and failed to recognize that because the project would constitute Scovil's sole source of income, payment on a piecework basis was unacceptable.[38]

Several additional letters failed to resolve the impasse, so Davis resigned as treasurer in a letter June 1, 1974, to the full AAVSO Council and all AAVSO officers.[39] In later correspondence, Davis made it clear that he had been considering that action before the emergence of the atlas project. The atlas project only crystallized his desire to be free of the responsibility. As later noted, even before the outbreak of the atlas affair, Davis had told another council member he intended to resign at the end of the fiscal year. So Davis' resignation occurred only a few months earlier than it would likely have occurred in any event.[40]

Mayall, acting as a belated peacemaker in the matter, called Scovil for a long telephone conversation and then dispatched Mattei to Stamford to discuss the matter face-to-face with Scovil. There is no extant record of the Mattei-Scovil discussion, in which Mattei found a way to mitigate her opposition to the atlas and ask for Scovil's help in regaining Ford's good will. The latter point, after all, loomed as the most urgent need that arose out of this situation. As a successful investor, Ford had already acquired a large estate. So in addition to the loss of his annual coverage of budget shortfalls, which had become an annual ritual that only the treasurer and the director really understood at this point, the AAVSO might no longer be a beneficiary of that estate if the situation was not resolved. Ford was a very strong contributor as an observer and already accrued long service as a Council member and president in addition to his service as secretary.

On Mattei's return to Cambridge, Mayall called Scovil again to suggest that Scovil rewrite the contract himself, using Davis' draft as a template, but causing it to comport with Ford's prospectus in a way that was acceptable to both him and Ford, and then submit that to the Council for approval.[41] Newton Mayall agreed to serve as an interim treasurer and, at the AAVSO Council meeting in Winnipeg, Canada, a few weeks later, was elected to finish out Davis' full term as treasurer.[42]

Following Mayall's suggestion, Scovil drafted a revised contract, following the format of Davis's proposal, but this time without the piecework payment schedule and eliminating some of the excessively protective clauses that Davis built into the original draft.

Scovil's agreement totaled only 5 pages rather than 11; it was signed by himself and by Mattei on June 24, 1974. Scovil continued work on the atlas, but within another year, Ford lost patience with the project and turned the administration over to Newton Mayall, who immediately produced a schedule for its completion.[43] Still, time had not been allocated in Newton Mayall's analysis and schedule for certain essential stages of checking that were proving very time-consuming, especially those steps being done at headquarters, outside Scovil's control from the Stamford Observatory.[44]

In early 1976, Scovil again addressed Newton Mayall in a letter submitting bills for payment. Scovil first reported that all the "masters" were completed and chided Newton Mayall for failing to return charts in a timely manner that permitted the work to proceed continuously. Scovil expressed the same concern for the charts that he had sent to Henry Specht, whose inclusion in the chart-checking process should have been recognized as a weak link with his ill health and aging problems.

These were not the worst of Scovil's problems. Newton Mayall wrote to Ford to complain about Scovil's productivity; Ford passed Mayall's letter to Scovil, who asked that any complaints be sent to him. Scovil noted, "Janet has made a very strong statement about the same thing as concerns AAVSO in general, and people talking to each other or worse, to outsiders, but not letting HQ know about the complaints."[45] Scovil then responded to Newton Mayall's complaints, as he paraphrased Newton Mayall's own phrase, apparently lifted from the letter to Ford that started the exchange: "There were important corrections to be made in the set of charts already 'finished' so I made them. I have completed the 'Master' charts, and am now going ahead with completing charts again, working on the ring at the equator. Again, if you have 'beefs' or questions *ASK ME*." Scovil then addressed a question apparently raised by someone at SAO about why the AAVSO felt the need to do this work in the first place since an atlas existed.[46] Scovil responded:

> The atlas I presume you are talking about would be the Eclipticalis-Borealis set. It is fine, but not for observing. It is huge. It is impossible to handle even in my protected dome, let alone in the open. It is out of date, and was incomplete when printed. That is one of the reasons that my project is taking extra long. We were depending on those atlas charts to show all variables. They don't! I have to check the catalogue [GCVS] and 2 supplements to be sure of being complete and up to date. The Ecl.-Borealis set does not show AAVSO variables, and any observer would have to go thru just about what I am doing to put in his own magnitudes. Talk about duplication and wasted effort!! – and still wouldn't have a handy atlas. Of course the people at SAO wouldn't know all this. They are not observers and have never struggled with the problems.[47]

Later in 1976, Ford's impatience began to show as he joined with Mattei, Mayall, and Scovil's other detractors and took the unusual step of publishing rather critical remarks about the lack of progress with the project in *JAAVSO*.[48] After the project had been extended for 1 year, Ford ceased making payments for the work in March 1977, but Scovil agreed to continue the project as a volunteer. Scovil interrupted work on the atlas in 1978 to make photographic plates of the northern skies for the *Papadopoulos All-Sky Photovisual Atlas* using the same camera that had been used in South Africa for the exposures of the southern and equatorial regions.[49]

The atlas project remained stalled until 1979, when Leif Robinson, editor of *Sky & Telescope*, suggested his firm might be interested in publishing the completed work. After a tense meeting with all parties involved, an agreement was reached on a schedule by which Scovil could resume and complete the work for publication. Robinson's offer was the push that finally got the whole organization over the hurdle and moving toward project completion.[50]

In late 1980, *The AAVSO Variable Star Atlas* was finally published by Sky Publishing Corporation. As Mattei stated in her annual report to the Association: "We can be proud of this Atlas and congratulate Charles Scovil for his excellent job. Final stages of the Atlas required a team effort, and this summer we had to make this project a priority and devote a major part of most of the staff's time."[51]

The published *Atlas* consisted of 178 charts, each printed on heavy stock, $11\frac{1}{2} \times 13\frac{3}{4}$ inches in size, together with a descriptive booklet of articles published in *Sky & Telescope*, all packaged in a sturdy box. The *Atlas* scored an instant success to the extent that by 1982, a second printing of an additional 3000 copies seemed reasonable.[52] By 1990, a second edition, this time published by the AAVSO, was scaled slightly differently to make publication of 178 standard $8^1/_2 \times 11$-inch notebook sized charts possible as hole-punched sheets in a ring binder. This proved equally popular as a more practical format for use at the telescope.

Clearly, this process, extended and painful as it might have been to all, had produced an asset of some value to the AAVSO as an organization, one in which its members might take pride. In the extended process in which the *Atlas* came into existence, however, there were a number of unintended consequences worthy of note. Ford, for his part, seems to have emerged from the *Atlas* project with a better appreciation of Mattei's intellectual toughness and willingness to scrap with anyone for what she regarded as the best interests of the AAVSO. Mattei must surely have learned a great deal about the complexity of interpersonal communication in such circumstances. She must also have realized how limited her powers were to control every detail of the Association's activities and what well-intentioned members might do as their part of the overall effort. Most importantly, however, Davis' resignation thrust Newton Mayall into an active role in the management of the day-to-day operations of the Association, a change that might not have occurred in the same way had Davis served out his term as the AAVSO Treasurer and been replaced in a Council election in the normal course of events.

The *AAVSO Journal*

Three issues of *The Journal of the AAVSO (JAAVSO)*, all edited by William Glenn with assistance from his wife Florence, had been published under Mayall's direction. *JAAVSO*, which came into existence with uncertainty about its intent and content, was still in its early stages in fall 1973 when Mattei replaced Mayall as AAVSO Director. For those first issues, all the mechanical work of typing, printing, assembling, folding and stapling, stuffing and addressing envelopes, and mailing the journal to members had been carried out at Stamford Observatory by Scovil and John Griesé, with limited help from other volunteers.[53] That arrangement continued for several more years.

As was the case with the *Atlas*, Mayall's agreement to go ahead with publication had been a bit too casual. That was made clear from the fact that, as outlined in the previous chapter, only two policy issues dominated her last Council meeting as the AAVSO Director at Lenox, Massachusetts – selection of her replacement and a re-hashing of the journal question. Was it intended to be primarily a journal of the Association or a journal of variable star astronomy? These issues remained unresolved through the next three issues, which under Glenn's editorial leadership began to mature nicely.

Although there was an Editorial Board, an editorial policy had not been established for *JAAVSO* through the first issues. Editor Glenn finally presented an editorial policy in a letter to the Council at the June 1974 Winnipeg Council meeting. *JAAVSO*'s editorial policy finally received Council approval after some discussion and modification. The principal point of contention in the Winnipeg discussions centered directly on Mattei's concerns about control over *JAAVSO* content. As the matter was left after the Winnipeg meeting, Mattei would receive copies of the unedited text of each article Glenn intended to publish. Mattei also insisted on the right to review and approve copies of all the edited text before the finalization of content and preparation for printing. Specifically, the issue boiled down to who had the final authority over *JAAVSO* content. At the 1974 Annual Meeting, Glenn again raised the issue. Owen Gingerich and Helen Hogg resigned from the Editorial Board. Furthermore, Mattei's questions about *JAAVSO* content and standards concerned Glenn. He discussed these issues with the Council at the Williams College meeting that fall 1974. After a contentious discussion, the Council agreed formally that the final authority on the content of any AAVSO publication, including *JAAVSO*, rested constitutionally with the director.

A petty dispute involving the Glenns' relationships within the Fairfield County Astronomical Society (FCAS) triggered Mattei's first opportunity to modify *JAAVSO*'s arrangements. The Glenns resigned from their memberships in the FCAS immediately. Furthermore, not wishing to continue his contact with Scovil and Griesé, Glenn resigned as *JAAVSO* editor at the completion of the second issue of Volume 3 (Fall 1974).[54] The Council reacted immediately by appointing Douglas S. Hall, an astronomy professor at Vanderbilt University, to the *JAAVSO* editorial board.[55] The Council deferred appointment of a new editor until Mattei could consider the options available and nominate an individual she felt confident could replace Glenn.

Up to that point, *JAAVSO* publication information on the inside front cover specified the editorial offices as "Dr. William H. Glenn, Department of Teacher Preparation, York College of the City University of New York, Jamaica, New York." In the next issue, the notice on editorial offices was changed to "The Editor of the *Journal of the AAVSO*, c/o AAVSO, 187 Concord Avenue, Cambridge, Massachusetts." That simple change satisfied one of Mattei's concerns about the loose control exercised over *JAAVSO* content by headquarters, namely her need for early understanding of the content that would appear, the process of selecting peer reviewers for submittals, and final approval of the edited content. In the future all articles would be submitted to the Cambridge address.

No record exists of the exact process by which Mattei went about considering, interviewing, and selecting a new *JAAVSO* editor. Nevertheless, she did so with dispatch. At the Council's 1975 Spring Meeting in Atlanta, Georgia, Mattei introduced her choice, SAO astronomer Charles A. Whitney. Whitney, a theoretical astronomer who had worked for decades on the problems of stellar evolution and pulsation theory, seemed an ideal choice. Furthermore, he was well known and respected among professional variable star astronomers, most of whom had surely studied his widely cited work on stellar evolution and pulsation, both alone and with J. P. Cox of the University of Colorado.[56]

Whitney spoke at length to the Council about his aims, purposes, and future plans for *JAAVSO*. These included:

(1) to encourage members to contribute papers;
(2) to increase direct communication to members through an informal section on observing hints;
(3) to add some younger members to the Editorial Board;
(4) to continue the existing format of *JAAVSO*;
(5) to continue the existing criteria for selection of papers to be published, including the use of referees to monitor original manuscripts; and
(6) to accept papers that would be useful and interesting to observers, would be acceptable to professional astronomers, and would contain new observational material and ideas.

Whitney agreed to print all this in *JAAVSO* and distribute it to Council members, referees, and other interested parties. The Council expressed its pleasure with Mattei's choice as well as Whitney's initial presentation.[57]

By 1978, Scovil's assistance with editing and preparation of *JAAVSO* was no longer needed.[58] Instead, a succession of employees at headquarters handled the production work for Whitney. By the end of 1980, excessive staff turnover began to take a toll on production. Problems getting *JAAVSO* published and distributed mounted; delays with individual issues became cumulative as more and more of the activities involved were transferred from Stamford to Cambridge.

AAVSO NEWS NOTES

The Council dealt with another publishing matter at the 1975 Spring Meeting in Atlanta, Georgia. The AAVSO had been publishing a periodical newsletter called *AAVSO News Notes* for a number of years. Undoubtedly, its publication constituted one more burdensome piece of work at headquarters, which seemed to offer little of substance to the AAVSO's progress in the field of variable star astronomy. Mattei presented the issue as an expense reduction measure and asked the Council for its agreement to terminate its publication. She suggested that members be encouraged to submit such information to former Vice President Carolyn Hurless, who since 1964 had been publishing a more frequent and very informal newsletter, *Variable Views* (*VV*), from her home in Lima, Ohio. Popular and eagerly awaited by its subscribers, who for the most part were also active observers, *VV* might well have already satisfied the intended need of *AAVSO News Notes*. The Council supported Mattei by approving her resolution to drop publication of *AAVSO News Notes*, with a warning, however, that no technical information or notes should appear in *VV*, only personal news.[59]

The audience for *VV*, as described in Chapter 16, consisted mainly of very dedicated observers, mostly US residents. They shared a camaraderie that stemmed, at least in part, from the gatherings, jokingly called "August Orgies," in Lima, Ohio, each summer for a number of years. First-time participants at the August gathering were required to spend one night at the Hurless home in the "weird room," a bedroom painted completely black with luminescent representations of stars applied to the ceiling.[60]

AAVSO Circular

In contrast to *JAAVSO*, the *AAVSO Circular*, introduced to the membership in late 1970 (see Chapter 16), continued to be edited by John Bortle, AAVSO observer, committee chair, and Council member. The first edition of the *AAVSO Circular* presented the observations of only three observers – Bortle, Ford, and Lowder – who reported their observations for 66 stars, including 13 classified as eruptive, 8 novae, 8 irregulars, and 37 Miras. By the time Mattei became AAVSO Director, the *Circular* featured up to 166 stars, including 30 eruptive, 30 novae, and 56 irregular variables. Observations of Miras no longer appeared in the *Circular* to make room for more of the truly unpredictable objects.

The arrangements for typing and printing the *Circular* from Bortle's typed drafts were the same as for *JAAVSO*. They were typed at Stamford Observatory, printed and assembled by a commercial printer in Stamford, and mailed from AAVSO Headquarters. Copies usually arrived in the mail to each subscriber within a month after the completion of any calendar month. The *Circular* attracted wide readership among both the AAVSO observers and within the professional astronomical community. With its widespread distribution, the mailings of the *Circular* became a vehicle for distributing new star charts of particular interest to that special audience.[61]

In the original arrangement, subscribers communicated their desire to participate directly to Bortle, sending him the nominal annual subscription fee. The list of subscribers grew regularly and by mid 1975 totaled 128 individuals by regular surface mail, 47 individuals or institutional subscribers requiring delivery overseas by air mail, and 32 institutional subscribers in North America (universities, observatories, and associations).

Observers were aware that Bortle attempted to maintain a fairly rigid schedule for *Circular* distribution to keep it timely, primarily because the events reported were transient. With regard to the data, it was understood that Bortle could only report what he found from his own observations and the observations reported to him by subscribers and others, with the understanding that AAVSO Headquarters would also be receiving observations that may or may not have been reported in a timely manner to Bortle. Thus Bortle was sometimes obliged to state that there was no report, that is to say, no observations, for a particular star that had been reported to him, and, in addition, he had not been able to observe it himself, a rare circumstance. Some observers preferred reporting their observations only to headquarters and taking their chances on the utility of their observations in the long run. Still others reported sudden brightening of cataclysmic variables or sudden fading of R Coronae Borealis stars by telephoning headquarters when their observations could confirm that a transient event had begun or was in progress. It was exciting work for many observers who enjoyed the thrill of observing such sudden and dramatic changes with the hope of receiving special recognition for being the first to report the event.

The operation of the *Circular* was not without its problems. Bortle handled the accounting and financial arrangements for a period of time. When the AAVSO developed the capability to add a subscription fee to the annual dues submittal form, the Council approved making an annual payment to Bortle for the sum of fees received.[62] By 1974, rising reproduction and postage costs outstripped the annual subscription fee and, as some observers were irregular about submitting the voluntary payment, the *Circular* began to experience cash shortfalls. By September 1, 1974, Bortle reported he was "flat broke." The Council had agreed a few years earlier that if the *Circular* subscription fees were insufficient to cover Bortle's expenses, the Treasurer was authorized to provide "emergency funds" of up to $200 to sustain the timely distribution of the publication. In his September letter, Bortle asked for $150 to cover his expense, assuming up to 50 subscriptions would possibly arrive in the interim.[63] Within a few days a check was mailed to Bortle without question.[64]

More importantly, though, there were significant process problems that Mattei saw in the *Circular* after becoming director. When she sent her late "Director's Requests" for the February 1974 issue of the *Circular*, Mattei called Bortle's attention to three problems.[65] First, the observer initials for a dozen overseas observers who had provided observations to the *Circular* did not agree with those observer initials assigned at Headquarters. The important difference might result in failure, on the part of the headquarters, to credit the observer with all the observations for which he or she was properly responsible in annual reports and historical records, an important issue from an observer's perspective. However, Mattei might have acknowledged that the reason the glitch had occurred had more to do with delays in assigning initials in headquarters than any fault on Bortle's part. In the absence of any "official" initials to use for coding observations reported to both the *Circular* and headquarters, Bortle exercised prudent editorial judgment and used the submitted observations for the *Circular* by assigning initials for the Bulletin's purposes only. By doing so, observations from longitudes well outside North American continental limits added a great deal of value to the *Circular* for its uses.

A related but even more important question is embodied in Mattei's second problem, that of the loss of observer confidence in the handling of their observations at AAVSO Headquarters. Comparing the observations reported to Bortle with the observations submitted to headquarters, Mattei recognized that there were observers who had lost confidence in the ability of headquarters to handle the data in an efficient and timely manner and had decided not to send their observations to Cambridge. In the short run that was an expedient approach, but it failed to recognize the importance of maintaining the completeness and integrity of the AAVSO's central database. Mattei's diplomatic approach, in effect, acknowledged that the problem was in headquarters and not with Bortle: "In order to benefit from these data for the records of the AAVSO, for special requests and reports would it be possible to have copies of them sent to HQ? I would really appreciate it if we could work out a method, and am anxious to hear your suggestions."

Third, another issue, not unrelated to the first and second, involved the fact that the chart-making activity at the Stamford Observatory by Ford and Scovil fell largely outside headquarters' knowledge or control. In another diplomatic request, Mattei suggested that needed corrections should be reported first to headquarters, "so that we will have a note of them before it goes to the Chart Committee." Perhaps realizing that there was no real reason to have that knowledge in headquarters first, Mattei added, as almost an afterthought, that Ford as chairman of the Chart Committee was the person who had the most immediate need for the information about poor or inadequate charts and that the information should properly be sent to him as well, adding what would become a standard Mattei phrase even at this early date: "Don't you agree?" She tagged on the real issue in this by noting, "Of course Chuck (Scovil) is very much involved with it and will probably do the follow up on the questions raised."

Mattei could not have helped but be impressed, however, by the acceptance that the *AAVSO Circular* had achieved under Bortle's editorship, but if she had missed that point it would soon be brought to her attention. Bortle had been working with Florida astronomer David Florkowski on plans for an observing run on the Greenbank Radio Telescope. Florkowski corresponded directly with Bortle on the necessary hot-line communications that he hoped could be maintained to him while he was observing in Greenbank.[66] When Bortle replied, providing Florkowski with the necessary telephone numbers, contact times, and agreement on the arrangements that Florkowski proposed, he prudently sent Mattei a copy of his response, adding a typewritten note across the top to Mattei directly:

> Hope you find these arrangements agreeable. As I mentioned in earlier correspondence on this project (early March), I am not trying to do your job but simply felt the work involved could be more easily handle[d] from my end because of the time factor and "hot line" individuals involved. If you should feel differently, please contact me immediately so that necessary changes can be made."[67]

In fact, Mattei did not find those arrangements agreeable. In a quick response to Bortle, she said as much very directly, and in effect scolded Bortle for not insisting that these arrangements be made through headquarters. She also insisted that all such communication of alerts originate at headquarters as well.[68] Her clearly stated reasons for staking out this position seem logical enough, but in addressing one of the Association's leading observers, who was also a Council member and an active committee chair, she might have been more congenial. She might have emphasized the external acceptance of the organization more and the role of headquarters slightly less. The Florkowski exchange offers yet another example of Mattei's strongly principled approach to her new position as the AAVSO Director, with only 6 months of service in that role to her credit. Her response also reveals her dedicated but impulsive approach to communicating that commitment.

Mattei obviously went right to work as the new AAVSO Director on headquarters control of the organization as she saw the needs. In retrospect, that single letter from the very new AAVSO Director reveals, with considerable clarity, the actual perceptions that had developed in the Association and its observers of the headquarters dysfunction because of its limited staffing and growing workload. Mayall and the Council had, for too long, ignored the growing discomforts that existed in the organization that were clearly reflected in the three independent initiatives discussed above, *The AAVSO Variable Star Atlas*, the *Journal of the AAVSO*, and the *AAVSO Circular*. It would take time, and the resolution of a few other problems, before the critical need for a different approach to the management of the Association at headquarters was clearly recognized. Her efforts gradually won grudging acceptance from the Council.

TECHNICAL ACTIVITIES

The technical work of committees within the AAVSO continued to mature. It would be inappropriate to engage in an exhaustive survey of those committees, which included ongoing work: Charts, Occultations, Solar, Nova Search, Eclipsing Binary Stars, RR Lyrae Stars, and Photoelectric Photometry. These committees existed at the time Mattei assumed the director's role and remained essentially the same during the following decade. Instead, we will touch on significant developments in each of these activities as they occurred.

Eclipsing Binary Committee

The work of the Eclipsing Binary (EB) Committee continued under the leadership of Chairman Marvin Baldwin. Taking advantage of the new *AAVSO Circular*, Baldwin began publishing notices of stars in need of observation, useful in also calling attention to the opportunities to observe in a very different mode than that typical for most observers. EB stars generally display such regularity in their periods that any deviation of the actual time of minimum from the predicted time may be significant. The regularity is such that after one minimum has been chosen as a starting epoch, future minima can be calculated using a linear mathematical relationship and standard elements of an epoch and a period published in international catalogs. This relationship is then used to project a table of calculated times of central eclipse to form an ephemeris of opportunities to observe the actual time of minimum. Observers use the ephemerides of all program stars to plan observing runs on individual stars to establish the observed time. By comparing that observed time with the calculated time of central eclipse (O–C) on a regular basis, it is possible to detect secular changes in the period of the star that reflect actual changes in the physical system. Stars that exhibit such changes in period represent opportunities to study the physical characteristics of the system more intently in order to understand the nature of the physical changes taking place, for example, mass transfer between the two stars, or a loss of mass from the system.

The established routine for EB Committee work included publication of charts identifying each of the EB program stars, annual ephemerides of predicted times of eclipse, and periodic reports of how the observed times of minimum deviated from the predicted times or O–C values. Chart preparation work during the 1970s by Halbach in Milwaukee, aided by Scovil at the Stamford Observatory, provided the basic tools for identifying the EB star and appropriate comparison stars nearby.

By the early 1970s, Baldwin, with help from a number of other committee members, notably Peter Taylor, Josefa Manella, and Mary Jane Taylor, computed annual ephemerides, published as *Eclipsing Binary Bulletins.* The eclipse timing observations, which were reported directly to Baldwin at his home in Indiana rather than through AAVSO Headquarters, were then reduced to central times for each observed eclipse, and an O–C computed for that particular epoch.

When the IAU's *Information Bulletin on Variable Stars* (*IBVS*) stopped publishing visual times of minima for eclipsing binary stars, Baldwin began publishing them in semiannual reports in *JAAVSO*. Those reports provide an interesting glimpse of the effort involved, indicative of an intense commitment by a number of individuals from the preparation of annual ephemerides to the final data reduction and analysis conducted by Baldwin. For example, in the first of these reports, Baldwin reported 441 heliocentric eclipse times for 50 eclipsing binary stars reported by 25 observers; as is usual in such activity, most of the eclipse timings were made by a few very active observers, including Baldwin, Bortle, Chapman, Cook, Cragg, Monske, and Hazel. Each timing represented a 2- to 4-hour run with magnitude estimates every 10 to 15 minutes. Baldwin singled out

Hazel for his contribution of timings for particularly difficult stars.[69] Several more of these compilations of observed minima for EB stars appeared in subsequent issues of *JAAVSO*.[70]

Following that impressive show of productive results, the next *JAAVSO* issue included a number of articles on EBs: a report on the little observed eclipsing binary V Sagittae by Ernst Mayer, a general article on techniques of observing EB stars by Baldwin, an analysis of short-term period changes in RZ Draconis by Anthony Mallama, and a period study on XZ Cygni by Mary Jane Taylor. Thus, of the eight contributed articles in this issue, four dealt with EB stars.[71]

Over time, Baldwin succeeded in attracting interest in the EB program among the emerging PEP observers, with Leonard Kalish, Thomas Renner, Howard Louth, Melanie Mitchell, and Howard Landis successfully timing minima for stars that were too difficult to observe accurately by the usual visual methods.

As the program matured, Baldwin reported EB stars that had been discovered to be significantly off their predicted times of eclipse. These reports provided an opportunity for others to study those stars intensely to determine why that shift occurred. Such gratifying discoveries provided the *raison d'être* for this whole activity.[72]

Through her contacts in the IAU, Mattei initiated a cooperative program for observing EB stars in the southern hemisphere. Other variable star observers in the southern hemisphere (e.g., the RASNZ) had not yet taken up the challenge of EB stars, which required a different approach for successful observing in comparison to the typical LPVs. Jan Hers of South Africa agreed to cooperate by providing a set of charts.[73]

With the active participation of Halbach in chart making as well as observing, the center of gravity for AAVSO's EB work gradually shifted to Milwaukee, Wisconsin. Several observers there became dedicated EB observers. Gerard Samolyk proved exceptionally committed to this program and became one of the leading observers in the effort. Halbach and the Milwaukee Astronomical Society (MAS) possessed eight identical f/5 Newtonian reflecting telescopes on convenient low-profile fork mounts at MAS's observatory. When these telescopes were not otherwise in use, Samolyk would arrange them in a circle, find a different eclipsing binary system in each telescope, and then walk around the circle recording observations every few minutes for each star for the evening, during which an eclipse had been forecast.[74] Samolyk soon led the EB committee in observations of minima per year.

Occultation Committee

At the 1974 Annual Meeting at Williams College, the Association finally matured to the point that it could consider shedding some of the work it had accumulated over the years. The work of the Occultation Committee, for example, had actually peaked in the late 1930s and through the 1940s. However, by the 1970s, the timing of simple lunar occultations was of little interest to AAVSO members. In 1974, Council member Bortle announced that he wished to resign as chairman of the Occultation Committee. After some discussion in the Council, Vice President George Fortier volunteered to serve as the Occultation Committee chair and to evaluate further opportunities that existed in this work for the AAVSO.[75] The work of the Occultation Committee was again raised at the St. Louis Council meeting in May 1976, this time in the context of a general discussion of all committees. Fortier recommended that the committee be disbanded; his motion was tabled because the previous discussion had closed when it was agreed to reconsider the matter again after a year, but less time than that had transpired.[76] Fortier then published a note in *JAAVSO*, an unusual measure for an announcement of a pending Council action, perhaps to survey the membership reaction to make sure there were limited or no objections to dropping the committee.[77]

Fortier again recommended that the committee be disbanded at the 1976 Annual Meeting. Fortier described the International Occultation Timing Association (IOTA) to the Council and the active participation in the work of that organization by some members of the AAVSO, including Halbach and other MAS members. Fortier surmised that the reason for the limited activity in the committee was that IOTA, an organization dedicated to occultation work, had replaced the AAVSO as the gathering point for such amateur activity. The AAVSO's occultation work had continued as a service to astronomy and a way of engaging amateur astronomers in active observational science for 50 years but was by 1976 no longer needed.[78] Leon Campbell's 1940s vision of shedding distracting activities not directly related to variable star astronomy had at last begun to be realized.

PROGRESS IN PHOTOELECTRIC PHOTOMETRY

At the time when Ohio businessman Art Stokes replaced Lewis J. Boss as the PEP Committee chair in 1967, the transition from vacuum tube to transistor electronics was nearly complete. Transistor-based electronics offered a number of advantages over vacuum tubes, especially for out-of-doors operation at night. Stokes took interest in the transistor and how it could be applied to the circuits required for PEP work. Over a number of years, Stokes published several papers describing progress in photometer amplifiers and power supplies based on transistors and a few observational papers. In the latter case, he seemed to be more intent on illustrating the technique than contributing to the longer term database of information. During Stokes' tenure, the PEP Committee remained in a holding pattern, full of potential but also loaded with technical complexity in both the electronics and in the data reduction work involved. Seemingly less interested in applied observation than he was in hardware-related problems, Stokes provided the AAVSO with an updated manual and circuit diagrams for several new transistorized devices that simplified observing setups. As awareness of the number of appropriate targets for PEP work mounted, the members continued to respond at a fairly low level of observations per observer per year. It wasn't lack of interest in PEP observing that retarded progress, but the fact that observers had to fabricate their own PE photometers, requiring both electronic expertise, mechanical skills, and shop tools, a combination that few possessed.

Figure 17.7. Arthur J. Stokes, Chairman, Photoelectric Photometry Committee 1967–1975.

Figure 17.8. Clinton Ford and Howard Landis, at Landis' observatory, Hampton, Georgia, 1975.

In May 1975, Stokes announced his retirement as the PEP Committee chair, having also served as AAVSO President and Council member. The Council appointed Howard Landis as Stokes' replacement as the PEP Committee chairman.[79] From then on, there was an increased and steady growth in the reported PEP observations. Those increases came partly with Landis' own observing, but more importantly, also with Landis' patient counseling and encouragement of new observers and the steady efforts of Raymond Thompson, Ken Luedeke, Howard Louth, and many others.

Within a few years, Landis realized that it would be more efficient to create a periodic newsletter than to continue individual correspondence. In November 1977 he published the *AAVSO Photoelectric Photometry Bulletin No. 1*. Printed and distributed at his own expense at first, the *Bulletin* rapidly became a useful substitute for letter writing and allowed Landis to solicit articles from others as well as writing his own. The *Bulletin* proved quite popular with PEP observers and with others in and out of the AAVSO membership. The publication proved useful in pulling together the small community of PEP observers and kept others informed of their progress as well.

After Landis published *PEP Bulletin No. 12*, he passed the editor's role to David Skillman, comparatively new to AAVSO as well as to PEP observing but who showed considerable promise as an observer. By 1978, Skillman had begun investigating the variability of several stars in the *Bright Star Catalogue* for Dorrit Hoffleit at Mattei's suggestion. Some, but not all, proved to be variable, as perhaps Hoffleit had suspected.[80]

Skillman, a very candid correspondent in his letters to Mattei, revealed that Douglas Hall of Dyer Observatory at Vanderbilt University had questioned him at a recent meeting about his observing. When Skillman showed Hall the list of stars he had been observing for Hoffleit, Hall became agitated, in fact irritated, that the stars were not on the list he had suggested for an AAVSO PEP observing program.[81]

Hall had been invited as a guest to participate in the 1976 AAVSO Council meeting in St. Louis, Missouri, and used that forum to make several points forcefully about the AAVSO's PEP program. Hall's contention then was that professional astronomers were not reading the information published in *JAAVSO* because the stars that seemingly attracted amateur observing efforts, as evidenced by the articles appearing in *JAAVSO*, were not sufficiently interesting. At that time, Hall suggested providing an opportunity for professional astronomers to comment on the AAVSO program by establishing a regular Letter to the Editors section in *JAAVSO*.[82] Hall's presentation at the Council meeting persuaded Mattei to place him in contact with Landis and allow him to work directly with some of the PEP observers on their program, as well as providing technical advice.

Hall's interaction with PEP observers evolved into problematic directions as far as Mattei was concerned. His suggestions for observing programs were well received by Landis and a few other observers who had, as a result, provided Hall with substantial data. All of this pleased observers, who reported feeling gratified with direct contact with a professional who wanted and would use their work product for contemporary research rather than having it filed in the AAVSO archives for possible future use at an indefinite date. The problem, as Mattei later described it to Skillman, involved Hall's publication of those observations in professional journals without crediting either the observers or the AAVSO as the source of his data.[83] Skillman, for his part, expressed willingness to work with Hall on programs that Mattei approved, but agreed with and appreciated her stand on the need for appropriate credits with publication of AAVSO observers' data. In a subsequent letter to Hall, Skillman staked out a strong position that he not only wanted his name associated with any of his data that Hall published, but he also wanted an asterisk by the name with a footnote on the bottom of the page identifying Skillman with the AAVSO.[84]

As other AAVSO observers joined Landis and Skillman in this crusade for recognition, Hall eventually turned sour on the relationship with AAVSO observers. He claimed their objections had delayed, unnecessarily in his view, publication of several papers involving AAVSO data while he attempted to reach a compromise wording that would satisfy all the separate observers. He closed a later letter to Skillman with a reflection of that attitude:

> If you feel very strongly about crediting the AAVSO in the maximum possible way, even though they don't actually play any role in your acquisition of or reduction of the data, let me suggest you send it to Janet, and have her forward it to me along with an invoice, a procedure you once suggested and which she has on one occasion done already.[85]

For Skillman, the whole experience was a sobering one. As he later confided to Mattei, his experience tempered his willingness to share his scientific results quite so freely. In an interesting letter that foreshadowed his resignation from AAVSO, Skillman advised Mattei that he planned to share some of his analytical work with Joe Patterson, a professor in the astronomy department at Columbia University. He provided a copy of his work to JAM in an envelope, asking her to hold it for him for a period of time, noting: "I'd like you to know of this work before I send it to Joe Patterson. I trust him, but let's be sure we are protected. I'm mailing the original to him today."[86]

Attached to Skillman's letter is a copy of a handwritten analysis of the V471 Tau modeling that he sent to Patterson. It is sophisticated work for an amateur astronomer, well beyond what most AAVSO members would attempt. Patterson recognized that talent immediately; he and Skillman soon decided to form their own organization, the Center for Backyard Astrophysics (CBA), which continues today. With participation from members located around the world, CBA produces sound scientific publications regularly.

Skillman served as the *AAVSO PEP Bulletin* editor for a short period and then handed that task over to Russell M. Genet, who served for another year and resigned. The editorial task was then turned over to John Percy, who edited the *PEP Bulletin* starting with Volume 4, Number 1, in June 1983. Percy continued as editor for more than 20 years, adding significantly to its usefulness and stature.

Mattei's rocky relationship with Hall was not over at this point. In 1980 Hall and Genet formed another organization, International Amateur-Professional Photoelectric Photometry (IAPPP). By 1982 they had published a resource manual for amateurs who wished to enter into this work. Charles Whitney proposed a rough division of labor that involved the IAPPP as a moderator of equipment and technique problems, while the AAVSO served as an archive of results and source of interpretation. Genet responded favorably to this suggestion, but the AAVSO never responded.[87]

Mattei, for her part, hung on tenaciously to what she perceived as her role in protecting the reputation of the Association she was charged with leading. Her approach at first irritated some professional astronomers and seemed perilous. However, member/observers found it refreshing that the director would stand up to the professional users of AAVSO data and demand that the observers be recognized, in some cases as co-authors of published technical articles. Hall's objections receded, as more and more frequently the compromise that seemed to work best was to list Mattei as one of the authors of a published paper and credit the AAVSO as an organization when data from multiple observers was used. She was not technically qualified, however, to solve many of the problems that emerged in PEP application to variable star observing in practice.

Problems with organizational conflicts like those discussed above were not simplified or helped by Mattei's inability to handle a huge volume of mail from members and others. Forced to prioritize her time spent in responding, and with limited staff to whom she could delegate this task, Mattei missed several opportunities to encourage the PEP observers, apparently leaving that to Landis, whom she trusted. An example of the consequences of this neglect may be found in the case of Thomas G. McFaul. Having equipped himself with a 14-inch Schmidt Cassegrain telescope and an all-solid state photometer, McFaul presented his initial evaluation of that setup as an article in *JAASVO*. By comparing his reduced observations of the brightness of a number of stars with those reported in the Arizona-Tonanzintla Catalogue, the very latest such effort by recognized authorities in the field of photometry, McFaul later developed PEP light curves for a series of RR Lyrae stars.[88] He reported his finding on SW Andromedae and asked a serious question to RR Lyrae Committee chair and local friend Baldwin who could not answer McFaul's question.[89] He passed the letter on to Mattei in headquarters. Unfortunately, Mattei never responded to the question, a common complaint in this period. A similar situation occurred with Harold Stelzer of MAS, who equipped himself with an observing setup identical to McFaul's and began PEP projects. After attending the AAVSO 1982 Spring Meeting in Milwaukee, he reported to Mattei how pleased he had been with the meeting, and especially with the opportunity to meet Landis in person.

Several letters and questions to Mattei about PEP observing never brought any response. He eventually found it more satisfying to observe for Hall rather than submit his observations to headquarters. Without ever having any acknowledgment of his work from Mattei, he apparently dropped out of PEP observing.[90]

SUMMARIZING A RISKY TRANSITION

As will be seen in the foregoing chapter, the AAVSO survived some major challenges in the first decade of Mattei's tenure as its director. Resource limitations proved a constant challenge as the membership increased, and the numbers of observations submitted to be archived and in some way – as yet undefined – analyzed for the benefit of variable star astronomy, increased even more rapidly.

All this occurred under the stress of constant financial problems. There was never enough money to take the most direct or logical route to solving a problem. The Council, for its part, was never willing to step up to the task of providing adequate financial resources for the organization on its own initiative and instead waited for its long-standing patron, Clint Ford, to rescue it; those rescues were occurring annually.

Late in the 1970s, newer members on the Council asked for more clarity in the financial reports provided by Treasurer Newton Mayall, but to no avail. The Treasurer's reports failed to provide a clear understanding of the perilous financial conditions that actually existed without Ford's donations to cover shortfalls. Newton Mayall's over-supervision of the AAVSO office and its activities and petty expense issues constituted yet another problem with his continued service as the AAVSO Treasurer. Mattei finally engineered his "retirement" and the election of a successful investment advisor, Theodore H. N. Wales, as the seventh Treasurer of the AAVSO.

Mattei's salary as AAVSO Director for this period was, in the words of then-President Carl Anderson, a disgrace. (She started at $8000 and had received two small raises over the 5-year period to a total of $10 000.) She worked for many years for a third of the salary received by her predecessor, which was, as determined by a later salary survey, more like what Mattei could have expected in other jobs. That she continued to do so and to proffer her efforts with cheerful vigor constitutes one of the mysteries surrounding her tenure that may never be fully understood.

Mattei could have commanded a much greater salary in many other types of employment in astronomy, and certainly in that era, in many other types of employment, with her background in physics and teaching. As the new treasurer, Wales recognized this problem immediately and, without any discussion with the Council, raised her annual salary by several thousand dollars. At the same time, other new Council members had recognized, even in Mayall's opaque financial reporting, that there had to be a problem with the director's salary and proposed a salary survey to establish a proper range within which it could be administered. When the results of a preliminary survey revealed how substantial the risk of Mattei's resignation had become, the Council aggressively increased the director's salary toward the appropriate benchmarks identified by the survey.[91] Mattei's relief was palpable when she learned of that first large increase made formally by the Council.

Mattei continued to experience difficulties dealing with many problems for the next decade. She did so as a tough-minded, conscientious representative of the AAVSO, but lacked the full credential of a Ph.D. degree. She had limited professional experience in dealing with such things as credit for published scientific work. Still, she stuck to her guns over a lot of opposition and criticism from the professional community for several years. Eventually, the AAVSO would produce results of interest to professionals. However, a demonstration of that potential from the archival data that had been accumulating for six decades or more could not yet be performed so convincingly. The amateur observers in the AAVSO proved their worth to the professionals in the next few years through their real-time contributions, not through the Association's archival resources, substantial as those resources had become.

The drastically improved salary for Mattei came at a fortuitous moment; it allowed Mattei to lay aside her concerns about whether the AAVSO would retain her as its director. By 1980 she had already begun making plans for the celebration of the AAVSO's 75th anniversary, still 6 years in the future. She hoped to have headquarters settled in a new building by then, an ambitious and challenging goal for someone with her limited experience, but entirely consistent with her drive and dedication to variable star astronomy, as well as to the AAVSO.[92]

18 · Learning the ropes the hard way

Twentieth century amateur astronomers have seldom had access to the advanced equipment that their talent and dedication deserves. That needs changing.

– Riccardo Giacconi[1]

In late 1973 and over the next 7 years, Janet Mattei was forced to deal with a multitude of problems that had been developing during the difficult years, after the eviction from Harvard. Membership continued to grow, and the annual total of observations submitted to headquarters grew even faster as international observers began to accelerate their contributions.

Although the commitment to modern data processing technology had been made, headquarters could not keep up with the growing volume of incoming data. Furthermore, observations still on keypunched cards needed to be converted to light curves so that the ever-increasing volume of questions from the professional community could be answered quickly. Arrangements for semiannual meetings and hand-plotting of light curves constituted an overwhelming workload for Mattei and left her too little time for membership communication and development of professional relations. A crisis erupted in the Council over the handling of *The AAVSO Atlas*, and then *JAAVSO* editors resigned. Seeing these crises as opportunities, Mattei began gaining control over the operation and gradually won the confidence of the Council.

This chapter considers the events of the next 7 years of development in the AAVSO, a period of very significant change for improvement.

THE LEADERSHIP FUNCTION IN THE AAVSO

For a number of years, Council meetings had typically lasted as little as 2 and as much as $4\frac{1}{2}$ hours, but rarely more than 3 hours. Meeting for such short sessions only twice each year was not enough time for the Council to consider the full range of problems that confronted the AAVSO. As a result, many problems that should have been handled by the Council, or with guidance from the Council, were simply dropped in the lap of the director, the treasurer, or both.

Secretary Clint Ford's minutes reflected his impatience and perhaps concern about the way the leadership avoided responsibility and failed to decide issues. Ford would summarize a discussion and frequently add the phrase "but no action taken." As a matter of tradition, a great deal of time in Council meetings was tied up in hearing committee reports. This frustrated the more action-oriented Council members, who realized they would be hearing the same reports in the general meeting. The same could be said of the semiannual director's report.

Mattei realized the weakness inherent in this mode of operation and expressed her frustration with the situation to some Council members. She could not get help on many problems she felt deserved extended discussion in the Council. Many members of the Council valued their time at each semiannual meeting for sharing technical information, discussing peculiar variable star behavior, and generally renewing acquaintances. Other Council members expressed impatience with the length of meetings that exceeded 3 hours. In effect, all too often the Council simply abdicated its responsibility to Mattei. This was especially troubling for her when it came to such important matters as assessing how the Association should be changing in the future. Among many other things, Mattei considered the relocation of headquarters an urgent priority.

At the 1979 annual Council meeting, the question of how the AAVSO planned for its future arose as the meeting neared its end. There was virtually no discussion of this topic before it was dismissed as the meeting adjourned, except for a letter with a proposal from a Council member to AAVSO President Carl Anderson,

Ford, and Mattei. The letter suggested how a discussion might be conducted on the Association's future plans. An AAVSO Futures Study could examine three cases for the AAVSO from 1985 to 2005: a Reduced Cost Case, an Amateur Society Case, and a Professional Service Case.[2]

Ford suggested almost immediately that the three possible cases did not match with the reality that the AAVSO "has always been, and probably always will be, an organization that will encompass, and struggle with, all those three categories." Ford sketched out answers, at least his answers, to the questions that were intended to stretch the vision of Council members to consider alternatives for the Association. His response considered the whole idea of Council planning for the future as a subversion of the constitutional mandate that such planning was the responsibility of the director.[3]

President Anderson, on the other hand, suggested that the matter be the topic of a special session at a future Council meeting. At the 1980 Spring Meeting in Houston, Anderson proposed adding an unusual 2-hour morning session to the Council's agenda. The Council agreed to a special session at the annual meeting in October 1980. Ford included a copy of the original proposal for each announcement of the special session for each Council member to consider before the meeting.[4]

The extra 2-hour Council meeting at the annual meeting proved insufficient to discuss the matter in any depth. Part of the session was consumed by discussion of the need for an in-house computer system to assist with processing variable star observations. It was apparent that the Council members, for the most part, lacked information about the detailed workings of the office and especially the financial status of the Association.

This lack of knowledge on the Council's part was, in effect, a cultural artifact of the several decades during which the AAVSO resided securely within the Harvard College Observatory (HCO). At that time, the Council had no budget to control and no responsibility for the ongoing support of the organization from an administrative point of view. After the eviction, and during the perilous times when Margaret Mayall supported the Association as an unpaid director, there was nothing to work with in terms of resources. Thus it is perhaps natural that the question of planning for the future would only arise after the problems of surviving the eviction and short-term coping with the new expense burden, and the transition to a new director, were all resolved. At that point it was clear the AAVSO had a future for which it should plan. In some Council member's minds, at least, such planning was overdue. Mattei expressed interest in extending the planning discussions, so the Council agreed to continue to work with the idea in future meetings.

When four new members were elected to the Council, they asked about the Futures Study mentioned in Council minutes. Newly elected President Art Stokes agreed that they should each have a copy, since the Council had agreed to continue the discussion.[5] Stokes wisely suggested that Council discussion of the subject be limited to only one or, at most, two items per meeting. "If too many ideas are discussed, the meeting can get to be a free-for-all exchange with no real action taken," he commented. Stokes agreed with Ford on the matter of the delegation of responsibility to the director for reporting the condition of the Association to the membership, but agreed the subject needed further discussion.[6]

Further discussion of the proposal at the 1982 Spring Meeting in Milwaukee benefited from an outline prepared by Mattei titled "The Present Status of the AAVSO." Perhaps at the suggestion of former president Anderson in 1980, Mattei had prepared a helpful five-page summary of the extant conditions for the AAVSO. Key elements of Mattei's summary included:

1. The significant value of current and archival data and of the current cooperation with satellite observatories, closing this part with the observation that "Astronomers have suggested to me that the AAVSO become the center of variable star data for the world, and of variable star information for the United States, because of our important long-term contributions, the quality of our observations, and our financial independence."
2. A very large turn-over in technical staff, noting that she had had five assistants in four years.
3. A need for a full-time assistant to help the director with the administrative aspects of her job, and of other staff in headquarters.
4. The need to relocate the headquarters to a larger, safer (fireproof) building to protect and preserve the archival records of observations as well as computing systems and other files and to provide adequate working space for the headquarters staff.[7]

Mattei's report also referenced the suggested Futures Study, in response to Ford's concept that the responsibility for planning for the future belonged to the Director and not the Council. In effect, Ford's view of the leadership functions in the AAVSO could be summarized as "The Director Proposes, the Council Disposes." What Mattei's response really implied was that, in her view, the responsibility for such planning was a joint effort involving both the director and the Council. Noting that the key resource missing was money, Mattei returned the problem to the Council since it was responsible for providing the necessary resources to operate the Association. In spite of this interest on the part of Mattei and some of the Council members, the Futures Study idea gained no momentum at the time and faded from view.

MAINTAINING THE ROUTINE

The leadership role shared by Mattei with the officers and Council of the AAVSO included the activity of the technical committees. This involved recruiting and motivating chairmen. As described in Chapter 16, a motivated chairman like Marvin Baldwin for the Eclipsing Binary and RR Lyrae Committees produced concentrated activity in specialized fields. In such cases, the rewards to the effort are found in the numbers of observations reported on a routine basis and, in the special case of the EB stars, in the number of reductions of those observations to establish whether or not a star is deviating from its ephemeris. The results are clear and measurable and reflect well on the chairman, as well as the program participants who contribute observations.

A different kind of problem exists in motivating observers in an area like Nova Search. In such a case, the success of the program may also be measured in terms of the number of observations submitted to the chairman, but those reports are, for the most part, negative reports. The opportunity to be the first to find and report a new star, whether it be a Galactic nova or an extragalactic supernova, creates a much different incentive for the observer and therefore a much different problem for the chairman. Finding and motivating observers to continue reporting negative results in the hope that they will eventually make a positive discovery is a difficult job. Making such a discovery, of course, is a matter of countless hours in routine searching and looking in the right place at the right time. The problems associated with supervising the technical committees were, at best, an annoyance for Mattei, but that work could easily have been assigned to one or more Council members and thus relieved the director of the task. Tradition ruled against that idea.

GOING INTO ORBIT

When the US space exploration program began in 1961, visionary astronomers began to anticipate the time when satellites above the Earth's atmosphere could observe the sky at wavelengths normally absorbed by the atmosphere. Satellites such as the X-ray satellites (HEAO-1and HEAO-2), International Ultraviolet Explorer (IUE), the Infrared Astronomical Satellite (IRAS), and others could observe variable stars along with many other targets of interest. Opportunities emerged for cooperation between the AAVSO and professional astronomers in charge of Earth-orbiting observatories, who needed predictions for eruptions of cataclysmic variables (CV) and then real-time alerts when eruptions actually occurred. Mattei accepted the challenge, in a limited way at first, but initial successes contributed to her confidence in the data as well as her ability to manage the effort. Mattei worked at first with John Bortle, *AAVSO Circular* editor, to learn techniques he had refined in more than a decade of regular work with these stars. Soon, though, she was prepared to handle this as a headquarters' task.

The accommodation of requests for information about the status of most variable stars still presented difficulties in 1976 when France Córdova, an astronomer then at the California Institute of Technology, requested help for scheduling time on a forthcoming X-ray satellite. The High Energy Astronomical Observatory (HEAO-1), scheduled to be launched in 1977, would include a number of dwarf novae among its targets. Córdova and the astronomers involved hoped some of these stars would erupt during the HEAO-1 flight. Simultaneous observations and alerts to the HEAO-1 flight controllers would allow observations of possible X-ray emissions during those outbursts. Her targets of interest included a number of dwarf novae and recurrent novae. Observers who participated in the alert program were given "hot line" telephone numbers and were asked to call Córdova directly if they detected a brightening of any of the program stars. In the case of SS Cygni, a last-minute call from Jim Morgan of Arizona came just as the observing window was about to close. Córdova persuaded the mission controllers to redirect HEAO-1 to the star at the beginning of an outburst. They successfully accumulated observations throughout the outburst and shortly afterwards.

Córdova attended the AAVSO's 1977 Annual Meeting to thank the AAVSO and RASNZ observers who had participated in this monitoring and provided the necessary alerts to the HEAO-1 astronomers. She announced to that meeting, and to the world at large, that an unexpected and substantial brightening of hard X-rays had been detected in U Geminorum during a maximum.[8] The astronomers involved in these missions were surprised by a flickering of hard X-rays emitted during a visually confirmed quiescent state of SS Cygni and soft X-ray emissions detected during the quiescent state of EX Hydrae.[9] These mission results, made possible by the continuous monitoring and alerts by AAVSO observers, gave ample evidence of the value of involving the AAVSO in satellite observing programs.

The consequences of Mattei's decision to become involved in the CV program were enormous for her personally. She needed to learn the astronomy and astrophysics in a hurry to catch up. She needed to be able to contribute to mission planning and the management of satellite observations on an international basis. A major complexity in that specialized type of observing was the constant attention required. The work proceeded on the schedule of the orbiting observatory rather than the usual daytime office routine or the usual observatory night-time routine; it was a 24-hour job.

A good example of what was involved is reflected in a cable to Mattei from Martine Mouchet of the Vrije Universiteit, Brussels. Mouchet needed guidance for a series of cataclysmic variables and asked for Mattei's prediction of which would be in outburst. Mouchet had been allocated simultaneous observing time with the IUE Satellite, the spectroscope on the 3.6-meter telescope, and the photometer on the 1.54-meter telescope, both at La Silla, Chile. Obviously, these telescopes could only see part of the northern hemisphere, whereas the satellite could see the whole sky. So in addition to the scanning of the possible targets that Mouchet and his colleagues had selected, Mattei had to sort those into northern visual observers and southern visual observers and notify Frank Bateson of the RASNZ of the need to alert his observers in the southern hemisphere. There is no indication that any other southern hemisphere association, whether in South Africa or Australia, was cooperating in this project.[10]

A similar request from James Pringle, Institute for Astronomy, Cambridge, England, involved setting up the simultaneous IUE observing run at Madrid (mentioned above). In that case, a computer failure wiped out the observing time available in four of the six windows of opportunity in the IUE schedule. It illustrates pretty dramatically the costs and frustrations of relying on a computer that might crash in the middle of a run and wipe out not only the data already acquired, but also the time allocated rather narrowly for this purpose.[11]

With a few successes under her belt and to meet the rapidly expanding interest in the professional community, Mattei elected to upgrade her astrophysical knowledge of the cataclysmic variable stars. With activities at headquarters somewhat stabilized, she attended the *Fifth Astronomy and Astrophysics Summer Workshop* in 1981 at the University of California, Santa Cruz. The 3-week plunge into astrophysics represented an intense dose of the science involved, but at the same time Mattei established many contacts in the CV community and also participated in the American Astronomical Society meeting. In her annual report, Mattei described a number of papers in the workshop that relied on or cited AAVSO observations, and quoted Sumner Starrfield, who, as chairman of one of the workshop sessions, called for a round of applause for the AAVSO and said, "We owe a great debt to your observers. I hope you will convey it to them."[12]

The following year Mattei attended an IUE Symposium in Haifa, Israel, and added a new CV to the AAVSO program. UW CMa (29 CMa) was a new type of X-ray binary in which the X-rays originate in the collision of the stellar winds between the two stars in the binary; a characteristic of the UW CMa stars is that the X-rays are not present during a primary eclipse.[13]

RESTORING MEMBER RECOGNITION

Although Mattei appeared to be preoccupied with technical matters, other aspects of the AAVSO's effectiveness began to receive her attention. For example, throughout most of the turbulent 1970s, the Council completely overlooked the value of recognizing the contributions of individual members. After presenting an AAVSO Merit Award to Richard Davis in 1971, there was no further recognition of individual members with

Figure 18.1. Edward Oravec and Clinton Ford at Mt. Holyoke College, October 1985.

AAVSO Merit Awards for nearly a decade. Finally, in 1980, at Mattei's suggestion, the Council recognized the outstanding service of Marvin Baldwin as the long-standing chairman of the Eclipsing Binary Committee and two-term president of the AAVSO. Baldwin richly deserved the recognition. In the following years, Merit Awards were presented to Charles Scovil, John Bortle, and, in 1985, to Edward Oravec, all outstanding observers who also served the Association in many other capacities over the years. The restoration of this important recognition provided a boost in morale among observers, in particular; a few commented that the time lapse between awards in the 1970s seemed to mean many other factors were more important to the Council than the visual observer and his observations. Mattei rightly saw the need to restore balance in this important area.

MATTEI'S 1983 ANNUAL REPORT

By 1983 Mattei had developed a definite style and modus operandi, such that the AAVSO finally began to show definite progress. Elements of this transition likely included, but were not limited to, the following: completion of her Ph.D. and graduation from the University of Ege in 1982[14]; Ford's gradual acceptance, as shown in a shift in his attitude from grudging cooperation to steady support; and a gathering of more supportive professional astronomers who facilitated and amplified Mattei's efforts in the professional community, as well as with the problems of Association management, including Charles Whitney, Lee Anne Willson, Martha Hazen, and John Percy, in addition to Mattei's oldest friend Dorrit Hoffleit. Mattei's effective participation in the workshops, IAU meetings, and presentations at AAS meetings brought a growing awareness of the existence and utility of the AAVSO in the professional variable star community. A continuing presence in all these venues proved very rewarding to her as that recognition began to materialize in the form of credits to her as well as to the AAVSO in oral presentations of other astronomers at scientific meetings as well as in published papers.

MATURING VARIABLE STAR ASTRONOMY

Variable star astronomy had matured as a scientific subfield within astronomy and astrophysics. Sophisticated modeling techniques using high-speed digital computers aided this process, facilitating studies of stellar evolution and its relation to stellar pulsation, the basic phenomenon studied in the earliest years of the AAVSO's focus on LPVs. Digital computers also facilitated the development of modeling for binary star systems that replaced the techniques developed by Henry Norris Russell and that were exploited so well by Harlow Shapley. Advanced observational techniques using photoelectric photometry, coupled with the growing numbers of moderate-to-large aperture telescopes, contributed to many discoveries after 1950 that significantly increased the understanding of stellar variability and added several categories of variable stars deserving study. Among the more interesting and important results was the understanding that cataclysmic variables (novae, dwarf novae, and related stars) are nearly always close binary systems in which mass may be moving between stars. Contemporary studies investigated the effect of stellar rotation and magnetic fields on stars, leading to eventual classification of even more distinct types of variable stars. The field was ripe with opportunities for amateur astronomers who, in the same time period, were accessing larger aperture telescopes at reasonable costs, making possible the routine study of ever fainter stars. The mid-1980s were ripe with potential for the AAVSO.[15]

Following are some of the key signs of progress visible in the Director's Tenth Annual Report to the Association.[16]

Grantsmanship

One area of evident progress in her first 10 years was the preparation of grant proposals and appeals to various funding agencies for the money with which to advance the work of the AAVSO, whether that progress came through investment in additional equipment for the office or through the employment of additional staff. The most important source of funding, of course, was the AAVSO's own secretary, Clint Ford. After nearly losing Ford as a donor through her clumsy handling of *The AAVSO Variable Star Atlas* (Chapter 17), Mattei worked for several years to regain Ford's trust and faith as the Association's director. To do so, she had to show she was making progress on several fronts as she found herself in a complex and demanding job that was, at the same time, changing rapidly. In addition to progress in the technical program – an obvious measure – it was also very important to find other sources of funding to meet the AAVSO's needs.

Mattei's efforts to raise money for direct support of the technical program came mainly through her dedicated effort to promote awareness of the AAVSO within the ranks of professional astronomers in the variable star community. Eventually, any application for a grant of funds from such sources as the National Aeronautics and Space Administration (NASA), the National Science Foundation (NSF), and similar governmental agencies, as well as private foundations like The Research Corporation, would be "peer reviewed" before favorable consideration could be given to the request. Inevitably, then, those reviewers had to be familiar with the science at stake, as well as the person and organization requesting funds.

Mattei waged a vigorous campaign to have the data used in the preparation of variable star research papers based on AAVSO observations be explicitly identified as the work of the AAVSO. Her earliest efforts in this regard were not always welcomed by the professional astronomers. In many cases, those astronomers welcomed the availability of the data, but at the same time were unsure about how their use of data from an association consisting mainly of amateur astronomers would be viewed by other professionals. Mattei's persistence in making her demands paid off in a gradual but steady increase in the presence of the Association in the scientific literature of variable star astronomy. The more frequently the AAVSO appeared in the credits on scientific papers, it seems, the more acceptable it became to credit the AAVSO.

Mattei also attended scientific workshops such as those discussed earlier in this chapter on the cataclysmic variable stars. Her purposes in doing so were always multiple: to learn how a field of variable star astronomy was changing, what problems were under consideration, and especially which problems might involve variable star observations already available in the AAVSO archives; to present AAVSO observational programs and results for the awareness of astronomers new to the field; to meet the participants and establish personal relationships within the professional community; and finally to draw more professional participation into the Association.

Success in winning grants then reflected success in a number of parallel efforts on Mattei's part to improve the visibility of the AAVSO in the professional variable star community. Gradually, her grant applications began receiving approvals. The progress visible in Mattei's annual reports is especially impressive in this regard. She reported with increasing frequency the grants from various funding agencies and how they funded specific projects. On Mattei's part, those grants were important for what they contributed to the needed improvements in AAVSO operations. Furthermore, each success reported in one of her annual reports witnessed Mattei's gradual maturation as an association manager. Finally, and perhaps most importantly, each success provided additional assurance to Ford that the Association would not rely solely on his support for survival.

Data processing

Mattei decided early in her tenure to shift emphasis on data processing in an important way.[17] Rather than attempting to bring all unprocessed data up to date, she decided to process – on a monthly basis – only the current observations. This change in emphasis created what came to be known as "The Gap" in the accessible observation files beginning with the October 1961 reports. Thus the backlog in unprocessed AAVSO variable star observations included not only the complete observational records from 1911 to 1961, but also the unprocessed post-1961 data. The plan, then, was to remain absolutely current on new observations so that these data could be available in a timely way to answer questions that arrived at headquarters with increasing frequency. Earlier observations would be processed as time allowed.

A major setback in the AAVSO program occurred in 1979 when the Harvard-Smithsonian Center for

Astrophysics (CfA) decided to replace their entire Control Data Corporation (CDC) high-speed computer with much faster Digital Equipment Corporation (DEC) PDP 11/60 and VAX 11/780 machines. Benefiting from a grant of computer time arranged by long-time AAVSO members Owen Gingerich and Barbara Welther through the CfA Director, the AAVSO relied on the use of the CfA high-speed computing facilities to convert data on punched cards to magnetic tapes that could be copied and archived in separate locations for data security. Those tapes then became the input data on which the CfA computers could be used to prepare light curves, an invaluable resource for the AAVSO. All of the computer programs necessary to read the cards into the computer, sort the observations by variable star and by date, and then merge those data with the previously archived data and record the consolidated data onto the magnetic tapes were specific to the CDC computer. The system change at CfA made it necessary to completely reprogram AAVSO data-processing routines in the new language specific to the DEC machines, FORTRAN, and facilitated new programs written to perform quality-control checks previously done laboriously by eye.

Punched cards – eventually some 1.5 million of them – accumulated at the Concord Avenue headquarters. The boxes of cards, invaluable as they were to the records of the Association, became a part of the furnishings – functioning as partitions and support for tables and gradually choking off space that was needed for other purposes. Evolution of the whole field of computing would eventually make it possible to bypass the production of cards and enter the data directly in a digital magnetic medium. As a first step in December 1981, a grant from the Charles M. Townes Fund of The Research Corporation made it possible to install an Ithaca Systems microcomputer at headquarters, later including two work stations on which data entry and verification could proceed separately but simultaneously. The data were recorded and stored on 8-inch magnetic diskettes that were compatible with the high-speed computing systems at CfA. Although it was still necessary to rely on CfA for the steps of sorting, merging, and processing the data and archiving them on magnetic tapes, the process was accelerated and the storage problem of cards was eliminated by the acquisition of the AAVSO's first in-house computer.[18]

By 1981, processing of the 1960s "Gap" had been completed, and data entry of the archival data for the period 1911 to 1960 had begun. During processing of the archival data, it was discovered that reports dating back to 1902 were also in the files. These pre-1911 reports included some from the HCO Corps of Observers, which were incorporated into the archived data. Mattei reported progress in each meeting in terms of the letter of the alphabet (the archival observations were filed by the observer's last name) on which data entry operators currently were working. Work proceeded slowly for a time as data entry and verification of the current month's observations took first priority.

The archival data entry continued over the next 15 years; it involved regular full-time staff who did data entry, programming, and data processing, as well as temporary part-time data entry help. A grant from Ford supported this effort, and AAVSO Treasurer Ted Wales funded the purchase of two high-capacity disk drives to further accelerate report preparation and headquarters administration.

International cooperation

As recognition of Mattei's progress with the AAVSO spread in the professional community, the International Astronomical Union (IAU) Commission 27 on Variable Stars proposed a resolution, adopted by the IAU at its 1982 meeting in Greece:

> The International Astronomical Union, recognizing the recent increased interest in cataclysmic variables, and the need for long term light curves to make possible the correlation of theoretical and observational research, and noting that data collected by AAVSO observers and other groups throughout the world are available on magnetic tape, RECOMMENDS the publication of this valuable material by the AAVSO.[19]

The implicit suggestion, of course, amounted to the idea mentioned by several individual astronomers over time that the AAVSO become an International Database for all variable star information and that associations that had not already done so should send their archived observations to the AAVSO for that purpose. After several attempts to convince European observers that the name of the database was not important – the word "American" in the Association's name was problematic for some – Mattei acknowledged the IAU resolution by announcing that the AAVSO database would henceforth be known as the AAVSO International Database (AID). The first case in which that recognition materialized in actual practice occurred in 1983 when the members of the Hungarian Astronomical Association sent several years of their variable star observations to the AAVSO for inclusion in the AAVSO International Database and the Association Française des Observateurs d'Étoiles Variables (AFOEV) began submitting data monthly.

Individual members of a number of variable star associations from around the world had been contributing observations in increasing numbers during this period. In her annual reports, Mattei listed the full names of the societies whose members were submitting observations, including those from France, South Africa, Austria, West Germany, England, Australia, Japan, The Netherlands, Norway, Brazil, Hungary, and Belgium. (A listing of all the international associations that have contributed observations to the International Database over the years is included in Appendix C.) Several societies were distributing AAVSO publications of interest to their members, including the annual *AAVSO Bulletin*, which provided predicted maxima and minima for long-period variable stars. Mutually beneficial exchanges of literature were taking place between the AAVSO and major universities, observatories, and astronomical societies with variable star sections around the world.

The *AAVSO Alert Notices* program attracted some international participation. The postal distribution of these notices, announcing the discovery of a transient object, a request for observations from a professional astronomer, or a special observing campaign, offered the opportunity for observers and professional astronomers around the world to learn of and observe in a timely manner a transient object, such as a nova, that was not a part of an AAVSO committee's regular work. A case in point is the search for extra-galactic supernovae. Australian amateur Reverend Robert O. Evans improvised his own finding aids, a series of galaxy images he copied onto 35-mm photographic slides at an appropriate scale, and began visually searching those galaxies. By 1983 Evans had made an impressive start on his program, having discovered four supernovae. The AAVSO announced his discoveries (eventually totaling 37 by 2010) in Alert Notices and recognized this

Figure 18.2. Rev. Robert O. Evans, of Australia, Nova Award recipient, Nantucket, 1983.

signal achievement by inviting Evans to attend the 1983 AAVSO Annual Meeting at the Maria Mitchell Observatory on Nantucket Island, Massachusetts. Evans presented a paper about his methods and the results; then AAVSO presented its Nova Award (a plaque) for his early supernova discoveries.[20] Evans' charming and self-effacing style won him many friends as he visited and spoke to several other groups around the United States before returning to Australia.

Publications

The preparation of *AAVSO Reports* summarizing the observed behavior of long-period and other types of variables had lagged behind the planned schedule for several years. In 1983, the availability of observations in the computerized database and using the in-house microcomputer began to show the benefits of this investment when *AAVSO Report 38*, observations of long-period variables from 1974 to 1977, was published, for the first time with computer-plotted light curves of individual observations of 490 stars. This represented a substantial improvement in the display of detailed information, while at the same time allowing the preparation of the light curves in a cost-effective manner.

Progress had also been made in preparing a follow-up report to Leon Campbell's *Studies of Long Period Variables*. Compilation of an extension to Campbell's study to include maxima and minima for 383 stars from 1950 to 1975 received support from a Small Research Program grant from the American Astronomical Society. Maxima and minima of long-period variables compiled by Mrs. Katherine Hazen as a volunteer effort neared completion as a further extension of this work. Thus the traditional work of the AAVSO showed signs of returning to health after a lapse of nearly a decade during which data processing problems were resolved at headquarters.

Handling special requests

A good measure of the extent to which the AAVSO gradually became more broadly visible in variable star astronomy due to Mattei's persistent efforts can be found in the number of requests coming from astronomers. In her 1983 Annual Report, Mattei noted that such requests for data to assist with either scheduling observing runs, making simultaneous observing efforts with ground-based observers, or for correlation with observational results had increased from zero in 1973 to 79 in 1983. Although she did not separate satellite observatories from similar requests for ground-based observatories (especially radio telescopes), it is apparent that satellite-based programs dominated these requests. The opportunities to observe with these scarce resources were in great demand, and lead times for scheduling runs were correspondingly long. With very narrowly defined windows of opportunity during which to observe with the satellite, it was vitally important to have ground-based support with the greatest possible geographical dispersion to guarantee that an observer with clear and dark skies could provide the comparison data. The *AAVSO Alert Notices* program set up by Mattei achieved many successes, as noted along the way in her annual reports, but she also noted a particular success when those alerts redirected not just one, but three orbiting observatories: the International Ultraviolet Explorer (IUE) and two new orbiting observatories, the European X-ray Observatory Satellite (EXOSAT) and the Infrared Astronomical Satellite (IRAS). Although observing time had been scheduled for visual light observations at several large observatories, all were clouded out at the critical time and only AAVSO data were available to confirm a brightening of SU Ursae Majoris.

The year 1983 also featured observations with EXOSAT while GK Persei was in outburst and a successful prediction and observation of an outburst by TY Piscium. The latter allowed one large photometric telescope to confirm that the star could be firmly identified as belonging to the SU Ursae Majoris classification with observations of the superhumps during this outburst.

Observations

The number of observations of variable stars from within the United States declined slightly over the years as prolific observers aged, but the number of observers outside increased dramatically in these 10 years. In Mattei's first annual report, for example, the number of international observers submitting observations to the AAVSO, as well as the number of observations they submitted, amounted to about one-third of the respective totals. By 1983, however, those ratios changed substantially. The number of US observers had increased slightly from 236 to 260, but the number of international observers had increased from 109 to 245, or more than doubled. Furthermore, although the total annual observations submitted by US observers had decreased slightly from 97 095 to 88 167, the corresponding totals from international observers nearly doubled from 55 637 to 106 428. At the end of the fiscal year 1982–1983, the total number of observations in the AAVSO International Database stood at just under 5 million.

Thus Mattei's 10th Annual Report reflected very significant changes occurring within her first decade as the AAVSO Director. Surely some of this change reflects a growth in interest in variable star astronomy throughout that decade. That change can be traced to the availability of such new tools as the high-speed computing capability with which theoretical astronomers began modeling stellar evolution and studying the conditions that might lead to pulsation or any one of

Figure 18.3. Cap Hossfield with Janet Mattei, Nantucket, 1983. In the background are Walter Scott Houston, Gerry Samolyk, and Charles Scovil.

the several other causes of stellar variation. The small growth in the numbers of US observers can likely be traced to a growing availability of inexpensive larger telescopes for amateur astronomers as well as increased interest in astronomy with the public because of NASA and the orbiting satellite observatories. Still, one cannot ignore the benefits of Mattei's international travel and the growing awareness of the AAVSO that is evident within the professional astronomical community. Mattei's comprehensive annual reports provided evidence over the years that the AAVSO and its membership were solid contributors to the variable star community at large. And finally, we note that all of this provides evidence of Mattei's growing maturity as an association leader. That should not be construed too broadly, however, in that she still lacked the confidence necessary to delegate effectively and to operate in anything other than a completely "hands-on" management style that limited both her effectiveness and the productivity of the headquarters staff.

MONEY MATTERS

The staffing problem

Progress on the issuance of *AAVSO Reports* and data entry of the archived observations continued to be disappointingly slow during the 1980s. In addition to technology limitations, one of the reasons for this slow progress could be traced to a continuing problem of staff turnover. At about the time when an individual hired on a work-study grant became truly efficient, those skills could be translated into a higher-paying job elsewhere or a job with paid benefits. At the same time, inflation in the general economy and the rising cost of living contributed to the same search for improved wages and benefits. Thus staff turnover contributed to increased costs and reduced efficiency. Fortunately, one area not affected by staff turnover was data processing and archiving. Elizabeth Waagen had been hired in 1979 to handle this area of operations, and it had remained the top priority among her diverse responsibilities, carrying out Mattei's mandate that current data processing must be kept up to date and that the backlog of observations be processed as soon as possible.

A salary survey undertaken by the Council established the apparent minimum wage that could be expected for this work at an institution like Harvard University. A large employer of such skilled workers, Harvard established the local pattern for wages and benefits for nonprofessional employees. A contact within the Harvard salary administration department proved valuable to Mattei in gradually understanding how the AAVSO with its budget limitations might become more competitive in the marketplace for data entry personnel.

At the same time, the services that the AAVSO provided to members and to the professional community in various ways continued to expand. The resulting increased demand for work in the AAVSO office, coupled with the staff turnover problem, frustrated Mattei's efforts to catch up. Progress on many fronts was visible, but the basic problems continued to accumulate and demanded a slow but continuous growth in the total number of AAVSO employees. The combined effect of general inflation and slowly rising staff levels as more services were provided led to an almost continuous concern in the Council about the Association's financial situation.

Mattei's own salary, in spite of generous increases for several years in a row, was still not comparable to the benchmark positions established in the 1982 salary survey and repeated in 1984. The Council tried to keep moving the director's salary to approximate those of four positions identified at the Harvard-Smithsonian CfA and McDonald Observatory that had similar supervisory and financial responsibilities. Although progress was being made in that regard, those salary ranges also increased in response to economic conditions, creating a constantly moving target.

Figure 18.4. Clint Ford, Ted Wales, and Janet Mattei at the Concord Avenue Headquarters, 1985.

Deficit spending

Examination of the earlier annual treasurer's reports for the AAVSO sheds little light on the resolution of the staff and salary problems. As mentioned in Chapter 17, when Ted Wales took over the treasurer's job in 1980, he reorganized the treasurer's report format so that the Council could better appreciate how money came to the AAVSO, including the small contribution to the overall budget made by the membership dues. That reorganization also allowed a better understanding of the ways in which money was spent. It became evident that staff salaries and benefits, limited as they were, constituted a major portion of the total expenses. As Wales came to understand the ways in which income and expenses had previously been accounted for, and how bank accounts had been managed, it was clear that the Association had been involved in deficit spending for some time. Each year, there was a draw-down on a fund identified as Accumulated Income that had been held in reserve by previous treasurers. Wales reported this candidly each year, along with his estimate of what was left in this reserve fund to cover the annual deficits. Each year, it seemed, the following year would see the reserve fund finally exhausted.

It was not evident to the Council members that behind the scenes, Ford watched the financial reports very carefully and established a separate confidential rapport with the treasurer. Each year Ford identified one or more major expense items that he would handle by making a special donation to the AAVSO for that identified purpose. The consequence of Ford's generosity on an ongoing basis was that the AAVSO always operated on the financial edge, never quite solvent but never quite broke. The effect of this continuous apparent peril was to limit the ambitions of the Council, but not always in a productive way.

Although Mattei and Wales understood these financial dynamics, others in the Council were slow to come to an understanding; that was as Ford intended – he did not want public recognition of his continuous generosity. Wales, though, recognized that the Council needed to do something more in the way of active involvement in the financial management of the Association. After all, that was an important role delegated to the Council by the Constitution. Discussions in the Finance Committee eventually highlighted the problems in a way that Ford found comfortable: by talking about possible ways to increase the Association's income.

One way of increasing income, though initially resisted by Wales, was to shift the balance between stocks and bonds in the AAVSO Endowment Fund. The traditional financial advisor's recommendation of 60 percent stocks and 40 percent bonds had been followed for this fund for many years. The wisdom of the approach can be seen in the fact that the Endowment Fund had grown to a value in excess of \$1 million from the original balances, nearly \$500 000 in the early 1960s. At the Council's meeting during the 1984 Annual Meeting, Wales reported that he had shifted the balance in the Endowment Fund to a ratio of 50 percent stocks and 50 percent bonds to produce more income from the Endowment. It was agreed that this was only a temporary solution to the problem and that other measures would be required. Wales estimated that the annual income from the Endowment Fund in this new arrangement would amount to \$84 000, which would not even cover staff salaries that by this time amounted to more than \$100 000 per year.

Fundraising by amateurs

The financial problems described above had been recognized to some extent for several years, but in a different guise. It was a matter of simple arithmetic to calculate that for current expenditures, the Endowment Fund would have to be doubled for the Association to become self-sufficient. The amount needed to do that – \$1 million, did not seem unrealistic to the Finance Committee members.

Beginning in 1982, efforts had been made to create fundraising activities in various ways. For example, a corporate fundraising campaign had been mounted, with very limited success. The ostensible purpose of the corporate campaign was to provide capital for a new building to house AAVSO Headquarters. After the organization of a list of possible donors, discussed within the Finance Committee, a letter-writing campaign was initiated to the candidate corporations most likely to respond. The corporations approached in this campaign were sympathetic, but did not respond with donations to the Association. The purposes of the Association, whether the fundraising was for a capital investment or for support to current operations, were too far outside their normal corporate guidelines for philanthropy. Variable star astronomy, and especially by an association of amateur astronomers, simply could not be presented in a way that matched corporate philanthropic intention.

Getting help on the financial problem

Finally, the Council acknowledged that something had to change in the way it approached its financial problems. Fundraising efforts showed some progress but too little to solve the problem in either a short- or long-term basis. In March 1984 the Finance Committee held a special meeting to address the funding problems and concluded that a consultant should be hired to assist the Council. The Council authorized Council member Keith Danskin, Wales, and Mattei to identify a candidate firm. By the spring 1984 Council meeting in Ames, Iowa, they had surveyed the consultants available in the Cambridge–Boston area. They recommended the firm of Jeffrey Lant Associates, a specialist in small, not-for-profit associations. Lant advised that the goal of raising $1 million on the timing needed for short-term relief was unrealistic, so based on Mattei's intermediate goals, the Council focused on raising $500 000 by the end of 1986. The Council authorized the Finance Committee to employ a fundraising consultant by September 1 and to report on its progress by the 1984 Annual Meeting. Jeffrey Lant's firm went to work with Mattei and Wales on planning for further fundraising.

PUBLICATION OF MONOGRAPHS

Lant suggested finding something to dedicate in honor of donors and at the same time attract money in the form of donations or grants to support the technical program. His idea crystallized in the form of publishing monographs containing observations of individual stars with data for an extended period. The IAU Variable Star Commission had endorsed the idea, and professional astronomers preferred to see a long data set on one star rather than a short data set on many stars. The monographs also made up, to some extent, for the lack of further AAVSO Reports. By publishing extended data on variable stars of high interest, the monographs helped to ease demand for publication of archived data on all the available stars. With a concerted effort on the part of Elizabeth O. Waagen and Michael Saladyga, the AAVSO published the first of an extended series of monographs toward the end of 1985. The first monograph, *SS Cygni Light Curves 1896–1985*, dedicated to Clinton B. Ford, recognized his many years of support to the AAVSO with personal services in various capacities and generous donations. Ford did make a substantial donation to support the whole monograph-publishing project.

Three and one-half pages of text were required to list all the observers who had observed this popular variable star. The next page detailed the known properties of SS Cygni, and a folded light curve on a single page presented data for the period from 1896 to 1903. Then the modern data were presented in 40 pages of light curves, with each page covering a span of 200 days of observations.

Every single observation in the AAVSO database is represented with a dot in these light curves, and no single line is attempted as a representation of the light curve, just the actual data points. That form of presentation, preferred by the professional astronomers who would be the primary users of the monographs, created one problem for the AAVSO staff. Because of this form of presentation, individual data point aberrations are not masked by averaging 10 days of observations, as was the practice for the long-period variables. Mattei reacted to the challenge by insisting that each individual data point be checked three times. First, an M.I.T. student, Charles Jones, employed part-time by the AAVSO, programmed a routine for screening the individual data points to identify obvious outliers. Waagen and Saladyga scrutinized the data in other ways by eye and using programs they had written. If an observation seemed clearly out of line, the original report from the observer could be checked to verify the accuracy of the data presented. Occasionally, obvious errors occurred in the way that an observer entered a Julian date, for example, and when rectified, the data point usually fit the remainder of the data well. After that detailed examination of the data, Mattei insisted that she had to examine every data point herself in what was, admittedly, usually a simple visual confirmation of the results of the work already done. Unfortunately, such triple redundancy occurred in many headquarters' activities; it would not be stretching the point to say that Mattei's stubbornness in this regard added yet another burden to her already full agenda of work and delayed not only monograph production, but other work as well.

A FIRE CHANGES EVERYTHING

In the spring of 1984, the AAVSO staff arrived for work one morning to find the Cambridge Fire Department mopping up after a fire in one of the third-floor apartments in the building at 187 Concord Avenue that housed the AAVSO's Headquarters. Fortunately, there was no damage to the AAVSO office and all that it contained. That was not the first fire to have been extinguished in that aging building over the years. However, recent developments for the AAVSO actually heightened concerns about the threat that a future fire in the building might actually result in asset losses for the AAVSO.

A whole series of issues were accumulating, including the risk of delays in computerization of the AAVSO historical archives, as well as current observations due to damage or loss of the in-house computing system. Also at risk were membership records and other current files, as well as books and serials, historical files, and the precious archival records of 2 million observations submitted to the AAVSO between 1911 and 1961.

Immediately before the May 1984 Council meeting in Ames, Iowa, Mattei briefed a few Council members on the fire and its possible consequences. The conclusion reached by that small group and later stressed as forcefully as possible to the whole Council emphasized the paramount importance of the security and preservation of the most important assets of the Association, the archived records of individual observers and their observations. Although not stated explicitly, this amounted to a declaration that the security of the archival records demanded a relocation of AAVSO's Headquarters. In an unusual full-day meeting, the Council reacted well by agreeing to authorize an immediate increase in staffing to accelerate the data-processing work to protect and preserve the Association's assets. This agreement to digitize all AAVSO observation records back to 1911 was an essential step for security, but in the process the Council expressed additional confidence in, and support for, Mattei.[21]

The idea of relocating headquarters remained in the background at the 1984 Annual Meeting in late October in Salem, Massachusetts. The Council discussed other aspects of the fund raising campaign. Wales had transitioned the investment portfolio to a stock/bond ratio of 1 to 1 to maximize current income, and David Skillman described the opportunities for participation in NASA's Space Telescope Project.

Although there was an urgent need for relocating the Association's headquarters, the matter was not formally discussed in the meeting. However, Mattei never lost sight of the problem or the urgent need. She and her husband Mike visited Ford on December 7, 1984, at his home in Wilton, Connecticut. The details of that meeting are not available, but evidently the topics discussed included the possibility of Ford making a substantial donation to the AAVSO in the following 2 years. One can imagine that Janet Mattei's dream of a permanent home for the AAVSO before the Association's 75th Anniversary in 1986 was discussed. In a letter to Mattei dated December 29, 1984, Ford mentions talking with his lawyer about "invading" his custodial account and confirmed that he planned to make an initial gift in 1985 and a second gift in 1986, but the timing or the amounts of these gifts was not mentioned. In a letter to President Ernst Mayer in January 1985, Ford acknowledged that the Matteis had visited on December 7:

> Yes, the topic of a "new Building" for HQ was discussed again by JAM and me when she was here in Wilton on 12/7. No conclusions reached. The present campaign for funds does not envision anything by way of a new building – much as I know it is needed! I think before we set up another *Committee* for a "New Bldg. Campaign," we better finish up with the campaign we have going now. Yes, I think a new *rental* location sounds like a feasible goal, if one can be found.[22]

It is unclear here whether Ford had agreed with Mattei to fund a new building, but elected not to share

Figure 18.5. At 25 Birch Street, summer 1985. From left: Tom Williams, Ted Wales, Ernst Mayer, Janet Mattei, Keith Danskin, Clint Ford, Charles Scovil.

that information with Mayer, or he was simply undecided in the matter himself. The matter of moving headquarters needed a bit more consideration.

By the time the Council met for the following spring meeting on June 21, 1985, in Seattle, the matter was still not resolved. It was necessary to first dispose of the idea of renting more space at 187 Concord Avenue. Clearly that rental would solve the space problems but would not deal in any way with the security issues. The Council then deputized Danskin to undertake a search for a more secure rental and a larger space to be placed under lease.[23] Danskin's search for such a property quickly proved fruitless, however, and Ford then signaled his willingness to underwrite the purchase of a building for a permanent home for the AAVSO if a suitable facility could be located.[24]

With new marching orders, Danskin quickly located an ideal property: a freestanding building at 25 Birch Street in Cambridge, Massachusetts, actually adjacent to a house owned by Sky Publishing and occupied by the editorial staff of *Sky & Telescope* on Bay State Road. The split-level concrete block building seemed to fit the Association's needs very well as far as floor space and location were concerned. After a hasty discussion with a few AAVSO officers to confirm that negotiations for the purchase of the property could proceed, Ford entered into negotiations with the property owners. In addition to bargaining over the price, it was necessary to clarify the use of parking space on a lot next to the building, but that issue was eventually resolved to Mattei's satisfaction.

When papers were finally drafted for the entire transaction, though, it involved a purchase by the AAVSO, with a downpayment provided as a gift to the AAVSO from Ford. The remainder of the agreed upon price was to be carried as a mortgage by the building owner, obligating the AAVSO for large monthly payments of principal and interest. A separate side agreement between Ford and the AAVSO provided that these monthly payments would be made by Ford as donations to the AAVSO. Both the mortgage papers and the AAVSO's side agreement with Ford were to be signed at the closing, scheduled for October 30, 1985, in Boston. At the last minute, however, AAVSO President Ernst Mayer balked at the prospect of committing the Association to the large indebtedness. Taking a principled and not unreasonable stand, Mayer refused to travel to Boston for the transaction closing. After Mattei's urgent appeal for a telephone conversation between Mayer and AAVSO's Vice President Tom Williams, it was agreed that Williams would represent the Association at the closing and sign the necessary papers over Mayer's objections. The closing, attended by Ford, Mattei, Wales, and Williams, then proceeded relatively smoothly. When the AAVSO Council met on October 31, 1985, the transaction had been finalized and the AAVSO had become a property owner for the first time.[25]

With this news in hand, the Council decided that both the Headquarters relocation and the planning for the 75th Anniversary celebration took priority over routine meetings. It was agreed that no full meeting would be scheduled for the spring of 1986. Instead, the Council would meet as normal, functioning with the Finance Committee, to make the necessary arrangements. Vice President Danskin agreed to act as Chairman of General Committee on Arrangements for the 75th Anniversary. The Council agreed on a tentative plan for a four-part meeting to include a dedication ceremony, a historical paper session, a scientific paper session, and a session on observing techniques. Council member John Percy then initiated a discussion of the future aims and purposes of the Association, but although the importance and timeliness of such a discussion was acknowledged by the Council, it was impossible under the circumstances to make much progress in the matter, and the discussion was finally postponed.[26]

The physical move from Concord Avenue to Birch Street took place in late January 1986. A number of

Figure 18.6. At Headquarters, 187 Concord Avenue, November 1985. From left: Tom Williams, Charles Scovil, Clint Ford, Janet Mattei, Keith Danskin, Mike Mattei.

Figure 18.7. At AAVSO Headquarters, 187 Concord Avenue, December 1985. From left: Janet Mattei, Director; Janet MacLennan Zisk, administrative assistant; Elizabeth O. Waagen, assistant to the director; Roy Lee, AAVSO member and observer.

Figure 18.8. AAVSO Finance Committee meeting at the new headquarters, March 1986. From left: Treasurer Theodore H. N. Wales, Jeffrey Lant, Mark Malmros, Secretary Clinton B. Ford, Vice President Keith Danskin, Director Janet A. Mattei, President Thomas R. Williams.

volunteers, not all from the Cambridge area, greatly aided the staff with the move. Although the building had been renovated in anticipation of the sale, there were other necessary modifications to be made. For example, the overhead sprinkler system received an overhaul from Danskin's father-in-law, a specialist in such systems. Other utilities were connected and upgraded in the process. Ed Halbach of Milwaukee and Roy Lee of Florida, for example, brought their own tools, supplied all the materials, and built bookcases and shelves, installed doors and workstations, and performed general handyman roles for 2 weeks. The physical move took $1\frac{1}{2}$ days, January 25 through 28 (the main move was on the 28th, the day of the space shuttle Challenger explosion), but the planning, packing, and unpacking that went into such a smooth move occupied much more time. There were, for example, 1500 boxes of computer cards and 55 filing cabinets and other furniture to relocate in the enlarged space available. The old furniture was, in many cases, inadequate, so new furniture was purchased – for example, a long conference table and chairs for the McAteer Library, moveable office partitions, and so on.

The first official meeting, actually a meeting of the Finance Committee, was held on March 7, 1986, at the Birch Street offices. As the practice had emerged in such meetings, the Finance Committee reviewed the state of the Association's finances, but also discussed a broader agenda of concerns about the health of the organization. The Association's fundraising consultant attended the Finance Committee meeting to review progress and go over his checklist of action items. The following day, the full Council continued the discussions in its regular meeting, but one that lasted into the early evening. The agenda included a diverse set of topics that ranged from new services for members and improvements to the *Journal* to plans for the celebration of the 75th Anniversary.

75TH ANNIVERSARY CELEBRATION

Vice President Danskin shared responsibility with Mattei in planning the 75th Anniversary celebration. Two committees assisted them – the Scientific Paper Session Committee chaired by John Percy and the Historical Paper Session Committee organized by Martha Hazen.

The AAVSO staff handled every other detail of the planning and execution of this important meeting.

The 75th Anniversary celebration included a formal ceremony to dedicate the new building at 25 Birch Street as the Clinton B. Ford Astronomical Data and Research Center on the first day, while a second phase, planned as an extended version of a normal 2-day annual meeting, lasted 4 days. Scientific and historical paper presentations took place at the Harvard University Law School. Registrants arrived from all over the world, including New Zealand, Canada, Australia, Spain, South Africa, Hungary, and Japan. Papers read in the meeting by AAVSO members from countries not represented included France, The Netherlands, Germany, Belgium, Austria, Finland, and Norway.

Figure 18.9. Dedication of a new headquarters and celebration of the AAVSO's 75th Anniversary, 1986. From left: Keith Danskin, Janet Mattei, Riccardo Giacconi of the Space Telescope Science Institute, Dorrit Hoffleit, Clinton B. Ford, Cambridge Mayor Walter Sullivan.

The building dedication

The Dedication Ceremony on August 6, 1986, at 25 Birch Street took place in a large tent erected in the parking lot next to the building. Although packed with meaning for those who understood the history of the AAVSO, the ceremony was simple enough. Talks by Danskin and Mattei opened the presentations, with each describing different aspects of the events leading up to the actual acquisition. Ford responded, recalling the earliest efforts by Newton Mayall to design a building specifically for the AAVSO in 1965. He indicated what great pleasure it gave him to be able to make this acquisition possible at a time when the needs of the Association cried out for change. Presented to Ford were letters of congratulations and gratitude from Massachusetts Governor Michael Dukakis and US President Ronald Reagan. Mattei then introduced the Dedication speaker, Dr. Riccardo Giacconi, director of the Space Telescope Science Institute.

Giacconi focused his remarks on the contributions of variable star observers to the progress of science. In the course of his talk, he noted that variable star observers of the Association continued to make valuable contributions to science, providing several anecdotal examples from recent experiences involving investigators on the Hubble Space Telescope. Giacconi concluded his remarks with an announcement:

> I noted that in the twentieth century, amateur astronomers have seldom had access to the advanced equipment that their talent and dedication deserves. That needs changing.

Figure 18.10. Florence Bibber, daughter of Leon Campbell, with Elizabeth O. Waagen and Margaret Mayall, at the AAVSO's 75th Anniversary celebration, August 1986.

> Tomorrow . . . I will be announcing plans whereby amateur astronomers will have observing time on the Hubble Space Telescope.[27]

With this announcement, those assembled gasped with surprise and delight, then warmly applauded Dr. Giacconi. It was then time for the official unveiling of the new sign on the side of the building identifying it as the Clinton B. Ford Astronomical Data and Research Center. Cambridge Mayor, the Honorable Walter S. Sullivan, remarked on how appropriate it was for the AAVSO to remain in Cambridge, a center of science and technology, before ceremoniously pulling down the fabric covering the sign. After the unveiling, Ford finalized the proceedings by cutting a ribbon at the front door and invited everyone inside for a glass of champagne and a tour of the spacious facilities. While a reception was held inside the building, the caterers laid out a full New England clambake for an outdoor supper to end a splendid day.

The 1986 Annual Meeting

The extended 1986 Annual Meeting began Thursday morning with papers discussing various historical aspects of variable star astronomy and the AAVSO, followed by afternoon presentations from various national organizations around the world. The following day featured a Variable Star Symposium during which experts described the state of the science in various categories of variable stars such as eclipsing binaries, cataclysmic variables, dwarf novae, symbiotic stars, and various classes of pulsating variables. These thoughtful presentations filled a long-felt need among many AAVSO members. Lacking time to read the astronomy journals or access to them, members fell further and further behind in understanding the latest observational and theoretical views of the many classes of stars in the AAVSO program. An informal pizza dinner and social program at the Harvard College Observatory finished off the second day on Friday evening. The following day's schedule returned to the traditional format for AAVSO annual meetings, in which Saturday morning is devoted to the annual business meeting, whereas the afternoon session is devoted to presentations on a wide range of topics.

The 75th Annual Banquet on Saturday evening proved a rich experience for all who participated. After dinner had been completed and a few introductions made, the speaker for the evening, University of Minnesota astronomer Willem J. Luyten, delivered a short but very interesting presentation. Luyten, who joined the AAVSO in 1917, was the only living Charter Member of the Association. A member for 69 years, Luyten had been with the AAVSO longer than anyone else in its history. In an anecdote he told about earning his doctoral degree, it appeared likely that he may have been the only person ever awarded the degree for a dissertation that included observations of variable stars made by a teen-aged high school student (himself). He left a copy of that dissertation for the AAVSO library.

AAVSO Merit Awards were then presented to three long-time member–observers: Thomas Cragg from Australia, M. Danie Overbeek of South Africa, and F. Lancaster Hiett of Virginia. In an unusual move, the Council also honored two senior participants in the meeting – Frank M. Bateson, the long-standing director of the Royal Astronomical Society of New Zealand's Variable Star Section, and E. Dorrit Hoffleit, former Maria Mitchell Observatory director and Senior Research Astronomer Emeritus of Yale University – as Honorary Members of the AAVSO, a rarely accorded honor.

A musical treat closed the 75th Anniversary Celebration Banquet. Distinguished AAVSO member–observer, Gerald P. Dyck, who earned his living as a high school music teacher and musician, presented his own composition, an elaborate oratorio featuring percussion instruments, mainly Asian or Indonesian, combined with a vocal chorus and a spectacular slide show to illustrate the passage of time from dusk to dawn through a typical night. Everyone present considered the presentation a fitting climax for the evening, and for the entire celebration.

The *Journal of the AAVSO*

One final aspect of the celebration of the 75th Anniversary deserves mention here. The 75th Anniversary issue of the *Journal of the American Association of Variable Star Observers (JAAVSO)*, Vol. 15, No. 2 (1986), provided a fitting reminder of the events. Since taking over as editor in 1975, Charles Whitney had upgraded the publication to a more professional standard of content, graphics presentation, and overall appearance – a considerable achievement since *JAAVSO* is neither solely a scientific journal nor a chronicle of the Association's activities. Combining these two functions in one journal requires a balancing act of presentation, selection, and organization of materials and recruitment of authors and referees who must become convinced that the scientific content of the journal meets the normal standards to be expected in a scientific journal. Scientific papers presented at AAVSO meetings constituted a major source of papers, so as the Association's meeting agenda became more scientifically oriented, so too did *JAAVSO*'s content. On the other hand, astronomers could, and eventually did, submit papers for consideration without prior presentation at a meeting.

The special 75th Anniversary Edition represented the culmination of a decade of effort on the part of Whitney, his editorial boards, and the AAVSO staff.

The Anniversary Issue contained complete texts for nearly every paper presented at the Anniversary Meeting, as well as the usual minutes for the meeting and the Director's Annual Report, and even the words to the songs sung after the pizza supper and the banquet. Both the papers and the descriptive materials were well illustrated with photographs of meeting participants and graphical materials from the scientific papers. The issue was expanded to 354 pages, considerably more than its usual 50 to 100 pages, and bound in an attractive cover. *JAAVSO* presented the Association in a new light to most individuals in the astronomical community.

TURNING POINTS

The dedication of the AAVSO's first solely owned building on Birch Street, the celebration of the Association's 75th Anniversary, and the publication of the *75th Anniversary Edition of the Journal of the AAVSO* all represented turning points for the Association as well as for its Director, Janet Mattei.

Over the first 12 years of her service as AAVSO Director, Mattei had matured a great deal; she developed a relaxed, warm style of dealing with associates and strangers alike. Her demeanor fostered confidence and a willingness, indeed desire, in all who worked with her to cooperate and facilitate the work of this unique association and its director. With these events, Mattei's position with the Association was secure; her salary treatment approached equity with other competitive positions, and she no longer had pause for concern about whether or not she had a position with the Association.

The challenges ahead focused Mattei's attention on a few critical issues: funding for the Association, enhanced availability and utility of the AAVSO's observational archives, increased participation in the database with observations archived by other variable star associations, and broadening the charter to include some elements of education for younger individuals. The Association and its director were well positioned to meet these challenges.

19 · Managing with renewed confidence

The observers are the heart of AAVSO. It is the dedication, the tireless efforts, and the valuable contributions of our observers that give the AAVSO its high reputation in the astronomical community. In this 75th Anniversary year and always we salute our observers!

– Janet A. Mattei, 1987[1]

After the exciting move to its new home at 25 Birch Street, the office routine at the American Association of Variable Star Observers returned to normal for most of the headquarters staff. But for Director Janet Mattei, the excitement didn't end with the building dedication.

Mattei had been working as an organizer for one session of International Astronomical Union (IAU) Colloquium 98 in June 1987 in Paris. The purpose of the Colloquium, *The Contributions of Amateurs to Astronomy*, was to present the many opportunities for amateurs to participate in the science of astronomy. Professional astronomers who worked on problems amenable to amateur participation were invited to present those opportunities, especially as examples of worthwhile work already underway.

In 1987, the year that included a return of Halley's Comet, there were many topics to discuss. The program included more than 120 papers covering observational methods and results, history, and popularization of astronomy. Among the astronomical topics were variable stars, supernovae, binary stars, the Sun, occultations by the Moon, asteroids, and comets – and of course Comet Halley. Techniques considered included visual, spectroscopic, photographic, and videographic means of capturing the observational data.[2] The Colloquium provided an excellent opportunity for Mattei to describe both variable star observing and the AAVSO to an international audience.

In addition to Mattei, other professionals included J.-C. Pecker, Brian Marsden, Adouin Dollfus, Paul Couteau, A. J. Meadows, and Frank B. Wood. One professional participant was Michel Grenon of Switzerland, who used the opportunity to informally present a project that was in the planning stages, The High Precision Parallax Collecting Satellite (HIPPARCOS). The support of amateur astronomers would be required for its success, and Grenon met with Mattei to discuss the possibility of AAVSO participation in the project to guide the HIPPARCOS observing schedule with respect to appropriate exposure times for photometric measurements of long-period variable (LPV) stars.[3]

The tentative HIPPARCOS pre-mission plan called for accurate photometry as well as astrometry for about 300 known LPVs. HIPPARCOS would also conduct a survey to identify possible LPVs not included in the AAVSO program. Preliminary light curves to guide mission planning would be required for those stars as well as the AAVSO program LPVs. The faintest brightness detectable for HIPPARCOS would be 12th magnitude, so it was necessary to predict the periods of time in which the brightness of any given LPV would exceed that limit. More efficient use of the HIPPARCOS capability was thus assured while also guaranteeing the most complete coverage of the light curves while the variables were above that threshold.

The AAVSO possessed the most complete archival records for LPV stars and, Grenon assumed, was likely to have the best understanding of LPV behavior patterns. Mattei was intrigued, but did not immediately accept Grenon's proposal. The HIPPARCOS project would require that she rapidly fine tune her skills on LPV stars to match the expertise she had acquired with the cataclysmic variable stars. AAVSO observers would have to be recruited to monitor an unknown number of additional LPVs for which there were no existing charts. Additional AAVSO staff time would have to be allocated to the coordination of the incoming information. The HIPPARCOS project, exciting as it seemed,

Figure 19.1. Margaret Mayall with Clinton Ford in the director's office at the AAVSO's new headquarters, August 1986.

posed substantial challenges for the AAVSO in the short term, but also would provide additional funding in the form of grants to support the effort dedicated to the project.

Colloquium 98 also provided a convenient venue for another invitation to Mattei and the AAVSO. Belgian astronomer Christiaan Sterken met with Mattei to propose that the AAVSO conduct a meeting at his university, the Vrije Universiteit Brussel, in 1990. It would be the first AAVSO meeting ever held away from the North American continent. The challenges for planning any meeting were substantial, as Mattei had long since learned, but doing so in a distant country posed many additional challenges. Sterken expressed confidence that his local organizing committee, drawn from the local society of amateur astronomers, Vereniging Voor Sterrenkunde, Werkgroep Veranderlijke Sterren (VVS-WVS), could successfully handle meeting details. It was an exciting idea, one that appealed to Mattei, but the approval of the AAVSO Council would be required.

In Paris, the Société Astronomique de France awarded Mattei their Gold Medal for her international leadership in variable star astronomy. Then, on June 27, 1987, Mattei traveled to Leiden Observatory to deliver an address in celebration of "Variable Star Day" and be recognized by the Dutch and Belgian variable star observers. After Leiden, Mattei gave her first address to the British Astronomical Association (BAA) at their July meeting in London. Thomas Williams, who also attended those meetings, observed the celebrity status accorded to the AAVSO Director in each of these international venues. There was no doubt that Mattei

Figure 19.2. Janet Mattei in her first address to a meeting of the British Astronomical Association, London, July 1987.

had matured in her job and had grown in international stature.

The Council at its 1987 Annual Meeting began considering a raise in membership dues. Treasurer Wales had worked out the kinks of his financial reporting system. His annual statements could reliably be used to consider, for example, exactly what fraction of the annual expenses were covered by the annual dues. Separating the annual expenses into those items that were comparatively fixed (not dependent on the number of members) or variable (for which the addition of one new member incurred additional costs) was now a comparatively simple matter. This allowed the officers to identify more clearly than ever the actual cost of supporting each member's needs with paper, ink, postage, and other costs considered variable. When these total annual variable costs were divided by the number of members to estimate the variable cost of maintaining an individual member on the rolls, the surprising result, at least to Mattei and most members of the Council, was that the annual dues covered less than half of the variable cost per member. A reasonable assumption, on the other hand, might be that the annual dues should, at a minimum, cover at least that variable cost of membership, and if possible, make a contribution to covering the fixed costs of the Association.

A substantial debate broke out in the Council over a motion to double the dues so that even the dues paid by juniors and students would cover the variable cost of keeping them on the mailing list. The main argument against the dues increase amounted to a fairness issue. In essence, those opposed argued that members who worked hard to provide useful observations should not also be taxed to maintain their membership. These arguments had been mounted successfully in the past to forestall previous attempts to increase dues, but failed in the face of the actual costs and the reasonableness of the implied policy. After discussion, the dues were actually increased by only 40 percent, a substantial increase, but still short of fully covering the costs involved.

With the dues matter settled, other means of increasing income for the year were considered and tabulated, amounting to about $75 000, if all were successful. The Council also authorized Mattei to employ an attorney and seek a property tax exemption for the AAVSO, since the Association confronted the tax for the first time. The City of Cambridge, home of a large number of not-for-profit organizations like the AAVSO, rejected the Association's appeal for an exemption, so that annual expense had to be covered.

The normal routine still pressed itself on the headquarters staff. By 1987, observations submitted to the AAVSO totaled approximately 200 000 per year as compared with only 175 000 per year a decade earlier. Only 3 years later, the total annual for observations submitted to AAVSO rose to about 250 000. Grants from NASA and the Astrophysics Research Fund provided for the addition of a new personal computer. With that new computer in 1988, early Internet capability allowed Elizabeth Waagen, in charge of data entry, processing, and archiving since 1979, to use a modem to transfer AAVSO raw data to the Harvard-Smithsonian Center for Astrophysics (CfA) for processing and storage, a vast improvement over carrying boxes of punched cards, and then diskettes, to CfA by hand.

The premature death of the AAVSO's leading woman observer, Carolyn J. Hurless (1934–1987) of Lima, Ohio, in February set an unhappy tone for the year. Hurless observed with an 8-inch Newtonian reflector she made for herself. With a very short focal length (near f/4), the compact tube assembly on a simple mounting could be moved around her residential property to find the right access to the sky and to obscure local lights behind intervening vegetation. The mounting, given to Hurless by her close friend Leslie Peltier, could easily be transported in the back seat of her car to Delphos, Ohio, for her monthly observing sessions with Peltier. Her lifetime total of more than 78 000 observations placed her in the top ranks of all AAVSO observers. Hurless served the AAVSO in many other ways, as an elected Council member and vice president and as an ambassador of goodwill to observers all over the world through her tireless correspondence.

Hurless' service in the leadership of the AAVSO came at an unfortunate time in the history of the organization. She might have been a very popular and effective president. Normally, after serving 6 years as second vice president, Hurless would have succeeded to the AAVSO presidency, replacing Scovil in 1974. She had maintained a long correspondence with Ford and others in New England, and, through her mid-summer gatherings in Lima, Ohio, which were well-attended by AAVSO members from Connecticut, New York, and other parts of New England as well as the Mid-West, she could be counted as a strong friend of the Stamford group.

However, as outlined in Chapter 17, Mattei found herself in conflict at an early date with Ford, Scovil, and others from the Stamford Observatory. Mattei, for her part, therefore saw Hurless' ascendancy to the presidency as a threat to her control over the Association. So, she engineered the election of Canadian physician George Fortier as the AAVSO first vice president who would later replace Scovil as president when his term expired.[4] Hurless retired to the sidelines gracefully and continued to serve the AAVSO in other ways, perhaps grateful that she would not be immersed in any conflicts.

Afflicted for many years with a neurological problem that caused her constant pain, Hurless nevertheless maintained a relentlessly cheerful disposition that concealed her condition from all but her closest friends. In part, she coped with the pain by remaining in ceaseless motion, with music and teaching occupying her daylight hours not spent in various astronomical endeavors, including the publication of the informal newsletter, *Variable Views*, that was circulated to a wide circle of friends and correspondents.

Her tragic death came as a shock to most AAVSO members.[5]

HIPPARCOS TAKES OFF

After returning from Paris, Mattei pondered the challenges she would have to confront if she were to decide to take on the HIPPARCOS project. After due consideration, Mattei decided that she could manage these challenges. Michel Grenon visited AAVSO Headquarters to make a formal appeal for assistance from the AAVSO. Grenon's role on the HIPPARCOS effort was on the input catalog preparation team. The catalog covered more than 100 000 objects of interest to be surveyed by the satellite for various reasons. The project needed light curves for LPV stars, current observations, and predictions of how the brightness of each LPV was likely to change in the short term. The AAVSO and its observers were best equipped to supply the necessary historical data and real-time mission support. With a scheduled launch in mid 1989, there was not a lot of room for misjudgments. With some trepidation, Mattei signaled her agreement to Grenon and began to lay out plans with the AAVSO staff for this complex and demanding activity.[6]

By early 1988, the AAVSO had sent 20 years of observations for 300 LPV stars to the input catalog preparation team in the form of light curve graphs. The staff in Europe was responsible for subjecting those data to Fourier analysis and, on that basis, preparing a preliminary prediction of the light curves for the 2-plus years of expected mission life. Mattei was then asked to comment on and adjust the predicted light curves based on her experience and judgment.

Questions arose regarding the validity of the AAVSO data; in subsequent discussions, Mattei explained the problems involved with poorly observed stars, including poor finding charts and uncertain sequences. Such circumstances discourage observers, so most under-observed stars can be attributed to those basic problems. Grenon agreed to conduct a substantial program of photometry to improve the sequences for many stars and field photography to guide preparation of improved finder charts by Charles Scovil. Many other procedural details had to be worked out.[7] Light curves needed to be obtained from photographic plates in the Harvard archives wherever the AAVSO database proved inadequate to support HIPPARCOS needs. Dates of observed maxima and minima for these stars, when available for the previous 75 years, were also provided. All this was necessary to predict the periods in which the LPVs in the study would be bright enough (brighter than 12th magnitude) to observe with HIPPARCOS.

The HIPPARCOS launch vehicle failed to achieve its planned orbit in August 1989. Its highly elongated orbit would produce more limited data and for an indeterminate period of time. In spite of the difficulties, the satellite successfully gathered data on a revised schedule. AAVSO observers cooperated fully with the project, headquarters sent data monthly to the HIPPARCOS team, and Mattei spent many hours on data analysis and in conference with team members. The AAVSO data were also critical in determining how the aging of the HIPPARCOS optics was affecting the calibration of the satellite data. The Association's participation succeeded in spite of the tight schedules and massive amounts of data involved. This would not, of course, even have been possible as little as 5 years before; success in the HIPPARCOS project participation confirmed the high value of the enormous effort invested in the processing of current as well as archival observations of variable stars, data quality-control procedures, ongoing upgrading of the headquarters data processing systems, and all that had been involved.[8]

DATA PROCESSING PROGRESS

Progress in data processing continued at a very rapid pace. By the end of 1992, for example, successful grant proposals had funded individual networked workstations for each staff employee at headquarters. Programming for the microcomputer had advanced to the point that each such networked station could call up the variable star database and display the light curve for any one of 3600 stars for any period of time. Furthermore, other grants had provided improved hard disk storage facility for headquarters, additional computer memory, and the software capability to handle all data processing, storage, and analysis at headquarters. Processing monthly data at headquarters had begun in 1989. By the end of 1990, the computers at CfA were no longer used for any aspect of AAVSO data processing or storage; the magnetic tape library kept there had been transferred to CDs and the magnetic tapes discarded. Increasing numbers of observers were submitting observations on diskette using the data-entry software created and distributed by headquarters.

Progress continued in the processing of old observations. By 1990, 90 percent of the archived data had been entered, but the project was far from complete. Still pending were several years of data preparation and processing requiring two headquarters staff dedicated to the project, Michael Saladyga and Elena Khan, with significant assistance from Sara Beck, who would collect, merge, and sort the hundreds of files from tapes and disks and perform a battery of quality-control checks.

FIRST INTERNATIONAL MEETING

While the HIPPARCOS effort unfolded and demanded a large investment of Mattei's time in analysis, travel, and communication, others worked with Chris Sterken on planning the Association's first international meeting, July 24–28, 1990, in Brussels, Belgium. AAVSO President John Percy and his staff at the University of Toronto shouldered much of the planning to relieve Mattei, while the local planning committee organized by Sterken handled their part of the planning and early meeting execution. Mattei and nearly the entire headquarters staff were involved as well, especially heavily as this exceedingly complex meeting drew near. More than 175 participants from 30 countries on 5 continents participated in the June meeting in Brussels. Importantly, there were almost as many professional astronomers attending as amateurs. The meeting provided a stellar example of both professional–amateur and international cooperation.

Adriaan Blaauw, former General Director of the European Southern Observatory and IAU president, provided an appropriate opening address in stressing both the historical and the contemporary necessity of cooperation in astronomy, a science dedicated to stretching beyond any previous limits, both scientific and technological. By design, Blaauw noted, it seems astronomy has also always stretched the financial means available to it over the centuries. Blaauw's point was simply to acknowledge that the AAVSO's first international meeting continued a long tradition of several forms of cooperation in astronomy.[9]

The following 4 days included paper sessions on history and organization, visual and instrumental photometry, analysis and interpretation of variable star information, and three separate sessions on various classes of variable stars: blue and yellow, red, and cataclysmic. In each session, there was an appropriate balance of amateur and professional papers. The professional papers tended to provide both an appropriate introduction to the state of knowledge in a particular field as well as insight into how amateur observations were being used in solving current problems.

At the Brussels meeting, many AAVSO members heard for the first time about an emerging technology that promised to revolutionize the instrumental observation of variable stars. Charge coupled devices (CCD) were being successfully applied astronomically in many different ways in orbiting satellites and at ground-based professional observatories. The considerable advantages claimed for the CCD included broad spectral response and quantum efficiency, linearity of response, and the availability of digitized output signals. The cooperation between amateur and professional astronomers in Europe was proving that with effort, amateurs could use CCDs with small telescopes. In comparison, there was no organized effort in the AAVSO to apply CCD technology in 1990. The meeting in Brussels showed that the technology had evolved rapidly. With basic commercial CCD cameras and programs for reducing CCD observations on personal computers becoming available, practical application of CCD photometry by amateur astronomers with smaller telescopes began to appear workable.

In addition to presentations on the cutting edge in areas like CCDs, the Brussels meeting provided exposure to the more traditional approaches to observing, and especially data analysis. John Isles of the British Astronomical Association-Variable Star Section (BAA-VSS) summarized that group's attempts to analyze the light curves of the 34 LPVs that had been observed by the BAA-VSS, up to a century in some cases. The conclusions and recommendations that could be drawn from the BAA-VSS effort included several points of particular interest. First, Isles recommended that the effort to observe the LPV stars should continue, even for those stars thought to be well understood. Second, the variable star groups should combine their data to provide longer term, better coverage for more stars. Third, the guidance of theoreticians would be welcome in the analysis of LPV data, taking into account the latest models of those stars.[10]

Isles' recommendations were, in a useful way, reinforced by a report of results for the very latest of technology.[11] A report on the progress of the HIPPARCOS project by Marie-Odile Mennessier provided an interesting contrast with respect to the LPVs. The HIPPARCOS project would provide high-quality data on a large number of LPV stars in a short period of time, but was dependent on the historical record of observations for planning purposes. Similarly, Margarita Karovska's summary of recent results in both the theoretical arena and with observations of LPVs in wavelengths other than visual and using interferometry, spectroscopy, and other tools also provided support for Isles' basic recommendations.[12]

The meeting organizers, Mattei, Percy, and Sterken, arranged to have the conference proceedings, including 31 longer or survey presentations, published as a book by Cambridge University Press.[13] Twenty-seven shorter scientific papers presented by both amateurs and professionals featuring the usual light curves, O–C curves, bar charts, and tables of data were published in an issue of *JAAVSO*.[14] At the Brussels meeting, each amateur organization described its unique practices. These presentations, mainly descriptions of variable star observing in various national contexts, were collected and published in 21 articles in another issue of *JAAVSO*.[15] Thus every presentation in the Brussels meeting was published in either the conference proceedings book or in *JAAVSO*.

The Brussels meeting seemed to energize Council members, who tackled Association problems with renewed vigor in the fall. President Percy set aside time in a Council meeting in November 1990 to reflect on the immediate past and brainstorm about the future. In the brainstorming session, Percy and the Council developed two lists for discussion, one that represented the main problems faced by the Association, and the other, a list of the opportunities to alter the Association's activities and programs in some useful new area. Percy invited Council members to focus not only on the continuing deficit spending, but also on ways in which meeting other Association goals might also be turned to advantage in solving that problem.

The three top priority ideas developed in that session were the need for an education initiative, the need to pursue the newly emerging CCD capability for amateur astronomers, and the need to increase AAVSO exposure to the professional astronomical community. The remainder of the list depended to some extent on the success of the decade-long project to computerize the AAVSO archive of observations, which was still a long way from completion.[16]

Hands-On Astrophysics

One of the ideas brought forth in Percy's brainstorming session seemed to have potential to reduce cash flow problems and at the same time begin a new program that Mattei had held in abeyance for some time. The idea was for the AAVSO to develop an education curriculum using variable star materials to teach astronomy and astrophysics to high school and lower-level college students. Quite apart from the other merits of an education program, Mattei felt that the idea could attract NSF or NASA funding to the Association without any increase in headquarters staffing.

As a long-term supporter of education initiatives, Percy volunteered to draft a grant proposal for such a project. He and Mattei submitted a preliminary proposal outline to the education office at the National Science Foundation (NSF). A seemingly continuous flow of new discoveries in astronomy appearing in the daily news validated the timeliness of such a project. After their review, the NSF staff encouraged the AAVSO to prepare a finished grant application.

Work on the project, by now identified formally as Hands-On Astrophysics (HOA), began in 1991. By the end of 1992, the NSF had accepted the AAVSO's completed proposal, and Hands-On Astrophysics emerged as a full project within the AAVSO. The funding proposed, \$303 943 over 3 years, included a generous allocation for Association overhead encouraged in all such proposals. Mattei assembled a team of individuals to work on the project. An experienced high school teacher, Jeff Lockwood, wrote a first draft of the curriculum manual. Charles Scovil agreed to prepare 45 finder charts for 15 variable stars in 5 constellations, and astrophotographer John Chumack of Dayton, Ohio, prepared sets of 35-mm slides of selected variable stars illustrating their variation. John Percy shared the leadership responsibility with Mattei, whereas headquarters staff, led by Michael Saladyga, provided office support for the whole project.[17]

By 1995, HOA development had achieved considerable progress, but was not yet completed. After Lockwood completed the first-draft manual, a team of three teachers – Kristine Larsen, Sharmi Roy, and Donna L. Young – took over to further develop and refine the

Figure 19.3. Participants in the second Hands-On Astrophysics Teachers' Workshop held at AAVSO Headquarters in 1995. Janet Mattei is at left in the second row. Staff members present are: Grant Foster (kneeling at left); Michael Saladyga (kneeling, third from left); Rebecca Turner (standing at left behind Dr. Mattei); Shawna Helleur (middle row, right). Directly behind Ms. Helleur is John Percy. Donna Young (second row, center) and Kristine Larsen (third from right in same row) were part of the curriculum development team.

curriculum. As the work proceeded to completion, Young became the manual's principal author and curriculum consultant.

One major stumbling block preventing completion of the project involved the need to assemble a database containing 27 years of observations for 46 variable stars, which amounted to a total of 607 458 observations. In the process of assembling these data, a number of errors in the AAVSO database were found and corrected. These errors were unique errors associated with data entry problems. The necessary corrections benefited the entire database.

Two workshops held in 1994 and 1995 at AAVSO Headquarters drew about 30 high school teachers from across the United States. The teachers evaluated the materials under development in a series of hands-on experiences. Their enthusiastic critique encouraged the project staff to complete and publish the resulting effort and also led to the preparation and distribution of a *Hands-On Astrophysics Newsletter*, which was received favorably by high school science teachers across the nation. In addition to the set of 35-mm slides and corresponding prints, preparations were well underway in 1996 for the production of a three-part video introducing variable stars and the process of making and interpreting variable star observations.[18]

ACCEPTING THE CCD CHALLENGE

After several decades, photoelectric photometry had emerged from the shadows of arcane technology practiced only by those with specialized knowledge of electronics. The new availability of standardized commercial photometers (and personal computers for data reduction) made it possible for many more amateurs to acquire the capability to contribute in this specialty, as the steady growth in the annual reports of observations from the AAVSO Photoelectric Photometry (PEP) Committee had demonstrated since 1983. As mentioned previously, at the Brussels meeting many AAVSO members heard about Charge Coupled Devices (CCDs) for the first time.

In response to growing interest in CCD technology expressed by some AAVSO observers, both before and especially after the Brussels meeting, Mattei scheduled the first of several CCD workshops for the 1991 Annual Meeting. This introductory session, conducted by Steve B. Howell of the Lunar and Planetary Institute in Tucson, Arizona, included several papers by Howell and a few AAVSO members, including George East, Priscilla J. Benson, and Gary P. Emerson, who were already practicing CCD photometry. The workshop oriented members to the potential for CCD photometry as well as the problems and properly introduced a subject that would continue to be of interest for the foreseeable future.

The 1991 Annual Meeting CCD workshop provided an excellent introduction to the topic so that those who were interested could absorb more detail in the second workshop, this time scheduled jointly with the American Astronomical Society (AAS). The second workshop in June 1992 preceded the AAVSO Spring Meeting in Columbus, Ohio. Many AAVSO members participated with interested AAS professionals.

In spite of these two sessions, however, Mattei felt uncertain about expanding the activity within the AAVSO. Certainly the complexity of the data processing and the lack of standardized data reduction software for amateur observers, as well as the vagaries of data acquisition at the telescope, left Mattei with some uncertainty about starting a new program in the AAVSO, which several members were urging her to do. To build her own confidence, Mattei attended a 1-week CCD workshop in February 1993 conducted by Steve Howell at the Lunar and Planetary Laboratory in

Figure 19.4. Priscilla Benson, Steve Howell, and Janet Mattei during an observing session to obtain AAVSO CCD finder charts. Kitt Peak National Observatory, February 1993.

Tucson, Arizona, and Kitt Peak National Observatory (KPNO). The workshop attracted a number of amateur astronomers, including John J. Briggs, Richard Berry, and Dennis di Cicco. Benson, a radio astronomer from Wellesley College who also attended the CCD workshop, had only recently become interested in variable stars because of maser emissions from water molecules in the outer atmosphere of some variable stars.[19] As a result, Benson agreed to organize and lead an AAVSO CCD committee and was appointed to that role at the 1991 Annual Meeting. Thus Howell's workshop helped orient the two key AAVSO leaders for this new activity.

A few weeks later, the AAVSO held its 1993 Spring Meeting in Berkeley, California. As in Columbus, the Berkeley AAVSO meeting coincided with the AAS semiannual meeting so that participants of both organizations could benefit from the second Joint AAS/AAVSO CCD Workshop. The technical content of the workshop continued to interest AAVSO members, but the most interesting thing about this meeting involved Sky Publishing Corporation. Rick Feinberg, the Chief Executive Officer of Sky Publishing, used the AAVSO meeting to announce that his firm had initiated a new magazine, titled *CCD*. Edited by Dennis di Cicco, *CCD* would feature articles of interest regarding important applications of the technology, recent developments and reviews of new hardware and software, and of course advertisements for the vendors of related products. Feinberg's announcement raised awareness of the growing interest in and importance of this new technology.

Benson had been unable to attend the Berkeley meeting, so Gary Walker, an active CCD practitioner, chaired the session at that AAVSO meeting. Because Benson's teaching schedule and research prevented her full participation in the AAVSO activity, Walker agreed to act as a co-chairman for the committee and eventually was appointed full chairman. In 1994 when the AAVSO held its fourth CCD Workshop in Houston, Texas, Benson was unable to attend, so Walker took charge of the activity, which was at the same time stimulating a great deal of interest among AAVSO observers. About this time, Berry, Kanto, and Munger published *The CCD Camera Cookbook*, which contained information a novice needed for building, testing, and operating a CCD camera and making images with it.[20] University Optics, Inc. offered a kit at a reasonable price, complete with all machined parts and directions such that a novice could understand and build a camera from scratch at a reasonable cost. The book and kit stimulated a great deal of additional interest among amateurs in CCD imaging and applications.

CLINTON BANKER FORD (1913–1992)

The AAVSO lost its strongest supporter on September 23, 1992, when Clinton Banker Ford passed away. Ford had suffered from diabetes most of his adult life, but had never taken the normal precautionary measures very seriously. His neglect finally resulted in gangrene in a leg that was therefore amputated. Confined to his bed, Ford apparently lost interest in life and died shortly afterward. The loss was acutely felt among variable star astronomers around the world.[21]

In her memorial obituary in *JAAVSO*, Dorrit Hoffleit, who had likely known Ford longer than anyone else alive at the time of his death, revealed that Ford had been the anonymous individual who saved the AAVSO in the days after the eviction from Harvard and had continued to support the AAVSO in the same manner. Although some of the Council members suspected that Ford was making more money available than was publicly known, no one except the director and the treasurer understood fully the extent of Ford's generosity.[22]

Quite apart from Ford's financial largess, his services to the Association over the years can hardly be overstated. His steadying and conservative influence on the Council, in spite of his likely impatience with the slow pace of the Council's development, must be

Figure 19.5. Brian Marsden of the Harvard-Smithsonian Center for Astrophysics with Margaret Mayall and Clinton Ford to celebrate the naming of minor planet 3342, "FIVESPARKS," in Mayall's honor in 1989.

credited with nearly equal importance as his contributions to the technical progress of the observing programs through his long-term chart development effort with Charles Scovil. Then, too, Ford's quiet encouragement of new observers, first time members of the Council, and others who clearly wanted to make a contribution of some sort to the Association's efforts reflected yet another aspect of his influence – helping members feel at home in the Association and welcoming their contributions to the overall program.

The AASVO recognized Ford's passing in the traditional way at the 1992 Annual Meeting with a moment of standing silence during the meeting. Near his birthday, which coincided with the date of his first variable star observation, a special memorial service was scheduled to honor Ford more formally. Also at that time, as Ford had planned if he had lived to do so, the last $75 000 of the mortgage on the Birch Street building was paid off. The money for that payment came from anonymous, no-interest loans to the Association from Ford's long-time friend Hoffleit and Treasurer Wales. It was very appropriate, therefore, that as part of the memorial service for Ford, the mortgage on the building at 25 Birch Street was burned as a final act of celebration and recognition of all that his support had meant to the AAVSO.

Over the years, Ford had mentioned to both Mattei and Wales that the AAVSO would be a major beneficiary in his will. Though the total was unclear, it was expected that the amount might exceed $7 or $8 million, but in the final settlement a total of more than $12 million flowed into the AAVSO coffers. Although payment of the principal from the Ford inheritance was delayed by a court challenge to the will, the executor of the estate began making distributions of the income from the Ford accounts available to the AAVSO. As cash from the Ford estate began to accumulate in the AAVSO's accounts, the Council quickly adjusted to the new financial realities that confronted the Association. The availability of the additional funds would make possible expanded services to the astronomical community as well as the AAVSO membership. However, inquiries from the staff regarding this new wealth made clear their expectation that historical compensation and benefit deficiencies would soon be rectified. By the mid 1990s, the professional staff had stabilized, but some staff turnover continued to limit the effectiveness of secretarial and clerical help. A need for retirement and improved medical benefits were among the earliest questions to arise among all the AAVSO employees. Such benefits were common from most employers and would now be appropriate for the AAVSO staff as well.

The availability of additional financial resources helped resolve another historical tension in the AAVSO leadership. Over the first 80 years of its existence, participation in the AAVSO Council as a volunteer always meant a financial sacrifice by the participants. In effect, only those who could afford to travel to meetings could participate in the Council, an unspoken requirement that limited the availability of talented individuals in the leadership of the AAVSO. With the advent of additional funding, it was agreed that Council members could be reimbursed for their travel expenses if they so desired. Submission of an expense statement for the meeting with the necessary receipts and tickets to justify the expenses would qualify the Council member for reimbursement. This allowed some broadening of the potential membership of the Council.

Modest staffing additions made possible by the availability of the new funds included a part-time bookkeeper to relieve Treasurer Wales of the routine data entry responsibility. Also, a limited expansion of technical and data entry staff accelerated the entry and analysis of the historical variable star observations and at the same time accommodated the simultaneous increase in requests for support to orbiting satellite observatories as well as various ground-based observing programs.

Figure 19.6. Professor Alice H. Farnsworth, Mount Holyoke College, South Hadley, Massachusetts, and her student, Martha Hazen, about 1953. Hazen would become AAVSO President in 1991. *Courtesy of Mount Holyoke College Archives and Special Collections.*

The AAVSO continues to benefit from Ford's generosity. About 90 percent of the endowment on which the Association now depends for a major part of the financial support of its activities was inherited from the Ford estate.

SUCCESSION PLANNING

Ford's death created a succession problem the Council had not faced for more than four decades – electing a new AAVSO Secretary. After private discussions of this problem with various Council members, President Martha Hazen expressed her own interest in assuming the secretary's role to Mattei. The two had evolved a close working relationship over many years preceding even Hazen's election to the Council for the first time. That choice only exchanged the problem of replacing a secretary with the new problem of replacing Hazen as president. Both Mattei and Hazen agreed that of the two problems, the naming of a new president was to be preferred.

Both Mattei and Hazen had reservations about the obvious choice to replace Hazen, that is, accelerating the election of First Vice President Wayne Lowder by 1 year. Lowder had not attended any Council meetings for several years. He could not be expected to understand the current status of the Association's business or to lead the Association's activities without a great deal of difficulty, both for him and for Mattei. On the other hand, Lowder was well known and widely respected as an observer of variable stars who also possessed broad knowledge of theoretical variable star astronomy. Mattei and Hazen agreed to temporize by asking former President Tom Williams to assume the role again for a 1-year term.

After discussing the problems presented by Lowder's poor participation in the Council and her concerns about working with Lowder, Mattei persuaded Williams to accept the position for 1 year and to work with Lowder to insure that he would be ready to serve as president at the end of that year. Mattei expressed confidence that this arrangement could bridge the necessary gap in leadership. In discussions of what she wanted to achieve in the following year, Mattei revealed that one of her motivations in asking Williams to fill this role was her desire to resurrect the idea of a Futures Study during his interim term as president. Mattei hoped this would help her clarify the direction the Association should take. Williams, in turn, discussed the matter with Lowder in a telephone conversation and secured his agreement to participate actively in the affairs of the Council thereafter. With that assurance from Lowder, Williams accepted the Mattei/Hazen proposal.

Confronted with the realization that he, too, was in line to become the president of the AAVSO, Second Vice President Paul Sventek expressed concerns about his readiness to accept that responsibility. He was persuaded to continue as an officer while he learned more about the workings of the Council and the Association. Recognizing Sventek's reservations and cautious reluctance, Mattei arranged to have Albert V. Holm nominated for first vice president and for Sventek to remain second vice president when Lowder finally assumed the presidency a year later.

AAVSO FUTURES STUDY REDUX

Ford's death freed Janet Mattei to initiate a detailed look at what the future held for the AAVSO – the first time such a detailed planning exercise had been undertaken on the Association's behalf. Recalling the effort to initiate a Futures Study that occurred a decade earlier, Mattei asked the Council to initiate the study. A major difference on this second attempt, however, was that instead of asking the Council to participate in the study, Mattei asked that the President appoint a group of AAVSO members to execute the study as an independent activity. The members asked to serve on the Futures Study Group represented a very broad cross-section of the AAVSO membership: professionals and amateurs, Council members, former Council members, and member/observers with no official portfolio. The group included Mark T. Adams, John Bortle, Martha Hazen, Albert Holm, John Isles, John Percy, David Williams, Thomas Williams, and Lee Anne Willson. Janet Mattei, AAVSO President Wayne Lowder, AAVSO Treasurer Ted Wales, and, when able, Second Vice President Paul Sventek sat in on meetings as an ex officio member of the study group.

The Futures Study Group convened three times in all-day meetings and twice in hour-long telephone conference calls. At its first meeting, October 13, 1993, the Study Group spent a half-day considering the strengths and weaknesses of the AAVSO. During the other half, the group gathered data on what it believed the future held for the AAVSO in terms of opportunities and risks. In the four remaining meetings, these issues were prioritized and discussed, options with which the Council might address the problems to enhance the Association's strengths were considered, and the final draft prepared and edited.

Several participants brought data they had gathered in preparation for the first meeting. Bortle, for example, presented a series of graphs of observing totals and numbers of observers spanning the previous 12 years. His concerns, graphically expressed, reflected the fact that although the observations received by the AAVSO each year continued to increase, those observations were increasingly made by only a few individuals and that those individuals were aging and could not be expected to contribute at the same level for many more years. Furthermore, Bortle observed the problem to be more severe as the gap was growing between the percentage of US observations as opposed to those from non-US sources, which in 1984 outweighed the US observations substantially.

John Percy had conducted a user survey of about 40 professional astronomers who had requested AAVSO data for use in their research work. About a third of the professionals responded, and most expressed satisfaction with the quality of the AAVSO data available to them. Not surprisingly, a strong preference emerged in Percy's survey for instrumental observations, either PEP or CCD, rather than visual observations.

The AAVSO staff conducted a membership survey to assess the general feelings that existed about the AAVSO and things that the membership believed might be improved. Headquarters staff tabulated the numerical results for this survey, whereas Lowder reviewed the verbal comments that members included with their numerical scoring of attitudes. In general, Lowder and the staff concluded that the membership felt the AAVSO was doing a good job but identified a number of areas in which some improvement would be desired, especially in communication and mentoring.

The final report of the Futures Study Group was organized in terms of problems in four major areas: demographics, technology, science, and funding. Statements, cast in the form, "By the year 2005, the AAVSO needs to . . . ," articulated the 10 recommendations of the Futures Study Group. Those recommendations did not address the possible solutions to problems. Instead, they simply recommended the capabilities that would apparently be needed in 2005 and left it to the Council and staff of the AAVSO to decide whether those future states could be achieved and, if so, how to proceed toward the goal.

Over a period of time after the completion of the Futures Study, the Council and officers achieved a great deal of what was recommended, and in the face of many unusual and unexpected problems. The Futures Study Group's acute perceptions of what conditions would be faced and how the leadership could position the Association to respond as effectively as possible to insure the continuity and growth of the AAVSO may have helped in that regard. However, the real importance of the Futures Study effort may have been that it marked Mattei's final release from the past and turned her focus to the enhancement of AAVSO research and services to its membership and the astronomical community in the present and for the future.

MARGARET MAYALL (1902–1995)

On December 6, 1995, the AAVSO lost another stalwart leader to whom it owed so much. Margaret Walton Mayall passed away in Cambridge, Massachusetts. The last formal AAVSO meeting Mayall attended was the 1992 Annual Meeting, though she also participated in the AAVSO staff Christmas party that year and in the

Figure 19.7. AAVSO Director Emeritus Margaret Mayall with Dorrit Hoffleit and Director Janet A. Mattei at the conclusion of a memorial service for Clinton B. Ford, February 1993.

Ford Memorial service early in 1993. Functionally blind during the last few years of her life, Mayall remained resolutely cheerful and supportive of the AAVSO to the end. Her husband Newton died 6 years earlier. They had no children, and so the AAVSO and her many friends in astronomy mourned the loss without the solace of continuing familial connections.[23]

Looking back from the year 2010, Mayall's choice of Mattei as her successor, though it seemed almost incomprehensible at the time, can now be seen as a very wise one. Mayall understood the difficulty that the Association faced in the coming years and likely appreciated better than anyone else the dedication that would be required to sustain the AAVSO. Mayall's unshakable dedication to the AAVSO through the difficult years after the Harvard College Observatory eviction was one of two key ingredients that insured that the AAVSO would survive, the other being Ford's anonymous financial support. But without Mayall's energy, enthusiasm, and loyalty to the Association, even Ford's funding could not have saved the Association from the fragile state that it experienced in 1954.

But raw survival is only one aspect of Margaret Mayall's legacy to the AAVSO. During her tenure as AAVSO Director, Mayall established a practice of welcoming and supporting students from the summer program at the Maria Mitchell Observatory (MMO) on Nantucket Island. The primary research programs carried out by these student astronomers were typically a combination of historical research using the plate collection on Nantucket with current observations of the same stars using the MMO telescopes and photographic tools. Mayall routinely invited these student astronomers to present the results of their summer's work in short papers delivered at the AAVSO annual meetings. For most of these students, the AAVSO talk was their first oral presentation of a scientific paper of any kind, a valuable experience for anyone. But the experience was especially meaningful for those students who continued their astronomical studies and became professional astronomers. Contact with MMO Director Hoffleit, Mayall, and the members of the AAVSO helped them crystallize their lifetime ambitions in astronomy. That was certainly true in the case of Janet Akyüz Mattei.

At the time of her retirement, the AAVSO Council established a Margaret Mayall Assistantship fund to support the employment of summer research assistants. The Mayall Fund is supported by member donations. A small inheritance from the Mayall estate was added to the general endowment of the Association.

NEW LEADERSHIP TAKES HOLD

Over the next four years – 1993 to 1997 – and during the tenure of Council presidents Wayne Lowder and Al Holm, the Council responded to the needs of the organization with renewed interest. The Futures Study seemed to revive interest in managing the enterprise for the good of variable star astronomy and for the benefit of members. It helped to have goals in sight toward which

Figure 19.8. AAVSO Headquarters staff in 1996. From left: Grant Foster, Michael Saladyga, Shawna Helleur, Lynn Anderson, Janet Mattei, Sara Beck, Bill Mackiewicz, Elena Khan, Barbara Silva, Elizabeth Waagen, John Beck, and Donna Eldridge.

the Council could gradually steer the organization, but more than anything else, perhaps, it gave the Council a renewed sense of its ability to do so.

Funds from the Ford estate gradually became available, first in the form of income from the undistributed principal and finally as large transfers of capital to AAVSO accounts. The Council and Mattei were finally able to address a number of issues that had troubled the operation for years. Adjustments to bring staff salaries more in line with local competition were helpful in stabilizing the office workforce to some greater degree. Modest benefit packages including a better medical plan, a retirement plan, and tuition reimbursement for night school courses relevant to the work in the office were all considered and adopted to some extent.[24]

NASA provided an Internet connection to AAVSO from the SAO, thus making it possible for AAVSO to become a node in the Internet with its own address (aavso.org). Council member Doug Welch led an effort to upgrade member communications to take advantage of that capability with various member communication utilities. Not least of those was a provision for an electronic file transfer system of chart delivery for variable star comparison charts. It would be possible to centralize all chart making and distribution into Cambridge at some time in the near future.

At the same time, the Council began taking action on what it saw as the implications of the Futures Study report. Presidents Lowder and especially Holm followed with a rigorous delegation of responsibility and a system to monitor progress being made against the specific Futures Study recommendations.

Efforts on the part of the Council to bring the AAVSO Constitution up to date took several years. The Council began to propose important changes in the way the Association operated. For example, a proposal to assign defined work to the two AAVSO vice presidents, specifically in their roles in providing some supervision for the existing committees, received favorable attention, and some work progressed toward its implemention by the elected Council members. Mattei, seeing this as an infringement on her rights and responsibilities, began an effort to undermine this proposal before it got too far.[25] Whether or not she had the time to do this work properly seemed irrelevent as she insisted on sharply delineating the actual work that the vice presidents might undertake with the Committee chairs. Over the next year, Mattei became ever more assertive in protecting what she viewed as the director's territory, eventually persuading the Council to back away from this initiative.

Apart from this conflict of roles that developed between Mattei and the Council over the implementation of the Futures Study Group recommendations, the AAVSO did move forward. President Holm's task assignments, spreadsheet tracking of results, and steady pressure on Council members to report progress regularly achieved beneficial changes in the way the AAVSO Council approached its role and in the way the Association operated.

EXTERNAL RECOGNITION

In the midst of all these signs of real organizational development, several AAVSO Council members were receiving some recognition, which added to the luster of the organization in general as well as those individuals specifically. For Mattei, the year 1993 started in an exciting way. At the American Astronomical Society annual meeting on January 22, she was awarded the AAS Georges Van Biesbroeck Award:

> . . . for her enthusiastic and unselfish leadership of the American Association of Variable Star Observers, and for making the AAVSO Data Base available to the astronomical community.[26]

In 1995, John Isles, a member of the AAVSO Council who had been a long-time member of the BAA and active in its leadership, received the BAA's senior award, the Walter Goodacre Medal and Gift. Also in 1995, the AAVSO membership rejoiced in two additional significant honors bestowed on its director. First, the Royal Astronomical Society of London awarded Janet Mattei their Jackson-Gwilt Medal for achievement in observational astronomy. Mattei had also been awarded the first Giovanni Battista Lacchini Award by the Unione Astrofili Italiani "for collaboration with amateur astronomers."

Figure 19.9. Janet Mattei with Carolyn and Albert Jones of New Zealand, at the Sion–St. Luc meeting, 1997.

A SPECIAL YEAR – 1997

At a special April 1997 Council meeting, the best news was that the final processing work for the long-delayed addition of the 1911–1961 observations to the AAVSO International Database was completed. Changes in place at headquarters for some time had meant that the database was essentially up to date for current observations, with many being received on computer disks and over the Internet, as well as on paper. It was a rare moment of relief and celebration for all concerned, but no one more than Mattei. She had suffered the trials and tribulations of this project for more than 20 years, supervising the transition from punch card operation to digital input on magnetic disks, followed by the various stages of upgrading headquarters software to validate and store the information, and the progressive movement of the whole data management system away from dependence on the Harvard-Smithsonian Center for Astrophysics computers to one handled completely on an in-house computing system. Although the AAVSO still had a long way to go to establish the full availability and credibility of its database to the astronomical community, it was a huge achievement to have reached this plateau in data management.[27]

Figure 19.10. M. Daniel Overbeek of South Africa with Frank Bateson of New Zealand at the AAVSO 1997 Spring Meeting held at Sion–St. Luc, Switzerland.

SECOND AAVSO INTERNATIONAL MEETING

The successful collaboration of the AAVSO with the HIPPARCOS project led to an invitation from Michel Grenon, the project's representative to the AAVSO, for the Association to hold a major conference in Switzerland. The meeting in Sion–St. Luc, Switzerland, in late May 1997 celebrated 10 years of fruitful collaboration between the AAVSO and the HIPPARCOS mission. Organized on the theme "Variable Stars: New Frontiers," the conference attracted participation from both amateur and professional astronomers from around the world for 5 days of intense science in a setting of incomparable scenic beauty.

A lengthy first session reviewed the relevant results from the HIPPARCHOS mission and other similar all-sky mapping projects. Then, the status of stellar variability became the major organizing principle for five sessions in which recent scientific results for various classes of variable stars were discussed. These sessions included pulsating stars,[28] cataclysmic variable stars,[29] symbiotic stars,[30] and other binary systems. Finally, other variable sources[31] and data analysis were combined in a final session in this part of the conference.

Each institution or association represented in the meeting came prepared to describe their particular contributions to variable star astronomy, featuring interesting presentations on observatories[32] and variable star organizations[33] from around the world. Finally, two workshops discussed data collections and distribution and astronomy education for those who had particular interests in these areas. The science papers presented in this meeting comprised a valuable snapshot of many branches of variable star astronomy as they existed in 1997.

AAVSO President Holm presented a noteworthy challenge to the AAVSO members present for his Sion talk, "Issues and Strategies for the Future." While recognizing the AAVSO's substantial progress over the previous decade, he identified a series of potential problems that confronted the Association and suggested possible solutions to those problems. "We have seen a number of the paths that we can follow to the future," he observed. "They all require that we change in some way or another, and that of course requires resolution and effort."[34]

The Astronomical Society of the Pacific (ASP) agreed to publish a volume dedicated to the Sion-St. Luc meeting in their Proceedings series. It remained to gather and edit the papers from those who made presentations at the Sion meeting and edit them for publication. There were 69 papers that summarized and detailed specific fields of variable star astronomy. An additional 67 papers were to be represented with abstracts summarizing the talks. The ambitious publication plan would require significant effort. Still, the mere fact that the AAVSO had led the effort to organize such a comprehensive review of the whole field of variable star astronomy established the Association as a leading participant in this interesting and important field of science. Recording that achievement in the widely read ASP Proceedings series would help secure that position.

Mattei and the AAVSO's leadership had much to be pleased about by the time the glow of the Sion meeting had diminished and normal work resumed.

20 · Expanding the scientific charter

The AAVSO web-site is one of my favorite astronomy tools, Thanks!

– Anonymous AAVSO website user, 2003[1]

After Janet Mattei's many contacts with international observers and her invitation for other national groups to submit their observations to the AAVSO International Database, its numbers exceeded 300 000 observations annually. Professional astronomers and educators were increasingly relying on these results for their studies of stellar phenomena.

The increase in observations tells us very little about the effort required to make those observations available for use by others. We discussed many aspects of the AAVSO's history as it evolved in the previous chapters, but said little about the staffing at headquarters organization other than for the recorder and/or the director. In the early years there was little more to consider than the position of secretary/receptionist, filled for many years by Helen Stephansky under Recorder Leon Campbell and then Director Margaret Mayall. Florence Campbell Bibber, who worked for Mayall from the 1950s through the 1970s, followed Stephansky. Mayall also began hiring work-study students and other part-time assistants to process observations and handle early programming efforts. With its limited financial resources, the Association could afford nothing more than temporary staff for this work.

During Mattei's tenure, the staffing situation changed radically as the workload increased. After experiencing the effects of excessive staff turnover on the productivity of the office, Mattei's strategy shifted. She hired more qualified individuals and, within the budget, provided compensation and benefits that were more competitive in the Cambridge employment market. Under this revised approach, the AAVSO enlarged its technical staff, including both astronomers and computer systems specialists, and support staff. In this chapter, we consider in somewhat more detail the staffing required for the AAVSO to function in the new age of information technology and science.

THE AAVSO INTERNATIONAL DATABASE

Activity remained high at AAVSO Headquarters with both routine and nonroutine work after the AAVSO's 1997 meeting in Sion–St. Luc, Switzerland. Annual totals of observations reported to and processed at headquarters for the AAVSO International Database (AID) regularly exceeded 300 000 per year. Visual observations from observers in the United States declined slightly, but observations from all other sources continued rising at a surprising rate. Similarly, total photoelectric photometric and CCD observations increased annually.

AAVSO directors had always emphasized quality, not quantity, of observations since the earliest days of the AAVSO. Thus they left the annual totals of observations in the last pages in their annual reports with some justification. Still, those totals provide a long evidence of progress that can't be ignored. In spite of all other difficulties, the AAVSO's observers continued to work in support of what they considered a valuable effort.

At the same time, the scientific scope of astronomy had gradually but continuously broadened over the previous five decades. By the year 2000, observational as well as theoretical interests in astronomy extended from the shortest wavelengths of gamma rays and X-rays to the longest wavelengths of radio astronomy within intermediate bands of the spectrum. It would be reasonable to believe, in view of wide-ranging interests, that there could be little interest in the work of a small group dedicated to making observations in the tiniest portion of the entire spectrum we identify as visual wavelengths. Yet, those observations in the AAVSO archives, representing more than 100 years of variable star activity, give valuable insights to astronomers about

Figure 20.1. AAVSO Council, 1998. Seated, from left: Dorrit Hoffleit, former President; Martha L. Hazen, Secretary; Kristine M. Larsen; Albert V. Holm, Past President; Janet A. Mattei, Director; Gary Walker, President; Daniel H. Kaiser, Vice President; William G. Dillon. Standing, from left: Mario E. Motta; Peter Garnavich; Lee Anne Willson, Vice President; Margarita Karovska; Theodore H. N. Wales, Treasurer; Roger S. Kolman.

the aggregate properties of classes of variable stars, as well as identifying the particular stars that might be most interesting scientifically. Mattei's struggle for 25 years to make the AAVSO database complete, accurate, and responsive to astronomers' needs finally began to pay off with real contributions to the science of astronomy.

STAFFING AAVSO HEADQUARTERS

Important changes in AAVSO staffing occurred as the Association gradually learned to manage salaries in a competitive manner and to selectively add benefit packages to remain competitive with the marketplace. For a period at the time of the 75th Anniversary in 1986, the staff numbers seemed to stabilize, but in 1987 resignations once again reduced the effectiveness of headquarters. The resignations forced Mattei to devote time and energy to hiring replacements. The balance point – that is, the right combination of a salary and a minimal benefits package sufficient to hire and hold younger, capable staff employees – eluded Mattei, immersed as she was in technical matters. The new balance seems to have been reached a decade later, but not without some difficulties.

From 1987 to 1997 and later, Mattei faced serious difficulties trying to focus on the technical and scientific aspects of her position while she repeatedly interviewed and selected staff members and then trained them for useful work. The AAVSO's location in an intellectual center like Cambridge no doubt made it attractive as an employer, but the cost of living in the Boston/Cambridge area ranks among the highest in the United States. Thus a pure intellectual attraction can only function as an incentive when the minimal considerations of salary and benefits have been accommodated for younger technical staff candidates.

Enhancing AAVSO science

For several years, members of the Council and others suggested that the AAVSO should employ a second Ph.D. staff astronomer who would engage, from a scientific perspective, the huge and growing database of variable star observations while assisting Mattei with questions from the scientific community. After resisting this idea for several years, Mattei finally agreed in 2000 and employed George Hawkins for 2 years as a postdoctoral fellow. Hawkins proved the value of having someone with high-level scientific competence on the payroll. Mattei co-authored several scientific papers with Hawkins. Nevertheless, as agreed at the time of his employment, Hawkins left at the end of 2 years.

The experiment of employing a staff astronomer having proven successful, Mattei replaced Hawkins with Matthew Templeton in 2002. Templeton proved equally helpful in headquarters and added significantly to the quality of communication between the AAVSO and the professional community. Each of these postdoctoral fellows handled challenging work appropriate to their degree while assisting with the evolution of AAVSO programs.

Over nearly a decade, the rest of the scientific and technical staff began to stabilize after salary administration and benefits considerations were worked out. Bearing titles as technical assistants, astronomical technical assistants, or in the case of Elizabeth Waagen, senior technical assistant, this professional staff was more fully qualified with their degrees in astronomy or related sciences, or prior employment in astronomy, than had been the case previously. Waagen, originally employed only for one year in 1979, was joined by Michael Saladyga (1985), Sara J. Beck (1991), Rebecca Turner (1995), Kerriann H. Maletesta (1996, resigned 2010), Gamze Menali (1998, resigned 2010), and Katherine Davis (1998, resigned 2010).[2]

Statistics and programming

Employment of non-astronomical scientific/ technology specialists offered another avenue for advancing the AAVSO's operations. Mattei's early experience in this regard had been very favorable. In 1989, in an amazing stroke of good luck, Mattei hired Grant Foster to perform routine data entry work. Foster had completed his coursework for a Ph.D. in statistics but had not yet written a dissertation. In the course of his undergraduate and graduate coursework, Foster had developed valuable skills in computer programming and technology that quickly became important at the AAVSO. Foster became interested in working with large data sets of variable star information over the next 5 years and introduced a number of valuable data analysis programs to the AAVSO repertoire, significantly enlarging the capability of the AAVSO staff to both evaluate and eventually exploit the AAVSO International Database.

Staffing for the web

As the urgent need to establish a presence on the World-wide web arose at the turn of the millennium, Mattei made several attempts to employ a systems analyst with appropriate skills. In 1998 she hired Aaron Price as systems administrator and webmaster. Like Foster, Price became interested in the analysis of large data sets but using web-based applications. Price resigned the following year to pursue other opportunities during the Internet Bubble (Web 2.0), but fortunately Price agreed to continue working part time on AAVSO problems, returning full time to the AAVSO in 2002 as the UNIX systems administrator. He earned an M.S. in computing systems technology and web applications in 2003.

Support staff

We are not enumerating here, for simplicity's sake, a continuing string of technical assistants who joined the AAVSO staff for a year or two and then left for greener pastures. There are simply too many of them. The same may be said of student assistants hired each summer for specific tasks. Over time the necessary data entry work associated with the backlogged historical observations was completed. In addition, more and more of the incoming observations were now being entered and transmitted electronically by the observers themselves. A diminishing number of paper reports was arriving at headquarters each month. Volunteers like Katherine Hazen, Dorothy Harvey, Frank McCorrison, Carl Feehrer, and Arthur Ritchie came to headquarters on a part-time basis to assist with mailings or routine clerical work, along with part-time data entry clerks Barbara Silva and Gloria Ortiz-Cruz.

The administrative staffing problem

Mattei struggled with a separate and nearly intractable employment problem that can be characterized as the routine management of a busy administrative office. Sharing this work with the technical staff was unavoidable because of the small staff size, and for many years, clerical work sessions were standard practice. Packaging and mailing two issues of *JAAVSO* annually would be a useful example. After the printer delivered the finished journals and the envelope labels had been printed from an internal database, the staff spent days collating the journals and extra materials, which involved folding, stuffing, sealing, affixing stamps and address labels, preparing a bulk mailing, and delivering dozens of sacks and trays of journals to the post office. Similarly, there were semiannual meeting announcements and other routine mass mailings. The staff was overqualified for such mindless work, but nevertheless it had to be done.

Mattei attempted for a number of years to hire and retain a more senior individual with the skills necessary to organize and maintain an efficient office and to help with her correspondence. The concept of an administrative assistant seemed clear enough in theory, but in practice it was difficult to implement for a number of reasons. The main problem was that the job of administrative assistant was difficult to describe well enough to establish an appropriate salary range compared with more normal administrative offices. Even if the salary would have been more generous than Mattei felt able to offer, the working environment was not one in which most experienced office managers or professional secretaries would feel comfortable. And finally, the job – as Mattei understood it and preferred to manage it – required an overly broad span of responsibilities ranging from that of a low-skilled receptionist to that of an experienced office manager and an executive secretary.

In several instances, very competent individuals, typically women, took the administrative assistant job and handled most parts of it very well. One of the first to tackle the job, Janet MacLennan Zisk proved the value of the right person in the job. Unfortunately she moved to Hawaii, and the others who followed her did not stay as long, retired, went to graduate school, moved away, or accepted positions with other employers. Frequently, they commented on the friendliness at the AAVSO, the interesting and challenging work environment, and the value of their friendships with the other staff members, especially Mattei, but the circumstances of those who changed employers demanded either higher pay, or better benefits, or both.

DATA MANAGEMENT AND PROCESSING

Over the period covered by this chapter, there was a steady improvement in the efficiency with which observers could submit their observations. By 1997, an average of 50 percent of all observations were submitted electronically. Previously those observations had been transmitted either by e-mail or mailed on a diskette. However, only 50 percent of those electronic reports were actually digitized in the format specified on the AAVSO's data entry form. Thus half of the observations still arrived at headquarters on paper, and an additional quarter required some manipulation before being added to the AID. Only the remaining quarter could be entered directly into the database. Even that was progress and offered encouragement to further develop the entry processes.

By 2004, popular web-based applications had been implemented so that observers could enter their observations directly into the AID, and at any time, not just monthly. Furthermore, within 15 minutes the observer could view his or her observations in a light curve of all observations for each star reported with that observer's contribution identified. Observers liked Foster's *QuickLook/Light Curve Generator*; it was a popular innovation. Many other innovations proved similarly popular and lightened the load for both headquarters and the observers. By 2004 less than 8 percent of all observations submitted for inclusion in the AID were received at headquarters on paper, freeing up staff time for many other important tasks.

Along with these data entry innovations, the final processing of the archival data from 1911 to 1961 had been completed in 1997, with all observations from 1911 to 2002 available online and updated regularly as observers submitted their observations. Mattei announced in 2002 that the AID had received and archived more than 10.5 million variable star observations. It was an important benchmark; 26 different associations from around the world provided observations to the AID in 2001 through 2002. As had been the case for many years, there were also many unaffiliated observers from around the world submitting their observations directly to the AAVSO on a regular basis.

THE AAVSO ARCHIVE PROJECT

At its October 2000 meeting, the AAVSO Council authorized partial funding for the organization of the AAVSO's collection of archival records and historical papers. The AAVSO Archive Project began later that month when Director Mattei asked AAVSO Technical

Assistant Michael Saladyga to take charge of all aspects of the work. Over the next 7 years, Saladyga evaluated, arranged, and cataloged about 80% of the archival holdings, storing the material in acid-free folders and boxes. The AAVSO Archive comprises some 400 linear feet of valuable correspondence and other papers from the nineteenth and twentieth centuries, official minutes for meetings, and other essential materials, including an extensive collection of Clinton B. Ford's secretary's files and the letters of Olcott, Campbell, Mayall, E. C. Pickering, and many others. The Archive Project was a major step toward preserving the AAVSO's institutional memory.

THE AAVSO AND THE WORLDWIDE WEB

The AAVSO first became involved in electronic mail handling in 1988. Acquisition of the necessary equipment to transmit messages across telephone circuits at voice frequencies in 1989 allowed AAVSO to transmit observations electronically to the mainframe computers at the Center for Astrophysics rather than carry them to the computing center on diskettes. By this time, many AAVSO members and observers who had acquired personal computers and were subscribing to Internet services were also corresponding with each other. The next step seemed natural, transmitting observations to headquarters as digital files recorded on magnetic media rather than as pencil marks on paper that had to be converted later to digital format. Headquarters quickly developed and distributed the necessary software for submitting observations on magnetic diskettes to interested members. By 1990, numerous observers were sending diskettes or attachments to e-mail messages, thus reducing the data entry workload at headquarters. Progress in the headquarters data processing systems was also evident in the fact that 1989 was the last year in which individual observations of variable stars were hand plotted on the pencil graphs of light curves. Thereafter, the computer plotted all light curves from the centralized database.

With sponsorship by several agencies within the National Aeronautics and Space Administration (NASA), the AAVSO was provided with the high-speed transmission line necessary to act as a node on the Internet. On August 1, 1994, the AAVSO posted its website to the Internet, with web pages designed by Technical Assistant William Mackiewicz. With numerous pages accessible through the AAVSO home page, the web site (http://www.aavso.org) was soon being used by members and nonmembers for alert notices, news, educational information about variable stars, light curves, and other published AAVSO materials. Over the course of the next year, 1600 unique computers accessed the AAVSO website nearly 34 000 times. The AAVSO website received many votes of appreciation from members and professional astronomers who valued the data and services provided by the AAVSO.

The new website thus proved its worth as a communication tool. More importantly, however, it was possible to use the same high-speed transmission facility to send more complex files through links provided on the website and through a separate file transfer protocol (FTP) site designed specifically for handling charts and other complex material. Thus the benefits of becoming active on the Internet were much larger than might be imagined from the above description, if use had been limited to communication of news and general information.

Growth of AAVSO website activity closely matched the more general phenomenon of growth in Internet activity. Only 3 years after the initial statistics mentioned above for the first full year of an established presence on the Internet, the AAVSO website was visited in 1997 more than 176 000 times from approximately 6000 unique computers over the Internet. The AAVSO website ranked as an unqualified success.

A NASA grant in 2001 made it possible to replace the AAVSO's Internet connection with a high-capacity T1 connection. The change was a major improvement because it made it possible to upload large files from the AAVSO computer to the Internet in a reasonable time period and to transfer large and complex documents such as AAVSO variable star charts and large data sets for everyone's benefit. Price, with webmaster Katherine Davis, expanded and significantly improved the AAVSO website, adding hundreds of features and changing software to speed delivery of files to the Internet with improved security.

What is impressive about all that, though, is not only that Price and Davis worked wonders, but that Mattei employed them and reacted quickly and supportively to their suggestions, reflecting her growing skills as a manager. The AAVSO's technical capability was enormously improved, and its reputation for service in the variable star community was greatly enhanced.

HIGH-ENERGY ASTROPHYSICS

The best example of Mattei's forward-looking leadership was in the AAVSO's involvement in high-energy astrophysics (HEA). The longer term observations and accumulation of light curves for the cataclysmic variables (CVs) proved invaluable as the complexity of these interesting stars came to be understood. Some CVs had been shown to emit X-rays and perhaps even higher energy particles. By the summer of 2000, that background, and the availability of a reliable corps of real-time observers, attracted the attention of those in the HEA program within NASA Marshall Space Flight Center. In an effort to interest the AAVSO and its observers in a cooperative effort, NASA offered a workshop for AAVSO members, which they hoped would attract observers to a more formal program of cooperation. That connection proved very productive when the AAVSO moved rapidly to support NASA's gamma-ray burst program.

Gamma-ray bursts of astronomical origin were first detected by the Vela satellites monitoring the Earth for the intense gamma radiation released from a nuclear detonation. When the source of one Vela detection of gamma-ray bursts had been localized, it was obvious the underlying object was in space and not on the Earth. This discovery prompted an entire new field of astrophysics and the launching of orbiting satellite observatories to observe and localize, as well as possible, the Galactic and extragalactic sources of gamma-rays.

Mattei scheduled the AAVSO 2000 Spring Meeting in Huntsville, Alabama, in cooperation with the NASA Marshall Space Flight Center. AAVSO members received a briefing by NASA scientists on both the technology and the science involved in the emerging field of gamma-ray astronomy. A joint program that resulted from the first HEA workshop benefited both the AAVSO and the NASA programs. NASA needed a way to quickly alert geographically dispersed observers in an effort to find the afterglow from the object that emitted a burst of gamma-rays once the burst was detected.

A joint program designed to search for and detect visual afterglows, if any, from gamma-ray burst events (GRB) originated from the Huntsville meeting. Response times measured in minutes were a critical need. Such afterglows, faint to begin with, were known to fade rapidly. NASA funded a special Alert System based on commercially available personal pagers that could be carried by AAVSO observers who participated in the program. When an alert was broadcast, observers who had dark skies agreed to drop everything and initiate a search for the object around the coordinates provided for the burst by NASA.

The first GRB afterglow discovered and observed by AAVSO observers occurred on March 23, 2003. Council member and USNO Astronomer Arne A. Henden of Flagstaff, Arizona, was engaged in online photometry in the southern hemisphere using a telescope at the Mt. John Observatory in New Zealand when the alert was posted. Alan Gilmore and Pam Kilmartin from the University of Canterbury physics department were also searching for the GRB with a CCD camera provided by the AAVSO and installed on the Optical Craftsman 24-inch telescope on Mt. John. The team located an object that did not appear on Henden's or other previous images of the region but fell within the error boxes for both the GRB detection and earlier detections of X-rays. In a short time it was apparent that the unknown object also faded at a rapid rate, thus confirming the object as the apparent afterglow of the GRB event. In less than a day after the original GRB detection, the afterglow faded from its discovery magnitude of approximately 18.6 to approximately 20.4, and within 10 days had faded to 24.3. The object received significant international attention, with many other astronomers attempting to gather photometric and spectrographic data.[3]

By October 2003, Mattei reported that the GRB network had expanded to include 185 active AAVSO observers, and several more GRB afterglows had been successfully observed and reported.[4] A trio of AAVSO observers in Finland subsequently discovered several other GRB afterglows,[5] and a number of afterglow observations after discoveries by others had also been reported. The joint AAVSO/NASA HEA program demonstrated the capability of AAVSO observers with remarkable clarity. The second and third HEA workshops extended AAVSO observer interest in the area. As a result of this interest, Mattei also expanded the highly successful GRB program to include Active Galactic Nuclei (AGNs), including blazars and magnetic cataclysmic variables. As a result of this broadening, the name of the network of committed observers was changed to the AAVSO High-Energy Network, soon abbreviated the AAVSO HEN. Aaron Price assumed the responsibility of coordinating HEN activities.[6]

Amazingly, once the alerts on GRB and X-ray transient events started coming, attentive AAVSO observers picked up other related effects. Long-time solar observer Cap Hossfield, chairman of the Solar Division and a former AAVSO president, noticed that his very low frequency (VLF) receiver detected the sudden ionization of the upper atmosphere caused by the arrival of GRB radiation. Hossfield and Solar Division observers had been detecting such events called sudden ionospheric disturbances (SID) for years without knowing their cause. Although not specific with respect to direction, the VLF receivers could at least confirm that an event was occurring and that an afterglow might be visible if any more specific target areas could be identified.[7]

MADISON, WISCONSIN – COUNCIL MEETING

With all of the excitement caused by the Internet and high-energy astrophysics, it would have been easy for the AAVSO Director to temporarily lose sight of the larger programs of the Association. Mattei, however, did not succumb to that attraction and instead kept equally focused on the long-term goals of the Association as well as the short-term advances and problems. Her dedication to long-term planning for the Association caused her to use the 2001 Spring Meeting in Madison, Wisconsin, to seek out additional areas in which the Association should be working. To some extent what Mattei wanted was an update of the Futures Study that had been conducted nearly a decade earlier, but with emphasis limited to the scientific aspects of the AAVSO program. Toward that end, she invited the Council to remain in Madison for an additional day and a half after the spring meeting to hear and discuss presentations by a panel of distinguished astronomers about their specialties. She described the meeting as a discussion of astronomy in the new century and the role of the AAVSO.

Among the panel members were Jay Gallagher (University of Wisconsin), who discussed development trends in land-based astronomy; Steve Beckwith (Space Telescope Science Institute) discussing future developments likely in space-based astronomy; John Good (California Institute of Technology, Infrared Processing and Analysis Center) discussing the proposed National Virtual Observatory and the AAVSO's potential role in it; and Henden of the U.S. Naval Observatory (USNO), Flagstaff, Arizona, who reviewed surveys and automated telescopes. Mattei, for her part, added her own assessment of the current roles and strengths of the AAVSO. Participants discussed these concepts well into the evening in two groups and reported to a joint session the following morning.

One new AAVSO initiative can be identified clearly as a consequence of this special Council meeting in Madison. After a great deal of discussion of potential new technologies that could be usefully adopted by AAVSO observers, the near-infrared spectral regions identified as the *H*, *J*, and *K* bands were clearly identified as opportunities for amateurs with modest-sized telescopes. Amplitudes of variation in the near *IR* bands might be smaller than those observed for the same stars in the visual band, but nonetheless would probably provide important information.

On learning about a new photometer using an indium-gallium-arsenide (InGaAs) detector, Mattei volunteered the necessary funds to purchase five photometers if there was sufficient interest among a small group of AAVSO observers. With appropriate filters, the Optec, Inc. SSP-4 photometer would be an ideal instrument to conduct studies in at least the *J* and *H* bands. Doug West, of Mulvane, Kansas, volunteered to lead the activity, and the AAVSO Infrared Photometry subcommittee of the PEP committee was organized. Since that time, *J*- and *H*-band observations have been accumulated in the AAVSO database. Interest in near-infrared photometry remains high, though recent efforts have emphasized a slightly different selection of filters.[8]

The second group at Madison focused on data mining and the National Virtual Observatory, with a recommendation that the AAVSO attempt to "be compatible with NVO" and seek opportunities to both contribute and to utilize the information thus made available.

THE PAN-PACIFIC MEETING

The Third International meeting of the AAVSO took place June 30–July 6, 2002, on the Big Island of Hawaii. The unusually long meeting included two extra events, the Fourth Annual Towards Other Planetary Systems

(TOPS) Teacher's conference and the Third HEA Workshop sponsored by NASA. The agendas of the AAVSO spring meeting and the TOPS conference were mingled and included alternating half-day sessions for each organization in addition to the normal AAVSO business meeting. For purposes of this discussion, we are treating them as separate meetings.

The TOPS Conference was organized and moderated by former AAVSO staff member and University of Hawaii astronomer Karen Meech. This meeting was the fourth annual TOPS conference organized by Meech and Mattei to promote better teaching of sciences in secondary schools through the use of astronomical observation. The participants presented their experience in teaching science; practical applications in papers on the observations of the Moon, Messier objects, and variable stars; and computer projects to interpret variable star observations. In addition to Meech and Mattei, the AAVSO was represented in the agenda by Hawaii resident Jim Bedient, a mentor in the TOPS program and co-author of a paper presented by three of his students on remote observing of AAVSO variable stars over the Internet, and by Mike Mattei, who taught telescope making and observing.[9]

In the Second HEA Workshop, the papers, mainly by professional astronomers, described a fascinating array of observing techniques and findings intended to both brief AAVSO members on the state of the science for various aspects of HEA and to attract AAVSO observers to support their programs. Topics included a background on detectors and history of the field, followed by papers on gamma-ray bursters (three papers and a panel discussion), neutron stars and magnetars, the Sun, and cataclysmic variables (four papers). Summary papers to introduce the sessions and highlight progress in the field helped integrate the diverse topics, whereas a paper on the fascinating Keck Telescope at the nearby Mauna Kea observatory held everyone's attention as a modest diversion from the main subject. Scientifically demanding at times, the sessions provided an attractive inducement for AAVSO observers to divert attention away from the main grist of LPVs to the more exotic and transient fields of high-energy astrophysics.

A DARK CLOUD LOOMS

As she had done for 4 years previously, Mattei returned to Hawaii to lead the fifth annual TOPS Conference, staying for 2 weeks in July 2003. She returned to Cambridge from Hawaii physically exhausted, a somewhat different experience since she usually returned exhilarated from the conferences. Along with the exhaustion, she experienced severe neck pains for which medication seemed to provide limited relief. It did not help that she had arranged a trip to Washington, DC, in August for which she had committed to a presentation.

Figure 20.2. Janet Mattei with observers Jan Hers and Alan Cousins at a meeting of the Variable Star Section, Astronomical Society of South Africa, December 2002.

Incorrect diagnoses of the neck, upper back, and spinal pains had resulted in unsuccessful treatment for muscle strain and polymyalgia rheumatica (PMR). Although unknown to AAVSO members, other than two key officers and the headquarters staff, in the 1990s Mattei had had a 2-month absence from work for surgery to remove a cancerous ovarian cyst. After the cyst removal, Mattei went for regular tests, at first every 3 months, and then every 6 months. Thus there was an ever-present risk, the magnitude of which only she and her husband Mike understood. The experience had a profound effect on Mattei, among other things deepening her religious faith. As a result she began regularly attending services at a Synagogue in Lowell, Massachusetts, and established a strong relationship with Rabbi Everett Gendler there.

Mattei went in for one of those tests on August 28, 2003, the Thursday before Labor Day. On Friday morning, Mattei appeared at the AAVSO office with her husband Mike, clearly frightened but still not willing to discuss the diagnosis. After a brief stay in her office checking messages and returning calls, Mattei announced she was going to the hospital for another test and left, saying only, "We shall see what this day brings." The AAVSO staff locked up the office on Friday evening for the long Labor Day weekend without any further feedback from either Janet or Mike about her condition.

Mattei spent the rest of Friday in her home office in Littleton, Massachusetts, making notes and sorting things for her husband to bring into the office for Elizabeth Waagen on the following Tuesday. On Friday evening, she returned to Boston and checked into the Brigham and Women's Hospital for further tests. By the next morning, tests had confirmed the preliminary diagnosis that Mattei suffered from acute myelogenous leukemia (AML).

On Saturday morning, Waagen received a telephone call at home from Mattei. From her bed at the hospital, Mattei advised Waagen that she had been diagnosed with AML and would begin chemotherapy treatments as soon as they completed their telephone conversation. Over the next hour, after formally appointing Waagen as the deputy director of the AAVSO, Mattei calmly outlined the threads of as many ongoing activities as possible, where to find files, what to look for in critical instances, who to rely on, and how to interpret clues and symptoms of various problems.

Waagen had additional worries. Her planned activity that weekend involved locating a new residence for herself and her mother Arline, who had moved from Virginia to the Boston area to establish a joint residence with Elizabeth. After some time to reflect on her conversation with Mattei, Waagen called AAVSO President Daniel H. Kaiser, Secretary Martha Hazen, and Treasurer Lou Cohen to advise each of this new dilemma the Association faced. Mattei's devastating illness, as well as the problems that Waagen would face as the deputy director of the AAVSO, continued to haunt her over the entire long Labor Day weekend.

On Tuesday morning, Waagen briefed the staff on Mattei's condition. For the entire staff, the absence of Mattei's smiling face, morning greeting, eager questions, and helpful suggestions came as a major dislocation. However, at that point each member of the staff had been in their respective assignments for a number of years. The tasks each person was responsible for were stable and reasonably well understood. The staff dedicated themselves to handling the Association's business with minimum disruption in honor of their now disabled director.

Mike Mattei visited AAVSO Headquarters later that morning to give Waagen several envelopes of materials that Mattei had assembled in her home office on Friday before leaving for the hospital. The material in the envelopes pertained to projects in progress at headquarters and administrative matters. Two file folders, for example, contained critically needed information, log-ins and passwords, for administration of grants supporting the AAVSO Data Validation project, with necessary instructions for requesting additional funds.

When the diagnosis of Mattei's illness became general knowledge, the AAVSO Council and its officers were faced with a difficult decision. The treatment for AML requires intensive chemotherapy in what may involve several attempts to destroy the leukemia cells. There was no way of knowing the length of time that would be required for the treatment and the recovery period that would surely be necessary. Adult survival rates for this drastic treatment were low; only those adults who, at the outset, were in excellent physical condition were likely to survive. Thus a realistic assessment of the odds of Mattei's survival seemed to point to a need to initiate a search for a replacement for Mattei as the AAVSO's Director, meaning the immediate appointment of a search committee.

Waagen, on the other hand, felt strongly that if any hint of a search reached the Matteis, the knowledge might adversely impact her will to survive. President Kaiser wisely weighed the odds involved. Could the AAVSO as an organization (both the staff and the officers) withstand the temporary buffeting that might occur while Mattei's treatment ran its course, and then if her treatment was not successful, the period in which a delayed search for a replacement for Mattei was mounted? That risk had to be weighed against the impact that knowledge might have on Mattei's own survival chances. Kaiser's decision, to wait and see, rested on his confident belief that the AAVSO as an organization enjoyed a maturity and stability that could carry it through a difficult period during which Mattei might be away from work. Therefore, Kaiser, with the concurrence of the Executive Committee, confirmed Mattei's last-minute commissioning of Waagen. Changing her title from Deputy Director to Interim Director of the AAVSO, the Council delegated the full director's responsibility to Waagen for the duration of Mattei's absence.

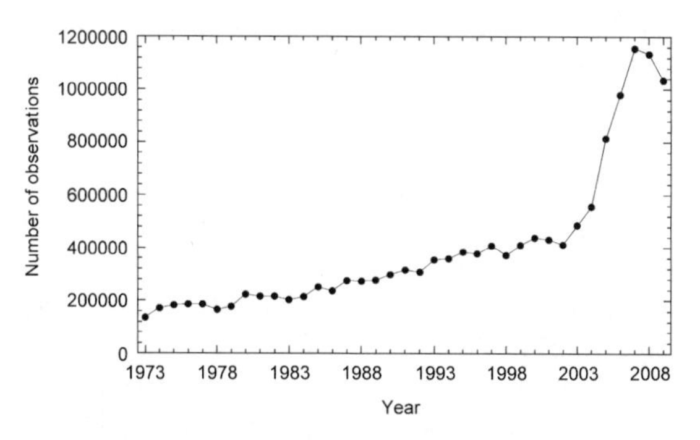

Figure 20.3. Annual totals of observations received at AAVSO Headquarters, 1973–2008.

MATTEI'S LAST ANNUAL REPORTS

Among the many uncompleted tasks on Mattei's desk at the time of her hospitalization was the *JAAVSO* version of her annual report for the fiscal year 2001–2002 (more complex than the oral report she delivered at the annual meeting) and, of course, after only 3 weeks of this interregnum, the annual report for the fiscal year 2002–2003 would also be due. Therefore, one of the early tasks that faced Waagen was the completion of not one but two annual reports. She published the 2001–2002 report under Mattei's name and published the 2002–2003 report under her own name, but with a note saying she had prepared it on Mattei's behalf. Taken together, those two reports provide a valuable summary of the condition of the AAVSO.

The Association's accomplishments during Mattei's 30th year as the director were impressive when compared with the early years of the Association. The annual total for observations added to the AAVSO International Database had exceeded 400 000 several years in a row compared with about 150 000 in Mattei's first year as director in 1973 (see Figure 20.3). Observations and technologies that were underreported, little used, or did not exist during those first years included PEP observations at about 3000 observations per year, annual CCD observing totals near 100 000 per year from more than 100 new observers, and observations in the multiple band passes (ultraviolet, blue, visual, red, and near infrared), most useful to astronomers in theoretical studies. The Eclipsing Binary, RR Lyrae, Supernova, and Nova Search Committees all continued to report valuable reduced eclipse minimum timings, maxima, and thousands of negative results in their nonetheless valuable nova search efforts. Sunspot observations and radio transmission enhancement events (SID events) rounded out an impressive array of observational contributions to astronomical knowledge from AAVSO members. All things considered, Mattei's tenure had shown impressive advances in these areas.

In her early years, Mattei (like her predecessors) had deliberately not emphasized the importance of observing totals as a way of emphasizing the need for quality observations. Still, recognizing that personal totals were important to many observers and that the overall totals of observers and observations was an indication of the health and international scope of the Association, the aggregate number of observations in the AAVSO International Database was another statistic that Mattei regularly included in her reports. By the 2001–2002 fiscal year, for example, she reported that the cumulative total observations exceeded 10.5 million for the first time. Annual reports each year after 1986 included both a graph of each individual year's observations and of the cumulative total, and both had increased in an exponential manner. Mattei called the latter graph the megasteps because it recorded not only the cumulative annual totals, but also the names of the individuals who contributed each millionth or each half-millionth observation to the total.

Another measure of progress during Mattei's 30 years might be found in the manner in which the users of variable star information, primarily professional astronomers, requested data from the AAVSO's archives. In Mattei's first annual report, she discussed 22 such requests, as compared with more than 530 in the summaries for each of the two final annual reports. Over the years, as the requests increased, Mattei shifted away from describing every request and instead described only a few she considered significant. Using pie charts, she characterized the uses to which the data were being applied, as well as the types of variable stars for which the information had been requested. These requests for information reflected a growing confidence in the user community that the Association reported reliable and appropriately characterized information based on more than a century of observations for some stars.

Mattei's annual reports had quickly fallen into a pattern that she maintained for her entire tenure. She first summarized progress on data processing, to which she had assigned her highest priority. Next she discussed the services the Association was providing in

Figure 20.4. AAVSO Council members Thomas R. Williams, John R. Percy, and Wayne Lowder at AAVSO Headquarters, about 1993.

terms of special requests for data. By 1995, however, the emphasis had shifted slightly, though the logic remained the same. With the advent of the Internet and all that it implied about the future, Mattei's annual reports emphasized the availability of observations and services through the AAVSO's Internet portal. The 2002–2003 fiscal report, her 30th report, reflected the growing wave of interpersonal, intra-organizational, and inter-organizational communication taking place on the Internet through the posting and management of information on web pages. She then noted the AAVSO's own "web presence." The number of times the AAVSO website was accessed in the year, usually referred to as the number of hits on individual site pages, soon became an important new measure of organizational effectiveness in addition to the annual observing totals. That statistic, when reported, provided a measure of public as well as member interest in the organization, its subject, and its results.

As far as data entry was concerned, the backlog of observations from 1911 to 1961 had been cleared up, but a never-ending array of other observations appeared from various sources. For example, more than 94 000 previously unreported observations had been entered from Wayne Lowder's observing notebook obtained in February 2003 after his death, and more additions came in from the prodigious southern hemisphere observer Albert Jones of New Zealand, who sent his observations in digital form.

In her annual reports, Mattei also called attention to other aspects of progress in the Association. In these years, for example, she reported on the completion of another 14 AAVSO variable star monographs summarizing data on individual stars of importance for a total of 26 monographs. Mattei also announced another data link in 2003 that would make astronomical data available: the Centre de Données astronomique de Strasbourg (CDS) SIMBAD catalog of astronomical data sources. CDS had agreed to provide a link in its catalog to the AAVSO light curves for stars in the AAVSO observing program.

In her two reports, she discussed several initiatives in updating charts used in making variable star observations. One involved international coordination in the effort to produce the charts from the best available sources and in a uniformly acceptable format. At the 2002 meeting in Hawaii, an international group assembled to discuss this problem, with participation from Japan, Australia, New Zealand, the United Kingdom, and Belgium. The goal of the International Chart Working Group was to provide guidelines for all variable star chart makers on selecting comparison star sequences and revising charts. Another change reflected the advance of technology. During his stay with AAVSO as its first post-doctoral fellow, George Hawkins developed a method of plotting charts from any electronic star catalog – for example, the Sloan Digital Sky Survey. The plotting program had wide applicability in AAVSO work.

In addition to these changes, a new AAVSO Charts Team had been formed, with active guidance from Mattei and Hawkins. Among its participants were Mike Simonsen as team leader, Charles Scovil as a senior advisor and mentor to the effort, USNO astronomer Henden as a technical advisor on photometry, and eight other AAVSO members. This team had a charter to maintain, revise, and create any comparison star chart necessary for any AAVSO observing program. The work of the new group was facilitated by a separate e-mail communication network. Significant results and annual summaries were to be communicated in the usual committee report formats. The group published notes on the Charts section of the AAVSO website, which also included an up-to-date catalog of all available charts. In addition, a separate team activity had been initiated with a mission to create a database of all comparison stars on all AAVSO charts. A catalog of these existing comparison stars was about 25 percent complete in 2003. This second team, soon including 16 team members, worked effectively under the direction of Vance Petriew as its leader. With the completed catalog as a tool, chart makers could easily add revised magnitudes or positions for comparison stars, which would then appear automatically on electronically generated charts.

These new initiatives were completely consistent with the whole idea of chart committees, going back to David B. Pickering and Dalmiro F. Brocchi in the 1920s and 1930s. The new programs relied on the very latest state-of-the-art in photometry, photometric catalogs, plotting technology, and graphic reproduction appropriate to AAVSO's standards and the fundamentals of web technology. In the first 4 months of the team's work, they had completed 222 new or revised charts, 24 blazar charts, and *Alert Notice* charts for three novae and three supernovae, a very impressive beginning fitting both the AAVSO tradition of member participation and newest styles of interactive cooperation over the Internet.

Figure 20.5. Janet Akyüz Mattei, AAVSO Director 1973–2004.

Taken together, these two annual reports of the AAVSO Director document a degree of progress in the Association that many members would have found difficult to imagine more than a decade earlier. In 1974, only 2 to 3 years of accumulated observations could be reviewed without a great deal of effort. In contrast, by 2004, available observations for any period of time could be selected and quickly and conveniently displayed in whatever detail an investigator required. It was an amazing transformation for anyone who had been in, or watching, the AAVSO for that entire period. As Waagen finished the one report and wrote the other for the absent director, she could not help but wonder whether these would be Mattei's last reports as well.

Part VI

ACCELERATING OBSERVATIONAL SCIENCE – THE ARNE HENDEN ERA

21 · Bridging the gap

Janet was enormously proud of the AAVSO and its members and observers. We were enormously proud of Janet, and we are grateful to her for giving herself to the AAVSO with such devotion, and for leaving us her personal legacy of striving for excellence, dedication, and compassion.

– Elizabeth O. Waagen, 2005[1]

An unexpected crisis occurred as the Association entered a new era in providing services to the community of variable star astronomy and reaping the benefits of its energetic director's international involvement in that science. The events involving Janet Mattei's illness occurred rapidly and caught everyone by surprise. Yet, with steadfast dedication and the strong support of the Council, the staff of the Association continued near normal operations and awaited the outcome of this crisis.

As Mattei left precipitously on September 6, 2003, to begin treatment for her rare and usually fatal blood disease, acute myelogenous leukemia, she appointed senior technical assistant Elizabeth Waagen as her deputy to manage operations while she received treatment, staff astronomer Matthew Templeton as science advisor, and technical assistant Aaron Price to be responsible for information technology. Within a few days the Executive Committee of the Council approved those choices and designated Waagen as the interim director, vesting her with all the authority that had been exercised by Mattei, and then waited to learn the outcome of the chemotherapy treatments already being administered to Mattei.

This chapter describes some of the key activities that engaged Waagen and the Council, including the selection of a new AAVSO Director, during those critical days of transition.

WAAGEN HOLDS IT TOGETHER

Determined to maintain the AAVSO and its services, Waagen had a better chance of fulfilling that challenge than anyone else. She was, after all, the most experienced member of the staff. Over the course of her work in the AAVSO offices, at that point spanning 25 years, Waagen had participated in virtually all of the technical activities involved in the enormous expansion of data-processing capabilities in the office, handled many of the technical alerts necessary to sustain the technical program, participated in meeting planning and execution, edited journals and other publications, assisted Mattei with grant and report writing and correspondence, and had routinely been responsible for "minding the store" during Mattei's travel absences. Those experiences would serve her well in the coming months, but her life experiences outside of AAVSO had also prepared her for such adversity.

Born in Buffalo, New York, on May 2, 1956, Waagen was the youngest of three children of Burton Selwyn Waagen and Arline Alice Otto Waagen. As a professor at the New York State College for Teachers at Buffalo (now Buffalo State College), Waagen had a good reputation in Industrial Arts Education. While raising Elizabeth and her two older brothers, Christopher and Douglas, Mrs. Waagen suspended her career in social work. After spending Burton's second 1-year sabbatical at Columbia University in New York City, the family returned to Buffalo for the next year. Then, Burton accepted a 3-year appointment in southern Africa in 1965. The family followed Burton, spending 6 weeks traveling in Europe along the way. Two years later, after a divorce, Arline Waagen and the children returned to Buffalo, where Elizabeth completed school and matriculated on scholarships to Smith College.

Although originally intent on a career in pediatrics, Elizabeth found the research-oriented focus of the

Figure 21.1. Elizabeth O. Waagen, AAVSO Interim Director 2003–2005.

biology major curriculum disappointing, whereas a course in astronomy in her junior year cemented a life-long love of the subject. Changing majors at the end of the junior year, she received an A.B. degree in astronomy in May 1979. One of Elizabeth's astronomy professors, Mrs. Krystyna Jaworowska, knew both Janet Mattei and Dorrit Hoffleit, the latter especially well. Near the end of Elizabeth's last semester, Mrs. Jaworowska insisted that she apply for a job opening at the AAVSO to process data, work on *JAAVSO*, and help the director with correspondence, even though it was advertised as available for only 1 year's duration.

By August 1979, Elizabeth was employed, presumably for only 1 year, as a technical assistant in the AAVSO offices at 187 Concord Avenue, Cambridge. The relationship proved a particularly happy one; the question of the 1-year limitation never again arose. Thus, on that fateful day in September 2003, Waagen, by now the senior technical assistant in the office, was best prepared to carry on operations. Everyone at the AAVSO knew their part in sustaining the routine in the office. They carried on business as usual while Waagen sorted out what needed to be done on Mattei's last-minute "to do" list, and the realization of her new responsibilities sank in.

AAVSO Treasurer Lou Cohen stepped in immediately on financial matters. He arranged check-signing privileges for Waagen at Cambridge Trust, the checking account for day-to-day matters, and at Bank of America, which handled the Ford Trust accounts. Trust money was left in Bank of America until it was needed for day-to-day expenses, then a check was written to transfer funds to Cambridge Trust. Because Cohen was already a signatory on the Bank of America account, he helped arrange the necessary documentation, and eventually the check-writing authority was straightened out. Similarly, Waagen was given access to the safety deposit box at Bank of Boston, where it was thought the deed to the Birch Street property would be found. The deed was later located in a locked file in headquarters along with the closing papers for that transaction.

The situation was complicated by the fact that Cohen had only been the AAVSO Treasurer for a year. During that first year, he had primarily directed his attention to finding a more reliable manager for the Endowment Funds. Cohen had started that search but was in mid-stream in the process when Mattei fell ill. Under the circumstances, it should surprise no one that the investment changes and the relocation of the Endowment were not completed for some time. Waagen also had to learn rapidly some of the other niceties of the financial end of the business, including cash management and understanding her role in management of the AAVSO assets. Fortunately she had written AAVSO checks, issued purchase orders, deposited incoming checks, and had worked with Mattei on membership dues and fundraising solicitations and other routine matters relevant to the operation of the AAVSO as a business.

The fiscal year ended on September 30, 2003, and created yet another set of problems for Waagen and Cohen. She had to prepare the Director's Annual Report for the Council and for the general membership meeting. Waagen had already started assembling the Director's Report for Mattei; she spent all of September and part of October finishing the report. One major complicating issue for Waagen, for example, was the huge influx of CCD data that had started in that reporting year, and continued. It was a new challenge for the director to decide how to handle that new stream of data in the annual statistics.

Waagen later commented that she could not have had a better Council to work with, representing experience and wisdom as good as any new director could want: Dan Kaiser, a successful small businessman, president; Bill Dillon, an executive in a large corporation, long-time member and observer, first vice president; Kevin Marvel, a professional astronomer and experienced association executive, second vice president;

Martha Hazen, professional astronomer and a Mattei confidant, secretary; Lou Cohen, an experienced engineering consultant with a lot of financial savvy, treasurer; Lee Anne Willson, a professional astronomer and long-time Mattei friend and colleague, past president; and Council member David Williams, also an experienced association executive, a long-term member, and observer. The remainder of the Council was equally strong and included Gary Billings, Doug West, and astronomers Geoffrey Clayton, Arne Henden, and Paula Szkody.

AAVSO's experienced staff easily handled the office routine on an ongoing basis. Mattei had already finalized plans for the forthcoming 2003 Annual Meeting. Waagen talked a lot with Kaiser, Hazen, and Cohen as the meeting approached on issues on which they could help. The usual crisis solving during the execution of the meeting had become, by now, a matter of routine for the staff, particularly for Rebecca Turner, the AAVSO's capable events coordinator.

The Executive Committee focused mainly on taking the pulse of the organization to make sure everything was functioning normally when it met before the annual meeting. It was Waagen's first Council meeting as a full participant. Completely prepared but not knowing what to expect, Waagen was gratified by the Council's acceptance of her reports and contributions to the meeting. A subdued general membership meeting itself acknowledged the difficult circumstances faced by the Association. The great uncertainty regarding Mattei's survival and the AAVSO's future hung in the air ominously. If that in itself were not enough to create a sober atmosphere, when Waagen announced, as was a common practice, the members who had passed away in the last year, her list was stunning for the number of well-known AAVSO leaders that it contained: past presidents Casper Hossfield, Wayne Lowder, and Ernst Mayer, and long-time treasurer Ted Wales.

At the general meeting on Saturday, committee reports were presented, but there was no scientific paper session or closing banquet – 2 years of abbreviated annual meetings had already been scheduled by Mattei and the Council in an effort to cut costs. Edward Halbach received the William Tyler Olcott award for his many years of distinguished service in outreach and education and, in a number of ways, to the AAVSO; and Council member Mike Mattei accepted a 30-year AAVSO service award for his hospitalized wife, AAVSO Director Janet Mattei. Kaiser ended his term as president and first vice president. Dillon began his term as president, along with Marvel becoming first vice president, and David Williams, second vice president. The meeting concluded with a buffet supper at headquarters.

THE FINAL LOSS

As soon as Mattei's hour-long conversation with Waagen had been completed on Saturday, September 6, 2003, the hospital staff started the chemotherapy. Mattei's immune system collapsed under the force of the treatment within days. Another acute health problem, a particularly virulent pneumonia that Mattei contracted from a resistant hospital-borne pathogen, nearly stopped her breathing. Mattei remained sedated, sustained by a respirator, for a number of weeks after the completion of the first round of chemotherapy.

On December 3, Mattei regained consciousness and eventually began to eat, carry on normal conversations, and recover some of her naturally cheerful personality. As her strength returned, she undertook a series of telephone calls to members of her family, some AAVSO members, and other close friends. None of the calls could last more than a few minutes, as her strength would not permit a sustained conversation, and she had a lot of names on her list to call. Most who received those calls expressed both astonishment and delight at hearing her voice again. It was a moment that all who had received such calls would later cherish. She was able to return home in mid-January 2004. Weak but responsive, Mattei remained at home for 5 weeks while undergoing treatment with a new experimental medicine. When that did not work, she was re-admitted to Brigham and Women's for the second round of chemotherapy in preparation for a stem-cell transplant from her brother Hayim. Unfortunately, Mattei did not survive the second round; she passed away on March 22, 2004, and was buried 4 days later.

Mattei's family wanted to return her remains to Turkey, where she would be buried alongside her father, but that Turkish cemetery was open only to Jewish males. More importantly, however, she wished to be buried in the Mount Auburn Cemetery in Cambridge, Massachusetts.[2] Furthermore, Janet was an internationally known figure of some importance, and Mike Mattei felt that her grave site should be accessible to

Figure 21.2. AAVSO Headquarters staff during the difficult interim period 2003–2005. From top left: Sara Beck, Katherine Davis, Carl Feehrer, Kerriann Malatesta, Gamze Menali, Gloria Ortiz-Cruz, Aaron Price, Arthur Ritchie, Michael Saladyga, Travis Searle, Sarah Sechelski, Barbara Silva, Matthew Templeton, Rebecca Turner, Elizabeth O. Waagen.

all at any time. Mike secured a double burial plot at the Mount Auburn Cemetery and a gravestone that was appropriate for Janet Mattei's faith, including an appropriate traditional message in Hebrew. On Friday, March, 26, 2004, a service was attended by family, AAVSO staff, and close personal friends at Temple Emanuel in Lowell, Massachusetts, and included a beautiful tribute by Lee Anne Willson. Attendees processed to Mount Auburn for a graveside service and traditional Sephardic burial. The service drew an overflow attendance of Janet and Mike Mattei's friends and supporters, although participation by family members only had been requested. The AAVSO Council and staff planned a Memorial Service in connection with the Annual Meeting in October 2004.

President Dillon called together the AAVSO Executive Committee for a telephone conference on March 24, 2004. Present on the telephone, in addition to Dillon, were past President Kaiser, First Vice President Marvel; Second Vice President Williams; Secretary Hazen; Treasurer Cohen; and Interim Director Waagen. Dillon limited the agenda to just four items of essential business: (1) Providing income and benefits for Mattei's husband Mike; (2) Dealing with Mattei's papers in the office and at her home office; (3) Planning for a search committee to replace Mattei; and (4) Communicating in response to questions – what to say to whom and when.

It was an effective and timely agenda for a sad but necessary meeting. On the matter of support for Mike Mattei, agreement was quickly reached to continue health care insurance for him until he qualified for Medicare insurance under his Social Security coverage, a matter of less than 2 years. Waagen assured the

Committee she had already been through most of Mattei's papers over the years and that accessing the files would not be a problem. The Executive Committee then commissioned a Search Committee and established a deadline for the selection of a new director.

The Mattei Memorial

The 2004 Annual Meeting was the first meeting for which the planning was entirely up to Waagen. She also involved Mattei Memorial Committee members Paula Szkody and Aaron Price and the entire headquarters staff. The meeting, complicated as it was by Mattei's death in March and the need to plan a memorial, could not in any sense be considered a "normal" meeting. One whole day, Friday, October 29, 2004, was devoted to a Memorial Service for Mattei at Brandeis University. The University was chosen for this meeting because Mattei was an alumna, a former Wein scholar, member of the Brandeis Recruiting Committee, and active in other ways with the university. The morning session concentrated on memorializing Janet Mattei in as many and as personal ways as her multifaceted personality deserved, and the afternoon included a Symposium that focused on specific aspects of variable star astronomy in which Mattei had been particularly involved; the presenters were all her long-time colleagues.

THE WORLD SPINS ON

On Saturday, after the Memorial Service and Symposium, the Council met for nearly the whole day, followed by a brief, late-afternoon general membership meeting. Members were not encouraged to stay for the afternoon meeting, but the formality was observed for Constitutional reasons. Committee reports were presented, and awards were announced, but there were no technical papers or banquet. The awards conferred by the Council included the William Tyler Olcott Award, given posthumously to Janet Mattei for her years of dedication and tremendous effort in outreach and education expended on behalf of the Association. The whole weekend was handled very well by the AAVSO staff, following in Mattei's footsteps in the process, an additional retrospective testimony to her success as an executive leader.

Waagen attempted to carry on professional relations and experienced full cooperation and no interference. She had the assistance she needed from the Council and Lee Anne Willson, and other professional astronomers on the Council volunteered to be available if needed. Also, Matthew Templeton, staff astronomer at headquarters, was a valuable resource with his expertise in pulsating stars and asteroseismology. Ongoing collaborations continued as usual, and grant administration was handled normally. The main grant project, validation of nearly 9.5 million variable star observations that was funded by NASA for 2 years, was in its first year and proceeding smoothly, thanks to the dedication of everyone involved – nearly the entire headquarters staff.[3]

The first in a series of AAVSO symposia on Mira Companions and Planets was organized by Aaron Price, dedicated to Janet Mattei, and held at the Harvard-Smithsonian Center for Astrophysics in April 2004. Planning for the next AAVSO meeting was well underway by the time Mattei went into the hospital. The 2004 Spring Meeting, to be held in July in Oakland and Berkeley, California, had been planned as a joint meeting with the Astronomical League (AL), the Astronomical Society of the Pacific (ASP), and the Association of Lunar and Planetary Observers (ALPO), in conjunction with three Bay Area amateur societies.[4] The AL, working with the local society's organizing committee, made all the logistical and scheduling arrangements for the meeting, so all that AAVSO and Waagen had to worry about was the meeting agenda and making sure the AL arrangements were adequate to meet the AAVSO's needs. In effect this was an easy meeting for Waagen, a real blessing in the face of the many other issues she was facing.

Chryssa Kouveliotou of NASA had obtained a grant in honor of Mattei when she became ill and had organized the whole proposal, listing Mattei as a co-investigator. Kouveliotou proposed a project to fund the purchase of three AAVSO CCD cameras to monitor gamma-ray bursters (GRBs) and to search for optical afterglows in a network to be called the AAVSO Joint Afterglow Network (JANET). JANET would expand the AAVSO High-Energy Network (HEN) from five sites to eight worldwide, providing dedicated hardware at headquarters for HEN activities and enabling a graduate student to work with the network and HEN data at headquarters. NASA funded the project and a later extension of the grant, all of which was formally administered by Waagen. JANET/HEN instruments are sited with experienced photometrists in Australia, Belgium, Finland, Hungary, India, New Zealand, and the United States, where they are used to follow up on GRBs and other high-energy events.

In terms of the routine, Waagen fielded requests for data and handled alerts in the, by now, well-established routine. An example of the vital contribution the AAVSO observers made was the request from Knox Long of the Space Telescope Science Institute and colleagues for coverage of VW Hyi in connection with their observations by the NASA satellite, Far Ultraviolet Spectroscopic Explorer (FUSE), to study the cooling of the white dwarf component following outburst and during quiescence. The FUSE observations had to be timed to monitor a VW Hyi outburst as it faded and through the subsequent quiescence until the next outburst started. However, the outburst following the quiescence could not be a superoutburst, which could damage the satellite should the variable become too bright. Thus AAVSO observers needed to monitor VW Hyi through a superoutburst and quiescence and then notify the astronomers as soon as the next, regular outburst started. Thanks to the AAVSO observers' diligence, Long was able to trigger the FUSE observations, which were carried out successfully. AAVSO data obtained before, during, and after the FUSE observations were used in the analysis of the satellite data. Waagen also co-authored two major papers, representing AAVSO observers and interpreting AAVSO data.[5]

THE SEARCH COMMITTEE

The Executive Committee quickly designated Vice President Marvel as chair of a Search Committee and included former presidents Gary Walker, Tom Williams, and Lee Anne Willson. In a later discussion with the whole Council on April 24, 2003, it was suggested that a member observer who had never served on the Council might provide a helpful leavening for the Search Committee's efforts. Jim Bedient of Honolulu, Hawaii, agreed to serve in that capacity.

At the same Council meeting, the history of salary administration for the AAVSO Director was presented with a view toward firming up the salary that should be offered to attract the right candidates for the position. Previous surveys, by then more than a decade out of date, had focused on positions that had either disappeared or changed radically; thus a more universal benchmark for the salary of the AAVSO Director was needed. After discussions with a number of qualified salary administrators, it was recommended that the Council adopt the Federal Civil Service grade GS15 as a benchmark salary for a very senior and well-qualified individual, in effect the ideal individual for the Association. The Council agreed and also accepted the recommendation that it use the grade GS14 for the first 4 years of a less experienced candidate, with a review in the fifth year.

Marvel was asked to arrange the Committee's activities in such a way as to present a rank-ordered list of three candidates for the Council's consideration at the 2004 Annual Meeting. President Dillon advised Marvel and the Council that he planned to sit in on the Search Committee proceedings as a nonvoting ex-officio member, and the search began. For only the second time in its more than 85-year history, the Council solicited applications for a new director. An advertisement was agreed upon and posted on the American Astronomical Society's Job Register, in the American Institute of Physics online job database, and in the *Chronicle of Higher Education* and was sent to all individual AAVSO members. Any astronomer who wished to be considered submitted not only a letter of application and curriculum vitae, but also a statement of the candidate's vision for the future of the AAVSO.

Marvel visited headquarters frequently during the interim. The staff regarded this as a very positive thing; their morale received a boost from that extra consideration in stressful circumstances. The staff wanted to provide their views of what was needed directly to the Search Committee, not filtered through the Council. Marvel conducted group meetings with the staff; then he held individual interviews to get a detailed view of each person's thoughts. After the group meeting and the one-on-ones, he held a feedback session with the whole staff. The process, from the staff's perspective, could not have been better.

A total of 13 applications were received by the Committee. Because the salary opportunity was attractive, the advertisement attracted a wide variety of candidates. Applicants included professional astronomers in all walks of scientific and academic life. Although there were a few less qualified candidates in the group, the Committee felt the general pool of submissions reflected a high level of respect for the Association and a generally favorable view of the potential for the position. USNO astronomer Arne Henden, a Council member who also wanted to be considered, recused himself from the further deliberations and submitted an application.

After discussion of the relative merits of each application, the Search Committee selected eight applicants for face-to-face interviews. Over a 2-day period in late August 2004, each of the seven applicants who agreed to interviews were allocated an hour and a half with the committee to both make a presentation and to answer questions. The meetings were scheduled with a minimum of 30 minutes between interviews to provide time for Committee discussion and to avoid embarrassing intersections of the candidates. Before the end of the second day, the Search Committee had narrowed the list to what it considered the four best candidates from the group for visits to the headquarters.

Each finalist spent a morning at the headquarters, toured the facility, made a presentation to the headquarters staff as a group, and spent time in discussions with individual employees. Marvel was present in headquarters for each visit, and spent the afternoon interviewing individual staff members, gathering their impressions of each candidate. After all four candidates had visited, the staff ranked them in the order of their acceptability. Marvel's summary of the entire process indicates that the staff participated in this process in a very realistic and grounded manner. Their comments generally supported, as well as supplemented, the views of the Committee members who had, by the nature of the process involved, even less time with each individual candidate than the AAVSO staff had collectively.

Marvel summarized the staff appraisals after the four candidate visits to headquarters for the rest of the Search Committee, and a final rank order of the candidates was agreed upon. Council members were provided copies of the four finalist candidates' applications together with the rank-order list for consideration at the regular Council meeting in October 2004. After discussion in that Council meeting, President Dillon conducted negotiations with the candidates. Henden emerged from the process as the candidate of choice for all involved. A short negotiation with AAVSO Secretary Walker finalized an employment contract on January 25, 2005, and the AAVSO had its new director. Henden reported to work at the headquarters on March 1, 2005, the fifth incumbent in that position – Olcott, Campbell, Mayall, Mattei, and Henden – one continuum of executive leadership. Henden and his wife Linda completed their move from Flagstaff to Cambridge after the 2005 Spring Meeting in Las Cruces, New Mexico.

One curious aspect of all this could be noted at this point. During this extended process as leader of the AAVSO Search Committee, Marvel had been a candidate for a new job, the position of Executive Director of the American Astronomical Society, a position occupied for a number of years by his boss, Robert Milkey. When notified by the AAS President that he had been chosen for the position, Marvel resigned as first vice president of the AAVSO, but first agreed to complete his assignment as the Search Committee chairman. David Williams moved up from second to first vice president, and in an impromptu election, Council member Chuck Pullen replaced Williams as second vice president.

A NEW REALITY

The 2005 Spring Meeting at New Mexico State University (NMSU) in Las Cruces, New Mexico, included the Third High-Energy Astrophysics Workshop for Amateur Astronomers. Many of the basic details of the meeting (but not for the workshop) had been worked out by Mattei and Rebbecca Turner before Mattei fell ill. Having been a graduate student at NMSU, AAVSO Staff Astronomer Matthew Templeton was able to provide assistance for the meeting with his familiarity of the region and the University. According to later musings by Waagen, for all these meetings, the Council did what it usually does: it came, met, accepted, rejected, decided, and then departed. Headquarters really ran the show, pretty much as normal.

The routine of the director's office was maintained, but many developmental activities were put on hold to await the new director's preferences. Also, it was Waagen's deliberate decision not to implement some of her ideas for new programs or procedures, again in deference to the preferences of the incoming director. As an interim director, Waagen was exemplary in maintaining current data processing operations, handling crises and alerts, keeping publications mainly on schedule, managing the Association's business with the help of the Council, and, to the greatest extent possible, keeping business as usual.

HONORING A LONG-TIME ALLY

With the negotiations to employ the new AAVSO Director in progress and everything else about the Association in apparent good order, there arose an activity in which the AAVSO should clearly be represented by a senior individual, but for which purposes the yet-to-be-named AAVSO Director would not be available. The Royal Astronomical Society of New Zealand (RASNZ) announced that it would be honoring its Variable Star Section's founder and only director, Frank M. Bateson, in December 2004. The event celebrated his 80th year of involvement in variable star astronomy. Bateson had emerged as a strong supporter and friend of the many AAVSO leaders with whom he came in contact and had been the one individual outside the AAVSO who first suggested that the AAVSO become the central repository for all variable star observations. For that reason, and in recognition of his contributions to variable star astronomy for the southern hemisphere, the AAVSO Council had singled out Bateson as an Honorary Member at the AAVSO's 75th Anniversary celebration.

Waagen was asked to represent the AAVSO at the Bateson celebration and to convey the Association's high opinion of its honoree as well as its thanks for the many years of valued cooperation and support.[6] At the meeting in Tauranga, New Zealand, Waagen spoke about the relationship between the AAVSO and the RASNZ-VSS since 1927, when Bateson contacted Leon Campbell to announce that he was starting a Variable Star Section in the RASNZ and to ask for finder charts and suggestions of additional stars for their program, and how that relationship matured and prospered through the Mayall and Mattei eras. Waagen also discussed how the RASNZ-VSS observers have contributed to variable star astronomy and the work of the AAVSO.

After 18 difficult months, Waagen and the AAVSO staff looked forward to a new era, expecting changes to come with a new director, but no one could imagine then how much change or how rapidly change might occur.

22 · Accelerating the science – the Henden era begins

Everyone who has contributed data, made a monetary donation, volunteered their time and energy, has made this organization the success that it is. We "stand on the shoulders of giants" – that came before us and built the foundation. . . . Without all of these people, the AAVSO would not exist.

– Arne A. Henden, 2007[1]

From September 2003 to March 2005, the AAVSO was without a director but was not rudderless as a result of the services of Elizabeth Waagen, who served as Janet Mattei's trusted senior technical assistant for many years. With careful attention from the AAVSO Council, Waagen maintained the business while the Council selected the next director for the Association.

In the first years after the appointment of Arne A. Henden as director, the Association displayed an outburst of activity in robotic observing, modernization of its database of observations, improvement of tools for observers, and general Internet presence. The variable star science community and amateur participation in the science benefited worldwide from these improvements. This chapter summarizes a few of the key developments in the AAVSO over this period. However, evaluation of the dramatic evolution that has taken place since Henden assumed the role of AAVSO Director will, of necessity, require historical perspective that at this early date would be impossible.

ARNE A. HENDEN

On March 1, 2005, Arne Anthon Henden assumed the role of AAVSO Director. He replaced Janet Akyüz Mattei after an 18-month period after her fatal illness and the time necessary for the search, candidate interviewing, and negotiations to select and employ a new director. At age 55, Henden assumed the AAVSO leadership as not only the oldest and most experienced, but also the most professionally qualified individual to assume the role.

A native of Huron, South Dakota, Arne A. Henden grew up in a family that moved frequently with his father's civil engineering career. In addition to a stint in Tehran, Iran, the family lived in a number of locations around the United States, primarily the western part. Henden received undergraduate and graduate training at the University of New Mexico, followed by a Ph.D. in astronomy at Indiana University in 1985. During his graduate school years, but after he completed his coursework, Henden's wife, Linda, was offered a significant job enhancement and training in computer languages in a position located in Washington, DC. They moved to accommodate her work opportunity; Henden took employment as a project manager supervising a small group at a systems engineering firm. He worked full time in that position while completing his doctoral research and writing a dissertation. His employment in software and equipment development during this period helped prepare Henden for his career in various aspects of astronomical instrument systems development and astronomical photometry.

After receiving his doctoral degree from Indiana, Henden accepted a post as a research astronomer at The Ohio State University, where he worked for 8 years. His work in Columbus involved development of instrumentation for their 1.8-m telescope in Flagstaff, Arizona, and participating in the initial stages of the Large Binocular Telescope. Henden then moved to the US Naval Observatory (USNO) in Flagstaff, Arizona, in 1993. At Flagstaff Henden engaged in astronomical photometry and worked on instrument development as well. He remained at USNO Flagstaff until he accepted the position as the AAVSO's Director in 2005.

Early in his graduate studies, Henden co-authored a well-respected book, *Astronomical Photometry*, with Ronald Kaitchuck that has remained in print for a

Figure 22.1. Arne A. Henden, AAVSO Director since 2005.

Figure 22.2. AAVSO Director Henden with former AAVSO President Al Holm at the 2006 Annual Meeting.

number of years.[2] In 1997, Henden began assisting AAVSO Director Janet Mattei by performing the photometry necessary to standardize comparison stars for AAVSO variable star observing charts. Henden's understanding of the organization increased with his presentation of workshops on CCD observing early in the AAVSO's evolution of a CCD observing program. He was elected to the AAVSO Council in 2000 and, at the time of Mattei's death, had been re-elected to a third term.

In his first year as AAVSO Director, Henden solidified and accelerated a number of existing program development projects to upgrade the AAVSO's data handling. The changes completed in this period included a sweeping conversion of the AAVSO International Database (AID) to a more powerful relational database structure using the MySQL program. This new, more efficient format provided many benefits, including faster access and more efficient storage. Having completed that transition, the entire headquarters computer network was upgraded to more modern computers. These changes vastly increased user access to data and charts for variable star observing. From a member's perspective, new web-based tools greatly simplified the planning for an observing program and submitting observations to the database. Progress in other areas was equally impressive.

A NEW HOME FOR THE AAVSO

Almost immediately after assuming his role as the AAVSO Director, Henden assessed the physical assets at 25 Birch Street. Quite a few of the problems evident in his assessment required further investments of one sort or another in the building that the AAVSO had owned for 20 years. Concerns included the fact that the basement of the split-level Birch Street building leaked badly during any serious storm. The basement housed, among other things, the valuable AAVSO archives; the potential for flooding created a hostile environment for the archives. Throughout the building, the carpets were worn out, and windows were neither secure nor energy efficient.

As he began to consider the condition of the Birch Street headquarters, Henden evaluated several solutions. Rehabilitation of the physical plant would require moving headquarters to a temporary facility while contractors completed the necessary repairs and remodeling, which would be expensive and disruptive. However, under the Cambridge city building codes, the maximum expenditure that could be made remodeling a building without triggering modern building codes was well below the necessary remodeling costs for the Birch Street building. Neither the electrical nor the plumbing services for the building would meet the modern code. Thus the AAVSO would require a fully-permitted reconstruction, adding to both the costs and the time delays. The split-level building would be especially

Figure 22.3. AAVSO Annual Meeting, 2006.

difficult and expensive to bring into compliance with the Americans with Disabilities Act (ADA), in addition to meeting the Cambridge city building code. Another consideration, one that could not be easily remedied, involved the limited parking space available at 25 Birch Street. Only six spaces were available, and those had to be leased from the former owner of the property.

Henden then considered the option to move out of state, thinking about relocating to one of the astronomical centers that had recently emerged in Arizona (e.g., Flagstaff or Tucson) or in New Mexico near Las Cruces. His findings were not comforting. Although the cost of living, and therefore operating, in either state was lower, the costs of building effective floor space in dollars per square foot were essentially equal to those in Massachusetts. Having eliminated the idea of moving out of state, Henden began the search for an alternative building in the Cambridge area.

In an unrelated but opportune development, Sky Publishing had decided to move to another facility and dispose of the three-building complex it owned at 49–50–51 Bay State Road. When Henden mentioned that the AAVSO might also want to move, Sky Publishing offered to lease AAVSO a floor in any of the buildings. Before this might have happened, Sky Publishing was purchased by New Track Media, and its three buildings were put up for sale. After long discussions on remodeling Birch Street or moving to a new building, the AAVSO Council purchased Sky Publishing's main building at 49 Bay State Road for $1 100 000. This represented a bargain price in the Cambridge real estate market but may have been indicative of several factors – Sky Publishing wanted the AAVSO to become the new owner, and New Track Media wanted to free itself from property management as quickly as possible. The Birch Street building was placed on the market and sold for $469 000. This was also a bargain price, but the Council wanted to free the director from the responsibility of maintaining and selling the old building, a distraction that might have lasted for months or even years. When both of the transactions were approved by the AAVSO Council, President David B. Williams signed the necessary documents. By April 2007, the AAVSO owned a much larger and more secure building for its new headquarters, just around the corner from the old facility on Birch Street.[3]

The new building was not without its problems, either, so many modifications were necessary to make it more comfortable for the headquarters staff. Essentially a three-part building, the facility included a two-story residential space, two stories of office space, and a one-story wing for a warehouse and shipping facility for Sky

Figure 22.4. AAVSO Headquarters, 49 Bay State Road, Cambridge, Massachusetts.

Publishing. Replacement of the windows with energy-efficient modern windows was a high priority. Also, the Hendens suggested, and after due consideration the Council agreed, that the original two-story residential building serve as a residence for the AAVSO Director. The modification of the residence also included a private guest apartment for visiting members and professional astronomers.

The building had been wired for the Internet, but additional upgrades on the wiring and air conditioning were required to meet the needs of the AAVSO's more extensive systems. The walls and woodwork all needed painting, and carpeting needed to be replaced, along with other revisions and upgrades throughout the building – all adding to the project's expenses.

As the case when the Birch Street building was purchased, several AAVSO members and staff volunteered on nights and weekends with painting and other work. The Hendens could be found in paint clothes on many weekends as they finished the residence beyond the contractor's structural work.

In total, another \$316 000 investment was required, in spite of all the volunteer efforts, to finish the building in an acceptable manner. Thus the total building cost was about \$1 416 000, including closing costs, remodeling, and permitting fees, and was only partly offset by the sale of 25 Birch Street, which covered about a third of that total cost. Rather than be saddled with a mortgage, the Council used the Endowment Fund for the actual building purchase. Funds for the remodeling were provided from other sources, including \$100 000 from a bequest from David Rosebrugh, \$92 000 from the Hendens, and \$37 000 from a bequest from Walter Feibelman. The residence is known as the David Rosebrugh Residence, whereas the guest apartment is the Feibelman Guest Suite. In return for these expenditures, of course, the AAVSO has added an asset of equal or greater value to the corporate capital accounts. In total, it is an addition in which all AAVSO members can take pride.

The AAVSO staff now has considerably larger and more comfortable working spaces in the new building, with the archives physically secure and actual parking space available and at no cost. Office space is available for visiting researchers as well. The former warehouse area is large enough to accommodate the annual meetings of the AAVSO or a moderate-sized technical symposium when appropriate. A generous bequest by Dorrit Hoffleit has been dedicated to refurbishing this space as a formal conference center. The first meeting held in the new building was the 2007 Annual Meeting, for which a Friday evening buffet dinner and informal gathering offered a pleasant introduction to the new facility for many members.

WRAPPING UP SION

One of the pieces of work left hanging by Janet Mattei's untimely death was the proceedings volume for the Sion–St. Luc meeting (see Chapter 19). The original plan was a good one: collect the papers presented in the meeting from the authors, edit them into a uniform format, and publish a proceedings volume for the meeting as quickly as possible. In contrast to previous AAVSO meetings, the breadth of the scientific presentations at Sion made this procedure particularly appropriate. A proceedings volume would be valuable to the astronomical community and in the process would add credibility for the AAVSO.

Unfortunately, the plan to publish the Sion meeting proceedings as a Conference Series volume by the Astronomical Society of the Pacific (ASP) went astray on Janet Mattei's desk. Mattei delayed the submittal well beyond the usual ASP requirements for a timely publication based on the dates of the meeting. At the time of Mattei's death, the book, in LaTeX coding and ready for publication, had still not received her final-final approval. Mattei's perfectionist outlook had, in the final analysis, proved the undoing of this good idea.

Confronted with this problem, Henden wisely decided that further delays would serve no purpose. ASP refused to accept the material as it was no longer timely. Under the direction of staff member Kerriann Malatesta, the authors were given an opportunity to update their articles, which many did. Elizabeth Waagen and Michael Saladyga prepared the articles for publication as an issue of *JAAVSO*. The updated Sion proceedings appeared nearly a decade after the meeting as *JAAVSO* volume 35, number 1 (2006).

TWO ADDITIONAL LOSSES

The holiday season was marred on December 23, 2006, by the death of former AAVSO President, Secretary, long-term Council member, and Mattei confidant Martha Hazen (1931–2006). Hazen's services to the AAVSO covered many years as the Curator of the astronomical plate collection at the Harvard-Smithsonian Center for Astrophysics, assisting Mattei and many AAVSO members with their investigations using historical plates to study variable stars. Hazen retired from the AAVSO Secretary's position after 13 years of service and many more on the Council. Her role as Mattei's confidant made her influence on the Association most significant. Also stricken with acute myelogenous leukemia and having witnessed Mattei's extended struggle to overcome that blood disease, Hazen rejected treatment and allowed the disease to run its course. Less than a year later, long-time AAVSO supporter Dorrit Hoffleit (1907–2007) passed away in New Haven, Connecticut, at the age of 100 years. Much has already been said in these pages about Hoffleit and her contributions to astronomy, and especially to the AAVSO.[4]

PROFESSIONAL–AMATEUR COLLABORATION

Mattei had agreed, on behalf of the Association, to support professional variable star observers using orbiting satellite observatories in the late 1970s (see Chapter 18). After that, professional–amateur collaborations increased, with AAVSO observers playing an increasingly important role in variable star astronomy. In addition to their contributions to the AAVSO International Database (AID), they actively participated in major research programs on invitation. For example, Bradley Schaefer of Louisiana State University invited AAVSO observers and others to participate in a 2-year (2008–2010) campaign in 2007 to monitor the recurrent nova U Scorpii. Schaeffer predicted U Scorpii would brighten sometime in that period after 10 years in a quiescent state. The campaign was approved by Henden and organized by Templeton, who issued frequent updates, starting with the campaign announcement on January 22, 2008. Seasonal activity summaries and a Variable Star of the Season article on the AAVSO website kept observer interest at a high level. Two years after Alert Notice 367 initiated this program, Schaeffer's prediction was fulfilled. Observing under difficult predawn conditions on January 28, 2010, AAVSO observers Barbara Harris and Shawn Dvorak discovered the predicted outburst. Their timely communications with Schaefer and AAVSO Headquarters alerted the international network of professional observers, who immediately redirected telescopes to U Sco. Schaefer and his colleagues had waited anxiously after Scorpio had disappeared into the sunset a few months earlier. They could breathe easier knowing the eruption had not occurred while the star was lost in the Sun's glare. The U Sco campaign, for which a number of amateur observers around the world contributed important observations, demonstrated once again both the utility and the spirit of professional–amateur collaboration at its best.

THE NEW AGE OF VARIABLE STAR ASTRONOMY

Automation and robotic control of telescope mounts had been increasingly common for professional telescopes for several decades; by 2005 it was an affordable reality for smaller telescopes. Sophisticated hardware, for example, digitally controlled stepping motors that moved telescopes on both axes, could be coupled to inexpensive digital computers, putting telescopes under remote pointing control. At the same time, CCD detectors produced the necessary images from which stellar photometry could be extracted. An assortment of software programs pointed the telescope; acquired a target field of view; took a series of CCD exposures suitable for photometry; changed filters for each exposure, if necessary; and transferred all the acquired information to the computer for further processing. Scientifically valid photometry could now be reduced through programs commonly known as a data pipeline. The raw observations can be massaged to correct for sky brightness, filter transmission, and the air mass through which the observation is executed. David Skillman predicted all this in 1979 (Chapter 17), and although it took a great deal longer than he anticipated, it did happen.

Related to this rapid progress in robotic telescopes and CCD photometry, one of Henden's more ambitious plans was to establish a fleet of robotic telescopes around the world that would be available for all AAVSO members. The telescopes in what became known as the AAVSOnet would be ideally dispersed by longitude and latitude for maximum flexibility and would avoid unfavorable weather in any one locale to the extent possible. In Henden's concept, an AAVSO member would submit an observing program of interest. After selection of the program by a Time Allocation Committee (TAC), the TAC would select a telescope with appropriate capability and place a request in the automated waiting list for scheduling on that telescope. At the scheduled time, the telescope is programmed to make the observation, process the results, and report the whole package to the "observer." In 2010, the photometric reduction program Photometrica was made available on the AAVSO website so that observers no longer needed either hardware or software to perform CCD photometry. The concept is impressive, and the execution even more so.

Using a mixture of new, decommissioned, or underused telescopes, the AAVSOnet project represents efficient use of resources otherwise unused. This new approach was possible because of advances in robotic technology, surely, but most of all to advances in communication made possible by the Internet. To set up a network, it is necessary to establish the availability of an appropriate telescope at a location where the conditions exist for effective use, including but not limited to dark skies, favorable climate, availability of appropriate utilities and individuals who will maintain the operation, and access to the Internet. If these conditions can be met, then the next step is to fund the changes, including the relocation, installation, automation, and operation of the telescope in the new robotic locations. These steps were accomplished with great success by AAVSO Director Henden for the AAVSOnet. Assets in the AAVSOnet in 2010 included the following telescopes:

1. A Celestron 14-inch telescope, owned and operated by John Gross at the Sonoita Research Observatory, 60 miles southeast of Tucson, Arizona; the first telescope in the AAVSOnet project, it was the proof-of-concept instrument and was used to obtain thousands of observations for the AAVSO comparison star sequences and program stars in the first few years of use. The AAVSO is allocated one-third of the usable telescope time.
2. Two automated telescopes with apertures of 11 inches and 12 inches, operated and maintained by AAVSO member Tom Krajci at Cloudcroft, New Mexico, and donated to the AAVSO by Paul Wright.
3. The Morgan 24-inch Tinsley telescope, donated to AAVSO by Lowell Observatory in Flagstaff, Arizona, maintained and operated by Tom Smith at his Dark Ridge Observatory near Weed, New Mexico.
4. A 24-inch Optical Craftsman telescope at the Mt. John University Observatory on New Zealand's South Island, a joint venture between the AAVSO and the Physics and Astronomy Department, University of Canterbury, Christchurch, New Zealand; it is one of four telescopes maintained by the university. The AAVSO refurbished and automated the telescope in return for two-thirds of the observing time. The South West Research Institute (SwRI) in Boulder, Colorado, paid for the refurbishment in exchange for one-third of the observing time, leaving the other one-third for AAVSO observers.

Professional astronomer Dirk Terrell of SwRI, an active participant in AAVSO eclipsing binary programs, manages the program for SwRI.

5. A 12-inch Meade telescope to be installed in the Cohen-Menke Observatory on the roof of AAVSO Headquarters in Cambridge, Massachusetts, donated by AAVSO members Lou Cohen and John Menke.
6. A 24-inch Boller & Chivens telescope, owned and operated by New Mexico State University at the Tortugas Mountain Observatory, Las Cruces, New Mexico. Funding for the automation of this telescope was provided through the AAVSO by private individuals, and the AAVSO will receive 75 percent of the available observing time.
7. A 60-mm Takahashi refractor in the Astrokolkhoz Observatory, Cloudcroft, New Mexico, known as the Bright Star Monitor, wholly owned by the AAVSO and operated by Tom Krajci. This telescope is dedicated to monitoring stars from second to eighth magnitude.

The larger grants for the AAVSOnet robotic telescope network included $40 000 from the Mount Cuba Astronomical Foundation, $25 000 from SwRI, $25 000 from SBIG (formerly The Santa Barbara Instrument Group), and $10 000 from Michael Kran.

Another robotic telescope project, but with a different mission, is identified as the AAVSO Photometric All Sky Survey (APASS). APASS consists of dual 8-inch ASA astrographic telescopes mounted together on a single mounting. Each telescope is equipped with a set of filters in various band passes, three on one telescope and two on the other. The APASS was placed in service in late 2009 at the Dark Ridge Observatory, Weed, New Mexico, to conduct an all-sky photometric survey of the northern hemisphere. A second system is being purchased and will operate at the Cerro Tololo International Observatory (CTIO) in Chile. After installation and testing, it will be used to compile a similar survey of the southern hemisphere. The entire APASS project has been funded through the Robert Martin Ayers Sciences Fund.

When the southern hemisphere exposures are completed, the APASS project will assemble and publish photometry in five accurately determined bands for every star in the sky down to a magnitude of $V = 16.0$.

Figure 22.5. Long-time member Tom Cragg with Jaime Garcia at the joint AAVSO–Society for Astronomical Sciences meeting held at Big Bear, California, May 2009.

As a reference catalog for future photometrists and chart makers, APASS will constitute a significant contribution to astronomy.

Henden traveled extensively in his first years as the AAVSO Director and enhanced AAVSO relations with other variable star organizations and independent observers around the world. His travels helped him organize the AAVSOnet and knit together the international variable star community. A fourth international meeting, this time organized by AAVSO President Jaime Garcia in Valle Grande, Mendoza, Argentina, in April 2010, allowed South American members and observers to experience the AAVSO firsthand rather than over the Internet.

MODERNIZING DATA MANAGEMENT

The reorganization of the AAVSO's data management systems that Aaron Price started in 1999 made substantial progress by 2005, but systems were still burdened by some basic problems. The AAVSO systems operated with legacy programs, which represented a veritable history of programming languages used in small computing systems. Many programs were based on Microsoft DOS and Microsoft Windows, which were, by modern standards, slow and inefficient. In the early 1990s, "open source" software came into widespread use within the astronomical community, and new programming languages added considerably to the ease of developing pages for data presentation on the Internet. And finally, converting the text-based file structure of the AID into a more modern relational database provided both faster access and more efficient storage of data for such a very large database. Price began converting all the headquarters computing systems to the newer open source languages and operating systems during the Mattei era. However, Henden placed a higher priority on that work, and by 2007, all the basic conversion work was completed. The Internet made it possible for the AAVSO to undertake several additional projects that Henden identified as collaborative computing.

MODERNIZATION OF DATA SYSTEMS

The modernization of all headquarters variable star data systems was revolutionary and will thus be described in greater detail than might otherwise be considered appropriate. These changes involved the evolution of the AAVSO comparison star chart system, the creation of a large database to catalog and make available information online for all known and suspected variable stars, and the addition of large collections of variable star observations that were previously inaccessible to the AID. Each of these changes is discussed in the following pages.

There have always been three parts to the problem of standardizing the charts necessary for variable star observations: (1) determination of the brightness of a sequence of stars to be used for making comparisons; (2) production of the charts, that is drafting, editing, and reproducing the charts; and (3) distribution of the charts to observers. These three tasks are further complicated by the fact that separate variable star associations around the world prepared their own charts, tailored to each group's preferred method of observation and reporting. Many associations had their own methodology for each step in this process, had trained their observers to use that methodology, and therefore resisted making changes. In spite of the best intentions of all involved in their separate evolution, different sequences existed for the same variable stars being observed by different organizations. Some of these sequences were close enough that the resulting light curves were essentially the same. In other cases though, in about 20% to 25% of the LPVs, for example, the sequences differed enough that observations could not easily be consolidated without introducing unacceptable scatter in the actual light curve.

Furthermore, many of the LPV early sequences had been developed at Leander McCormick Observatory by methods that rendered the magnitude scale for stars fainter than 12th magnitude increasingly nonlinear. Thus the light curves for many LPVs had undefined or flat minimums that were obviously in error. Quite apart from the differences between associations, this need to extend the sequences for fainter stars became apparent as more and fainter stars were added to Association programs.

Changes in the AAVSO's method of producing comparison star charts illustrate the evolution of technology as well as Association processes over this period. Traditionally, charts had been drafted by hand in a format that evolved slowly over the decades.[5] Beginning in the 1970s, the process had been modernized by Charles Scovil and Bob Leitner at the Stamford Observatory, who were later assisted by Marc Biesmans of Belgium. Working with digitized sky surveys to populate the variable star fields, they used those files as the basis for the charts, but the comparison star sequences were assigned manually by the person making the chart. When they completed a chart, it was not quite ready for publication. Headquarters had to approve each chart, and that process always involved a final approval by Janet Mattei. Hence new charts were approved at the rate of four or five per month. After receiving the director's approval, the charts were made available on the AAVSO website to be printed by the individual observers on demand, which solved the third problem, the task of distribution.

The first of the three comparison star chart problems, the selection of appropriate sequences of comparison stars, caused the most problems for the professional users of variable star light curves. Poor sequences and discrepant sequences in use by different associations created excessive scatter in the light curves for individual variable stars. At the suggestion of John Toone, BAA-VSS Chairman Roger Pickard proposed a cooperative effort to resolve the inter-organizational sequence conflicts to Janet Mattei during the AAVSO's first international meeting in Brussels in 1999. Pickard suggested that the AAVSO take the lead in the effort.[6] Mattei agreed and called on other variable star observing organizations to participate. The first meeting of the International Chart Working Group (ICWG) took place in

Huntsville, Alabama, the following spring in connection with the High Energy Astrophysics Workshop and AAVSO meeting. At that meeting, representatives of the AAVSO and BAA-VSS were joined only by representatives from the Vereniging Voor Sterrenkunde, Werkgroep Veranderlijke Sterren (VVS-WVS) of Belgium.

A broader constituency came together for the second meeting of the ICWG in the autumn of 2000 at Cambridge, Massachusetts. Associations represented in that meeting included not only the AAVSO, BAA-VSS, and VVS WVS, but also the Royal Astronomical Society of New Zealand (RASNZ) and independent observers from Belgium, Finland, and Argentina. Representatives from VSNET, another international organization based in Japan, joined the ICWG at a third meeting in connection with the Second High Energy Astrophysics Workshop and AAVSO meetings in Hawaii. At that third meeting, the ICWG adopted and prioritized criteria for the selection of sequence stars to limit the discrepancies between international sequences to less than 0.2 magnitude.

The most important agreement, the one that accounted for the most problematic differences between sequences, limited the range of colors for comparison stars in any given sequence. Other criteria dealt with the issues of using double stars or slightly variable stars as comparison stars and eliminating the use of multiple comparison stars of the same magnitude in one sequence, among other considerations. The new criteria constituted a breakthrough in international cooperation in variable star astronomy and were made possible by the widespread availability of high-precision multicolor photometry over the Internet.

Observers became impatient with the slow rate of progress in this process as well as AAVSO chart production. They were anxious to expand their observing programs as interest in the cataclysmic variable stars ramped up. Mike Simonsen of Imlay City, Michigan, made his own charts using the *Digital Sky Survey* with sequences prepared by Bruce Sumner from photometry originally generated by Henden at the USNO in Flagstaff. Simonsen made his charts available on his personal website; those charts found some acceptance from other observers. Simonsen then approached Mattei with a proposal to organize and lead a new chart committee of observers and members. The chart committee would replace the entire AAVSO chart catalog with charts produced with his all-digital technique, using standardized precision multicolor photometry and plotted to the AAVSO standards and make those available to everyone over the Internet. But Mattei was not ready to let work proceed on the basis that Simonsen proposed, since she was concerned with quality-control issues.

That is where things stood when the ICWG convened in Hawaii in July 2002 for its third meeting to resolve the problems involved in sequence selection. Mattei's principal concerns centered on the selection of the best available photometry for each sequence. After ICWG agreed on a set of ground rules that could be uniformly applied in the selection of sequence stars from available photometry, the way was cleared for an alternate approach to chart making. Still, she rejected Simonsen's proposal at the 2002 Annual Meeting to assemble a team of observers and members to accelerate the chart-making process.

By the 2003 Spring Meeting, Mattei finally accepted Simonsen's proposal. Simonsen assembled the Chart Team that summer and began preparing variable star charts for each of about 4000 variable stars in the AAVSO program. When Mattei's illness removed her from all activities, Henden stepped in to provide additional scientific as well as Council leadership for the effort.

Around the same time, George Hawkins, the first AAVSO postdoctoral intern, developed a semiautomated chart-plotting program in Interface Description Language (IDL). One of the early steps taken by Henden after he became Director was to replace this program with a more modern version written in a nonproprietary language. For that purpose he contracted with AAVSO observer Michael Koppelman's company, Clockwork Active Media, to write a better chart-printing program using open-source software. The resulting automated plotting program, Variable Star Plotter (VSP), called the necessary comparison star sequences from a separate comparison star database that was under development so that the charts would be automatically updated whenever a comparison star magnitude was updated without redrawing the entire chart. The user interface for the software was designed by Christopher Watson and programmed by Aaron Price.

The necessary table of comparison star magnitudes was created by a separate team led by Vance Petriew that first created a database called CompDB that contained basic information such as accurate coordinates and chart magnitudes for all of the comparison stars for every chart on the AAVSO program. CompDB was completed in the fall of 2007. Henden then matched the stars in CompDB with the best available standardized photometry and updated all comparison stars used by the AAVSO. The merged database, now called the Variable Sequence Database (VSD), provides accurate comparison star magnitudes to VSP.

The next piece of this puzzle, access to an accurate and up-to-date catalog of the data describing a variable star (position, classification, spectral class, color, magnitude range, etc.), represents the most revolutionary of all the changes in the AAVSO systems. Historically, the most authoritative information has been located in the *General Catalogue of Variable Stars* (*GCVS*), published in Moscow in book form on behalf of the International Astronomical Union (IAU). The fourth edition of the *GCVS* was published in four volumes beginning in 1985 and completed in 1990. Thereafter, an online catalog of the *GCVS* was posted to the Internet and maintained by the IAU office in Moscow. The online *GCVS* was updated periodically as the fully validated data were approved by those responsible – a consensus process that was, by its nature, a slow one. At the same time, the *GCVS* office was overwhelmed by the large number of new variable stars discovered after the advent of CCD imaging and application of that technique to astronomical research, especially in variable star photometry. Thus, even the online *GCVS*, last published in 2008 with 41 483 variable stars, was ill equipped to provide authoritative information rapidly on a large scale.

Concurrently, AAVSO observers were rapidly expanding their horizons and looking for newer and comparatively under-observed variable stars to add to their programs. The *GCVS* offered little help in this regard. Furthermore, variable star astronomy literature was littered with misidentified and misclassified stars. Far too frequently the same star was given several different names or classifications, or both, in the absence of authoritative information. Finally, there was no convenient way to report the discovery of a new or suspected variable star so other astronomers could contribute observations without confusing name problems.

For the AAVSO, these issues had arisen in both the AAVSO Council and in active e-mail discussions among interested variable star and asteroid observers. Responding to a suggestion by Council member Lew Cook, the Council formed an ad hoc group chaired by David Williams to consider the problems. The e-mail group included Price, Simonsen, Petriew, John Greaves, Brian Skiff (Lowell Observatory, in Flagstaff), Bill Gray, and Henden. After considerable discussion, Simonsen drew in Christopher Watson, a variable star observer who had worked on other AAVSO system problems, including VSP.

Figure 22.6. AAVSO President David B. Williams at the 2005 Annual Meeting.

Watson identified with the problem and within a short period conceptualized a framework for an online database of variable star information that could be updated frequently to reflect the latest information on any number of variable stars. He mounted such a system on a private Internet site and invited Henden to examine it. Their dialogue quickly drew in David Williams' ad hoc group. Although Henden agreed that Watson's system met their requirements, he insisted that the system had to be coded in open-source software. Watson agreed to convert his proprietary system to meet AAVSO's requirements. That file of variable star information, described by some as the "*GCVS* on steroids," went online to enthusiastic acceptance by AAVSO observers

Figure 22.7. The 2006 AAVSO chart team. From left: Richard Huziak, Jim Bedient, Roy Axelson, Mike Simonsen, Robert Stine, Chris Watson, Aaron Price, and Vance Petriew.

in 2006. Still not connected in any way to other systems, the new database, identified as the Variable Star indeX (VSX), seemed to fulfill most of the requirements.

The missing piece of the puzzle at that point was a way to identify each object in VSX, any objects added to VSX, and all variable star data in the AID so as to eliminate, to the maximum extent possible, the rampant confusion of any multiple identities. Petriew, who was developing a new database schema for comparison star photometry (VSD), championed a new object identification protocol that could be used throughout the AAVSO's systems and processes as a "primary key," identifying and linking individual objects unambiguously across all systems. That identifier, called the AAVSO Unique Identifier (AUID), was essential to completing the goal of linking the systems that existed in 2006 – the variable star observations (AID), the variable star catalog (VSX), and the comparison star database (VSD). Watson developed a new system to assign AUIDs automatically to every existing object in each of these three systems and for any new information, including variable star observations, to be added to any AAVSO system. A series of quality-control procedures added to the AUID system insured that any new information was vetted before being admitted to the system.

The process of merging the AUID information with the existing databases was initiated in 2008 and completed in 2009. In addition to Watson, the team that managed the process included Katherine Davis, Matthew Templeton, and Patrick Wils, with assistance from Price and Richard Kinne. This team was responsible for the assignment of AUIDs to every piece of data in any of the databases. Under the control of the AUID system, observations are added to the system, variable star and comparison star data are updated, and up-to-date comparsion star charts can be plotted with VSP.

It was a remarkable achievement. VSX contained 133 662 variable stars when the system was officially released for online access on May 10, 2006.[7] In announcing the public availability of VSX itself, Watson described VSX as:

> a comprehensive relational database of known and suspected variable stars gathered from a variety of respected published sources and made available through a powerful Web interface. It provides tools for visitors to search and view the data, registered users to revise and add to the data, and authorized moderators to vet the data, creating a consistently reliable "living" catalog of the most accurate and up to date information available on these objects.[8]

In practice, a variable star observer was free to call up the charts for each observing session on a computer monitor in the observatory, never relying on a printed copy. The user could also customize all aesthetic aspects of the chart so it would reflect what was seen in the eyepiece.

Since the merger to create VSD, it has been under constant review by Mike Simonsen's Sequence Team, correcting the residual sequence errors using problems reported by observers and cataloged online. The process of correcting and creating new sequences was greatly enhanced by a program called Seqplot created by Sara

Beck of the AAVSO staff. It displays color and magnitude information and uses magnitudes from several photometric catalogs including Tycho2 photometry and more than a million stars that have been calibrated using the robotic telescopes in the AAVSOnet, as well as earlier work done by Henden at the USNO.

All of these new facilities for variable star observing are available to the international community, with sequences established with the best available precision photometry and selected according to the criteria published by the ICWG. This accomplishment, with volunteers and AAVSO staff, open-source software, and publicly available photometry in a comparatively short period (less than 7 years), stands as a monumental achievement in collaborative computing.

PUBLIC OUTREACH AFTER HANDS-ON ASTROPHYSICS

The Hand-On Astrophysics (HOA) grant to the AAVSO from the National Science Foundation in 1993 totaled $384 000 over a 3-year period and had, to that point, been the largest grant ever received by the Association. The HOA project, when completed, achieved its ambitious goals of spreading astronomical education and awareness of variable stars and the AAVSO through secondary education teachers across the United States and beyond. When the last of the assembled HOA sets was sold or distributed by the AAVSO, the entire project was updated by its principal author Donna Young. Now called *Variable Star Astronomy* and residing on the AAVSO website, the project continues to be accessed frequently and is spreading in use by secondary school teachers as well as others who are curious about variable star astronomy. This education project could therefore be described as an unqualified success.

Aaron Price of the AAVSO staff conceived a new public outreach project to attract the public to the observation of an astronomical event taking place in real time that illustrated variable star astronomy. Doing so would place the participant in the role of a scientist performing research. As a graduate student in education at Tufts University, Price had researched the methodology for making such a study.

A natural forum existed for establishing such a project. The IAU designated the year 2009 as the International Year of Astronomy (IYA 2009). The American Astronomical Society (AAS) and the AAVSO were both participants in the United States IYA Program Committee for which Price served as the chair of its Working Group on Citizen Science. Rick Feinberg of *Sky & Telescope* suggested the forthcoming eclipse of the bright star ε Aurigae as the astronomical event to study. With a visual magnitude range of 3.0 to 3.7, the star is easily visible to the naked eye, even in fairly light-polluted skies, and undergoes an eclipse only once every 27 years. Each eclipse lasts about 18 months.[9] That project idea was further developed by the committee with Lucy Fortson of Adler Planetarium, Michael J. Raddick at Johns Hopkins University, Robert E. Stencel of the University of Denver, and Ryan Wyatt, Director of the Morrison Planetarium and Science Visualization at the California Academy of Science in San Francisco.

The AAVSO and the AAS embraced the Committee's proposal as a logical vehicle for their IYA 2009 efforts. With that endorsement, Price recruited co-investigators in early 2008 to help develop the project, manage the process, and supervise the analysis of the results. His co-investigators included Fortson, Raddick, and Stencel, with AAVSO Director Henden as Principal Investigator. The AAS Executive Director Kevin Marvel, as well as the IYA Committee chairs, gave their approval to the project.

Price wrote and submitted a proposal for NSF funding in June 2008, with a requested start in October 2008. The project was approved by the NSF on September 1, 2009, with a total approved funding over 3 years totaling $794 000. It was a new high for grants received by the AAVSO. The entire project was implemented by a team led by AAVSO's Rebecca Turner, assisted by Price and Brian Kloppenborg, a graduate student at the University of Denver. Throughout the project, Stencel at the University of Denver acted as scientific advisor to the project and reviewed each step of the activity for accuracy.

The Citizen Sky project first established an interactive website. That complex activity, led by AAVSO webmaster Katherine Davis, developed a useful tool through which interested parties could learn more about the project, register to participate in the project, submit observations of ε Aurigae, follow the news about the star and related activities, and participate in the scientific analysis of all the data submitted.[10] Turner narrated two videos for the website and YouTube, so those interested might become better informed about the project.

The project also created two press releases used by the National Geographic Society and *Wired Magazine*.

The AAVSO worked with Ryan Wyatt of the California Academy of Sciences to produce a professional planetarium show trailer about ε Aurigae and the Citizen Sky project. Narrated by the well-known author and astronomy popularizer Timothy Ferris, the trailer was distributed to planetariums around the world. In 2010 the planetarium show was entered in an NSF science visualization competition. As one measure of the project's success, the NSF chose Citizen Sky to be one of its annual "NSF Highlights," a list of exemplary NSF projects provided to Congress and used in various NSF promotional materials.

The first year of Citizen Sky was dedicated to building the infrastructure, promotion, and training observers. An initial workshop for those signed up for the project was held August 4 through 7, 2009, at the Adler Planetarium. Turner planned and handled the workshop with support from Adler Planetarium staff. The workshop attracted about 60 persons, including 50 participants in the project, and the project staff. The actual eclipse of ε Aurigae started on schedule in August 2009. Data documenting the decline in brightness and progress of the eclipse were displayed on the website as the results were reported.

As of August 1, 2010, the Citizen Sky project had more than 3300 registered participants; about three-quarters were male with a mean age of 41 years. One-fourth of the participants reported no prior experience in amateur astronomy. Typically well educated with a bachelors or higher degree, the participants represented 18 countries, with 42% of the participants outside the United States.[11] By the time ε Aurigae reached mid-eclipse in summer 2010, more than 3400 observations had been received, and the project proved itself an unqualified success in meeting the NSF goals.

SOCIAL NETWORKING AND THE AAVSO'S INTERNET PRESENCE

First Vice President Simonsen introduced another aspect of work on the Internet through social networking, a series of programs that emerged during the Internet bubble (frequently referred to as Web 2.0) in the early 2000s and developed an enormous audience among young Internet users. Simonsen considered both Facebook and Twitter presence on the Internet necessary for the Association to attract younger new members who might only get involved initially through such forms of communication. Simonsen created an AAVSO page on Facebook, added comments to the page periodically, and advertised it in his notes to members of the AAVSO, other variable star astronomers, and other correspondents. Those visiting the page could become a "fan" of the AAVSO and link to their own Facebook page. Those links show up not only on the AAVSO Fans List but on other pages linked to the Fans creating the social network. Such networks grow exponentially, and that proved to be the case with the AAVSO fans list, with well over 2000 fans. Simonsen also took advantage of Twitter, another social networking program. Although Facebook pages are complex, often with graphics, music files, and other items, Twitter is limited to short (128 characters maximum) text files known as Tweets. Twitter provides some convenience for real-time inter-observatory coordination of observing runs. Short text messages can be transmitted by speaking into a telephone that converts voice to text, leaving hands free to manipulate telescopes and record data. Thus it has become possible to exchange information effortlessly without approaching a keyboard.

THE COUNCIL AND ORGANIZATIONAL CONSIDERATIONS

The by-laws were once again revised and approved by the membership in 2009. Among the changes implemented was a slight shrinkage of the number of Council members and a provision for the Council to change classes of membership without requiring a change of the by-laws, thus obviating the need for a full membership vote on such a small issue. A number of other issues involving the organization were dealt with during this period, as follows.

Treasurer turnover and financial considerations

Over many years the AAVSO had enjoyed stability in the Council brought about by the long tenure in two positions, the treasurer and the secretary. In addition to being a long-term financial benefactor for the AAVSO, Clinton B. Ford served as secretary of the Association for 44 years. With comparatively long terms served by Olcott, Rosebrugh, and Hazen as well, the average tenure of AAVSO secretaries had been 16 years.

AAVSO Treasurer Ted Wales finally resigned after 19 years in the office. After Wales, however, the treasurer changed several times very quickly: former AAVSO President Wayne Lowder died unexpectedly less than 3 years after taking responsibility as the AAVSO Treasurer; then his replacement, Louis Cohen, resigned after 5 years because of health problems. David Hurdis replaced Cohen for 2 years but never became comfortable with the responsibility and so asked to be replaced; the Council then elected Gary Billings.

For most of his tenure as treasurer, Wales performed all the bookkeeping needed by the AAVSO. As the AAVSO matured as an organization with more resources and flexibility after Ford's endowment, the bookkeeping was finally turned over to a part-time employee with appropriate accounting skills, along with an external accountant who reconciled accounts monthly. With annual and independent audits, this was a reasonable and safe move for the Council, particularly in view of the fact that the individuals elected to the treasurer's position after Wales did not live in close proximity to AAVSO Headquarters. However, the bookkeeping is only one of the responsibilities of the treasurer. Management of the AAVSO's assets – that is, the endowment accounts, cash bank accounts, and payroll-related activities – constitutes other concerns. The series of limited-tenure treasurers served to highlight the necessity for attention to these other aspects of the treasurer's role in an important way.

When he took over the treasurer's position, Cohen recognized those other responsibilities immediately. His examination of the financial statements from the investment bank that had been managing the AAVSO's endowment funds revealed that those funds had been performing poorly. Cohen began an examination of the alternatives available to the AAVSO and eventually planned to move the funds to another investment broker. Mattei's unexpected illness and death resulted in a short delay in those plans. As the market rose over the next 2 years, the value of AAVSO investments swelled to more than $17 000 000. The growth was all in stocks. In the hands of more astute fund managers and with an appropriate stock/bond ratio, some of that growth might have resulted in significant gains in the AAVSO's long-term financial position. Instead, stocks were not sold to maintain a targeted 60/40 stock-to-bond ratio and capture some of the market-based asset appreciation in less volatile securities. When the market collapsed in 2008, the paper gains vanished into losses in the Endowment. (With surprisingly good luck, the new building acquisition in 2007 required a significant withdrawal from the funds, but at the peak of the market.) The whole problem demonstrated the necessity of the treasurer's attention to all aspects of the responsibilities delegated in the by-laws to the treasurer's position.

Organizational changes

Normally such a financial swing would have little effect on the AAVSO. In this particular period, however, the shrinkage in the market value of the stock portion of the portfolio occurred at the time when the AAVSO's expenses were peaking. Although many of the additional programs and activities were funded by grants, the stock market retreat left the Council with less than the necessary income from the investments to handle current expenses, even with the grants. As a consequence, the Council felt constrained in 2009 and 2010 to sharply limit expenses and reduce the AAVSO staff positions. At the same time it became apparent that the work of the organization had evolved significantly. An additional level of management was considered appropriate to divide and share the direct supervisory responsibilities and work of the director. Thus the AAVSO staff was reorganized into three major areas of activity: science, systems, and administration. Matthew Templeton became the scientific director with oversight of staff related to the AID and the research-oriented activities on variable stars, including *JAAVSO* and other technical publications. Aaron Price was named assistant director to manage staff and those activities involving the maintenance and development of headquarters computing systems and presence on the Internet, as well as the day-to-day activities in meeting planning, publicity, and other outward-looking activities. Director Henden will continue to supervise other activities essential to the internal functioning of AAVSO headquarters. As always, the director remains responsible to the Council for the Association's overall operation, including the annual financial audit, required state and federal reporting, and all other external affairs of the Association.

AAVSO sections

Over several years, the Council discussed other aspects of the AAVSO organization as well. The existing committee structure did not seem particularly well designed to meet current needs. Although there was a continuing interest in various forms of instrumental observation including PEP and CCD, the Council discontinued these technique-oriented committees. Instead, new individuals who wished to get started in these activities would be mentored within several new "sections" focused on individual astronomical topics.[12]

Under the new plan, the AAVSO's scientific work was divided into eight topical sections: Cataclysmic Variables, Data Mining, Eclipsing Variables, Long Period Variables, Supernova and Nova Search, Solar, Short Period Pulsating Variables, and the High Energy Network including Gamma Ray Bursters and other phenomena. Each topical section has both a leader and a scientific advisor. Together, they manage the AAVSO program for a specific class of objects. That responsibility includes a mentoring role for the specific observing techniques, whether visual, instrumental, photometric, or spectrographic. This new approach was intended to attract a larger number of professional astronomers into the direct work of the Association and therefore strengthen the AAVSO's support to ongoing science.

Also missing in this new section structure are specific functions, such as chart making. The automation of the chart-making function makes a section for this activity unnecessary. By the same token, other similar areas of activity not related to the mainstream of the AAVSO's activities will be handled by the Council on an ad hoc basis when any need develops in the future.

JAAVSO

Before *The Journal of the AAVSO* (*JAAVSO*) was initiated in 1972 (Chapter 16), each semiannual meeting was followed by *AAVSO Abstracts*, which contained the minutes, reports, and abstracts of papers presented at that meeting. When the semiannual *JAAVSO* replaced the abstracts, traditional thinking gave the journal a split personality. Each issue contained the reports and papers from a semiannual meeting, as well as submitted research papers. Because *JAAVSO* was considered as a record for each meeting, it was often delayed while missing reports and papers were solicited. Finally, when the question came up regarding whether to continue *JAAVSO* on paper or as an electronic-only publication, Henden realized that the two *JAAVSO* missions could be separated. *JAAVSO* could focus on research papers, whereas organizational matters such as minutes, the director's and committee reports, and similar items could be consolidated in a separate *Annual Report*. Research papers could be posted on the website as soon as they were accepted (the posting of such electronic "preprints" began in 2005), eliminating the delays in publication that resulted from the semiannual *JAAVSO*. Thus the decision to go digital would not only save steadily increasing postage and printing costs, but also speed up the publication of research papers, making the *JAAVSO* more attractive to potential scientific contributors. Beginning in January 2009, *JAAVSO* was published only in electronic form, downloadable from the AAVSO website. Printed and bound *JAAVSO* copies were made available "on demand" from a separate online vendor.

A FINAL LOOK AT THE OBSERVATIONS

Through this first century of its existence, the primary mission of the AAVSO has been to observe, record, report, archive, and make observations of variable stars available to the scientific community. For the first 40 years or so when the Association was closely controlled by Harvard College Observatory, that mission was limited mainly to the LPVs. In the last 60 years, AAVSO interest has broadened to include all types of variable stars within the means of the Association's observers. Over time those observers have become more sophisticated instrumentally speaking, but the visual observations of a century still form the backbone of the AAVSO's credibility and ability to serve the variable star astronomy community. Therefore, in closing this history, it seems appropriate to reflect on the total accumulated observations now included in the AID, the one statistic above all others that stands as a measure of the success of the venture founded by William Tyler Olcott in 1911.

There is, of course, an obvious contradiction built into any discussion of the contents of the current AID in the context of the centennial history of the AAVSO. The total observations in the database are not solely those of the AAVSO and its observers. Instead, observations included in the AID were provided by members of many other organizations as well as the AAVSO. Many individual observers from outside North America have reported their observations to the AAVSO without formally joining the Association. Many other observations have been archived through agreements with national variable star associations.

The AID's value as a scientific database is directly proportional to its coverage in terms of both its spatial (both celestial hemispheres observed from all longitudes) and chronological completeness. The AID is therefore fully acknowledged as representing the efforts of the international community of variable star observers. The AAVSO is honored by the trust expressed in its enduring stewardship of the world's variable star data by the various individuals and associations that have generously contributed their observations to the AID.

It is also necessary to recognize that over time, but especially in the last decade, more effort has been devoted to pulling as many historical collections of observations of variable stars into the AID as possible, and that effort is expected to continue. For example, the AID has acquired the invaluable collection of observations of southern hemisphere variable stars by Alexander William Roberts of Lovedale, South Africa, who observed from about 1890 to 1916. The Roberts observations were reduced by Tim Cooper and his father, the late Dennis Cooper, in South Africa during the period from 2003 to 2007. From their effort, a total of 70 053 observations by Roberts were added to the AID, with support from a small research grant provided to the AAVSO by the American Astronomical Society.

Henden also arranged for the variable star observations of Olin J. Eggen made during his long professional career at a number of observatories and each recorded on a 3 × 5–inch index card. The cards were scanned and posted on the AAVSO website and have gradually been reduced by volunteer efforts as a "cloudy night" project for eventual inclusion in the AID. A similar project, the digitization of the AAVSO data published in the *Harvard Annals*, was also initiated. The AAVSO archives also include the observing logs of Paul S. Yendell and several other early observers. These will eventually be reduced to modern photometric standards and added to the AID. It may also be possible to add observations of specific variable stars included in research journals. The intent is to enhance the value of AID for research purposes, and for those purposes, any and all such observations converted from paper to digital form will help achieve that goal.

With the above context clearly in mind, it was still gratifying to recognize that as of August 2010, the AID included more than 18 500 000 observations of variable stars. The progressive growth in the annual totals added to the AID can be seen in Chapters 8, 16, and 20.

Epilogue · A new century

Each age spells its own advance, and the all-important present soon fades into the shadowy and forgotten past . . . but let us not ridicule past ages for their crude notions and quaint fancies, lest some of the cherished ideas of which we boast be transmuted by the touch of time into naught but idle visions.

– William Tyler Olcott, from *Sun Lore of All Ages*, 1914

ADVANCING VARIABLE STAR ASTRONOMY – THE FUTURE?

In April 1911, a lawyer in Norwich, Connecticut, announced that he was founding the American Association of Variable Star Observers (AAVSO) and invited others, both amateur and professional astronomers, to join him in this interesting scientific pursuit. By December of that same year, William Tyler Olcott was able to publish the first compilation of 198 observations of 69 stars from the earliest 7 members of this new organization. Since 1911, the AAVSO has grown to more than 1200 members scattered around the world and has archived well over 19 million observations for a list that has grown to include more than 10 000 known variable stars, and the AAVSO is arguably the leading international organization of variable star observation. This signal record of achievement has been the work of many individuals over the century, so in this history we celebrate the contributions of members and observers, as well as the administrative leaders who contributed to that growth.

The AAVSO has much to be proud of in considering its achievements over the first century of its existence. As the AAVSO enters its second century, the future for the Association and the constructive involvement of its members in the science of astronomy has never seemed brighter, though there are many challenges to be confronted in the process.

Tremendous progress has been made in institutionalizing good science and modern technology in the AAVSO's systems and procedures over the past two decades, fitful as that process may have been. The Association has the resources to serve astronomy: a large, growing database of observations; a director and staff astronomer who can help interpret that data; a technically adept staff; and so many member observers who are eager to apply their talents toward making it all work.

Change is inevitable. But having a clear sense of what direction change will take is another matter. By the early 1900s, it was thought that observations of newly discovered variable stars were not necessary – all possible permutations seemed to be accounted for. In 1954 when the AAVSO had accumulated less than 40 years of variable star observations, nearly everyone was saying that long-period variable stars were well-understood, further observations were of little use, and the AAVSO ought to be disbanded. Years from now what will be said about the changes happening today?

The AAVSO's eviction from HCO was a blessing in disguise, as it freed the Association to expand its observing programs in new directions. It took Mayall's dedication and sacrifice to save the AAVSO, but once the organization stabilized, due in a large part to the Ford endowment, the field was clear for it to go far and wide. Mattei saw that opportunity and devoted her career to steering the Association on a forward course. She embraced the new and advanced the science methodically while staying up to date with the technology; that was how Mattei exercised her commitment to perfection and in so doing protected the reputation of the AAVSO.

There are also blessings in disguise in Henden's iconoclasm as well as his vision. In 5 short years, he has led the Association to astonishingly new and innovative ways to collect, preserve, disseminate, and understand variable star observations. Efficient database technology, robotic telescope networking, an up-to-date emphasis on photometry – these are just a few ways in which Henden has demonstrated his willingness to change everything to better the scientific contribution.

In all cases, the AAVSO has been blessed with directors who were – and are – capable, dedicated, and willing to work extraordinarily long hours to advance variable star astronomy as well as the work of the Association. Last, but not least, is the dedication of the AAVSO's observers and members who serve the Association in so many ways as volunteers, usually but not always suffering in silence as the Association struggled with the various challenges we have presented here.

More than 100 years ago, Edward C. Pickering envisioned a cooperative relationship between the amateur and professional astronomer. What Pickering had in mind was for the amateur to serve the practical needs of the professional. This was the model that Olcott followed when he founded the AAVSO. Through the decades, thousands of individuals have contributed to that amateur–professional relationship by faithfully observing, recording, and reporting the brightness of variable stars for the benefit of some future scientific use.

With that model of cooperation, the AAVSO has through the years served astronomy well. But increasingly over the last few decades, advances in technology have enhanced the value of the amateur's work. The mostly one-directional "co-operation" of the past is being rapidly displaced by a "convergence" of amateur and professional interest toward a truly collaborative relationship. More than ever before, amateur and professional variable star astronomers are working together, and the AAVSO will continue to advance variable star astronomy.

Appendix A • AAVSO historiographic notes

In this book, we have attempted to correct some mistaken views of the history of the AAVSO that have developed over a number of years. Those problems have crept into the historical record in a number of ways, but all seem to have their origin in the Pickering memorial booklet. That document, crafted to honor Edward C. Pickering after his death in 1919, asserts that he was "a founder." We have attempted to show that Pickering was so busy with other more important matters of both research and observatory administration that he had little time to spend on William T. Olcott and his variable star observers. Also, the outcome of Olcott's initiative was far from certain, and Pickering had good reasons to limit his interest to providing the necessary technical support to Olcott in parallel to two other groups of variable star observers who also provided variable star observations for his use. That technical support hardly qualifies him as a founder.

Thus the history of the AAVSO appears to have gotten off track in that honest attempt to honor Pickering. Olcott's modest and self-effacing personality may have prevented him from speaking up in the matter when confronted with two strong professional women who were the other members of this committee – and likely strong advocates of the "Pickering as Founder" thesis. We have no records of that committee's work, only their final report. We also know, however, that Olcott made one further effort to distance the AAVSO from HCO by building a separate observatory in which AAVSO Headquarters could be located. That initiative was also warped into a fund to endow a separate chair in the HCO organization named in honor of Pickering. At that point in 1922, Olcott's attitude clearly changed and he distanced himself from the organization even more. It took about 8 more years of progress in the AAVSO for Olcott to finally reconcile himself to the fact that others recognized Pickering, however inappropriately, as a founder. His 1930 poem "Fait Accompli" seems to be our clearest evidence that such a shift of resignation had occurred.

Leon Campbell's 1935 attribution of the founding not only to Pickering and Olcott, but also to Herbert C. Wilson, the editor at *Popular Astronomy*, only adds to the confusion in the matter. It was very much in Campbell's style to mention as many people as possible in his many public talks, so the insertion of Wilson in the list of "founders" was consistent with Campbell's style but only further compounded the origin problems.

William Tyler Olcott, sometime shortly before his death in 1936, wrote a short article for *Variable Comments*, "History of the American Association of Variable Star Observers." It is interesting to note that in Olcott's description of his first entry into variable star observing, he makes no mention of Campbell's visit to Norwich to teach him how to observe, only that the first observation was difficult and only achieved after several attempts. Olcott's description of the history of the Association includes, as a separate item, only the first meeting in 1914 in New York City and omits all subsequent meetings, including the Boston meeting and the visit to HCO. It seems likely that the Olcott history was written as a rebuttal to Campbell's careless attempts to expand the number of founders once again.

In his 1941 book, *The Story of Variable Stars*, coauthored with Luigi Jacchia, Campbell includes a small (one-sixth of a page) portrait of Olcott with an inscription that reads "Wm. Tyler Olcott, 1873–1936 Co-founder and 1st Secretary of the A.A.V.S.O." There is no explanation here of who the other co-founder or co-founders might have been. We understand Campbell and Jacchia's intent to reduce Olcott's importance in the matter by turning to the front of the book, where the page facing the title page is a full-page portrait with a caption "Edward Charles Pickering (1846–1919) Founder of the American Variable Star Astronomy" with no mention of the AAVSO. Campbell and Jacchia do not address the matter of co-founder, nor where AAVSO Headquarters was located, in the short description of the AAVSO. Thus the matter of who were the other co-founders was left for the reader to discern.

In 1951, Margaret Mayall, Clinton Ford, and David Rosebrugh were appointed by the AAVSO Council to bring the history of the Association up to date and publish it. Their effort, never successfully concluded, fell victim to the events then beginning to impact HCO. Under leadership of Harvard President James B. Conant, the university's administration began to focus on HCO and its problems by appointing an external review committee. Thus a final draft of the updated AAVSO history, prepared by Rosebrugh in January 1952, never saw the light of publication under the pressures Mayall began to feel from Donald H. Menzel and the Observatory Council.

Although the Mayall, Ford, and Rosebrugh update was to cover 1936 to 1951, the authors did include a few paragraphs about the previous years, based on a 1943 letter from Leon Campbell to one of them, probably Rosebrugh. In the direct quotation they provide, Campbell describes his understanding of the early history. A copy of that letter is not in the Campbell papers in the AAVSO archives. It is important to recognize that Campbell was in Arequipa, Peru, at the time of the founding in 1911. In fact, Campbell did not return from Arequipa to HCO until August 1915, by which time the Association was well established and had been publishing observations under its own name in *Popular Astronomy* for 4 years. We cite the existence of this unpublished Mayall, Ford, and Rosebrugh history as valuable in two respects: first, Campbell's own bias is evident in the information quoted directly from his letter, in that he attempts to minimize the nature of the Association until after its first meeting in Boston in November 1915, during which the members first visited HCO. Campbell describes that meeting as the first AAVSO meeting to be held at HCO, whereas it was neither the first meeting nor was it held at HCO. It is also from that letter that we derive our best understanding of Campbell's role in those early years after his return from Peru. Before his formal recognition as the AAVSO Recorder in 1925, that work was only a part of his formal HCO work and was in combination with work as an elected AAVSO officer and as a volunteer, and all done after hours from his home. There is no denying that Campbell was deeply involved in AAVSO activities after 1915; the only question is one of formal roles. Second, Campbell's letter demonstrates that without an adequate archive on which to rely, even the membership of the Association was dependent on limited and erroneous or distorted prior written histories.

In 1961, Margaret Mayall's husband, R. Newton Mayall, published a two-part article titled "The Story of the AAVSO" in *The Review of Popular Astronomy*. In that essay, likely intended to promote the AAVSO for fundraising purposes as well as membership recruitment, Mayall wisely skirted around the issues outlined above, by sticking to Olcott's own words and ignoring Campbell's various texts. Perhaps this also reflects the lack of archival research, but very likely also recognizes that the relationship between the AAVSO and HCO was already on the mend after the eviction but still tender and that the Association could only lose by reopening such issues.

Using Campbell and Jacchia as their sole source, Jones and Boyd in their 1971 book *The Harvard College Observatory – The First Four Directorships, 1839–1919* offer an explanation of these events:

> [Pickering] offered full cooperation when in 1911 the American Association of Variable Star Observers was formed with headquarters at the Observatory. William Tyler Olcott, a gifted amateur, became its president and received help from and advice from both Pickering and Leon Campbell. (page 381)

Their assessment of the situation confuses the incorporation of the Association at HCO in 1918, with its founding in 1911, but we believe also recognizes the ambiguity of both Pickering and Campbell's roles in that founding.

The AAVSO staff was understandably confused by all this as successive retellings of the AAVSO history added more and more layers of misinformation. In connection with the 1986 celebration of AAVSO's 75th Anniversary, Janet Mattei prepared a talk that was later published in *JAAVSO* as "The founding of the AAVSO and its first seventy-five years." She described the November 1915 Boston meeting, during which the members watched the Harvard-Yale football game and visited HCO, as "the first official meeting of the AAVSO took place at" HCO. She further asserted that at that time, Pickering appointed Campbell "Recorder of Observations," though that is not factual. It appears that her effort was to make the best sense she could out of the previously published articles, without access to the archival resources that would later become available. Her effort only supports the prior distortions already described. By 2007, the story of the AAVSO's founding was so confused that the *Biographical Encyclopedia of Astronomy* entry for Anne Sewell Young described her as one of eight founders of the AAVSO. No doubt the source of that error is traceable to an obituary for Young written under the pressure of time and without access to archival records outside those available at Mount Holyoke College.

There is a comparable problem of resource quality for even this history. Although we have an excellent archive of AAVSO materials including Olcott, Campbell, Mayall, and Mattei papers, these collections are compromised in ways all too familiar to historians with any experience in archival research. The archive was organized beginning in 2000 from materials that had been stuffed into file folders in filing cabinets from previous administrations or dropped off in boxes at the various AAVSO Headquarters over the years. These materials were arranged – as found – into an archive of AAVSO correspondence and other collections. Problems exist with the previous stewardship of these records, especially of the Campbell papers, which appear to have been severely excised. For example, there is little or no correspondence with David B. Pickering, who was a prolific writer and deeply involved with the AAVSO for many years. It is true that, at the time of the AAVSO's eviction from HCO, Mayall and others had to dispose of a great many files, but one wonders what was lost. Also, the Mattei papers, though voluminous, appear to be missing in key areas. Hopefully the existence of this note will key historians of astronomy to the availability of these archives as well as the service they can perform by calling the attention of the AAVSO archivist, Dr. Michael Saladyga, to the availability of substantial materials elsewhere that are relevant to this history. His address is c/o AAVSO, 49 Bay State Road, Cambridge, Massachusetts 02138.

As a final comment, we should note that, as authors, we have made a conscious effort not only to access available archival materials, but also to understand the context within which those archival documents were created. We present in this book our best interpretation of those archival materials in the context of all that surrounded their creation as we understand it. However, we make no claim for infallibility and hope that in the future, additional archival materials will come to light, and that where those resources offer additional understanding of this history, future historians of the AAVSO will review all available resources, including those cited in this history, before coming to their own judgments and writing a new history of the Association. Only thus will the best available history of the AAVSO gradually emerge.

Appendix B · Top AAVSO observer totals

TOP 100 VISUAL OBSERVERS: TOTALS AS OF FY 2007–2008

Observer initials	Group affiliation	Name	Country/state	Total
JA	14	Jones, Albert	New Zealand	448 449
OB	10	Overbeek, M. Daniel	South Africa	292 711
LX		Lowder, Wayne	NY	208 630
BRJ		Bortle, John	NY	180 800
SRX	14	Stubbings, Rod	Australia	173 988
OV		Oravec, Edward	NY	170 663
VED	01	Vedrenne, Paul	France	168 149
PYG		Poyner, Gary	England	166 908
CR	14	Cragg, Thomas	Australia	166 062
DK		De Kock, Reginald	South Africa	161 745
DGP		Dyck, Gerald	MA	155 854
BM		Baldwin, Marvin	IN	153 372
MOW		Morrison, Warren	Canada	152 039
CMG	04	Comello, Georg	Netherlands	149 943
FE		Fernald, Cyrus	ME	132 448
P		Peltier, Leslie	OH	130 608
HE		Hiett, F. Lancaster	VA	115 918
AB		Albrecht, William	WI	115 403
WPX	14	Williams, Peter	Australia	107 814
MUY	05	Muyllaert, Eddy	Belgium	105 533
SHS		Sharpe, Steven	Canada	102 590
HK		Halbach, Edward	MN	100 715
VET	01	Verdenet, Michel	France	91 393
FJH	04	Feijth, Henk	Netherlands	88 738
CGF		Chaple, Jr., Glenn	MA	77 879
DPA	05	Diepvens, Alfons	Belgium	77 602
SAH		Samolyk, Gerard	WI	77 371
OJR	30	Ripero Osorio, Jose	Spain	77 062
YRK		York, David	NH	75 494
KOS	03	Kosa-Kiss, Attila	Romania	75 396
HR		Hurless, Carolyn	OH	75 279
VFK	02	Vohla, Frank	Germany	74 255
MYR		Mayer, Ernst	OH	73 937
PPS	03	Papp, Sandor	Hungary	73 023

Observer initials	Group affiliation	Name	Country/state	Total
RAE		Roberts, Alexander	South Africa	69 050
AD		Adams, Robert	MO	64 888
GUN	01	Gunther, Jean	France	63 951
FD		Ford, Clinton	CT	61 839
KRS		Kolman, Roger	IL	61 758
H		Hartmann, Ferdinand	NY	60 102
DZS	12	Dominguez, Sergio	Argentina	59 971
MED	20	Medway, Kenneth	England	59 001
AC		Anderson, Curtis	MN	58 196
KIS	02	Krisch, Gunther	Germany	57 669
GRL	08	Granslo, Bjorn	Norway	57 288
VAM		Vattuone, Mario	Argentina	57 047
MRX	02	Marx, Harald	Germany	55 283
L		Lacchini, Giovanni	Italy	54 560
HSG		Hanson, Gene	AZ	53 634
HGC	14	Herdman, Gordon	New Zealand	53 141
MMI	02	Moeller, Michael	Germany	52 650
SXN		Simonsen, Michael	MI	52 466
SSW		Swierczynski, Stanislaw	Poland	50 967
MTH		Matsuyama, Hiroshi	Australia	50 631
SQR		Schmude, Jr., Richard	GA	50 167
CH		Chandra, Radha	India	49 667
MZS	03	Mizser, Attila	Hungary	49 528
TNX	14	Taylor, Noel	Australia	49 168
AAP	27	Abbott, Patrick	Canada	48 063
VDL	05	Van der Looy, Johan	Belgium	47 085
KMA		Komorous, Miroslav	Canada	45 623
SJZ		Speil, Jerzy	Poland	45 367
DPV	99	Dubovsky, Pavol	Slovakia	45 226
J		Jones, Eugene	MA	44 764
SCZ	01	Schweitzer, Emile	France	44 712
RB		Rosebrugh, David	NY	44 152
HOX	14	Hull, O.	New Zealand	43 724
MGE		Mavrofridis, George	Greece	41 748
GBZ	21	Gabzo, Ofer	Israel	41 034
AH		Ahnert, Paul	Germany	40 846
RMQ		Reszelski, Maciej	Poland	40 718
FMR		Fonovich, Marino	Croatia	40 391
LVY		Levy, David	AZ	37 935
STI		Steffey, Philip	FL	37 608
SCE		Scovil, Charles	CT	37 131
HMR		Ham, Ronald	CO	37 113
SJQ		Sajtz, Andrei	Romania	37 108

Continued

TOP 100 VISUAL OBSERVERS (*Continued*)

Observer initials	Group affiliation	Name	Country/state	Total
BL		Baldwin, James	Australia	36 872
SET		Stephan, Christopher	FL	35 729
BIC	01	Bichon, Laurent	France	34 972
HUZ		Huziak, Richard	Canada	34 591
VMT	05	Vanmunster, Tonny	Belgium	34 272
HCS	03	Hadhazi, Csaba	Hungary	34 231
GOP		Goodwin, Paul	LA	33 997
HEF		Heifner, Mark	CO	33 384
PAW		Plummer, Alan	Australia	33 366
SNO	03	Szentasko, Laszlo	Hungary	33 197
SC	27	Spratt, Christopher	Canada	33 069
HIR		Hirasawa, Yasuo	Japan	32 280
MBB	14	Menzies, Barry	New Zealand	31 443
WSN		Wilson, Thomas	WV	30 761
STF		Stefanopoulos, George	Greece	30 593
MJ		Montague, Allen	MI	30 187
WEF		West, Frederick	MD	30 174
MLF	10	Monard, Libert	South Africa	29 419
IML		Idem, Michael	NY	29 097
MKJ		McKenna, Jerome	MN	28 630
ANN		Annal, Robert	CA	28 484
GPX	14	Goltz, William	Australia	28 449
HHU	05	Hautecler, Hubert	Belgium	28 426

Note: Group affiliation codes indicate observers are also affiliated with the groups listed in Appendix C.

TOP 50 CHARGE-COUPLED DEVICE (CCD) OBSERVERS: TOTALS AS OF FY 2007–2008

Observer initials	Group affiliation	Name	Country/state	Total
MXT	10	Middleton, Christopher	South Africa	375 784
JM		James, Robert	NM	295 097
TRE		Tomlin, Ray	IL	256 209
DKS		Dvorak, Shawn	FL	226 552
SAH		Samolyk, Gerard	WI	217 316
SBL	05	Staels, Bart	Belgium	185 724
MLF	10	Monard, Libert	South Africa	172 913
PVA	27	Petriew, Vance	Canada	151 994
COO		Cook, Lewis	CA	151 346
VMT	05	Vanmunster, Tonny	Belgium	150 686
HUZ		Huziak, Richard	Canada	126 418

Observer initials	Group affiliation	Name	Country/state	Total
DSI		Di Scala, Giorgio	Australia	122 267
ASAS3		All Sky Automated Survey 3	Chile	97 497
CTX		Crawford, Tim	OR	94 245
BDG	20	Boyd, David	England	81 674
MEV	01	Morelle, Etienne	France	77 042
GKA		Graham, Keith	IL	68 036
BIZ		Bialozynski, Jerry	AZ	65 485
BXS		Brady, Steve	NH	63 779
MIW	20	Miller, Ian	Wales	61 309
OAR	17	Oksanen, Arto	Finland	60 437
BIW		Butterworth, Neil	Australia	60 334
DPP	05	De Ponthiere, Pierre	Belgium	59 939
KRV		Koff, Robert	CO	57 321
GFB		Goff, Bill	CA	56 319
NLX	14	Nelson, Peter	Australia	54 971
ASAS2		All Sky Automated Survey 2	Chile	53 088
RIX	29	Richards, Thomas	Australia	40 967
SDB		Starkey, Donn	IN	40 568
JJI		Jones, James	OR	36 532
ROE		Roe, James	MO	34 928
JLT	02	Jensen, Lasse	Denmark	34 475
SFY	20	Shears, Jeremy	England	34 187
KTC		Krajci, Tom	NM	28 848
MDW	27	MacDonald, Walter	Canada	28 507
ZRE		Zissell, Ronald	MA	27 831
SWIL		Stein, William	NM	27 788
MZK		Menzies, Kenneth	MA	25 634
PRX		Poklar, Rudy	AZ	23 673
RCW		Robertson, Jr., Charles	KS	22 297
MXL	20	Miles, Richard	England	20 233
ATE		Arranz, Teofilo	Spain	19 027
SRIC		Sabo, Richard	MT	18 703
OYE		Ogmen, Yenal	North Cyprus	18 330
SRIH		Schwartz, Richard	WA	18 160
NMR	20	Nicholson, Martin	England	16 874
GCO		Gualdoni, Carlo	Italy	16 216
GOT	06	Gomez, Tomas	Spain	16 111
KGE	08	Klingenberg, Geir	Norway	16 013
HES		Hesseltine, Christopher	WI	15 037

Note: Group affiliation codes indicate observers are also affiliated with the groups listed in Appendix C.

TOP 50 PHOTOELECTRIC PHOTOMETRY (PEP) OBSERVERS: TOTALS AS OF FY 2007–2008

Observer initials	Group affiliation	Name	Country/state	Total
APOG		Auckland Photometric Observers Group	New Zealand	12 324
THR		Thompson, Raymond	Canada	8231
LKA		Luedeke, Kenneth	NM	4243
WJM		Wood, James	CA	3236
JRW	10	Jones, Raymond	South Africa	2987
STQ		Stoikidis, Nick	Greece	2726
HPO		Hopkins, Jeffrey	AZ	2407
DSG	18	Dallaporta, Sergio	Italy	1806
SMQ		Smith, Michael	AZ	1800
MBE		McCandless, Brian	MD	1656
LOT		Louth, Howard	WA	1636
FXJ		Fox, James	MN	1203
CLX		Cox, Louis	FL	1187
MPH		Manker, Phillip	NM	1098
CCB		Calia, Charles	CT	1029
WN		Wing, Robert	OH	1010
HKB		Hakes, Brian	IL	979
WJD		West, J. Doug	KS	943
LND		Landis, Howard	GA	812
BTU		Beresky, Ted	MO	744
CLK		Clark, Wayne	MO	696
MTL		Milton, Russell	CA	664
FBN	10	Fraser, Brian	South Africa	650
WI		Williams, David	IN	623
SOH	11	Sorensen, Hans	Denmark	609
ZRE		Zissell, Ronald	MA	562
KFE		Kameny, Franklin	MA	547
DFR	27	Dempsey, Frank	Canada	545
CUO		Curott, David	AL	479
WSI		Wasatonic, Richard	PA	466
KPL		Kneipp, Paul	LA	422
DVI	10	De Villiers, Fanie	South Africa	418
KLG		Kohl, George	AZ	411
PDO		Pray, Donald	MA	359
VBR		Van Bemmel, Henri	Canada	339
HWD		Hodgson, William	Australia	294
CRR		Crumrine, Robert	NY	267
PPK	17	Paakkonen, Pertti	Finland	259
LTW		Langhans, Thomas	CA	256
BWK		Barksdale, William	FL	244
CJH		Crast, Jack	NY	243
KRK		Krisciunas, Kevin	TX	237

Observer initials	Group affiliation	Name	Country/state	Total
RRC		Reisenweber, Robert	PA	180
RTH		Rutherford, Thomas	TN	171
FT		Fortier, George	Canada	171
GBI		Grim, Bruce	UT	158
WPK		Wiggins, Patrick	UT	150
ILS	20	Isles, John	MI	147
PHD		Powell, Harry	TN	140
GGL	18	Galli, Gianni	Italy	132

Note: Group affiliation codes indicate observers are also affiliated with the groups listed in Appendix C.

TOP 50 PHOTOGRAPHIC/PHOTOVISUAL OBSERVERS: TOTALS AS OF FY 2007–2008

Observer initials	Group affiliation	Name	Country/state	Total
DL		Dahlmark, Lennart	France	2349
AUB	01	Aubaud, Jean	France	1677
DMI	02	Dahm, Michael	Germany	1648
TRF		Trefzger, Charles	Switzerland	1418
SUZ	03	Szutor, Peter	Hungary	1273
KIL	03	Kiss, Laszlo	Hungary	856
BSF		Barnhart, Stephen	OH	738
BJS		Bedient, James	HI	408
AU		Aurino, Salvatore	Italy	371
SIV		Sergey, Ivan	Belarus	370
BMP		Bemporad, Azeglio	Italy	338
ERU	02	Rudolph, Eyck	Germany	314
SGU	03	Sari, Gyula	Hungary	244
PAH		Price, Aaron	MA	237
LVY		Levy, David	AZ	208
LIW		Liller, William	Chile	196
RR		Royer, Ronald	CA	170
HAI		Hastings, Allen	MA	170
VAQ		Vacquier, Victor	CA	121
AH		Ahnert, Paul	Germany	115
TMT		Templeton, Matthew	MA	104
S58		Swope, Henrietta	CA	103
CQS		Cheng, Simiao	China	101
SCK		Schaefer, Bradley	LA	99
CZT	03	Csiszar, Tibor	Hungary	94
MCC		Macris, Constantinos	Greece	93
MAJ		Moorhouse, A.	MI	85

Continued

TOP 50 PHOTOGRAPHIC/PHOTOVISUAL OBSERVERS (*Continued*)

Observer initials	Group affiliation	Name	Country/state	Total
SCZ	01	Schweitzer, Emile	France	76
SH		Specht, Henry	CT	73
SCE		Scovil, Charles	CT	67
CSI	03	Csoti, Istvan	Hungary	62
OGJ	01	Ortega, Juan Carlos Gil	Spain	61
WJD		West, J. Doug	KS	60
LEG	01	Leger, L.	France	56
KBJ		Kaufman, Robert	Australia	54
LLR	01	Letellier, Georges	France	52
SZK	03	Szitkay, Gabor	Hungary	51
FEO	03	Farkas, Erno	Hungary	47
REN	01	Renault, Jean	France	47
MW		Miller, Robert	FL	46
FD		Ford, Clinton	CT	35
ELW		Elwin, S.	Australia	28
MGY	03	Mogyorosi, Imre	Hungary	28
DUP	01	Dupasquier, Paul	France	26
BTP		Bhaskaran, T.	India	25
MJK		May, Jack	AR	25
PAR		Patterson, Jr., Russel	NY	19
CZJ		Canelhas, Jorge	Portugal	16
HZL		Hazel, Lawrence	NY	16
FC		Focas, Joannis	Greece	15

Note: Group affiliation codes indicate observers are also affiliated with the groups listed in Appendix C.

Appendix C • Variable star observing groups represented in the AAVSO International Database

15	Agrupacion Astronomica de Sabadell (Spain)
00	American Association of Variable Star Observers (AAVSO)
29	Asociacion Amigos de la Astronomia (Buenos Aires, Argentina)
07	Asociacion de Variabilistas de Espagne (Spain)
01	Association Française des Observateurs d'Étoiles Variables
16	Association of Variable Star Observers "Pleione" (Russia)
10	Astronomical Society of Southern Africa, Variable Star Section
24	Astronomischer Jugendclub (Austria)
11	Astronomisk Selskab (Scandinavia)
20	British Astronomical Association, Variable Star Section
02	Bundesdeutsche Arbeitsgemeinschaft für Veränderliche Sterne e.V. (Germany)
31	Center for Backyard Astronomy
32	Groupe Europeen d'Observations Stellaires
23	Grupo Astronomico Silos (Zaragoza, Spain)
21	Israeli Astronomical Association, Variable Star Section
04	Koninklijke Nederlandse Vereniging Voor Weer-en Sterrenkunde, Werkgroep Veranderlijke Sterren (Netherlands)
12	Liga Iberoamericana de Astronomia (South America)
06	Madrid Astronomical Association M1 (Spain)
03	Magyar Csillagàszati Egyesület, Valtózocsillag Szakcsoport (Hungary)
08	Norwegian Astronomical Society, Variable Star Section
26	Red de Observadores (Montevideo, Uruguay)
13	Rede de Astronomia Observacional (Brazil)
27	Royal Astronomical Society of Canada
14	Royal Astronomical Society of New Zealand, Variable Star Section
22	Sociedad Astronomica "Syrma" (Valladolid, Spain)
19	Svensk Amator Astronomisk Forening, variabelsektionen (Sweden)
09	Ukraine Astronomical Group, Variable Star Section
18	Unione Astrofili Italiani (Italy)
17	URSA Astronomical Association, Variable Star Section (Finland)
33	Variable Star NETwork
25	Variable Star Observers League in Japan
29	Variable Stars South (New Zealand)
05	Vereniging Voor Sterrenkunde, Werkgroep Veranderlijke Sterren (Belgium)

Appendix D · AAVSO Awards

AAVSO MERIT AWARD

The valuable data processed and distributed by the AAVSO to astronomers throughout the world are obtained from volunteer observers. Their constant watch over variable stars is a satisfying activity that results in a genuine contribution to astronomy. There are many observers whose contributions to the AAVSO have gone beyond observing to other activities as well. The Association shows its gratitude to such individuals by presenting them with a Merit Award, an illuminated scroll and citation listing their contributions. This award is presented to any member who, by majority vote of the AAVSO Council, is deemed to have made an outstanding contribution as an observer, or in other ways. The following is a list giving the names of the AAVSO Merit Award recipients, home state or country, and year of award.

1. Leslie C. Peltier, Ohio, 1934
2. Rev. Tilton C. H. Bouton, Florida, 1936
3. William Tyler Olcott, Connecticut, 1936
4. Eugene H. Jones, New Hampshire, 1937
5. David B. Pickering, New Jersey, 1938
6. Dalmiro F. Brocchi, Washington, 1942
7. Charles W. Elmer, New York, 1943
8. Leon Campbell, Massachusetts, 1944
9. Cyrus F. Fernald, Maine, 1948
10. Harlow Shapley, Massachusetts, 1951
11. David W. Rosebrugh, Florida, 1951
12. Percy Witherell, Massachusetts, 1954
13. Roy A. Seely, New York, 1954
14. Margaret W. Mayall and R. Newton Mayall, Massachusetts, 1956
15. Reginald P. de Kock, South Africa, 1961
16. Clinton B. Ford, Connecticut, 1961
17. Richard W. Hamilton, Connecticut, 1962
18. Curtis E. Anderson, Minnesota, 1965
19. Ralph N. Buckstaff, Wisconsin, 1965
20. Richard H. Davis, Massachusetts, 1971
21. Marvin E. Baldwin, Indiana, 1980
22. Charles E. Scovil, Connecticut, 1981
23. John E. Bortle, New York, 1983
24. Edward G. Oravec, New York, 1985
25. Thomas A. Cragg, Australia, 1986
26. M. Daniel Overbeek, South Africa, 1986
27. F. Lancaster Hiett, Virginia, 1986
28. *no award*
29. Arthur J. Stokes, Ohio, 1987
30. Howard J. Landis, Georgia, 1987
31. Carolyn J. Hurless, Ohio, 1987
32. Edward A. Halbach, Colorado, 1988
33. Theodore H. N. Wales, Massachusetts, 1991
34. Charles A. Whitney, Massachusetts, 1993
35. John R. Percy, Canada, 1993
36. Thomas R. Williams, Texas, 1995
37. Martha L. Hazen, Massachusetts, 2005
38. Louis Cohen, Massachusetts, 2006
39. Gerard Samolyk, Wisconsin, 2007
40. Lee Anne Willson, Iowa, 2008
41. Albert F. Jones, New Zealand, 2008
42. Lewis M. Cook, Texas, 2009

AAVSO WILLIAM TYLER OLCOTT DISTINGUISHED SERVICE AWARD

Initiated in 2000, this award is presented to a member of the AAVSO for outstanding contributions in promoting variable star astronomy through outreach and education.

1. David H. Levy, Arizona, 2000
2. Leif J. Robinson, Massachusetts, 2001
3. Dorrit Hoffleit, Connecticut, 2002
4. Edward A. Halbach, Colorado, 2003
5. Janet A. Mattei*, Massachusetts, 2005
6. John R. Percy, Canada, 2007
7. Vladimir Strelnitski, 2008
8. Karen Meech, 2009

*Posthumous award

AAVSO DIRECTOR'S AWARD

The Director's Award is awarded at the director's discretion in the form of a plaque to an outstanding observer who contributes to special observing projects.

1. M. Daniel Overbeek, South Africa, 1994
2. John E. Bortle, New York, 1995
3. William G. Dillon, Texas, 1996
4. Thomas A. Burrows, California, 1997
5. Michel Grenon, Switzerland, 1997
6. Albert F. Jones, New Zealand, 1997
7. Gene A. Hanson, II, Arizona, 1998
8. Charles E. Scovil, Connecticut, 1999
9. Ronald E. Zissell, Massachusetts, 2000
10. Stephen D. O'Connor, Canada, 2001
11. William B. Albrecht, Hawaii, 2002
12. Thomas A. Cragg, Australia, 2002
13. Rod Stubbings, Australia, 2002
14. Richard E. Wend, Illinois, 2002
15. Richard Huziak, Canada, 2003
16. Gary Poyner, England, 2003
17. Christopher Stephan, Florida, 2004
18. Arto Oksanen, Finland, 2004
19. Michael A. Simonsen, Michigan, 2005
20. Vance Petriew, Canada, 2007
21. Sebastian A. Otero, Argentina, 2009

AAVSO NOVA AND SUPERNOVA AWARDS

At the Annual Meeting in October 1928, a Nova Award Medal was instituted to encourage and recognize the visual discoverers of novae. The Medal was not restricted to members of the AAVSO, but would be awarded to the original discoverer of a nova by direct visual methods. Initially, the medals were made of gold and were donated by AAVSO member and jeweler David B. Pickering. The Nova Award Medal was discontinued, and the Nova Award Plaque was awarded in its place on subsequent occasions. This award is also awarded to the visual discoverer of a supernova; until the late 1990s such a discoverer received a Nova Award, but now receives a Supernova Award. The following is the list of AAVSO Nova and Supernova Award recipients.

Note: listing gives the year an award was made, which is not necessarily the same year that the discovery was made.

George E. D. Alcock, England, 1976 (three awards), 1981, 1991
Samantha Beaman, Australia, 1995, 1996
Kenneth C. Beckmann, Michigan, 1988
Jack Bennett, South Africa, 1976
Robert Benton, Australia, 1996
Nicholas J. Brown, Australia, 1995
Peter Challis, Massachusetts, 1995
Peter L. Collins, Arizona, 1985, 1988, 1993
Giancarlo Cortini, Italy, 1995
Thomas A. Cragg, Australia, 1995
Bernhard Dawson, Argentina, 1942
Robert O. Evans, Australia, 1983 (four awards), 1984 (five awards), 1986, 1988 (three awards), 1991 (six awards), 1993 (three awards), 1995 (two awards), 1996 (three awards), 1997, 2000, 2001, 2002, 2003 (four awards), 2005, 2007
Francisco Garcia Diaz, Spain, 1993
Alan C. Gilmore, New Zealand, 1999
Kazuaki Gomi, Japan, 1936
Olaf Hassel, Norway, 1960
John G. Hosty, England, 1979
K. Ikeya, Japan, 1988
John Jarman, Australia, 1995
Gus E. Johnson, Maryland, 1979
Wayne Johnson, California, 1995
Stephen P. Knight, Maine, 1991
Reiki Kushida, Japan, 1993, 1995
Piero Mazza, Italy, 1996, 1997
Doug Millar, California, 1995
Larry Mitchell, Texas, 1995
Libert A. G. Monard, South Africa, 2001
Warren C. Morrison, Canada, 1979
Gary T. Nowak, Vermont, 2000
Sigeki Okabayasi, Japan, 1936
Kentaro Osada, Japan, 1975
Dana J. Patchick, California, 1988
Leslie C. Peltier, Ohio, 1963
Alfredo Pereira, Portugal, 2000, 2001 (two awards)
Stefano Pesci, Italy, 1996, 1997
James Peters, Massachusetts, 1995
J. H. M. Prentice, England, 1935
John Shobbrook, Australia, 1995
Mirko Villi, Italy, 1995
Brett White, Australia, 1998, 2001
Peter Williams, Australia, 1999, 2006
William Wren, Texas, 1993, 1995

Appendix E · Officers of the AAVSO

RECORDER

1911–1918 William T. Olcott (Assisted by Leon Campbell after 1915)

Note: from 1918 to 1925 this function was handled by the Recording Secretary and Corresponding Secretary.

1925–1949 Leon Campbell
1949–1956 Margaret W. Mayall

DIRECTOR

1956–1973 Margaret W. Mayall
1973–2003 Janet A. Mattei (d. 22 March 2004)
2003–2005 Elizabeth O. Waagen (Interim Director September 2003–February 2005)
2005– Arne A. Henden

CORRESPONDING SECRETARY

1911–1931 William T. Olcott

RECORDING SECRETARY

1918–1924 Howard O. Eaton (Assisted by Harvard College Observatory staff in most years)
1924–1925 Florence Cushman

SECRETARY

1931–1936 William Tyler Olcott (d. 6 July 1936; Charles W. Elmer filled Olcott's unexpired term)
1936–1937 Charles W. Elmer
1937–1945 David W. Rosebrugh
1945–1948 Herbert M. Harris
1948–1992 Clinton B. Ford (d. 23 September 1992)
1992–2005 Martha L. Hazen
2005– Gary Walker

TREASURER

1917–1921 Allen Burbeck
1921–1929 Michael Jordan (d. 22 Dec. 1929; Willard J. Fisher filled Jordan's unexpired term)
1929–1931 Willard J. Fisher
1931–1961 Percy W. Witherell
1961–1973 Richard H. Davis (resigned mid–term)
1973–1979 R. Newton Mayall
1979–1998 Theodore H. N. Wales
1998–2001 Wayne M. Lowder
2001–2006 Louis Cohen
2006–2008 David A. Hurdis
2008– Gary W. Billings

PRESIDENT

1917–1918 David B. Pickering
1918–1919 Harold C. Bancroft, Jr.
1919–1922 Leon Campbell
1922–1924 Anne S. Young
1924–1926 J. Ernest G. Yalden
1926–1927 Charles C. Godfrey
1927–1929 David B. Pickering
1929–1931 Alice H. Farnsworth
1931–1933 Harriet W. Bigelow
1933–1935 Ernest W. Brown
1935–1937 Harlow Shapley
1937–1939 Charles W. Elmer
1939–1941 Helen S. Hogg
1941–1943 Dirk Brouwer
1943–1945 Roy A. Seely
1945–1947 Charles B. Smiley
1947–1948 Marjorie Williams
1948–1949 David W. Rosebrugh
1949–1951 Neil J. Heines
1951–1954 Martha S. Carpenter
1954–1956 Cyrus F. Fernald

1956–1958	Richard W. Hamilton
1958–1960	Ralph N. Buckstaff
1961–1963	Dorrit Hoffleit
1963–1965	George Diedrich
1965–1967	Edward G. Oravec
1967–1969	Frank J. DeKinder
1969–1971	Casper H. Hossfield
1971–1973	Charles M. Good
1973–1975	Charles E. Scovil
1975–1977	George L. Fortier
1977–1979	Marvin E. Baldwin
1979–1981	Carl A. Anderson
1981–1983	Arthur J. Stokes
1983–1985	Ernst H. Mayer
1985–1987	Thomas R. Williams
1987–1989	Keith H. Danskin
1989–1991	John R. Percy
1991–1992	Martha L. Hazen
1992–1993	Thomas R. Williams
1993–1995	Wayne M. Lowder
1995–1997	Albert V. Holm
1997–1999	Gary Walker
1999–2001	Lee Anne Willson
2001–2003	Daniel H. Kaiser
2003–2005	William G. Dillon
2005–2007	David B. Williams
2007–2009	Paula Szkody
2009–	Jaime Garcia

Appendix F · AAVSO Council members

Name	Years of service	Name	Years of service
Mark T. Adams	1992–1997	Claude B. Carpenter	1939–1943
Robert M. Adams	1960–1964; 1979–1981	Martha E. Stahr Carpenter	1951–1956; 1970–1973
Leah B. Allen	1930–1934		(also see Stahr)
Joseph A. Anderer	1961–1963	Kenneth E. Chilton	1970–1972
Carl A.Anderson	1964–1967; 1974–1983	Morgan Cilley	1942–1944
Curtis E. Anderson	1965–1969	Geoffrey C. Clayton	2001–2003
Luverne E. Armfield	1934–1939	William G. Cleaver	1959–1963
Joseph Ashbrook	1949–1953; 1976–1980	Inez L. B. Clough	1931–1933
Solon I. Bailey	1918–1919	Louis Cohen	2001–2006
James M. Baldwin	1940–1942	Peter L. Collins	1979–1981
Marvin E. Baldwin	1969–1981; 1983–1988	Lewis M. Cook	1985–1989; 2000–2007
Harry C. Bancroft	1917–1919	Louis B. Cox	1989–1993
James C. Bartlett, Jr.	1951–1953	Thomas A. Cragg	1951–1953; 1962–1966
Barry B. Beaman	2006–2009	John J. Crane	1917–1919
Margaret H. Beardsley	1950–1954	Florence Cushman	(served but not elected)
James Bedient	2007–2009		1924–1925
Priscilla J. Benson	1991–1995	Keith H. Danskin	1982–1991
Raymond B. Berg	1999–2002	Richard H. Davis	1961–1974
Priscilla A. Bibber	1970–1974; 1977–1978	Frank J. DeKinder	1963–1970
Harriet W. Bigelow	1929–1933	George Diedrich	1957–1967
Gary W. Billings	2003–2009	William G. Dillon	1997–2007
Dorothy W. Block	1918–1919	Howard O. Eaton	1923–1925
Harry L. Bondy	1956–1958	Charles W. Elmer	1926–1930; 1933–1943
Larry C. Bornhurst	1966–1968	Luis E. Erro	1942–1944
John E. Bortle	1970–1974	Alice H. Farnsworth	1923–1931; 1933–1935
Lewis J. Boss	1943–1947; 1962–1964	Charles A. Fausel	1985–1989
Tilton C. H. Bouton	1934–1938	Helen S. Federer	1944–1948
Robert F. Brady	1960–1962	Cyrus F. Fernald	1947–1962; 1967–1972
Dalmiro F. Brocchi	1936–1938; 1940–1941	Willard J. Fisher	1925–1931
Harold G. Brock	1975–1977	Clinton B. Ford	1947–1992
Dirk Brouwer	1938–1947	George L. Fortier	1969–1979; 1988–1990
Ernest W. Brown	1929–1935	Caroline E. Furness	1918–1920
Ralph N. Buckstaff	1945–1949; 1954–1963	Jaime R. Garcia	2000–2009
Allan B. Burbeck	1917–1921	Peter M. Garnavich	1996–2000
Arthur W. Butler	1927–1931	Pamela L. Gay	2005–2009
Leon Campbell	1918–1949	Charles H. Giffen	1963–1965
Annie J. Cannon	1918–1922	Jocelyn R. Gill	1951–1955

Name	Years of service
Owen Gingerich	1965–1969
William H. Glenn	1959–1961
Charles C. Godfrey	1925–1927
Charles M. Good	1957–1961; 1966–1975
Paul N. Goodwin	1980–1982
Robert M. Greenley	1952–1954
Casper R. Gregory	1939–1941
John W. Griesé III	1985–1990
Edward F. Guinan	2008–2009
Phoebe W. Haas	1937–1939
Edward A. Halbach	1940–1943
Richard W. Hamilton	1949–1960; 1968–1970
Herbert M. Harris	1944–1948
Ferdinand Hartmann	1936–1938; 1948–1952
Margaret Harwood	1925–1929; 1934–1936; 1942–1944; 1954–1955
Michael B. Hayden	1995–1997
Lawrence E. Hazel	1974–1976
Martha L. Hazen	1984–2005
Neal J. Heines	1945–1954
Arne A. Henden	1999–2009
William Henry	1928–1932
Richard E. Hill	1991–1992
Sarah J. Hill	1956–1960
E. Dorrit Hoffleit	1943–1945; 1954–1965; 1972–1974; 1977–1981; 1989–1993
Helen Sawyer Hogg	1933–1945; 1950–1952
Albert V. Holm	1990–1999
William L. Holt	1943–1947
Casper H. Hossfield	1964–1975
Nancy Houk	1971–1973
Walter S. Houston	1961–1965; 1981–1985; 1989–1993
Stephen C. Hunter	1921–1923
David A. Hurdis	2001–2009
Carolyn J. Hurless	1965–1973; 1983–1985
Katherine Hutton	2008–2009
John A. Ingham	1932–1934
John E. Isles	1993–1997
Eugene H. Jones	1932–1936
Michael J. Jordan	1918–1919; 1921–1930
Daniel H. Kaiser	1993–2005
Margarita Karovska	1997–2001
Winifred M. Kearons	1939–1943
Donald S. Kimball	1940–1944
Jeremy H. Knowles	1976–1978
Roger S. Kolman	1997–1998
Michael Koppelman	2008–2009
Giovanni B. Lacchini	1930–1932; 1961–1963
Howard J. Landis	1975–1979
Arlo U. Landolt	2006–2009
Kristine M. Larsen	1997–2001
William Liller	1978–1980
Jeffrey F. Lockwood	1995–1996
Wayne M. Lowder	1971–1975; 1985–2003
Richard Lynch	1979–1980
Mark K. Malmros	1985–1986
Kevin B. Marvel	2000–2005
Janet A. Mattei	1973–2004
Michael Mattei	1972–1976; 1979–2007
Margaret W. Mayall	1948–1973
R. Newton Mayall	1957–1979
Ernst H. Mayer	1973–1987
Charles Y. McAteer	1917–1918
Karen J. Meech	2002–2008
Joseph W. Meek	1938–1940
Donald H. Menzel	1949–1953
Paul W. Merrill	1933–1935
William A. Miller	1955–1957
Peter M. Millman	1947–1949; 1958–196
Walter L. Moore	1968–1970
Francis P. Morgan	1962–1964
James E. Morgan	1981–1983
Charles S. Morris	1981–1985
Mario E. Motta	1996–2000
George S. Mumford	1963–1967
Paul H. Nadeau	1946–1950
Virginia McKibben Nail	1955–1957
Arto Oksanen	2005–2007
William Tyler Olcott	1917–1936
Charles P. Olivier	1935–1939
Edward G. Oravec	1955–1960; 1963–1969; 1973–1975
M. Daniel Overbeek	1977–1979
Samuel R. Parks	1944–1946
Arthur E. Pearlmutter	1969–1971
Leslie C. Peltier	1932–1936; 1967–1970
John R. Percy	1982–1993
Arthur C. Perry	1922–1924
David B. Pickering	1917–1918; 1919–1926; 1927–1931

Name	Years of service	Name	Years of service
Edward C. Pickering	1918–1919	Bertram J. Topham	1942–1944
A. Charles Pullen	2001–2006	Theodore R. Treadwell	1946–1948
William H. Reardon	1918–1919	Georges Van Biesbroeck	1926–1928; 1931–1932
Walter P. Reeves	1953–1957	Elizabeth O. Waagen	2003–2005
Francis H. Reynolds	1953–1955	George C. Waldo	1931–1935
Marguerite A. Risley	1974–1976	Theodore H. N. Wales	1977–2001
David W. Rosebrugh	1937–1951	Gary Walker	1993–2001; 2005–2009
Ronald E. Royer	1996–2000	George Wallerstein	1990–1994
John J. Ruiz	1952–1956; 1958–1962; 1967–1969	Christopher Watson	2007–2009
Gerard Samolyk	1986–1987; 1988–1990	Harold B. Webb	1933–1937; 1950–1952
Leo J. Scanlon	1936–1940	Douglas L. Welch	1995–1999; 2007–2008
Charles E. Scovil	1969–1977; 1981–1983; 1994–1996	John E. Welch	1955–1959
Roy A. Seely	1940–1948	Barbara L. Welther	1987–1991
Alan H. Shapley	1956–1958	Jerry Doug West	2003–2005
Harlow Shapley	1931–1939	Charles T. Whitehorn	1919–1921
Emil A. Sill	1944–1946; 1951–1953	Winston S. Wilkerson	1977–1981
Michael A. Simonsen	2005–2009	C. Wilkerson–Montout	1980–1981
David R. Skillman	1984–1986	David B. Williams	1968–1970; 1990–1994; 1998–2009
Charles H. Smiley	1939–1949; 1953–1955	Marjorie Williams	1941–1949; 1975–1977
Martha E. Stahr	1946–1951 (also see Carpenter)	Thomas R. Williams	1978–1989; 1992–1996
Arthur J. Stokes	1964–1968; 1976–1990	Isabel K. Williamson	1961–1965
Richard J. Strazdas	1980–1984	Lee Anne Willson	1980–1984; 1993–2003
Paul L. Sventek	1983–1985; 1990–1995	Robert F. Wing	1994–1996
Helen M. Swartz	1917–1918	Percy W. Witherell	1930–1967
Paula Szkody	2003–2009	Ida E. Woods	1917–1919
Domingo Taboada	1952–1956	Joseph L. Woods	1938–1940
Peter O. Taylor	1974–1978	J. Ernest G. Yalden	1920–1929
		Anne S. Young	1917–1926

Appendix G • AAVSO Scientific committee, section, division, and program chairs

Committee, section, division, or program	Years	Chair; co-chair
Auroral	1940–1948	Edward A. Halbach
	1951–1954	Donald S. Kimball
Cataclysmic Variable	2008–2009	Mike Simonsen; Gary Poyner
Charge-Coupled Device (CCD)	1992–1994	Priscilla J. Benson
	1994–1996	Priscilla J. Benson; Gary Walker
	1996–1999	Gary Walker; Priscilla J. Benson
	1999–2008	Gary Walker
Classical Cepheids	1967–1994	Thomas A. Cragg
Chart Curators	1928–1935	Helen S. Hogg
	1936–1947	Ferdinand Hartmann
	1948–1961	Richard W. Hamilton
Charts	1918–1920	John J. Crane
	1921–1934	David B. Pickering
	1934–1948	Dalmiro F. Brocchi
	1948–1955	Harold B. Webb
	1955–1969	Richard W. Hamilton
	1971–1992	Clinton B. Ford
	1992–2006	Charles E. Scovil
Data Mining	2008–2009	Michael Koppelman
Eclipsing Binary	1965–1969	David B. Williams
	1969–2007	Marvin E. Baldwin
	2007–2008	Gerard Samolyk
	2008–2009	Gerard Samolyk; Gary Billings
Education and Outreach	2006–2009	Pamela Gay
Long Period Variable	2008–2009	Katherine Hutton; Mike Simonsen
Moon	1935	Ernest W. Brown
Nova Search	1935–1939	Luverne E. Armfield
	1939–1948	Roy A. Seely
	1953–1955	Margaret W. Mayall
	1955–1972	George Diedrich
	1972–1983	Carmine V. Borzelli
	1983–2009	Rev. Kenneth C. Beckmann
Occultations	1928–1937	J. Ernest G. Yalden
	1937	Dirk Brouwer
	1937–1938	Anne S. Young
	1938–1951	Alice H. Farnsworth

Continued

Continued

Committee, section, division, or program	Years	Chair; co-chair
	1951–1956	Margaret H. Beardsley
	1956–1958	Sarah J. Hill
	1958–1968	Roy A. Seely
	1968–1971	Charles M. Good
	1971–1976	John E. Bortle
	1976–1977	George Fortier
Photoelectric Photometry	1955–1966	Lewis J. Boss
	1966–1975	Arthur J. Stokes
	1975–2003	Howard J. Landis
	2003–2004	J. Phillip Manker
	2004–2006	AAVSO Headquarters
	2006–2009	James H. Fox
Photographic	1936–1938	Lynn H. Matthias
	1938–1941	J. L. Woods
	1941–1942	A. E. Navez
	1942–1949	Leon Campbell
	1961–1963	Charles H. Giffen
	1963–1964	Richard H. Davis
Red Variables	1936–1937	Leon Campbell
RR Lyrae	1968–2007	Marvin E. Baldwin
	2007–2008	Gerard Samolyk
Satellite Tracking	1956–1957	R. Newton Mayall
Short Period Pulsator	2008–2009	David A. Hurdis; Gerard Samolyk
Solar	1944–1954	Neal J. Heines
	1955–1963	Harry L. Bondy
	1963–1978	Casper H. Hossfield
	1981–1984	Robert B. Ammons
	1984–1999	Peter O. Taylor
	1999–2000	Elizabeth Stephenson
	2000–2001	Joseph Lawrence
	2001–2006	Carl E. Feehrer
	2006–2009	Paul Mortfield
Supernova Search	1988–2007	Rev. Robert O. Evans
	2007–	AAVSO Headquarters

Note: Years shown before 1949 are as found in available archival records and may not be definitive.

Notes

1 THE EMERGENCE OF VARIABLE STAR ASTRONOMY – A NEED FOR OBSERVATIONS

1 Friedrich W. A. Argelander, "Aufforderung an freunde der astronomie," *Schumacher's Jahrbuch 1844* (1844), 123–254; and Friedrich W. A. Argelander, "The variable stars [in 3 parts, part 2]," *PA* **20** (1912), 151.

2 Helen Lewis Thomas, "The early history of variable star observing to the XIX century," Ph.D. dissertation, Radcliffe College, 1948, p. 105. Thomas was apparently unaware that Wolfgang Schüler of Wittenberg and others had observed the new star before Tycho. For more recent discoveries of observations in the Orient, see David H. Clark and F. Richard Stephensen, *The Historical Supernovae*, Oxford, UK and New York: Pergamon Press, 1977, p. 173 and 14–39.

3 Argelander, "The variable stars," 92.

4 Thomas, dissertation, p. 2 and pp. 34–40; E. Dorrit Hoffleit, "History of the discovery of Mira stars," *JAAVSO* **25** (1997), 115); Argelander, "The variable stars," 92; and Agnes M. Clerke, *A Popular History of Astronomy during the Nineteenth Century*, London: Adam and Charles Black, 1908, p. 10.

5 Argelander, "The variable stars," 93.

6 Michael A. Hoskin, *Stellar Astronomy*, Chalfont St. Giles, Bucks, England: Science History Publications, 1982, 41–3.

7 Caroline E. Furness, *An Introduction to the Study of Variable Stars*, Boston: Houghton Mifflin Company, 1915, pp. 269–70.

8 Furness, *Study of Variable Stars*, p. 270; Michael Hoskin, "Goodricke, Pigott and the quest for variable stars," *JHA* **10** (1979), 34; and Carolyn Gilman, "John Goodricke and his variable stars," *S&T* **56** (1978), 400–3.

9 Agnes Mary Clerke, "An historical and descriptive list of some double stars suspected to vary in light," *Nature (London)* (1888), 55–8; and John E. Isles, *Variable Stars*, Hillside, New Jersey and Aldershot, Hants, UK: Enslow Publishers, Inc., 1990, p. 3.

10 William Herschel, *The Scientific Papers of Sir William Herschel (in 2 volumes)*, London: The Royal Society and the Royal Astronomical Society, 1912, Vol. 1, pp. 530–58. By the time of his death, Herschel had compiled two additional catalogs – these were published by Harvard College Observatory Director Edward Charles Pickering in the *Harvard Annals*. Edward C. Pickering, "Observations with the meridian photometer during the years 1879–82," *AnHar* **14** (1884), 345–56.

11 Seth C. Chandler, Jr., "Third catalogue of variable stars," *AJ* **16** (1896), 145–72.

12 Cited in John R. Percy, *Understanding Variable Stars*, Cambridge, UK: Cambridge University Press, 2007, p. 3. Herschel's statement appeared originally as part of Dionisius Lardner's *Cabinet Cyclopedia* in 1833. John Frederick William Herschel, *A Treatise on Astronomy*, London: Longman, Rees, Orme, Brown, Green & Longman, Paternoster Row; and John Taylor, Upper Gower Street, 1833.

13 Friedrich W. A. Argelander, *Uranometria Nova*, Berlin: Simon Schropp, 1843; Argelander, "Aufforderung," 122–254; Friedrich W. A. Argelander, *De Stella Lyrae Variabili Disquisitio*, Bonn: Formis Caroli Georgii, 1844; Friedrich W. A. Argelander, "Beobachtungen und Rechnungen über Veränderliche Sterne," *Bonner Beobachtungen* **7** (1869); and Thomas, dissertation, p. 10. In 1883, Eduard Schönfeld, Argelander's successor at Bonn, brought to the attention of Pickering a series of observations made by Argelander after 1867, but never published. E. C. Pickering published about 4000 observations made from 1859 to 1871 in this series in the *Harvard Annals* in 1900. Edward C. Pickering, "Observations of variable stars by Argelander," *AnHar* **33** (1900), 29–74.

14 E. Dorrit Hoffleit, "A history of variable star astronomy to 1900 and slightly beyond," *JAAVSO* **15** (1986), 78.

15 Furness, *Study of Variable Stars*, pp. 88–9.

16 Argelander, "The variable stars," 91.

17 Friedrich W. A. Argelander, *Bonner Durchmusterung (volumes 1–3)*, Bonn: 1859.

18 Furness, *Study of Variable Stars*, pp. 42–50.

19 Eduard Heis, *Atlas Coelestis Novus*, Cologne: M. Dumont-Schauberg, 1872; and Furness, *Study of Variable Stars*, pp. 38–9 and pp. 252–3.
20 Edwin Dunkin, *Obituary Notices of Astronomers*, London: Williams & Norgate, 1879, pp. 40–6; and Rev. Johann Georg Hagen, S. J., *Die Veränderlichen Sterne (vol. 1, part 1)*, Freiburg, Germany: Herdersche Verlagshandlung, 1913, 13.
21 Rev. Johann Georg Hagen, S. J. (Editor), *Beobachtungen Veränderlicher Sterne von Eduard Heis aus den Jahren 1840–1877 und von Adalbert Krueger aus den Jahren 1853–1892*, Berlin: Felix L. Dames, 1903.
22 HCO director Pickering published a selection of Schmidt's observations in the *Harvard Annals* in 1900. Edward C. Pickering (Compiler), "Observations of variable stars by Schmidt," *AnHar* **33** (1900), 95–134.
23 Eduard Schönfeld, "Beobachtungen von veränderlichen sternen," *Sitzungsberichte der Kais. Akad. der Wissensch., Math.-Naturw.* **42**; **44** (1861), cited in *Harvard Annals* 33.4, 1900; and Hagen, *Var. Stars I*, p. xiv. E. C. Pickering converted these to the Harvard photometric scale and published the results in the *Harvard Annals* in 1900. Edward C. Pickering, "Observations of variable stars by Schönfeld," *AnHar* **33** (1900), 75–94.
24 Furness, *Study of Variable Stars*, pp. 260–1; John A. Parkhurst, "Variable stars. V.," *PA* **1** (1894), 315; and W. Valentiner, "Nachgelassene beobachtungen von E. Schönfeld," *Veröiffentlichungen der Grossherzoglichen Sternwarte zu Heidelberg, Astronomisches Institut* **1** (1900).
25 Chandler, "Third catalogue," 145–72.
26 John Toone, "British variable star associations 1848–1908," *JBAA* **120** (2010), 135–51.
27 Toone, "British variable star associations," 136–8.
28 J. B. Hearnshaw, *The Measurement of Starlight: Two Centuries of Astronomical Photometry*, Cambridge, UK: Cambridge University Press, 1996, pp. 74–6.
29 Chandler, "Third catalogue," 145–72.
30 Rev. Johann Georg Hagen, S. J., *Supplementary Notes to the Atlas Stellarum Variabilium I. The New Star in Perseus II. Some Engraved Charts of Pogson's Proposed Atlas*, Washington, DC: Georgetown College Observatory, 1904, 25–33.
31 Joseph Baxendell, "Baxendell's observations of variable stars. Edited by H. H. Turner," *MmRAS* **73** (1913), 73; and George Knott, "Observations of twenty-three variable stars by the late George Knott. Edited by H. H. Turner," *MmRAS* **52** (1899), 1–318.
32 Chandler, "Third catalogue," 145–72.
33 H. H. Turner, "Observations of variable stars made at the Rousdon Observatory, Lyme Regis, under the direction of the late Sir C. E. Peek. Edited by H. H. Turner," *MmRAS* **55** (1904), 55; and P. M. Ryves, "Mary Adela Blagg," *Monthly Notices of the Royal Astronomical Society* **105** (1945), 65–6.
34 Richard Baum, "Historical note: The Observing Astronomical Society – birth of a legend," *JBAA* **108** (1998), 42–3.
35 Gerard Gilligan *et al.*, *The History of the Liverpool Astronomical Society*, Liverpool, England: Liverpool Astronomical Society, 1991; and Thomas R. Williams, "Getting Organized: A history of amateur astronomy in the United States," unpublished Ph.D. dissertation, Rice University, 2000, 73–7.
36 Toone, "British Variable Star Associations," 142–9; and E. E. Markwick, "Variable stars and how to observe them. I–XII," *English Mechanic and World of Science* **84** (1906–1907).
37 David S. Evans, *Under Capricorn: A History of Southern Hemisphere Astronomy*, Bristol, UK and Philadelphia: Adam Hilger, 1988.
38 John Tebbutt, "On the desireability of a systematic search for and observations of variable stars in the southern hemisphere," *Transactions of the Philosophical Society of New South Wales*, 126–39.
39 John Tebbutt, "On the variable star R Carinae, read before the Astronomical Section of the Royal Society of N.S.W. on 1 July 1881," *RSNSW* **15** (1881), 380–5; Wayne Orchiston, "John Tebbutt of Windsor, New South Wales: pioneer southern hemisphere variable star observer," *The Irish Astronomical Journal* **27** (2000), 47–54; and Toone, "British variable star associations," 43.
40 Wayne Orchiston, "John Tebbutt and observational astronomy at Windsor Observatory," *JBAA* **114** (2004), 47–8.
41 Orchiston, "Tebbutt pioneer VSOer," 51–2.
42 Alexander William Roberts, "Density of Close Double Stars," *ApJ* **10** (1899), 308–14; Henry Norris Russell, "Densities of the variable stars of the Algol type," *ApJ* **10** (1899), 315–18; and Donald G. McIntyre, "Alexander William Roberts (1857–1938)," *Journal of the Astronomical Society of South Africa* **4** (1938), 116–24.
43 D. B. Herrmann, "B. A. Gould and his *Astronomical Journal*," *JHA* **2** (1971), 98–108.
44 Marc Rothenberg, "The education and intellectual background of American astronomers, 1825–1875," Ph.D. dissertation, Bryn Mawr College, 1974, pp. 127–8.

45 Friedrich W. A. Argelander, "Minima Algol and T Cancri," *AJ* **4** (1856), 185–6.

46 Benjamin A. Gould, "Observations of the variable stars," *AJ* **4** (1856), 190.

47 Benjamin A. Gould, "Preamble to the seventh volume," *AJ* **7** (1886), 1; Furness, *Study of Variable Stars*, pp. 270–2; and Herrmann, "B. A. Gould and *AJ*," 104–6.

48 Benjamin A. Gould, "Uranometria Argentina," *Resultados del Observatorio Nacional Argentino en Córdoba* **1** (1879), 1–20.

49 Benjamin A. Gould, "Catalogo general argentino: posiciones medias de estrellas australes determinadas en el Observatorio Nacional," *Resultados del Observatorio Nacional Argentino en Córdoba* **14** (1886), 141–650.

50 Bill Carter and Merri Sue Carter, *Latitude: How American Astronomers Solved the Mystery of Variation*, Annapolis, Maryland: Naval Institute Press, 2002, 56; and Gould, "Preamble," 1.

2 A NEED FOR OBSERVERS

1 Edward C. Pickering, *A Plan for Securing Observations of the Variable Stars*, Cambridge, Massachusetts: privately printed, 1882, p. 4.

2 Stillman Masterman, "Observations of Algol-minimum," *AJ* **5** (1857), 31; Donald deB. Beaver, "The American scientific community, 1800–1860: A statistical-historical study," Ph.D. dissertation, Yale University, 1966, p. 319; and Michael Saladyga, "The 'pre-embryonic' state of the AAVSO: amateur observers of variable stars in the United States from 1875 to 1911," *JAAVSO* **27** (1999), 155.

3 "Our amateurs in science," *The Sunday Herald,* Boston, Massachusetts, 7 August 1898, p. 27; Solon I. Bailey, *The History and Work of Harvard Observatory*, New York: McGraw-Hill, 1931, pp. 266–8; Saladyga, "The 'pre-embryonic' state," 155; and Bill Carter and Merri Sue Carter, *Latitude: How American Astronomers Solved the Mystery of Variation*, Annapolis, Maryland: Naval Institute Press, 2002, pp. 81–9.

4 Paul S. Yendell, "S. C. Chandler," *Vierteljahrschrift der Astronomischen Gesellschaft* **49** (1914), 240–2; and Carter and Carter, *Latitude*, pp. 39–42 and 69–70.

5 Edward C. Pickering, "Observations with the meridian photometer during the years 1879–82," *AnHar* **14** (1884), 1–512.

6 Edwin F. Sawyer, *Scrapbook of Newspaper Clippings*, EFS, 1, Scrapbook, AAVSO.

7 Seth C. Chandler, Jr., "On the methods of observing variable stars," Cambridge, Massachusetts: [*Astronomical Journal*], 1887, pp. 3, 5. Reprinted from the *Science Observer* of January, February, and March, 1878. With an introduction by B[enjamin] A[pthorp] G[ould].

8 A. Searle, "S. C. Chandler," *PA* **22** (1914), 271–4.

9 Pickering, "Observations 1879–82," 71–5.

10 Searle, 271–4; Edward C. Pickering, "Observations with the meridian photometer during the years 1879–82," *AnHar* **14** (1885), 431–76; and Seth C. Chandler, Jr., "Catalogue of variable stars," *AJ* **8** (1888), 90.

11 Horace A. Smith, "Popular Astronomy Magazine and the development of variable star observing in the United States," *JAAVSO* **19** (1980), 41. Yendell's unpublished observations are in the AAVSO archives.

12 Edwin F. Sawyer, "Catalogue of the magnitudes of southern stars from 0° to –30° declination, to the magnitude of 7.0 inclusive," *Memoirs of the American Academy of Arts and Sciences, New Series* **12** (1893), 1–6.

13 Paul S. Yendell, "Suggestions to observers of variable stars," *PA* **2** (1894), 12.

14 Helen Lewis Thomas, "The early history of variable star observing to the XIX century," Ph.D. dissertation, Radcliffe College, 1948, pp. 14–15.

15 AAVSO, *Memorial to E. Charles Pickering: 1846–1919*, Norwich, Connecticut: AAVSO, 1920, p. 9. See also Dorrit Hoffleit, "Preludes to the founding of the AAVSO," *JAAVSO* **15** (1986), 110.

16 Edward C. Pickering, "Variable stars of short period," *Proceedings of the American Academy of Arts and Sciences* **16** (1881), 257.

17 Edward C. Pickering, *A Plan for Securing Observations of the Variable Stars*, Cambridge, Massachusetts: privately printed, 1882, p. 4.

18 William W. Payne, "Editorial notes," *SidM* **1** (1882), 237; and William W. Payne, "Editorial notes," *SidM* **2** (1883), 25.

19 Edward C. Pickering, "A plan for securing observations of the variable stars [part 1]," *Obs* **6** (1883), 46–51; and Edward C. Pickering, "A plan for securing observations of the variable stars [part 2]," *Obs* **6** (1883), 79–82.

20 Edward C. Pickering, "Co-operation in observing variable stars," *PA* **9** (1901), 148–51; Edward C. Pickering, "Cooperation in observing variable stars," *HarCi* No. 112 (1906), 1–3; and Edward C. Pickering, "Cooperation in observing variable stars," *HarCi* No. 166 (1911), 1–6.

21 W. J. Youmans, ed., "Editorial note," *The Popular Science Monthly* **32** (1887), 282–3.

22 Edward C. Pickering, "Aid to astronomical research," *Sci* **41** (1915), 82–5; and Catherine Elaine Nisbett, "Business practice: The rise of American astrophysics, 1859–1919," unpublished Ph.D. dissertation, Princeton University, 2007, pp. 9–12.

23 *ibid.* Though her title is misleading, Nisbett's Princeton University dissertation deals exclusively with Pickering's many and diverse interests and distractions, as well as his drive for efficiency during this period.

24 J. B. Hearnshaw, *The Measurement of Starlight: Two Centuries of Astronomical Photometry*, Cambridge, UK: Cambridge University Press, 1996, pp 88–94. Tables 3.2 and 3.3 are especially informative about the magnitude of Pickering's personal efforts as an observer.

25 Howard Plotkin, "Pickering, Edward Charles," in John A. Garraty and Mark C. Carnes (General Editors), *American National Biography,* 1st Ed., New York and Oxford: Oxford University Press, 1999, 476–8. See also John Lankford and Ricky L. Slavings, "The Industrialization of American astronomy, 1880–1940," *PhT* **49** (1996), 34–40.

26 William W. Payne, "How to observe variable stars," *SidM* **10** (1891), 249.

27 William W. Payne and George Ellery Hale, *Astronomy and Astro-Physics* **11–12** (1892). See also Donald E. Osterbrock, "Founded in 1895 by George E. Hale and James E. Keeler: The *Astrophysical Journal* Centennial," *ApJ* **438** (1995), 3.

28 Yendell, "Suggestions to observers," 12–17.

29 Paul S. Yendell, "On the variable stars of short period [I–IV]," *PA* **2** (1894–1895): I, pp. 160–8; II, pp. 202–9; III. pp. 269–74; IV. pp. 362–6.

30 Annie J. Cannon, "Williamina Paton Fleming," *ApJ* **34** (1911), 314–17.

31 Owen Gingerich, "Henry Draper's Scientific Legacy," A. E. Glassgold *et al.*, Editors, *Symposium on the Orion Nebula to Honor Henry Draper*, December 1981, New York University, New York City (New York, New York: New York Academy of Sciences, 1982), pp. 315–18; and Howard Plotkin, "Henry Draper, Edward C. Pickering, and the birth of American astrophysics," in A. E. Glassgold *et al.*, Editors, *Symposium on the Orion Nebula to Honor Henry Draper*, December 1981, New York University, New York City (New York, New York: New York Academy of Sciences, 1982), pp. 326–8.

32 J. B. Hearnshaw, *The Analysis of Starlight: One Hundred and Fifty Years of Astronomical Spectroscopy*, Cambridge, UK: Cambridge University Press, 1986, p. 127.

33 Seth C. Chandler, Jr., "Second catalogue of variable stars," *AJ* **13** (1893), 90–1.

34 Chandler, "Catalogue of VS," 94.

35 Mrs. M. Williamina Paton Fleming, "A field for woman's work in astronomy," *A&A* **12** (1893), 688.

36 Seth C. Chandler, Jr., "Third catalogue of variable stars," *AJ* **16** (1896), 145.

37 John Lankford, "Amateur versus professional: The transatlantic debate over the measurement of Jovian longitude," *JBAA* **89** (1979), 574–82; John Lankford, "Amateurs versus professionals: The controversy over telescope size in late Victorian science," *Isis* **72** (1981), 11–28; and Marc Rothenberg and Thomas R. Williams, "Amateurs and the society during the formative years," in David H. DeVorkin (Editor), *The American Astronomical Society's First Century,* Washington, DC: American Astronomical Society, 1999, 40–52.

38 Yendell, "Suggestions to observers," 17.

39 Paul S. Yendell, "The study of the variable stars. II.," *PA* **4** (1896), 18.

40 Pickering, "Variable stars of short period."

41 John A. Parkhurst, "Variable stars. VIII.," *PA* **1** (1894), 460–4.

42 John A. Parkhurst, "Variable stars. V.," *PA* **1** (1894), 316.

43 Parkhurst, "Variable stars. V.," 316.

44 Chandler, "Third catalogue," 145.

45 AAVSO, *Memorial to ECP*, 10.

46 Joseph Ashbrook, "J. G. Hagen and his cosmic clouds," *S&T* **42** (1971), 215.

47 Alice H. Farnsworth, "Rev. John G. Hagen, S. J. 1847–1930," *PASP* **42** (1930), 281–4; and J. Stein, "Johann George Hagen S. J.," *PA* **39** (1931), 8–13. Denver Applehans called our attention to Hagen's engagement of his Prairie du Chien students in variable star observing at the History of Astronomy Workshop, Notre Dame University, July 2007.

48 Rev. Johann Georg Hagen, S. J., "An atlas of variable stars in preparation," *PA* **5** (1897), 20.

49 Caroline E. Furness, *An Introduction to the Study of Variable Stars*, Boston: Houghton Mifflin Company, 1915, pp. 55–6.

50 O. C. Wendell, and E. C. Pickering, "Observations of fifty-eight variable stars of long period during the years 1890–1901," *AnHar* **37** (1902), 2, 189.

51 Furness, *Study of Variable Stars*, 52–7.

52 O. C. Wendell, and E. C. Pickering, "Observations of circumpolar variable stars during the years 1889–1899," *AnHar* **37** (1900), 1; and Edward C. Pickering, *Variable Stars of Long Period*, Cambridge, Massachusetts: privately printed, 1891.

53 Wendell and Pickering, "Fifty-eight variables," 204.
54 AAVSO, *The Practical Observing of Variable Stars*, Norwich, Connecticut: AAVSO, 1918, 32.
55 Pickering, "Co-operation," 1901, 148–51.
56 William M. Reed, and Annie J. Cannon, "A provisional catalogue of variable stars," *AnHar* **48** (1903), 1–123; and Annie J. Cannon, "Second catalogue of variable stars," *AnHar* **55** (1907), 1–94.
57 Pickering, "Cooperation," 1906, 1–3.
58 Edward C. Pickering, "843 new variable stars in the Small Magellanic Cloud," *HarCi* No. 96 (1905), 1–4; and Henrietta Swan Leavitt, "New variable stars in the Small Magellanic Clouds," *Publications of the Astronomical and Astrophysical Society of America* **1** (1910), 257–8.
59 Edward C. Pickering, "A durchmusterung of variable stars," *HarCi* No. 116 (1906), 1–3.
60 Pickering, "843 new variables," 1–4; and Pickering, "A durchmusterung," 1–3.
61 The undated lists of active and inactive variable star observers are found in the Harvard University Archives, Observatory Director's Correspondence, UAV 630.17.5, File II, Box 37 VAL-WEN, Folder Variable Star (Observers and Work). Plotkin's list is in personal correspondence in the author's files.
62 Pickering, "Cooperation," 1911; and various, "AAVSO archival observation reports (1902–1961)," 1902, various reports, AAVSO.
63 Pickering, "Cooperation," 1911, 1–3.
64 Furness, *Study of Variable Stars*, 255–6.
65 Henry M. Parkhurst, "Observations of variable stars," *AnHar* **29** (1893), 29–170.
66 Edward C. Pickering (compiler), "Observations of variable stars by Argelander," *AnHar* **33** (1900), 29–74; Edward C. Pickering (compiler), "Observations of variable stars by Schönfeld," *AnHar* **33** (1900), 75–94; and Edward C. Pickering (compiler), "Observations of variable stars by Schmidt," *AnHar* **33** (1900), 95–134. Pickering's copies of the Argelander, Herschel, and Schmidt observations are preserved in the AAVSO Archives.
67 Paul S. Yendell, "The Heis-Krueger observations of variable stars," *PA* **18** (1910), 516–17.
68 A nearly complete collection of the *Monthly Register of the Society for Practical Astronomy* exists in the archives of the US Naval Observatory in Washington, DC, as QB1.S75. We are unaware of any other collection of this journal. See Thomas R. Williams, "Reconsidering the history of AAVSO-Part 1," *JAAVSO* **29** (2001), 132–45.
69 Frederick C. Leonard, "The Society for Practical Astronomy," *PA* **19** (1911), 455–6.
70 Russell Williams Porter, and Robert H. Bowen, "Special notice of the proposed International Conference of the S.P.A. in 1914," *MRSPA* **6** (1914), 32.
71 Robert H. Bowen, "Report on the first international conference of the Society for Practical Astronomy," *MRSPA* **6** (1914), 49–53.
72 "Meeting of the Society for Practical Astronomy at Chicago," *MESM* (1915), unpaginated; and "Conference of the Society for Practical Astronomy," *PA* **23** (1915), 536. In personal communication in the author's files, Dr. Roy Clarke has pointed out that Francis P. Leavenworth was a friend of the Leonard family from their Indiana days and visited them regularly when traveling from Indiana to Wisconsin each summer to work at Yerkes. The young Leonard is found in several group pictures taken at Yerkes. Thus Leonard was well known among professional astronomers when his first announcement of the formation of the SPA was published.
73 S. F. Maxwell, "Minutes of the second annual conference [of the Society for Practical Astronomy] held at The University of Chicago, Ill. 1915 August 16, 17, and 18," *MRSPA* **7** (1915), 46–56.
74 "Dr. Edward Gray," *PA* **21** (1913), 375.
75 Ursula B. Marvin, "The Meteoritical Society: 1933 to 1993," *Meteoritics* **28** (1993), 281–2.
76 Roy S. Clarke, Jr., "The AAVSO," to Thomas R. Williams, 12 July 2000, copy in author's files.

3 THE AMATEUR'S AMATEUR

1 William Tyler Olcott to Harriet W. Bigelow, 4 October 1911, Org ECP-WTO-LC, Box 1, Early Correspondence folder, AAVSO.
2 William Tyler Olcott, "The first monthly report of the American Association of Variable Star Observers," *PA* **19** (1911), 655–8.
3 "William M. Olcott," *Norwich Bulletin*, Norwich, Connecticut, 24 June 1902, p. 1. Microfilm Newspapers Collection, OLA; and Orderly Sgt. James R. Haydon, "U. S. Zouave Cadets Govenors [*sic*] Guard of Illinois," EEE, Box 125, folder 1, CHS. The Haydon notebook, to which both the ownership and attributed authorship are only "probable," includes a roster of members of the unit in the order in which they were enlisted. The pages of this notebook are unnumbered.

4 Charles A. Ingraham, *Elmer E. Ellsworth and the Zouaves of '61*, Chicago: The University of Chicago Press, 1925, pp. 69, 89; and Orderly Sgt. James R. Haydon.

5 "William M. Olcott." The fact that Olcott had been in the coal business before going into the Army is documented in several places in the Ellsworth papers, and in fact, E. B. Knox, a former Zouave and later Colonel Knox, referred to him as "Billy Olcott, the coal man." Olcott was apparently well liked among the Zouave Cadets and participated in their reunions. Ingraham, *Ellsworth and the Zouaves*, p. 67.

6 There is, for example, in the Norwich Township Land Records, a recorded loan by Capt. Roath to Samuel Tyler in 1830. Samuel Tyler, and Lyman Brewer, "Lien on Property on Church Street to Edmund D. Roath," 28 July 1830, Norwich Land Records, Volume 43, pages 49–50, Norwich Township Clerk's Office, Norwich, Connecticut. Land adjacent to the Glebe House was mortgaged to meet demands against the estate of Rev. John Tyler, but the mortgage was quickly released as the estate was settled. Edmund D. Roath, "Release of Lien on Property on Church Street," 7 September 1830, Norwich Land Records, Volume 43, pp. 50–1, Norwich Township Clerk's Office, Norwich, Connecticut.

7 "Obituary: The death of Mr. Wm. S. Tyler," *Norwich Morning Bulletin,* Norwich, Connecticut, 1 October 1864, p. 2. OLA; (contributed), "Mrs. E. Olivia Tyler Olcott," *Norwich Bulletin,* Norwich, Connecticut, 23 February 1915, p. 8. Microfilm Newspapers Collection, OLA; and John Plummer, "Glebe House," Hartford, Connecticut, April 1981, G 728 Nor, OLA.

8 Colbert and Chamberlain's standard period reference on the Great Chicago Fire included a "Map of Chicago; showing Parks, Boulevards, and Burnt District." Elias Colbert and E. Chamberlin, *Chicago and the Great Conflagration*, Cincinnati and New York: C. F. Vent, 1872, Map following p. 10.

9 "Removal," *Chicago Tribune,* Chicago, Illinois, 20 November 1871, p. 5. ProQuest Historical Newspapers, Chicago Tribune (1849–1986); "Consumers of Coal," *Chicago Tribune,* Chicago, Illinois, 17 December 1871, p. 7. ProQuest Historical Newspapers, Chicago Tribune (1849–1986); "The court record. The record of Aug. 12, Bankruptcy. New petitions," *Interocean,* Chicago, Illinois, 5 January 1875, p. 3. Newsbank, American Antiquarian Society, 2004; General Items, W. M. & J. F. Olcott 1875 #74150; "Lakawanna Coal," *Chicago Tribune,* Chicago, Illinois, 4 February 1877, p. 1. ProQuest Historical Newspapers, Chicago Tribune (1849–1986); and "The courts – superior court in brief," *Chicago Tribune,* Chicago, Illinois, 26 August 1879, p. 9A.

10 "In General," *Chicago Tribune,* Chicago, Illinois, 3 May 1888, p. 8; and Olivia Mahoney, *Douglas/Grand Boulevard: A Chicago Neighborhood*, Chicago: Arcadia, 2001, p. 17.

11 (contributed), "Mrs. E. Olivia Tyler Olcott"; and "William M. Olcott."

12 Robert P. Keep to Faculty of Trinity College, 12 July 1892, AFC, folder William Tyler Olcott, Class of 1896, TCA.

13 Edmund J. Churchill, "Received of Olivia A. T. Roath," Norwich, Connecticut: 5 January 1892, Norwich Land Records, Vol. 101, p. 126, Township Clerk's Office, Norwich, Connecticut; George W. Hill, "To All Whom It May Concern," Norwich, Connecticut: 5 January 1892, Norwich Land Records, Vol. 101, page 126, Township Clerk's Office, Norwich, Connecticut; and Olivia Ann Tyler Roath, "Inventory of Edmund D. Roath Estate," Norwich Probate Records, Volume 40, p. 77, Norwich, Connecticut Township Clerk's Office. The inventory of Capt. Roath's estate, valued at about $58 000, included real estate in Norwich; shares of stock in railroads, insurance companies, and banks; and a mortgage payable to the estate. Olivia Ann Tyler Roath was designated executrix in Capt. Roath's will. Olivia sent W. Tyler Roath's share of Capt. Roath's estate to the executor of Tyler's estate in Cheyenne. "The Late W. Tyler Roath," *Norwich Morning Bulletin,* Norwich, Connecticut, 23 June 1891, p. 6. Microfilm records, Archives of Otis Library, Norwich, Connecticut; and "Churchill, Receipt."

14 Laurence R. Veysey, *The Emergence of the American University*, Chicago: The University of Chicago Press, 1965, p. 263.

15 Veysey, *The American University*, pp. 270–1.

16 Veysey, *The American University*; pp. 276–8.

17 Registrar of Trinity College, "*Record of Student Standings, 1894–1900*," Administrative Records, Archives 17.3.1, Archives of Trinity College, Hartford, Connecticut. A large hand-written ledger book, this volume reveals the relative weight placed on such courses as bible study and themes, in contrast to what might today be considered the core courses. There is a separate unnumbered pair of facing pages for each class of students for each academic year, showing all the courses taken by every student in the class. Thus the Record of Student Standings affords the interested researcher not only a view of the individual, his course undertakings, and his relative performance in that course, but also of the differences in course undertakings by each member of the class. With a class of only 26, a reasonable understanding of an individual in the context of the whole class is clearly displayed.

18 *Trinity College Ivy for 1896,* Hartford, Connecticut: Trinity College, 1895, SC, TCA. Seven of the 26 members of the

class of 1896 were elected to Medusa, or slightly more than a quarter of the class. Mr. Peter Knapp, Trinity College Archivist, was very helpful during an all too brief visit to the Trinity University campus and archives. Mr. Knapp offered this explanation of the role of the Medusa society as a campus function. Olcott remained a member of the Kappa Delta Epsilon, supporting a membership in the fraternity's club in New York City as well as participating in Trinity College events from time to time.

19 *Trinity College Ivy for 1897,* Hartford, Connecticut: Trinity College, 1896, SC, TCA; and "A Norwich Boy writes a song for Rice's new comic opera," *Norwich Bulletin,* Norwich, Connecticut, 12 February 1896, p. 5. Microfilm Newspapers Collection, OLA.

20 Alfred Zantzinger Reed, *Present Day Law Schools in the United States and Canada*, New York: Carnegie Foundation for the Advancement of Teaching, 1928, p. 111, and pp. 131–4. Yale University Law School's admission requirements and curriculum were no more rigorous than those of the NYLS, and Yale ranked well below Harvard and Columbia in national stature in 1897. See Robert W. Gordon, "Professors and Policymakers," Anthony T. Kronman (Editor), *History of the Yale Law School, the tercentennial lectures,* New Haven, Connecticut and London, UK: Yale University Press, 2004, p. 81. One can therefore only assume, as we have here, that the attraction of NYLS included not only the opportunity of leaving law school after only 2 years, but also the prospect of an entertaining 2 years in mid-town Manhattan. William Tyler Olcott, "William Tyler Olcott," Hartford, Connecticut, 11 December 1919, AFC, folder Olcott, William Tyler, Class of 1896, TCA.

21 Julius Goebel (Compiler and Editor), *A History of the School of Law, Columbia University,* New York City: Columbia University Press, 1955, pp. 151–2.

22 Alfred Zantzinger Reed, *Training for the Public Profession of the Law*, New York City: The Carnegie Foundation for the Advancement of Teaching, 1921, pp. 192–6, and p. 452.

23 "Died – Roath," *Norwich Bulletin,* Norwich, Connecticut, 7 April 1899, p. 7. Microfilm Newspapers Collection, OLA; and "Funerals-Mrs. Edmund D. Roath," *Norwich Bulletin,* Norwich, Connecticut, 11 April 1899, p. 5. Microfilm Newspapers Collection, OLA. Olivia Ann Tyler Roath's will bequeathed an estate estimated by her executors, William Marvin Olcott and Elizabeth Olivia Olcott, to be valued at $82 854, in 2008 dollars equivalent to about $2 000 000. *Norwich Death Records*, Volume 50, p. 51. Olivia Ann Tyler Roath had inherited a substantial estate from her first husband, William Samuel Tyler, as well as the Roath estate. Thus this inheritance received by William Tyler Olcott was a combination of the estates of the Tyler family and that of Capt. Roath.

24 Michael McCarthy, "Memorandum of a telephone conversation with Thomas R. Williams re William Tyler Olcott," 5 November 2008 (Copy in Author's file); and Allon Schoener, *New York,* New York and London: W. W. Norton & Company, 1998, p. 108. At NYLS, a third year was required for completion of a degree, thus Tyler never "graduated" from law school.

25 Dale Plummer, *Norwich,* Charleston, South Carolina: Acadia Publishing, 2003.

26 Clara Hyde Olcott and William Tyler Olcott, *Social Scrapbook*, Norwich, Connecticut: Unpublished; Diane Norman, "Arcanum Club," to Thomas R. Williams, 11 November 2008 (copy in Author's file); and Diane Norman, "Olcott," to Thomas R. Williams, 18 November 2008 (copy in Author's file).

27 *History of Miss Dana's School, Finding Aid to Miss Dana's School Records, 1873–1938*, undated, Morristown and Morris Township Library Finding Aids, *http://www.morristownmorristwplibrary.info/HCFindingAids/MissDanas.xml.*

28 "Untitled newspaper fragments re Olcott wedding," unknown newspaper, June 1902, p. Unknown. Org ECP-WTO-LC, Box 2, file Olcott Social Scrapbook, AAVSO; "Funeral – William M. Olcott," *Norwich Bulletin,* Norwich, Connecticut, 26 June 1902, Microfilm Newspapers Collection, OLA; "Sympathy for Mrs. William M. Olcott," *Norwich Bulletin,* New London, Connecticut, 25 June 1902, p. 5. Microfilm Newspapers Collection, OLA; and "William M. Olcott."

29 Olcott and Olcott, *Social Scrapbook*; and Mary Backus Hyde, "Incidents in the life of an astronomer," Undated, but likely after W. T. Olcott's death, folder Olcott, Tyler, OLA. In this charming and affectionate unpublished tribute to her brother-in-law, Mary Hyde provides a great deal of the background that follows. The pages are unnumbered. Norman, "Olcott."

30 Hyde, "Incidents in the life of an astronomer."

31 William Tyler Olcott, *Star Lore of All Ages; a Collection of Myths, Legends, and Facts Concerning the Constellations of the Northern Hemisphere*, New York: G. P. Putnam, 1911.

32 We are grateful for the help of AAAS Archivist Dr. Amy Crumpton for digging out the meeting schedules for the AAAS 1909 meeting. She clarified not only the location of the meeting, but also the fact that Pickering was not scheduled to make a formal presentation in this meeting.

Also, although all laboratories on both campuses were open for the duration of the meeting and welcomed visitors, there was no scheduled tour of the Harvard College Observatory, which is located about 1 mile west of Harvard Square. In all likelihood, the HCO, like other laboratories on the campuses, held itself in an open house mode. So it seems likely that what Tyler Olcott saw was a poster or display on a wall of the observatory during his informal visit there. It is also unclear exactly where the Mathematics and Astronomy Division of the AAAS actually met. Thus this explanation of the first communication between Pickering and Olcott must be taken as a "best guess" in the absence of any formal evidence to the contrary.

33 William Tyler Olcott, "Variable star work for the amateur with a small telescope," *PA* **19** (1911), 129–42. Olcott had taken care to have his article reviewed for accuracy at HCO. Pickering had Leon Campbell review the article and then indicated to Olcott that the article was acceptable. Edward C. Pickering to William Tyler Olcott, 31 January 1911, Org ECP-WTO-LC, Box 1 Pickering-Olcott letters, AAVSO.
34 Herbert C. Wilson, "Notes for Observers – What an amateur can do," *PA* **19** (1911), 456–7.
35 Herbert C. Wilson to Edward C. Pickering, 11 September 1911, HCO-ECP, UAV 630.17.5, Box 44, folder Popular Astronomy 1911, HUA.
36 Edward C. Pickering to Herbert C. Wilson, 7 September 1911, HCO-ECP, UAV 630.17.5, Box 44, folder Popular Astronomy 1911, HUA; and Wilson, 11 September 1911.
37 William Tyler Olcott to Edward C. Pickering, 29 September 1911, HCO-ECP, UAV 630.17.5, Box O-Pai, folder William Tyler Olcott 1911–1915, HUA.
38 William Tyler Olcott, "Association of Variable Star Observers," *Scientific American* **105** (1911), 484.
39 Edward C. Pickering to William Tyler Olcott, 2 October 1911, HCO-ECP, UAB630.17.5, Box O-Pai, folder William Tyler Olcott 1911–1915, HUA.
40 William Tyler Olcott to Edward C. Pickering, 29 September 1911, HCO-ECP, UAV 630.17.5, O-Pai, William Tyler Olcott 1911–1915, HUA.
41 William Tyler Olcott to Harriett W. Bigelow, 4 October 1911, ECO-WTO-LC-Org, 1, Early Correspondence folder, AAVSO.
42 Pickering, 7 September 1911. On the matter of professional control of amateurs, see Marc Rothenberg, "Organization and control: Professionals and amateurs in American astronomy, 1899–1918," *Social Studies of Science* **11** (1981), 305–25.
43 Edward C. Pickering to William Tyler Olcott, 7 October 1911, HCO-ECP, UAV 630.16, HCO Letters, July-December 1911, HUA.
44 David P. Todd to Edward C. Pickering, 4 October 1911, HCO-ECP, UAV 639.17.5, File II, box 51, TIE-VTI folder Todd, David, HUA.
45 Edward C. Pickering to David P. Todd, 5 October 1911, HCO-ECP, UAV 630.16, HCO Letters, Jul-Dec 1911, HUA. The Amherst operation was valuable to Pickering in that with a larger than average telescope, they were able to follow most long-period variables through minima that most amateur observers could not reach with smaller telescopes. Pickering was anxious not to offend Todd or create double work for him.
46 David P. Todd to Edward C. Pickering, 14 October 1911, HCO-ECP, UAV 630.17.5, File II, box 36 TIE-VTI folder Todd, David, HUA.
47 William Tyler Olcott to Helen M. Swartz, 29 October 1911, Org ECP-WTO-LC, Box 1, Early Correspondence folder, AAVSO.
48 R. Newton Mayall, "The story of the AAVSO," *RPA* **55** (1961), 4–9.

4 AMATEURS IN THE SERVICE OF SCIENCE

1 William Tyler Olcott to Edward C. Pickering, 30 December 1912, HCO-ECP, UAV 630.17.5, 41, Olcott, Tyler Wm., HUA.
2 J. Rode Jacobson, "Sigurd Enebo, discoverer of 'Nova Geminorum' and his observarory," *PA* **20** (1912), 328; and Herbert C. Wilson, "A new star (Nova Geminorum No. 2)," *PA* **20** (1912), 448–9.
3 Edward C. Pickering, "Nova Geminorum, No. 2," *HarCi* No. 176 (1912), 1–6.
4 Edward C. Pickering, "Photometric measurements of Nova Geminorum, No. 2," *HarCi* No. 175 (1912), 1–3.
5 John A. Parkhurst, "Changes in the early spectrum of Enebo's Nova in Gemini," *PA* **20** (1912), 236–8.
6 Edward C. Pickering to William Tyler Olcott, 18 April 1912, HCO-ECP, UAV 630.16, Outgoing Letters, Jan-Jun 1912, HUA.
7 William Tyler Olcott, "The monthly report of the American Association of Variable Star Observers," *PA* **20** (1912), 323.
8 Frederick C. Leonard, "Observations of Nova (2) Geminorum," *PA* **20** (1912a), 310–13; Frederick C. Leonard, "Observations of Nova (2) Geminorum," *PA* **20** (1912b), 379–80; and Frederick C. Leonard, "Observations of Nova (2) Geminorum," *PA* **20** (1912c), 449–50.

9 (form letter) HCO-ECP; UAV 630.17.5, File II, 52 VAL-WEN Variable Star (Observers and Work), HUA.

10 William Tyler Olcott to Edward C. Pickering, 24 February 1912, HCO-ECP, UAV 630.17.5, 41, Olcott, Tyler Wm., HUA; Edward C. Pickering to William Tyler Olcott, 26 February 1912, HCO-ECP, UAV 630.16, Outgoing Letters, Jul-Dec 1911, HUA; Edward C. Pickering to William Tyler Olcott, 1 May 1912, HCO-ECP, UAV 630.16, Outgoing Letters, Jan-Jun, 1912, HUA; and Edward C. Pickering to William Tyler Olcott, 10 June 1912, HCO-ECP, UAV 630.16, Outgoing Letters, Jan-Jun 1912, HUA.

11 Edward C. Pickering to William Tyler Olcott, 28 October 1912, ECP-WTO-LC-Org, 1, Early Correspondence, AAVSO.

12 "Dr. Edward Gray," *PA* **21** (1913). Wilson's thinly veiled cheerleading for Olcott's AAVSO in preference to Leonard's SPA is apparent in this announcement.

13 William Tyler Olcott to Edward C. Pickering, 14 December 1911, HCO-ECP, UAV 630.17.5, 41 O-Pai, Olcott, Tyler Wm., HUA; William Tyler Olcott to Edward C. Pickering, 30 December 1911, HCO-ECP, UAV 630.17.5, 41 O-Pai, Olcott, Tyler Wm., HUA; and William Tyler Olcott to Edward C. Pickering, 20 January 1912, HCO-ECP, UAV 630.17.5, 41, Olcott, Tyler Wm., HUA.

14 Although the blueprint medium of reproduction remained standard, the format of charts continued to evolve over time. See Kerriann H. Malatesta, and Charles E. Scovil, "The history of AAVSO charts, Part I: The 1880s through the 1950s," *JAAVSO* **34** (2005), 81–101.

15 Edward C. Pickering to William Tyler Olcott, 4 November 1913, ECP-WTO-LC-Org, 1, Pickering-Olcott letters, AAVSO; and William Tyler Olcott to Edward C. Pickering, 9 November 1913, HCO-ECP, UAV 630.17.5, 41, Olcott, Tyler Wm., HUA.

16 Edward C. Pickering to William Tyler Olcott, 18 December 1914, HCO-ECP, UAV 630.16, HCO Letters, July-December 1914, HUA.

17 See for example Frederick C. Leonard, "Results of a preliminary investigation of the light variations of 202946 *SZ* Cygni," *MRSPA* **8** (1916), 37–9.

18 Source: "Variable Star Observers, 1912–13," HCO-ECP, UAV 639.17.5, File II, 52 VAL-WEN Variable Star (Observers and Work), HUA; and "Variable Star Observers, 1913–14," HCO-ECP, UAV 639.17.5, File II, 52 VAL-WEN Variable Star (Observers and Work), HUAU.

19 Edward C. Pickering to William Tyler Olcott, 9 February 1913, ECP-WTO-LC-Org, 1, Pickering-Olcott letters, AAVSO.

20 Edward C. Pickering to William Tyler Olcott, 22 January 1913, ECP-WTO-LC-Org, 1, Pickering-Olcott letters, AAVSO.

21 William P. Sheehan, and Anthony Misch, "Ménage à Trois: David Peck Todd, Mabel Loomis Todd, Austin Dickinson, and the 1882 Transit of Venus," *JHA* **35** (2004), 123–34.

22 Frank E. Seagrave to Edward C. Pickering, 29 May 1913, HCO-ECP, UAV 630.17.5, 33, Sco-Sme, Seagrave, F. E. folder 1, HUA; and Frank E. Seagrave, Boston, Mass: 13, HCO-ECP,UAV 630.17.5, 48, Sco-Sme, 3 Seagrave, F. E., HUA.

23 Edward C. Pickering to William Tyler Olcott, 22 January 1913, ECP-WTO-LC-Org, 1, Pickering-Olcott letters, AAVSO.

24 David H. DeVorkin, "Community and spectral classification in astrophysics: The acceptance of E. C. Pickering's system in 1910," *Isis* **72** (1981), 29–49; Howard Plotkin, "Henry Draper, Edward C. Pickering, and the birth of American astrophysics," in A. E. Glassgold *et al.*, Editors, *Symposium on the Orion Nebula to Honor Henry Draper*, December 1981, New York University, New York City (New York, New York: New York Academy of Sciences, 1982), 404; and Howard Plotkin, "Pickering, Edward Charles (1846–1919)," in John Lankford (Editor), *History of Astronomy: An Encyclopedia*, New York and London: Garland Publishing, Inc., 1997.

25 Howard Plotkin, "Edward C. Pickering and the endowment of scientific research in America, 1877–1918," *Isis* **69** (1978), 44–57; Bessie Zaban Jones and Lyle Gifford Boyd, *The Harvard College Observatory, The First Four Directorships, 1839–1919*, Cambridge, Massachusetts: Harvard University Press, 1971, pp. 176 and ff; and Plotkin, Lankford, *History*, p. 404.

26 David H. DeVorkin (Editor), *The American Astronomical Society's First Century*, Washington, DC: The American Astronomical Society, 1999.

27 For an elaboration on the whole topic of patronage in American astronomy, see John Lankford, *American Astronomy – Community, Careers and Power, 1859–1940*, Chicago and London: The University of Chicago Press, 1997, especially chapters 7 and 8.

28 An interesting glimpse of the activities at HCO in this busy period may be had by scanning the annual reports that were published each year by Pickering.

29 Edward C. Pickering to William Tyler Olcott, 4 November 1913, ECP-WTO-LC-Org, 1, Pickering Olcott letters, AAVSO.

30 "Sixty-seventh annual report of the director of The Astronomical Observatory of Harvard College for the year

ending September 30, 1912," Cambridge, Massachusetts: 1913, p. 7.

31 William Tyler Olcott, "Untitled table," Varied, Org ECP-WTO-LC, 2, Meeting Minutes, General and Council, AAVSO, inside cover. In this volume, in which Olcott kept the records of both meetings of the Association and of its Council, he also maintained, on the inside of the cover, a list of the meeting dates and locations. R. Newton Mayall, "The story of the AAVSO," *RPA* **55** (1961), 6.

32 Edward C. Pickering to William Tyler Olcott, 11 September 1915, ECP-WTO-LC-Org, 1, Pickering-Olcott letters, AAVSO; Edward C. Pickering to William Tyler Olcott, 23 September 1915, ECP-WTO-LC-Org, 1, Pickering-Olcott letters, AAVSO; and Edward C. Pickering to William Tyler Olcott, 28 September 1915, ECP-WTO-LC-Org, 1, Pickering-Olcott letters, AAVSO.

33 Samuel Eliot Morison, *Three Centuries of Harvard, 1636–1936*, Cambridge, Massachusetts: Harvard University Press, 1936, p. 414; and Edward C. Pickering to William Tyler Olcott, 6 October 1915, HCO-ECP, UAV 630.16, Outgoing Correspondence, Jul-Dec, 1915, Olcott, W. T., HUA.

34 Edward C. Pickering to William Tyler Olcott, 22 November 1915, HCO-ECP, UAV 630.17.5, 41, Olcott, W.T., HUA.

35 William Tyler Olcott to Edward C. Pickering, 6 December 1916, HCO-ECP, UAV 630.17.5, 41 O-PAI, Olcott, William Tyler, HUA.

36 Edward C. Pickering to William Tyler Olcott, 29 November 1916, ECP-WTO-LC-Org, 1, Pickering-Olcott letters, AAVSO.

37 Roy S. Clarke, Jr., "The AAVSO," to Thomas R. Williams, 12 July 2000.

38 Edward C. Pickering to William Tyler Olcott, 22 January 1916, WSH, William Tyler Olcott, AAVSO; and William Tyler Olcott to Edward C. Pickering, 24 January 1916, HCO-ECP, UAV630.17.5, 26, Olcott, William Tyler, HUA.

39 Joel Hastings Metcalf *et al.*, "Report of the Committee to visit the Astronomical Observatory," 8 October 1917, Harvard University Visiting Committee Report No. 103, Cambridge, Massachusetts: Harvard University.

40 Leon Campbell to H. W. Vrooman, 24 May 1917, LC-Cor, 8, Vrooman, H. W., AAVSO.

41 "All points about the draft," *Norwich Bulletin,* Norwich, Connecticut, 14 July 1917, p. 13. Microfilm Newspaper Collection, OLA.

42 Leon Campbell to Forrest H. Spinney, 23 November 1916, LC-Cor, 7, Spinney, F. H., AAVSO.

43 William Tyler Olcott, and William Tyler Olcott to Edward C. Pickering, 9 July 1917, HCO-ECP, UAV 630.17.5, 41 O-PAI, Olcott, William Tyler, HUA.

44 AAVSO, *The Story of the AAVSO and AAVSO Observers and Observations 1911–1993*, Cambridge, Massachusetts: American Association of Variable Star Observers, 1993. See the historiographic notes in the back matter for this book.

45 William Tyler Olcott, "Minutes of the American Association of Variable Star Observers," 10 November 1917, ECP-WTO-LC-Org, 2, Meeting Minutes, General and Council, AAVSO.

46 William Tyler Olcott, "Minutes of the Council of the American Association of Variable Star Observers," 10 November 1917, ECP-WTO-LC-Org, 2, Meeting Minutes, General and Council, AAVSO.

47 Copies of the constitution and by-laws, articles of incorporation, and the certificate of incorporation are preserved in the AAVSO archives.

48 The two-sided yellow cards by which Schulmaier gathered the information from AAVSO members remain in the AAVSO archives and form a valuable resource for twenty-first century historians.

49 For more on the history of the chart committee see Malatesta and Scovil, "AAVSO Chart History," 81–101.

50 William Tyler Olcott, "Minutes of the American Association of Variable Star Observers," 23 November 1918, ECP-WTO-LC-Org, 2, Meeting Minutes, General and Council, AAVSO.

5 LEON CAMPBELL TO THE RESCUE

1 Leon Campbell, "Looking backwards, or A. A. V. S. O. in retrospect," *VC* **2** (1931), 45–52.

2 S. B. Sutton, *Cambridge Reconsidered: $3\frac{1}{2}$ Centuries on the Charles*, Cambridge, Massachusetts, and London, England: The MIT Press, 1976, pp. 55–9.

3 Sutton, *Cambridge Reconsidered*, 59–72.

4 Cambridge city directories published by W. A. Goodnough & Company, Publishers and titled Cambridge City Directory, Annual Population and Business Addresses. The tragically short marriage of William J. Campbell to Leonora Rawding and the birth of twins to the couple is documented on the Hoenecke-Cannon, Parks-Williston-Sawicki, Couts-Rawding, and Monohan Family Trees published on http://Ancestry.com and queried for the last of several times on May 15, 2010.

5 Margaret Harwood, "Fifty years at HCO," *S&T* **8** (1949), 191.

6 Edward C. Pickering, Director, Harvard College Observatory, *53rd Annual Report of the Director of the Astronomical Observatory of Harvard College for the Year Ending September 30, 1898*, Cambridge, Massachusetts: Harvard College Observatory, 1898, pp. 5–6.

7 Harwood, "Fifty years," 191.

8 Ibid, 191–2; Leon Campbell, "Leon Campbell," Undated, LC-Cor, 2, Campbell, L., AAVSO.

9 Leon Campbell, "Leon Campbell," LC-Cor, 2, Campbell, L. Seth Carlo Chandler provided a detailed explanation of the Argelander step method. See Seth C. Chandler, Jr., *On the Methods of Observing Variable Stars*, Cambridge, Massachusetts: [*Astronomical Journal*], 1887.

10 Edward C. Pickering, Director, Harvard College Observatory, *58th Annual Report of the Director of the Astronomical Observatory of Harvard College for the Year Ending September 30, 1903*, Cambridge, Massachusetts: Harvard College Observatory, 1903, p. 4.

11 Pickering, Director, Harvard College Observatory, *HCO 58th Annual Report*, p. 4; and Edward C. Pickering, Director, Harvard College Observatory, *60th Annual Report of the Director of the Astronomical Observatory of Harvard College for the Year Ending September 30, 1905*, Cambridge, Massachusetts: Harvard College Observatory, 1905, p. 9.

12 Lyman R. Martineaux, Jr., "1910 United States Federal Census," 20 April 1910, http://search.ancestry.com, accessed 2 July 2009.

13 Leon Campbell, "Leon Campbell," LC-Cor, 2, Campbell, L., AAVSO.

14 Campbell, "Looking backwards," 45.

15 Catherine Elaine Nisbett, "Business practice: The rise of American astrophysics, 1859–1919," unpublished Ph.D. dissertation, Princeton University, 2007, 52–96.

16 Nisbett, dissertation, 97–150; David S. Evans, *Under Capricorn: A History of Southern Hemisphere Astronomy*, Bristol, UK, and Philadelphia: Adam Hilger, 1988, pp. 137–8; and Solon I. Bailey, *The History and Work of the Harvard Observatory 1839 to 1927*, New York and London: McGraw-Hill Book Company, 1931, 58–63.

17 Bessie Zaban Jones and Lyle Gifford Boyd, *The Harvard College Observatory: The First Four Directorships, 1839–1919*, Cambridge, Massachusetts: Harvard University Press, 1971, pp. 297–324; and *Forty-Sixth Annual Report of the Director of The Astronomical Observatory of Harvard College for the Year Ending September 30, 1912*, Cambridge, Massachusetts: 1891, pp. 6–7. During this period of time, Bailey made the important discovery that the globular clusters contained many variable stars and began his longer-term study of these clusters. Bailey, *History and Work*, 183.

18 Leon Campbell, "Weekly Report 165," Arequipa, Peru: 2 June 1911, HCO-BS, UAV 630.100, Box 6, folder 6, HUA.

19 Edward C. Pickering to Leon Campbell, 15 September 1911, HCO-ECP, UAV 630.100, 3, 7, HUA.

20 Edward C. Pickering to Leon Campbell, 22 July 1914, HCO-BS, UAV 630.100, Box 3, HUA.

21 Leon Campbell to Edward C. Pickering, 7 February 1913, HCO-BS, UAV 630.100, 6, 7, HUA.

22 Edward C. Pickering to Herbert C. Wilson, 7 January 1912, HCO-ECP, HUG 1690.6.5, 2, 352 Replies of Thanks, HUA.

23 E. E. Markwick, "Section for the Observation of Variable Stars Sixth Report of the Section 1900–1904," *MmBAA ROS* **XV** (1906); and E. E. Markwick, "Seventh Report of the Section for the Observation of Variable Stars 1905–1909," *MmBAA ROS* **XVIII** (1912).

24 Brook actually provided 10-day means for all the BAA-VSS long-period variables which were included in the AAVSO/Harvard Corps of Observers programs.

25 Leon Campbell to H. W. Vrooman, 8 January 1917, LC-Cor, 8, Vrooman, H. W., AAVSO; Leon Campbell to Forrest H. Spinney, 15 February 1917, LC-Cor, 7, Spinney, F. H., AAVSO; Charles L. Brook, "Section for the Observation of Variable Stars Eighth Report of the Section 1910–1914," *MmBAA ROS* **XXII** (1918), 1–352; and Charles L. Brook, "Section for the Observation of Variable Stars Ninth Report of the Section 1915–1919," *MmBAA ROS* **XXV** (1924), i–x; 1–554.

26 The journal sheets of 10-day means computed by volunteers exist in the AAVSO archives but are unfortunately not identified in terms of which volunteers actually delivered the results in this program.

27 Leon Campbell, 26 October 1916, LC-Cor, 8, Vrooman, H. W., AAVSO; Leon Campbell to H. W. Vrooman, 12 November 1916, LC-Cor, 8, Vrooman, H. W., AAVSO; and Leon Campbell to Forrest H. Spinney, 26 October 1916, LC-Cor, 7, file Spinney, F. H., AAVSO.

28 John Toone of the BAA-VSS has undertaken studies of this same issue in the last decade, indicating that indeed "observer personal equation" with respect to red sensitivity remains an issue.

29 J. Ernest G. Yalden to Leon Campbell, 16 December 1919, LC-Cor, 8, Yalden, J. E. G., AAVSO.

30 Leon Campbell to Howard O. Eaton, 3 April 1916, LC-Cor, 3, Eaton, H. O., AAVSO.

31 Anne Sewell Young to Leon Campbell, 3 June 1919, LC-Cor, 8, Young, A. S., AAVSO.

32 Anne Sewell Young *et al.*, Cambridge, Massachusetts: 8 November 1919, ECP-WTO-LC-Org, 1, Misc. org. papers and reports, AAVSO.

33 The evidence for this survey consists mainly of the few scattered replies that Campbell received from the membership, of which we cite as an example Yalden's detailed reply. Yalden, 16 December 1919.

34 Leon Campbell to Anne Sewell Young, 19 November 1921, LC-Cor, 8, Young, A. S., AAVSO.

35 Leon Campbell to Donald H. Menzel, 21 May 1920, LC-Cor, 6, Menzel, D. H., AAVSO.

36 Donald H. Menzel to Leon Campbell, 21 May 1920, LC-Cor, 6, Menzel, D. H., AAVSO; and Donald H. Menzel to Leon Campbell, 7 August 1920, LC-Cor, 6, Menzel, D. H., AAVSO.

37 Leon Campbell to Donald H. Menzel, 23 April 1922, LC-Cor, 6, Menzel, D. H., AAVSO; Donald H. Menzel to Leon Campbell, 29 April 1922, LC-Cor, 6, Menzel, D. H., AAVSO; and Donald H. Menzel to Leon Campbell, 13 May 1922, LC-Cor, 6, Menzel, D. H., AAVSO.

38 J. Ernest G. Yalden, "Minutes of the Council of the American Association of Variable Star Observers, 20, ECP-WTO-LC-Org, 2, Meeting Minutes, General and Council, AAVSO. "The Old Guard" was, at first, a title given to those enrolling as sustaining members. Willard J. Fisher, "Story of the 1928 Fall Meeting, A. A. V. S. O.," *VC* 1 (1929), 89–93. The term, if not already in general usage, quickly caught on and included anyone who was a part of the AAVSO's earliest years.

39 "Dr. Edward Gray, ship surgeon and student of Spanish, drops dead," *Berkeley Observer*, Berkeley, California, 20 January 1920, Obituary File, AAVSO.

6 FORMALIZING RELATIONSHIPS

1 Solon I. Bailey, Acting Director, Harvard College Observatory, *74th Annual Report of the Director of the Astronomical Observatory of Harvard College for the year ending September 30, 1919*, Cambridge, Massachusetts: Harvard College Observatory, 1920, 5.

2 The HCO Director's Annual Reports mentioned here – all titled *The Annual Report of the Director of the Astronomical Observatory of Harvard College for the year ending September 30, Year* – were accessed digitally in the SAO/NASA Astrophysical Data System, http://adsabs.harvard.edu/historical.html.

3 Campbell's actual role, in all these cases, consisted merely of ledgering the observations as they came in to the observatory and perhaps commenting to the observer on an occasional observation he might, as a result of that ledgering process, have noticed was discordant. There was literally no "supervision" of these volunteer observers in the managerial sense of the word common in the late twentieth century and after.

4 Radhagobinda G. Chandra to William Tyler Olcott, 26 November 1919, LC-Cor, 2, Chandra, R. G., AAVSO.

5 Leon Campbell to Henry N. Russell, 12 June 1920, LC-Cor, 7, Russell, H. N., AAVSO.

6 The Committee on Charts, "Second Announcement Chart Committee of the American Association of Variable Star Observers," Cambridge, Massachusetts: 1 February 1921, LC-Org, 1 Chart Program (4) Chart Committee Announcements & Notices, AAVSO; Chart Committee, "Third Announcement Chart Committee of the American Association of Variable Star Observers," Cambridge, Massachusetts: 1 May 1922, LC-Org, 1 Chart Program (4) Chart Committee Announcements & Notices, AAVSO; and Kerriann H. Malatesta, and Charles E. Scovil, "The history of AAVSO charts, Part I: The 1880s through the 1950s," *JAAVSO* **34** (2005), 89–101.

7 Warren K. Green to Leon Campbell, 7 March 1922, LC-Cor, 3, Green, W. K., AAVSO; Leon Campbell to Warren K. Green, 27 March 1922, LC-Cor, 3, Green, W. K., AAVSO; and Warren K. Green to Leon Campbell, 14 June 1922, LC-Cor, 3, Green, W. K., AAVSO.

8 Warren K. Green to Leon Campbell, 27 October 1924, LC-Cor, 3, Green, W. K., AAVSO; Leon Campbell to Warren K. Green, 28 October 1924, LC-Cor, 3, Green, W. K., AAVSO; Leon Campbell to Warren K. Green, 7 May 1925, LC-Cor, 3, Green, W. K., AAVSO; Warren K. Green to Leon Campbell, 26 June 1926a, LC-Cor, 3, Green, W. K., AAVSO; and Warren K. Green to Leon Campbell, 26 June 1926b, LC-Cor, 3,Green, W. K., AAVSO.

9 William Tyler. Olcott, Secretary, Minutes of the Council of the American Association of Variable Star Observers, William Tyler Olcott, *Record of the Minutes of the Council of the American Association of Variable Star Observers*, Cambridge, Mass. HCO, 18 October 1929, ECP-WTO-LC-Org, 2, Meeting Minutes, General and Council, AAVSO; Jane Opalko, "Helen Sawyer Hogg," in Benjamin F. Shearer and Barbara S. Shearer (Editors), *Notable Women in the Physical Sciences, A Biographical Dictionary*, Westport, Connecticut: Greenwood Press, 1997, 191–6; and Leon Campbell, "Looking backwards, or A. A. V. S. O. in retrospect," *VC* **2** (1931), 51. See also Kerriann H. Malatesta and others in two important articles in the *Journal of the AAVSO*. Malatesta and Scovil, "AAVSO Chart History," 81–101; and Kerriann

H. Malatesta *et al.*, "The history of AAVSO charts, part II: the 1960s through 2006," *JAAVSO* **35** (2007), 377–89.

10 Two things are important about the Julian Day system of denoting time. First, days are numbered continuously from day 1, without years or months. Day One was defined as January 1, 4713 B.C. Second, the Julian date starts, by definition, at exactly 12 noon at Greenwich, England, and is always reckoned in decimal fractions of a day. P. Kenneth Seidelmann, *et al.*, "Time," P. Kenneth Seidelmann (Editor), *Explanatory Supplement to the Astronomical Almanac,* Mill Valley, California: University Science Books, 1992, 55–6.

11 Conversion of civil times to JD and decimal is now provided on the AAVSO Web site.

12 See, for example, Harlow Shapley, "Eclipsing binaries," *PA* **20** (1912), 573–9.

13 William Tyler Olcott, "Minutes of the American Association of Variable Star Observers," 6 November 1920, ECP-WTO-LC-Org, 2, Meeting Minutes, General and Council, AAVSO.

14 Ernest W. Brown, "Request for more observations of occultations," *VV* **1** (1927), 69–71.

15 Albert Van Helden, "The birth of the modern scientific instrument, 1550–1700," John G. Burke (Editor), *The Uses of Science in the Age of Newton,* Berkeley and Los Angeles, California: University of California Press, 1983.

16 J. B. Hearnshaw, *The Measurement of Starlight: Two Centuries of Astronomical Photometry*, Cambridge, UK: Cambridge University Press, 1996, 79–80 and 89–96.

17 Hearnshaw, *Measurement of Starlight*, 185–9.

18 Owen Gingerich and Kenneth R. Lang, *A Source Book in Astronomy and Astrophysics, 1900–1975*, Cambridge, Massachusetts, and London, England: Harvard University Press, 1979, pp. 39–45.

19 Lewis J. Boss, "The use of selenium cells in astronomy," *PA* **28** (1920), 154–6; and Lewis J. Boss, "Some results obtained with the selenium cell," *PA* **29** (1921), 213–6.

20 Lewis J. Boss to Cyrus F. Fernald, 31 January 1955, CFF, 1, AAVSO Committee Appointments, AAVSO; Lewis J. Boss to Leon Campbell, 30 November 1922, LC-Cor, 1, Boss, L. J., AAVSO; Howard O. Eaton, "Monthly Report of the American Association of Variable Star Observers, September 20 to October 20, 1922," *PA* **30** (1922), 659; Leon Campbell to Lewis J. Boss, 10 November 1922, LC-Cor, 1, Boss, L. J., AAVSO; and Leon Campbell to Lewis J. Boss, 6 December 1922, LC-Cor, 1, Boss, L. J., AAVSO.

21 Leon Campbell to Lewis J. Boss, 6 December 1922, LC-Cor, 1, Boss, L. J., AAVSO. Correspondence fell off for about 20 years, presumably because Boss was too busy with his professional and family life, and war work. He resumed correspondence with Campbell in 1946 and maintained a very active correspondence with Margaret Mayall once she became the recorder.

22 Morgan Cilley to Leon Campbell, 4 June 1924, LC-Cor, 2, Cilley, M., AAVSO.

23 Leon Campbell to Anne Sewell Young, 14 May 1923, LC-Cor, 8, Young, A. S., AAVSO.

24 Anne Sewell Young to Leon Campbell, 10 May 1923, LC-Cor, 8, Young, A. S., AAVSO; and Anne Sewell Young to Leon Campbell, 12 June 1923, LC-Cor, 8, folder Young, A. S., AAVSO.

25 Leon Campbell to C. E. Kenneth Mees, 31 August 1923, LC-Cor, 3, Eastman Kodak, AAVSO; and Loyd A. Jones to Leon Campbell, 10 September 1923, LC-Cor, 3, Eastman Kodak, AAVSO. See also Hearnshaw, *Measurement of Starlight*, especially pp. 57–9.

26 Leon Campbell to Eastman Kodak Company, 15 January 1926, LC-Cor, 3, Eastman Kodak, AAVSO; Loyd A. Jones to Leon Campbell, 21 January 1926, LC-Cor, 3, Eastman Kodak, AAVSO; and Loyd A. Jones to Leon Campbell, 8 February 1926, LC-Cor, 3, Eastman Kodak, AAVSO.

27 Henry Norris Russell, "Variable stars," *Science* **49** (1919), 127–39.

28 Jean-Louis Tassoul and Monique Tassoul, *A Concise History of Solar and Stellar Physics*, Princeton and Oxford: Princeton University Press, 2004, 105–8.

29 Cecilia H. Payne, *Stellar Atmospheres; A Contribution to the Observational Study of High Temperature in the Reversing Layers of Stars,* Cambridge, Massachusetts: Harvard University Press, 1925, QB871.P3; and Cecilia H. Payne, *The Stars of High Luminosity,* New York and London: McGraw-Hill Book Company, Inc., 1930, QB815.P33. Payne's problems with Henry Norris Russell and the hydrogen abundance anomaly are elaborated well in David H. DeVorkin, *Henry Norris Russell: Dean of American Astronomers*, Princeton, New Jersey: Princeton University Press, 2000, 201–4.

30 Lee Anne Willson, "Miras," in A. G. Davis Philip *et al.*, Editors, *Anni Mirabiles: A Symposium Celebrating the 90th Birthday of Dorrit Hoffliet*, March 1997, New Haven, Connecticut, Schenectady, New York: L. Davis Press, Inc., 1999, 89.

31 Sidney Dean Townley *et al.*, "Harvard catalogue of long period variable stars," *Harvard College Observatory Annals* **79** (1928), Preface. On Townley, see also Robert Dale Hall, "Education of American Research Astronomers, 1876–1941," Ph.D. diss., Oregon State University, 1999, 920.

32 William Tyler Olcott, "Minutes of the Council of the American Association of Variable Star Observers," 26 May 1923, ECP-WTO-LC-Org, 2, Meeting Minutes, General and Council, AAVSO. This business was one of several that Campbell ran on the side over the years to supplement the meager income he received from Harvard College Observatory.

33 Leon Campbell to J. H. Skaggs, 21 April 1922, LC-Cor, 7, Skaggs, J. H., AAVSO. A loyal member and variable star observer, Skaggs not only printed *Variable Comments*, he also printed Julian Day Calendars and other specialized documents like the Pickering Memorial Fund brochure. At about the same time he was joined by another printer, Charles E. Barns of Morgan Hill, California, who volunteered to typeset and print a number of pamphlets, circulars, and catalogs for the AAVSO.

34 Anne Sewell Young to Leon Campbell, 16 February 1924, LC-Cor, 8, Young, A. S., AAVSO; and Leon Campbell to Anne Sewell Young, 18 February 1924, LC-Cor, 8, Young, A. S., AAVSO. One aspect that Campbell highlighted in his letter pertains to the fact that under the extant by-laws, Olcott, was responsible, as editor of everything, for whatever the AAVSO published, but Campbell also observed that the rule was pretty much ignored in the breech.

35 William Tyler. Olcott, "Minutes of the Council of the American Association of Variable Star Observers," 19 October 1929, ECP-WTO-LC-Org, 2, Meeting Minutes, General and Council, AAVSO. A total of four volumes, published between 1923 and 1949, were bound and constitute a fruitful resource for historians.

36 William Tyler. Olcott, "Settin' Elmer's Glass – The Ridicule Method," *VC* 1 (1924), 6–8.

37 Margaret W. Mayall, "Well-known eastern amateur astronomer dies," *S&T* 4 (1955), 189. Although Mayall's comment with respect to variable star observations is correct (his actual lifetime total never exceeded 33), Elmer succeeded very well in photographing variable star fields with very good quality plates, which he then donated to HCO. Campbell congratulated him on the fine quality of the plates and states that "The variable star group here under Miss Mohr should feel gratified with such a fine contribution of plates." Leon Campbell to Charles W. Elmer, 25 January 1939, LC-Cor, 3, Elmer, Charles W., AAVSO.

38 "The Old Guard" was, at first, a title given to those enrolling as sustaining members. Willard J. Fisher, "Story of the 1928 Fall Meeting, A. A. V. S. O.," *VC* 1 (1929), 89–93. The term – if not already in general usage – quickly caught on and included anyone who was a part of the AAVSO's earliest years.

39 "Dr. Edward Gray, ship surgeon and student of Spanish, drops dead," *Berkeley Observer*, Berkeley, California, 20 January 1920, Obituary File, AAVSO; and J. Ernest G. Yalden, "Minutes of the Council of the American Association of Variable Star Observers," 21 December 1920, ECP-WTO-LC-Org, 2, Meeting Minutes, General and Council, AAVSO.

40 "Veteran Rail Engineer Killed by Truck as He Dashes for Car," Pittsburgh, Pennsylvania, 20 October 1924, WTO, 2, Misc. Clippings, AAVSO; and "Engineman by day and an observer of stars in the night," Pittsburgh, Pennsylvania, 29 May 1924, Biographical Pamphlets-M, AAVSO Archive.

41 William Tyler Olcott, "Minutes of the Council of the American Association of Variable Star Observers, 22 October 1926, ECP-WTO-LC-Org, 2, Meeting Minutes, General and Councill, AAVSO; William Tyler. Olcott, "Minutes of the Council of the American Association of Variable Star Observers," 21 October 1927, ECP-WTO-LC-Org, 2, Meeting Minutes, General and Council, AAVSO; and George Waldo, Jr., "Charles Cartlidge Godfrey," *PA* 36 (1928), 221–6.

42 Leon Campbell to Alice H. Farnsworth, 13 October 1927, LC-Cor, 3, Farnsworth, A. H., AAVSO.

43 William Tyler. Olcott, "Minutes of the Council of the American Association of Variable Star Observers," 22 October 1927, ECP-WTO-LC-Org, 2, Meeting Minutes, General and Council, AAVSO.

44 Leon Campbell to Alice H. Farnsworth, 23 December 1929, LC-Cor, 3, Farnsworth, A. H., AAVSO; and Leon Campbell, "Michael J. Jordan In Memoriam," *VC* 2 (1930), 21–2.

45 John A. Ingham, "The 1929 Harvard Meeting," *VC* 2 (1930), 17–20.

46 Leon Campbell, "Minutes of the Council of the American Association of Variable Star Observers," 11 October 1930, ECP-WTO-LC-Org, 2, Meeting Minutes, General and Council, AAVSO.

47 David Bedell Pickering, "The astronomical fraternity around the world, part I," *PA* 35 (1927), 157–68. This article began a series of almost 40 articles in *Popular Astronomy* over the next 9 years in which Pickering described his travels around the world, including his meetings with astronomers in each country he visited, their observatories, and personalities, along with other interesting commentary.

48 William Tyler Olcott, "Minutes of the Council of the American Association of Variable Star Observers," 6

November 1920, ECP-WTO-LC-Org, 2, Meeting Minutes, General and Council, AAVSO. At the same time, the Council elected S. A. Mitchell, the Leander McCormick Observatory director, as a patron, recognizing his continuing interest and support.

49 Leon Campbell to Henri Grouiller, 23 November 1920, LC-Cor, 3, Grouiller, H., AAVSO.

50 Stephen Crasco Hunter to Leon Campbell, June 1921, LC-Cor, 4, Hunter, S. C., AAVSO.

51 William Tyler Olcott, "Minutes of the Council of the American Association of Variable Star Observers," 28 May 1921, ECP-WTO-LC-Org, 2, Meeting Minutes, General and Council, AAVSO; Stephen Crasco Hunter to Leon Campbell, June 1921, LC-Cor, 4, Hunter, S. C., AAVSO; and Henri Grouiller to Leon Campbell, 15 July 1921, LC-Cor, 3, Grouiller, H., AAVSO.

52 Amalendu Bandyopadhyay, and Ranatosh Chakraborti, "Radha Gobinda Chandra – A pioneer in astronomical observations in India," *Indian Journal of History of Science* **26** (1991), 104–13.

53 Radhagobinda G. Chandra to Leon Campbell, 22 January 1920, LC-Cor, 2, Chandra, R. G., AAVSO; Radhagobinda G. Chandra to Leon Campbell, 26 February 1920, LC-Cor, 2, Chandra, R. G., AAVSO; Leon Campbell to Radhagobinda G. Chandra, 15 April 1920, LC-Cor, 2, Chandra, R. G., AAVSO; William Tyler Olcott, "Minutes of the Council of the American Association of Variable Star Observers," 23 May 1925, ECP-WTO-LC-Org, 2, Meeting Minutes, General and Council, AAVSO; and Helen Swartz, 'The A. A. V. S. O. meeting at Yale, 1927," *VC* **1** (1927), 73.

54 Willard J. Fisher, "Story of the 1928 Fall Meeting, A. A. V. S. O.," *VC* **1** (1929), 89; and David Bedell Pickering, "New observers in 1928," *VC* **2** (1929), 1.

55 Leon Campbell, "Minutes of the Council of the American Association of Variable Star Observers," 10 November 1930, ECP-WTO-LC-Org, 2, Meeting Minutes, General and Council, AAVSO.

56 Anne Sewell Young, *VC* **2** (1931), 35.

7 THE PICKERING MEMORIAL ENDOWMENT

1 Leon Campbell, "Leon Campbell," Undated, LC-Cor, 2, Campbell, L., AAVSO.

2 Anne Sewell Young to Leon Campbell, 10 December 1920, LC-Cor, 8, Young, A. S., AAVSO; and Leon Campbell to Stephen Crasco Hunter, 21 December 1920, LC-Cor, 4, Hunter, S. C., AAVSO.

3 "Charles Alfred Post," *Monthly Notices of the Royal Astronomical Society* **72** (1922); and "Attorney and Astronomer," unidentified newspaper, unknown, Obituary File, AAVSO.

4 Charles Alfred Post (1844–1921) wrote a long article in *Astronomy & Astro-Physics* with details of his observatory, including the 6-inch Post telescope. The telescope is a rare example of an instrument made by Gregg. The refractor's lens cell was threaded to accommodate a photographic corrector plate. C. A. Post, "The balance roof for telescope buildings," *A&A* **2**, 400.

5 Arthur C. Perry to Leon Campbell, 17 June 1921, LC-Cor, 6, Perry, A. C., AAVSO.

6 Leon Campbell to George Waldo, 21 January 1921, LC-Cor, 8, Waldo, G., AAVSO. One of the other telescopes acquired from Post by the AAVSO at the same time, a 5-inch f/4/9 doublet camera, was loaned to the Texas Observers of Fort Worth, Texas, and used by AAVSO members James Logan, Sterling Bunch, and Oscar Monnig of that group, who were also very involved in occultation timings for the AAVSO in the late 1920s and early 1930s. Robert Brown *et al.*, Members, Texas Observers, *Texas Observers Souvenir Scrapbook*, Fort Worth, Texas: Texas Observers, 1929, LC-Org, 3, Texas Observers, AAVSO.

7 William Tyler Olcott, Secretary, Minutes of the Council of the American Association of Variable Star Observers, William Tyler Olcott, *Record of the Minutes of the Council of the American Association of Variable Star Observers,* Harvard College Observatory, Cambridge, Massachusetts, 4 November 1921, ECP-WTO-LC-Org, 2, Meeting Minutes, General and Council, AAVSO.

8 Richard B. Morris *Encyclopedia of American History,* New York: Harper & Row, Publishers, 1965, pp. 538–40.

9 Morton Keller and Phyllis Keller, *Making Harvard Modern: The Rise of America's University*, Oxford, UK and New York: Oxford University Press, 2007, pp. 87–9 and 102–3.

10 Samuel Eliot Morison (Editor), *The Development of Harvard University since the Inauguration of President Eliot, 1869–1929,* Cambridge, Massachusetts: Harvard University Press, 1930, pp. 303–6. As outlined by Morison with help from Solon I. Bailey, astronomy education actually crept into Harvard through a back door. As late as 1890–91, and in spite of the existence of the HCO for nearly 50 prior years, Harvard offered no separate course on astronomy. The only course that included astronomy was Applications of Spherical Trigonometry to Astronomy and Navigation. Robert W. Willson (1873–), a native of Salem and inspired by the seafaring nature of his home community, received an

appointment as Instructor in Astronomy in 1891 and began offering courses then. Originally housed in the Engineering department, by 1897 Willson had broken away, joined the movement to astronomy and moved to the physics department. He taught a variety of astronomy courses, including both survey courses and detailed professional level instruction. Eventually a separate astronomy department was established, which led to the granting of a degree in astronomy. Harlan True Stetson replaced Willson in 1919. The president of Harvard and the Dean of Liberal Arts no doubt planned to eventually consolidate the Observatory and the department, but that would prove difficult unless a director was appointed from outside the existing observatory staff.

11 Bart J. Bok, "Harlow Shapley 1885–1972," *Biographical Memoirs of the National Academy of Sciences* (1978), pp. 231–45.

12 Bok, *NAS Biographical Memoirs*, pp. 241–3; Horace A. Smith and Virginia Trimble, "Shapley, Harlow," Thomas A. Hockey (Editor in Chief), *Biographical Encyclopedia of Astronomers*, New York: Springer, 2007, pp. 1048–9; and Joann Palmeri, *An Astronomer beyond the Observatory: Harlow Shapley as a Prophet of Science*, University of Oklahoma, 2000, pp. 27–8.)

13 Henry Norris Russell, "On the determination of the orbital elements of eclipsing binary variable stars. I," *AJ* **35** (1912), 315–40; Henry Norris Russell, "On the determination of the orbital elements of eclipsing binary variable stars. II," *AJ* **36** (1912); Henry Norris Russell, and Harlow Shapley, "On darkening at the limb in eclipsing variables. I," *AJ* **36** (1912), 54–74; and Henry Norris Russell, and Harlow Shapley, "On darkening at the limb in eclipsing variables. II," *AJ* **36** (1912), 385–408. Russell had been working in this direction for over a decade. In 1899, both he and South African amateur Alexander William Roberts came to the conclusion that some binary stars were tenuous and extended objects of low density, but it took another decade to establish conclusively that Cepheids were not eclipsing binaries but single stars. Alexander William Roberts, "Density of close double stars," *AJ* **10** (1899), 308–14; and Henry Norris Russell, "Densities of the variable stars of the Algol type,"*AJ* **10** (1899), 315–18. See also Josef Kallrath and Eugene Milone, *Eclipsing Binary Stars: Modeling and Analysis*, New York: Springer, 2009.

14 Bok, B., *NAS Biographical Memoirs*, p. 245. Shapley's mentor at Missouri, Frederick Seares, by then an astronomer at Mount Wilson, arranged for Shapley to meet George Ellery Hale.

15 Kyle M. Cudworth, "Galactic structure, globular structures," Stephen P. Maran (Editor), *The Astronomy and Astrophysics Encyclopedia*, New York: Van Nostrand Reinhold, 1992, pp. 200–2. Shapley used the few nearby Cepheids for which distances were known for his calibration of the period-luminosity relationship. He also used the assumption that the diameters of all globular clusters are approximately the same, as a secondary method of determining the distances of the galactic globular clusters. In his application of these techniques, he ignored warnings from Russell and others that interstellar absorption of light could introduce an error in his distance estimates, an oversight that resulted in his distance estimates being overstated by a factor of nearly two.

16 Harlow Shapley, "The scale of the universe: Part 1," *Bulletin of the National Research Council of the National Academy of Sciences* **2** (1921), 171–93; and Heber Doust Curtis, "The scale of the universe: Part 2," *Bulletin of the National Research Council of the National Academy of Sciences* **2** (1921), 194–217.

17 Michael A. Hoskin, "The 'Great Debate': what really happened," *Journal for the History of Astronomy* **7** (1976), 169–82; and Robert W. Smith, *The Expanding Universe: Astronomy's 'Great Debate' 1900–1930*, Cambridge, UK: Cambridge University Press, 1982, pp. 77–96.

18 Russell strengthened and supported his former student's contentions that the spiral galaxies could not be outside the Milky Way Galaxy.

19 David H. DeVorkin, *Henry Norris Russell: Dean of American Astronomers*, Princeton, New Jersey: Princeton University Press, 2000, 169–70.

20 Paul W. Merrill to Leon Campbell, 19 January 1920, LC-Cor, 6, Merrill, P. W., AAVSO; and Paul W. Merrill to Leon Campbell, 7 December 1921, LC-Cor, 6, Merrill, P. W., AAVSO.

21 William Tyler Olcott, Secretary, Minutes of the Council of the American Association of Variable Star Observers, William Tyler Olcott, *Record of the Minutes of the Council of the American Association of Variable Star Observers*, Leonia, New Jersey, residence of JEG Yalden, 23 May 1925, ECP-WTO-LC-Org, 2, Meeting Minutes, General and Councill, AAVSO. See also Campbell's letter to Anne S. Young after he returned, full of vigor, from the July 1928 meeting of the IAU Variable Star Commission in Leiden, The Netherlands, Leon Campbell to Anne Sewell Young, 13 August 1928, LC-Cor, 15, Young, A. S., AAVSO.

22 Samuel Eliot Morison, 299–300.

23 William Tyler Olcott to J. S. Plaskett, 19 November 1921, JSP, RG 48, Vol. 49, file Correspondence, Misc. "N," "O," "P," LAC; and William Tyler Olcott to J. S. Plaskett, 19 November 1921, JSP RG 48, RG 48, Vol. 49, file 6–25, Misc. "N," "O," "P," LAC, LAC. We take these identical letters to J. S. Plaskett, from the Dominion Astrophysical Observatory archives, and to E. B. Frost, from the Yerkes Observatory archives, to be typical of the letters Olcott sent to many other observatory directors. We are grateful to Peter Broughton of Toronto for providing us with access to the Plaskett documents.

24 William Wallace Campbell to William Tyler Olcott, 2 December 1921, HU-PO-ALL, UA I 5.160 Series 1919–1922, folder 14, HUA; and J. S. Plaskett to William Tyler. Olcott, 9 December 1921, LAC, RG 48, Vol. 49, file Correspondence, Misc. "N," "O," "P."

25 Edwin B. Frost to William Tyler Olcott, 12 December 1921, Off of Dir, 84, folder N-O, Yerkes.

26 William Wallace Campbell to Harlow Shapley, 24 January 1922, HCO-HS, UAV.630.22, 3, folder 25, HUA.

27 Henry Norris Russell *et al.*, "The Edward C. Pickering Memorial Endowment," 1921, LC-Org, 1, E. C. Pickering Memorial Endowment, folder 1, AAVSO.

28 William Tyler Olcott to Harlow Shapley, 5 December 1921, HU-PO-ALL, UA I 5.160, Series 1919–1922, Folder 14, HUA.

29 Harlow Shapley to William Wallace Campbell, 1 February 1922, HCO-HS, UAV.630.22, box 3, folder 25, HUA. See note 24 in this chapter for the Plaskett response to Olcott's letter.

30 Harlow Shapley to A. Lawrence Lowell, 13 December 1921, HU-PO-ALL, UA I 5.160 Series 1919–1922, folder 14, HUA.

31 A. Lawrence Lowell to Harlow Shapley, 21 December 1921, HU-PO-ALL, UA I 5.160 Series 1919–1922, folder 14, HUA.

32 Leon Campbell to Anne Sewell Young, 19 November 1921, LC-Cor, 8, Young, A. S, AAVSO; Leon Campbell to Anne Sewell Young, 1 April 1922, LC-Cor, 8, Young, A. S, AAVSO; and Harlow Shapley to Anne Sewell Young, 26 October 1923, HCO-HS, UAV.630.22, 19, folder 146 Young, A.S., HUA.

33 David Bedell Pickering to AAVSO Membership, LC-Org, 1, E. C. Pickering Memorial Endowment, folder 2, AAVSO.

34 Olcott's actual donations, small in amount and frequent, added up to only $1000 by the time the campaign finally ended. Thus his actual contribution was less than a fifth of what he had implied to Campbell he would do, $5000 and the land for the observatory, and even that would have been substantially below his capacity to give. At this point in the mid 1920s, the US economy prospered, and economic conditions cannot explain his reluctance. Willard J. Fisher, "List of subscribers to the Edward C. Pickering Memorial Fund of the A.A.V.S.O. (1921–1922)," 1930, LC-Org, 1, E. C. Pickering Memorial Endowment, folder 1, AAVSO.

35 William Tyler Olcott to Harlow Shapley, 21 October 1925, HCO-HS, UAV.630.22,box 14, folder 106, HUA.

36 Shapley's attitude in this case could be no different from those he later exhibited in hiring and underpaying Radcliffe graduates to work as computers. See John Lankford, *American Astronomy – Community, Careers and Power, 1859–1940*, Chicago and London: The University of Chicago Press, 1997, p. 341. Campbell's scientific limitations as well as his "beloved" status "throughout the astronomical world" were later memorialized by Donald H. Menzel in an unpublished autobiography and cited by Leif Robinson, who holds the only extant copy of Menzel's manuscript. Leif J. Robinson, "Enterprise at Harvard College Observatory," *Journal for the History of Astronomy* **21** (1990), 91, and note 13, p. 102.

37 Harlow Shapley, acting Director, Harvard College Observatory, *77th Annual Report of the Director of the Astronomical Observatory of Harvard College for the Year Ending September 30, 1922*, Cambridge, Massachusetts: Harvard College Observatory, 1923.

38 Harlow Shapley, Director, Harvard College Observatory, *78th Annual Report of the Director of the Astronomical Observatory of Harvard College for the Year Ending September 30, 1923*, Cambridge, Massachusetts: Harvard College Observatory, 1924.

39 Harlow Shapley, Director, Harvard College Observatory, *79th Annual Report of the Director of the Astronomical Observatory of Harvard College for the Year Ending September 30, 1924*, Cambridge, Massachusetts: Harvard College Observatory, 1925.

40 Lankford wrongly identifies variable stars as one of Shapley's "new" initiatives at Harvard, but correctly emphasizes the extent to which Shapley expanded and accelerated variable star research at HCO by employing additional low-wage "computers" to examine photographic plates from the HCO plate stacks and develop light curves to be interpreted later by the professional astronomers (Lankford, p. 341). This work on variable stars was not supervised by Campbell. See also John Lankford, and Ricky L. Slavings, "The Industrialization of American astronomy, 1880–1940," *Physics Today* **49** (1996), 34 (40).

41 During the first half of the twentieth century, at least, an important part of any university president's role was in regulating the flow of requests for funds from the university faculty to important sources of funds like the IEB. Thus it seems likely that even if Shapley independently identified the IEB, a subsidiary of the Rockefeller Foundation, as a possible source of money to endow his emerging scheme for the HCO modernization, he had first to gain the confidence of Lowell in both the soundness of his proposal for the observatory and in his ability to represent Harvard in negotiations with the IEB. Wickliffe Rose, the director of the IEB, had risen to prominence beginning in 1909 with his appointment as director of the Rockefeller Sanitary Commission. He led that organization in a very successful effort to eradicate the hook worm. The IEB absorbed part of the scientific and medical programs of the Sanitary Commission. A former philosophy professor, Rose valued science with "an amateur's fervor," an outlook that colored the approach of the foundation profoundly; see Waldemar A. Nielsen, *The Big Foundations*, New York and London: Columbia University Press, 1972, pp. 53–5. Rose established patterns of giving from the IEB that benefited Harvard enormously in future years. See, for example, references to the foundation scattered throughout the later history of the university (Keller and Keller).

42 Shapley's one-sentence descriptors for the other elements of his program include, for example, "A. A plate vault to house the world's greatest collection of stellar photographs," "E. Equipment and endowment for the study of faint nebulae and of variable stars in the Milky Way," as well as "F. The establishment of a bureau for the measurement of plates in the Harvard collections," all of which would include and support the study of variable stars. Thus variable star astronomy constituted a major element of Shapley's program for HCO. Seen in this context, the request for the Pickering Memorial Endowment funds (i.e., Shapley's item C.) is an important element of his overall plan. The Harvard Observatory, "The Harvard Observatory Development," undated, LC-Org, 1, E. C. Pickering Memorial Endowment, folder 3, AAVSO.

43 Harlow Shapley to Wickliffe Rose, 30 October 1926, HCO-HS, UAV 630.22, 16, folder 120 Robinson-Rose.W 120, HUA; and Ernest W. Brown, "Request for more observations of occultations," *VV* 1 (1927), 69–71. Apparently energized by this opportunity, which obviously had not occurred to him before, Brown also chose to publish his request to other astronomers through the Astronomical Society of the Pacific. Ernest W. Brown, "Request for more observations of occultations," *Publications of the Astronomical Society of the Pacific* **39** (1927), 37–41.

44 Harlow Shapley to Wickliffe Rose, 12 November 1926, HCO-HS, UAV 630.22, 16, folder 120 Robinson-Rose.W, HUA; and Harlow Shapley to Wickliffe Rose, 16 November 1926, HCO-HS, UAV 630.22, 16, folder 120, Robinson-Rose, W., HUA.

45 Michael Saladyga, "A history of AAVSO's headquarters," *JAAVSO* **35** (2007), 390–406.

46 Leon Campbell, Secretary Pro-Tem, Minutes of the Council of the American Association of Variable Star Observers, William Tyler Olcott, *Record of the Minutes of the Council of the American Association of Variable Star Observers,* Nantucket, Mass, 13 June 1930, ECP-WTO-LC-Org, 2, Meeting Minutes, General and Council, AAVSO.

47 Leon Campbell, Secretary Pro-Tem, Minutes of the Council of the American Association of Variable Star Observers, William Tyler Olcott, *Record of the Minutes of the Council of the American Association of Variable Star Observers,* Nantucket, Mass., 13 June 1930, ECP-WTO-LC-Org, 2, Meeting Minutes, General and Council, AAVSO. Harvard's receipts for each of the three transfers are also in this same folder in the AAVSO archives.

48 Harriet W. Bigelow to AAVSO Council, 2 April 1931, LC-Org, 1, E. C. Pickering Memorial Endowment, AAVSO, and other letters in this folder.

49 Leon Campbell, Secretary Pro-Tem, Minutes of the Council of the American Association of Variable Star Observers, William Tyler Olcott, *Record of the Minutes of the Council of the American Association of Variable Star Observers,* Nantucket, Mass., 11 October 1930, ECP-WTO-LC-Org, 2, Meeting Minutes, General and Council, AAVSO.

50 Leon Campbell to William Tyler Olcott, 28 January 1931, ECP-WTO-LC-Org, 2, Campbell/Olcott letters, AAVSO.

51 Leon Campbell, Secretary Pro-Tem, Minutes of the Council of the American Association of Variable Star Observers, William Tyler Olcott, *Record of the Minutes of the Council of the American Association of Variable Star Observers,* Nantucket, Mass., 2 January 1931, ECP-WTO-LC-Org, 2, Meeting Minutes, General and Council, AAVSO.

52 Nielsen, *The Big Foundations*, pp. 56–8.

53 A. Lawrence Lowell to Trevor Arnett, 26 January 1931, HU-PO-ALL, UA I 5.160 International Education Board [Series 1928–1930], folder 205, HUA; Max Mason to A. Lawrence Lowell, 23 January 1931, HU-PO-ALL, UA I 5.160 International Education Board [Series 1930–1933], folder 97, HUA; and A. Lawrence Lowell to Max Mason, 26 January 1931, HU-PO-ALL, UA I 5.160 International Education Board [Series 1930–1933], folder 97, HUA.

54 Harlow Shapley to F. W. Hunnewell, 22 December 1930, HU-PO-ALL, UA I 5.160 Funds – Wyeth [Series 1930–1933], folder 275, HUA; and F. W. Hunnewell to Harlow Shapley, 27 December 1930, Records of the President of Harvard University, Abbott Lawrence Lowell, 1909–1933, UA I 5.160 Funds – Wyeth [Series 1930–1933], folder 275, HUA.

55 Office of the Comptroller Harvard University, "Astronomical observatory, gift from the Rockefeller Foundation for buildings, equipment and endowment, statement of account, fiscal years 1931–32 to 1940–41 inclusive," HCO-DHM, UAV 630.37, 6, HCO Council Budget, HUA.

56 William Tyler Olcott, Secretary, Minutes of the Council of the American Association of Variable Star Observers, William Tyler Olcott, *Record of the Minutes of the Council of the American Association of Variable Star Observers,* Cambridge, Mass HCO, 16 October 1931, ECP-WTO-LC-Org, 2, Meeting Minutes, General and Council, AAVSO.

57 William Tyler Olcott, Secretary, Minutes of the Council of the American Association of Variable Star Observers, William Tyler Olcott, *Record of the Minutes of the Council of the American Association of Variable Star Observers,* Cambridge, Mass., HCO, 16 October 1931, ECP-WTO-LC-Org, 2, Meeting Minutes, General and Council, AAVSO. The minutes quoted above are ambiguous well beyond the norm even for AAVSO minutes. Absent here is any reflection of discussion of who, for example, was to pay for the clerk mandated as an assistant for the Recorder. The notes are not signed by Olcott, but instead are merely labeled "Attest William Tyler Olcott."

58 Clara Hyde Olcott to Leon Campbell, 11 June 1929, LC-Org, 6, Olcott, C. H., AAVSO; and Leon Campbell to Clara Hyde Olcott, 24 June 1929, LC-Org, 6, Olcott, C. H., AAVSO.

59 Leon Campbell to William Tyler Olcott, 8 January 1931, LC-Org, 6, Olcott, W. T., AAVSO.

60 Leon Campbell, "Looking backwards, or A. A. V. S. O. in retrospect," *VC* **2** (1931), 45–52.

8 FADING OF THE OLD GUARD

1 William Tyler Olcott to Leon Campbell, 2 October 1934, ECP-WTO-LC-Org, 1, Olcott-Campbell, AAVSO.

2 Edwin H. Hall to Leon Campbell, 24 December 1931, LC-Cor, 4, H misc. letters, AAVSO. It was through a donation of funds that long-time AAVSO member and patron Dorrit Hoffleit was originally hired as part of this program.

3 Olcott, to Campbell, 2 October 1934.

4 See, for example, William Tyler Olcott to Leon Campbell, 9 March 1933, ECP-WTO-LC-Org, 1, Olcott-Campbell, AAVSO.

5 Ronald Florence, *The Perfect Machine*, New York: Harper Collins Publishers, 1994, 197–206 and ff.

6 George W. Gray to Leon Campbell, 16 March 1930, LC-Cor, 3, G misc. letters, AAVSO; and (Leon Campbell to George W. Gray, 18 March 1930, LC-Cor, 3, G misc. letters, AAVSO.

7 G. E. Ensor to Leon Campbell, 14 October 1932, LC-Cor, 3, Ensor, G. E, AAVSO.

8 The decade of the 1930s witnessed five major planetarium openings, including Adler in Chicago, 1930; Fels in Philadelphia, 1933; Griffith in Los Angeles, 1935; Hayden in New York, 1935; and Buhl in Pittsburgh, 1939. Marché provides an excellent analysis of this movement in the social context of the time and of the later evolution of planetaria to simpler and less expensive projectors that served equally well and facilitated both public education and entertainment. Jordan D. Marché II, *Theaters of Time and Space*, New Brunswick, New Jersey, and London: Rutgers University Press, 2005, p. 26.

9 Leon Campbell to William Tyler Olcott, 21 August 1931, ECP-WTO-LC-Org, 1, Campbell-Olcott, AAVSO; and Leon Campbell to William Tyler Olcott, 18 June 1935, ECP-WTO-LC-Org, 1, Campbell-Olcott, AAVSO.

10 Thomas P. Fahy, *Richard Scott Perkin and the Perkin–Elmer Corporation*, unknown: privately published at Perkin–Elmer Corporation, 1987, pp. 36–9.

11 William Tyler Olcott, "Minutes of the Council of the American Association of Variable Star Observers," 20 October 1934, ECP-WTO-LC-Org, 2, Meeting Minutes, General and Council, AAVSO.

12 William Tyler Olcott, "Minutes of the Council of the American Association of Variable Star Observers," 23 November 1918, ECP-WTO-LC-Org, 2, Meeting Minutes, General and Council, AAVSO.

13 William Tyler Olcott, "Minutes of the Council of the American Association of Variable Star Observers," 8 May 1920, ECP-WTO-LC-Org, 2, Meeting Minutes, General and Council, AAVSO.

14 William Tyler Olcott, "Minutes of the Council of the American Association of Variable Star Observers," 28 May 1921, ECP-WTO-LC-Org, 2, Meeting Minutes, General and Council, AAVSO.

15 "Engineman by Day And an Observer of Stars in the Night," Pittsburgh, Pennsylvania, 29 May 1924, Biographical Pamphlets, AAVSO.

16 David W. Rosebrugh, *VC* **3** No. 13 (1937), 60–5; and David W. Rosebrugh, "Minutes of Council meeting A.A.V.S.O, October 18, 1940," 18 October 1940, ECP-WTO-LC-Org, 1, Council and Meeting Minutes (folder 1), AAVSO.

17 Leah B. Allen, "Minutes of the Council of the American Association of Variable Star Observers," 6 May 1932, ECP-WTO-LC-Org, 2, Meeting Minutes, General and Council, AAVSO.

18 Leon Campbell to Harold B. Webb, 29 July 1932, LC-Cor, 8, Webb, H. B., AAVSO.

19 William Tyler Olcott to Leon Campbell, 11 August 1933, ECP-WTO-LC-Org, 1, Olcott-Campbell, AAVSO.

20 William Tyler Olcott and Charles W. Elmer, "Minutes of the Council of the American Association of Variable Star Observers," 19 October 1935, ECP-WTO-LC-Org, 2, Meeting Minutes, General and Council, AAVSO.

21 A list dated February 1936 is filed with William Tyler Olcott, "Minutes of the Council of the American Association of Variable Star Observers," 22 May 1936, ECP-WTO-LC-Org, 1, Council and Meeting Minutes (folder 1), AAVSO.

22 William Tyler Olcott, "Minutes of the Council of the American Association of Variable Star Observers," 14 October 1932, ECP-WTO-LC-Org, 2, Meeting Minutes, General and Council, AAVSO.

23 Curvin H. Gingrich to Leon Campbell, 30 September 1935, LC-Cor, 6, Popular Astronomy, AAVSO.

24 Leon Campbell, "Variable star notes from the American Association of Variable Star observers," in *PA* **43**, No. 12 (1935), 571–94; and in *PA* **44**, No. 1 (1936), 34–6.

25 William Tyler Olcott, "Minutes of the Council of the American Association of Variable Star Observers," 20 October 1935, ECP-WTO-LC-Org, 2, Meeting Minutes, General and Council, AAVSO.

26 Leon Campbell, "Minutes of the Council of the American Association of Variable Star Observers," 11 October 1930, ECP-WTO-LC-Org, 2, Meeting Minutes, General and Council, AAVSO.

27 Leah B. Allen, "Minutes of the Council of the American Association of Variable Star Observers," 6 May 1932, ECP-WTO-LC-Org, 2, Meeting Minutes, General and Council, AAVSO.

28 William Tyler Olcott to Leon Campbell, 4 October 1934, ECP-WTO-LC-Org, 1, Olcott-Campbell, AAVSO.

29 Olcott, 4 October 1934. See also William Tyler Olcott to Leon Campbell, 3 May 1935, ECP-WTO-LC-Org, 1, Olcott-Campbell, AAVSO; William Tyler Olcott to Leon Campbell, 25 June 1935, ECP-WTO-LC-Org, 1, Olcott-Campbell, AAVSO; William Tyler Olcott to Leon Campbell, 7 October 1935, ECP-WTO-LC-Org, 1, Olcott-Campbell, AAVSO; and William Tyler Olcott to Leon Campbell, 25 October 1935, ECP-WTO-LC-Org, 1, Olcott-Campbell, AAVSO, in which Olcott complains he is out of the prospectus and various stationery items and asks Campbell why the latter has not already supplied his needs.

30 David W. Rosebrugh, and Clinton B. Ford, "History of the American Association of Variable Star Observers – 1936–1951," MWM-Org, 5, Papers and talks, non-technical, historical (AAVSO), AAVSO, 14.

31 Harlow Shapley, *Through Rugged Ways to the Stars*, New York: Charles Scribner's Sons, 1969, 116–19.

32 Inez L. B. Clough, "The A. A. V. S. O. at the American Museum of Natural History in New York City," *VC* **2**, No. 10 (July 1931), 41–4.

33 Claude B. Carpenter, "The twenty-ninth annual spring meeting of the A.A.V.S. O. at Toronto," *VC* **3** (40), 105–8.

34 Caroline E. Furness, *VC* **2**, No. 18 (July 1933), 78.

35 Leon Campbell, *VC* **2**, No. 19 (October 1933), 84; Leon Campbell, *VC* **2**, No. 21 (April 1934), 91; and Félix Eugène Marie de Roy, "Le system de Z Andromedæ," *Ciel et Terre* (1942), 63–5. de Roy replaced C. L. Brook as director of the BAA-VSS in 1921 and served in that capacity for a number of years. In 1929, de Roy summarized what was then known about LPVs in a preface to the "Tenth report of the Variable Star Section, 1920–1924. Félix Eugène Marie de Roy, *MmBAA* **28** (1929), vii–xxix.

36 Lois T. Slocum, *VC* **2** (January 1933), 69–70. Since the time that Mitchell began providing positions and magnitudes to the BAA and AFOEV committees, those stars had been reflected on AAVSO charts as well. Kerriann H. Malatesta, and Charles E. Scovil, "The history of AAVSO charts, Part I: The 1880s through the 1950s," *JAAVSO* **34** (2005), 97.

37 Malatesta and Scovil, "AAVSO chart history," 96–7.

38 William Tyler Olcott to The Council of the AAVSO, 1934, CBF, 6, Secretary File, 1932–1938, AAVSO; and Charles W. Elmer, "Meeting of the Council of the AAVSO (Minutes)," 24 May 1935, CBF, 6, Secretary File, 1932–1938, AAVSO.

39 Harold B. Webb to Leon Campbell, 12 October 1934, LC-Cor, 8, Webb, H. B., AAVSO. The first formal black and white charts were not issued from headquarters until 40 years later when the AAVSO began to issue charts as xerographic reprints made at headquarters.

40 Frederick C. Leonard, "Observations of Nova (2) Geminorum 1913–1914," *PA* **22** No. 8 (1914), 479–86; and Leon Campbell, "A systematic search for bright novae," *PA* **22** No. 8 (1914), 493–5.

41 Campbell, "A systematic search," 493–5.
42 David B. Pickering to Issei Yamamoto, 25 November 1920, LC-Cor, 8, Yamamoto, I., AAVSO.
43 Olcott and Elmer, "Minutes of the Council of the American Association of Variable Star Observers," October 19, 1935, ECP-WTO-LC-Org, 2, Meeting Minutes General and Council, AAVSO. A total of four awards of the medal were made: Prentice in 1934; Kazuaki Gomi of Japan for Nova Lac 1936; Sigeki Kabayashi of Japan for Nova Sgr 1936; and Bernhard Dawson of Argentina for Nova Pup 1942. There were no further awards of the medal; apparently, the cost of having each medal made became prohibitive. See LC-Org, 2, Nova Award, AAVSO.
44 Carpenter, "Twenty-ninth Spring Meeting," 106.
45 John Birmingham, "The red stars: observations and catalogue," *RIAT* **26** (1879), 249–354; T E. Espin, "Astronomical Fine Objects," *English Mechanic*, London, 23 January 1880, in files; T. E. Espin to W. H. Wesley, 31 December 1889, RAS Letters, AAVSO; John Birmingham, "The red stars: observations and catalogue," *PRIAA* **5** (1890); and A. F. Kohlman, "The double and colored star section," *MRSPA* **5** (1913), 67–8.
46 David W. Rosebrugh and Clinton B. Ford, "History"; and Malatesta and Scovil, "AAVSO Chart History," 96.
47 Leon Campbell to Edward A. Halbach, 24 May 1940, LC-Org, 2, Photographic variable fields program, AAVSO; and Edward A. Halbach to Leon Campbell, 10 June 1940, LC-Org, 2, Photographic variable fields program, AAVSO.
48 C. E. Kenneth Mees, "Astronomical photography looks to the red," *The Photographic Journal* **58** (1934), 448–56; and C. E. Kenneth Mees, "Astronomical photography looks to the red," *The Telescope* **Series II, vol. 1** (1934), 102–13.
49 George Waldo to Leon Campbell, 24 May 1933, LC-Cor, 8, Waldo, G., AAVSO.
50 William B. Albrecht, "Milwaukee Astronomical Society Chronology," MAS; R. Newton Mayall and Margaret W. Mayall, *Skyshooting: Photography for Amateur Astronomers*, New York: Dover Publications, 1968.
51 Leon Campbell to Lynn H. Matthias, 5 June 1935, LC-Org, 2, Photographic Variable Fields Program, AAVSO.
52 Lynn H. Matthias to Leon Campbell, 10 October 1935, LC-Org, 2, Photographic Variable Fields Program, AAVSO; and Lynn H. Matthias to Leon Campbell, 11 October 1936, LC-Org, 2, Photographic Variable Fields Program, AAVSO.
53 Leon Campbell to Lynn H. Matthias, 29 December 1936, LC-Org, 2, Photographic Variable Fields Program, AAVSO; and Leon Campbell to Lynn H. Matthias, 6 January 1937, LC-Org, 2, Photographic Variable Fields Program, AAVSO.
54 Lynn H. Matthias to Leon Campbell, 7 March 1937, LC-Org, 2, Photographic Variable Fields Program, AAVSO.
55 Leon Campbell to Lynn H. Matthias, 19 March 1937, LC-Org, 2, Photographic Variable Fields Program, AAVSO.
56 Leon Campbell, "The amateur in astronomy," *The Telescope* **Series II, vol. 5** (1938), 91, 94.
57 David W. Rosebrugh and Clinton B. Ford, "History," 8.
58 David W. Rosebrugh, "Minutes of the Council of the American Association of Variable Star Observers," 28 May 1938, ECP-WTO-LC-Org, 1, Council and Meeting Minutes (folder 1), AAVSO.
59 John A. Ingham, "The 1929 Harvard meeting," *VC* **2** No. 4 (January 1930), 19.
60 Olcott reported this comment by Hunter in the obituary he wrote after the latter passed away: William Tyler. Olcott, "Stephen Crasco Hunter," *PA* **41** (1933), 378.
61 Leon Campbell, "Harvard Observatory and the amateur astronomer," *Harvard Alumni Bulletin* (24 April 1930), 836–9, 838.
62 Leon Campbell to Harriet W. Bigelow, 30 November 1931, LC-Cor, 1, Bigelow, H., AAVSO.
63 Leon Campbell, "Some interesting A. A. V. S. O. variable stars," *VC*, **2** No. 19 (October 1933), 81.
64 Alice H. Farnsworth, "Minutes of the Council of the American Association of Variable Star Observers," 20 October 1933, ECP-WTO-LC-Org, 2, Meeting Minutes, General and Council, AAVSO. It later emerged that another member, Eppe Loreta, had also observed RS Oph in its outburst, a day or so earlier than Peltier. Incredibly, Loreta could not interest anyone in Italy in his discovery. After this obvious miscarriage came to light, the Council recognized Loreta's contributions by approving funds to send him new charts. See Leon Campbell, "Second annual report of the Pickering Memorial Astronomer," *VC*, **2** No. 19 (October 1933), 81; and David W. Rosebrugh, "Minutes of the Council meeting," 14 October 1938, ECP-WTO-LC-Org, 1, Council and Meeting Minutes (folder 1), AAVSO.
65 William Tyler Olcott, "Minutes of the Council of the American Association of Variable Star Observers," 26 May 1934, ECP-WTO-LC-Org, 2, Meeting Minutes, General and Council, AAVSO.
66 David W. Rosebrugh, Minutes of the Council of the American Association of Variable Star Observers, 15 October 1937, CBF, 6, Secretary File, 1932–1938, AAVSO. Unfortunately, if the notion that Loreta might deserve to be nominated again at some later date was part of the decision to

defer an award for this dedicated variable star worker, that became impossible when Loreta died sometime during or after World War II. Amédée Dermul, "M. Eppe Loreta," *Gazette astronomique, Bulletin de la Société d'Astronomie d'Anvers* **28** (1946), 5–6.

67 David W. Rosebrugh, "Minutes of joint meeting with the Bond Astronomical Club," 14 October 1938, ECP-WTO-LC-Org, 1, Council and Meeting Minutes (folder 1), AAVSO; and Louise E. Ballhaussen, "A.A.V.S..O. meeting, October 14–15, 1938," *VC* **3** No. 16 (1938), 88.

68 The distinction is outlined in detail in Thomas R. Williams, "Getting Organized: A history of amateur astronomy in the United States," unpublished Ph.D. dissertation, Rice University, 2000.

69 "Amateur star gazers plan an organization," *Milwaukee Journal,* Milwaukee, Wisconsin, 18 September 1932, p. 5. Armfield scrapbook, Aug 1932 to Oct 1933, MAS Archives; and Albrecht, MAS Chronology.

70 Marjorie B. Leavens, *VC* **2**, No. 20 (January 1934), 85.

71 Cyrus F. Fernald, "Soliloquy of a variable star observer," *S&T* **2** (August 1943), 10–11. In this article Fernald not only describes the telescope and its mounting in some detail, but also provides a useful discussion of his observing practices.

72 Cyrus F Fernald to Leon Campbell, 11 November 1945, LC-Cor, 3, Fernald, C. F, AAVSO; and Fernald, "Soliloquy," 10–11.

73 Thomas R. Williams, "Three AAVSO leaders: De Kock, Fernald and Peltier," *JAAVSO* **15** (1986), 135–8.

74 William Tyler Olcott, *VC* **2** No. 22 (July 1934), 92; and Leo J. Scanlon, "Efficiency in amateur variable star observing," *VC* **2** No. 22 (1934), 95–6.

75 Leo J. Scanlon, *Address to the Amateur Astronomer's Association of Pittsburgh*, Pittsburgh, Pennsylvania: 16 Mar 1994, in author's files; "Leo J. Scanlon amateur astronomer, built first aluminum-domed observatory," *Pittsburgh Post-Gazette,* Pittsburgh, Pennsylvania, 29 November 1999, p. A15; and Dennis di Cicco, "Leo J. Scanlon, 1903–1999," *S&T* **99** (2000), 92.

76 Jean-Louis Tassoul and Monique Tassoul, *A Concise History of Solar and Stellar Physics*, Princeton and Oxford: Princeton University Press, 2004, pp. 120–32.

77 Leon Campbell to Joy G. Goodsell, 26 August 1933, LC-Cor, 3, G misc. letters, AAVSO.

78 Jesse L. Greenstein to Leon Campbell, 2 July 1941, LC-Cor, 3, G misc. letters, AAVSO; and Leon Campbell to Jesse L. Greenstein, 12 July 1941, LC-Cor, 3, G misc. letters, AAVSO. See also Donald E. Osterbrock, "AAS meetings before there was an AAS: The pre-history of the society," David H. DeVorkin (Editor), *The American Astronomical Society's First Century,* Washington, DC: American Astronomical Society, 1999, pp. 3–19.

79 Neal J. Heines to Leon Campbell, 5 November 1937, LC-Cor, 4, Heines, N. J, AAVSO.

80 Neal J. Heines to Leon Campbell, 31 January 1934, LC-Cor, 4, Heines, N. J, AAVSO.

81 Neal J. Heines to Leon Campbell, 7 August 1935, LC-Cor, 4, Heines, N. J, AAVSO; and Harry L. Bondy, "Desiderata," 6 October 1972, MWM-Org, 5, Papers and Talks, non-technical, historical (AAVSO), AAVSO.

82 For example, a 1938 report from David Rosebrugh, who observed aurora from his home near Poughkeepsie, New York, was simply filed away. David W. Rosebrugh to AAVSO, 3 September 1938, LC-Org, 1, Auroral Program, AAVSO.

83 Carl W. Gartlein, "Request for auroral observations," *JRASC* **33** (1939), 46–50.

84 Elizabeth Wight, "Minutes of the Observatory Staff Meeting," 19 March 1939, File 39, MAS.

85 Catherine Stillman Pierce *et al,* "Twenty-Eighth annual spring meeting of the A. A. V. S. O, Ann Arbor, Michigan, May 19–21, 1939," *VC* **3** (1939), 89–92.

86 Roy A. Seely, "A. A. V. S. O. meeting, October 13–14, 1939," *VC* **3** No. 18 (1939), 95; and John Ball, Jr., "Minutes of the Observatory Staff Meeting," 21 January 1940, File 40, MAS.

87 Edward A. Halbach, "Report of the Auroral Committee," 1 June 1940, LC-Org, 1, Auroral program, AAVSO; and Carpenter, "Twenty-ninth Annual Spring Meeting."

88 Clinton B. Ford, *VC* **4** No. 2 (1941), p.

89 Samuel Eliot Morison, *Three Centuries of Harvard, 1636–1936*, Cambridge, Massachusetts: Harvard University Press, 1936.

90 Morton Keller and Phyllis Keller, *Making Harvard Modern: The Rise of America's University*, Oxford, UK, and New York: Oxford University Press, 2007, pp. 3–10. Russell's comment appears on page 5. Russell delivered his address to the Harvard Tercentennial more than a decade before Fred Hoyle uttered his now famous sobriquet, "the Big Bang," which achieved instant widespread usage in spite of the fact that Hoyle intended it as a slur. Helge Kragh, "Gamow's game: The road to the Hot Big Bang," *Centaurus* **38** (1996), 336.

91 Keller and Keller, *Making Harvard Modern*, pp. 102–3; and Jeanne E. Amster, "Conant, James Bryant," John A. Garraty and Mark C. Carnes (General Editors), *American National*

Biography, 1st Ed., New York and Oxford: Oxford University Press, 1999, 313–16.

92 We are indebted to University of Michigan historian Rudi Paul Lindner for this clarification of Shapley and Bok's motivation.

93 Peggy Aldrich Kidwell, "Harvard astronomers and World War II – Disruption and opportunity," Clark A. Elliott and Margaret W. Rossiter (Editors), *Science at Harvard University: Historical Perspectives,* Bethlehem, Pennsylvania: Lehigh University Press, 1992, p. 290; Michael Saladyga, "Jacchia, Luigi Guiseppe," Thomas A. Hockey (Editor-in-Chief), *Biographical Encyclopedia of Astronomers,* New York and Berlin: Springer, 2007, p. 582; and Luigi Jacchia, "Le Stelle Variabili," *Publicazioni dell'Observatorio Astronomico di Universitá Bologna* **2** (1933), 174–265.

94 Leon Campbell to Charles W. Elmer, 4 April 1941, LC-Cor, 3, Elmer, C. W., AAVSO.

95 The original scheduled date for publication was June 1941. Leon Campbell and Luigi Jacchia, *The Story of Variable Stars,* 1941.

96 David S. Evans, *Under Capricorn: A History of Southern Hemisphere Astronomy*, Bristol, UK, and Philadelphia: Adam Hilger, 1988, p. 163; and G. E. Ensor to Leon Campbell, 4 September 1939, LC-Cor, 3, Ensor, G. E., AAVSO.

97 Albert Graham Ingalls, "Telescoptics," *SciAm* **163** (1940), 108–9.

98 Clara Hyde Olcott to Leon Campbell, 11 June 1929, LC-Org, 6, Olcott, C. H., AAVSO.

99 In 1938, Carpenter formally replaced Andrew Ellicott Douglass as director of the Steward Observatory, although he had effectively managed that institution since his arrival in 1925. Carpenter was made chairman of the University's astronomy department in 1936. See Robert Dale Hall, "Education of American Research Astronomers, 1876–1941," Ph.D. dissertation, Oregon State University, 1999, p. 862; and Donald E. Osterbrock, "Young Don Menzel's amazing adventures at Lick Observatory," *JHA* **33** (2002), p. 102.

100 Leon Campbell to William Tyler Olcott, 20 May 1931, ECP-WTO-LC-Org, 1, Campbell-Olcott, AAVSO.

101 William Tyler Olcott to Leon Campbell, 25 November 1931, ECP-WTO-LC-Org, 1, Olcott-Campbell, AAVSO.

102 William Tyler Olcott to Leon Campbell, 5 December 1931, ECP-WTO-LC-Org, 1, Olcott-Campbell, AAVSO.

103 William Tyler Olcott to Leon Campbell, 5 January 1932, LC-Cor, 6, Olcott, W. T., AAVSO.

104 Leon Campbell to William Tyler Olcott, 14 January 1932, LC-Cor, 6, Olcott, W. T., AAVSO.

105 "Note to be added to the minutes of the Oct. 1932 Meeting of the A.A.V.S.O.," CBF, 6, Secretary File, 1927–1932, AAVSO; and Leon Campbell, *VC* **2**, No. 17 (April 1933), 73.

106 The memorial bronze tablet has since always been displayed in AAVSO Headquarters.

107 William Tyler Olcott to Leon Campbell, 20 January 1932, LC-Cor, 6, Olcott, W. T., AAVSO; Leon Campbell to William Tyler Olcott, 3 February 1933, LC-Cor, 6, Olcott, W. T., AAVSO; and William Tyler Olcott to Leon Campbell, 13 February 1932, LC-Cor, 6, Olcott, W. T., AAVSO.

108 William Tyler Olcott to Leon Campbell, 27 February 1932, ECP-WTO-LC-Org, 1, Olcott-Campbell, AAVSO.

109 William Tyler Olcott to Leon Campbell, 25 April 1932, ECP-WTO-LC-Org, 1, Olcott-Campbell, AAVSO.

110 Leon Campbell to William Tyler Olcott, 1 November 1933, ECP-WTO-LC-Org, 1, Campbell-Olcott, AAVSO.

111 William Tyler Olcott to Leon Campbell, 6 December 1933, ECP-WTO-LC-Org, 1, Olcott-Campbell, AAVSO.

112 Leon Campbell to William Tyler Olcott, 19 January 1935, ECP-WTO-LC-Org, 1, Campbell-Olcott, AAVSO; and Leon Campbell to William Tyler Olcott, 28 January 1935, ECP-WTO-LC-Org, 1, Campbell-Olcott, AAVSO.

113 Leon Campbell to William Tyler Olcott, 10 April 1935, ECP-WTO-LC-Org, 1, Campbell-Olcott, AAVSO.

114 William Tyler Olcott to Leon Campbell, 27 April 1935, ECP-WTO-LC-Org, 1, Olcott-Campbell, AAVSO.

115 William Tyler Olcott to Leon Campbell, 28 May 1935, ECP-WTO-LC-Org, 1, Olcott-Campbell, AAVSO.

116 William Tyler Olcott, *The Romance of the Southern Cross,* Miami, Florida: Privately published, 1935.

117 William Tyler Olcott to Leon Campbell, 26 June 1936, ECP-WTO-LC-Org, 1, Olcott-Campbell, AAVSO. Olcott's confusion on the issue of handedness in the southern hemisphere is a common one for all astronomical travelers. The confusion is easily understood in the following explanation: in the northern hemisphere, when we face the North Celestial Pole (NCP), we observe that the Sun, Moon, planets, and stars all rise to our right and set to our left. Upon crossing the Earth's equator, we can no longer see the NCP and so turn around to face the South Celestial Pole (SCP) for which, regrettably, there is no convenient marker like the star Polaris. Nevertheless, the Sun, Moon, planets, and stars now are found to rise to our left and to set to our right. Had Olcott not been pressed to leave for a vacation in New Hampshire, he would doubtless have puzzled this through.

118 "Dies suddenly while lecturing on astronomy," *Norwich Bulletin*, Norwich, Connecticut, 8 July 1936, p. 12. WSH, 1932–1994, AAVSO, 12; and David Bedell Pickering, "William Tyler Olcott," *PA* **44** (1936), 409–12.

119 William Tyler Olcott, "Fait accompli," 21 May 1930, LC-Cor, 6, Olcott, W. T., AAVSO.

9 GROWING PAINS AND DISTRACTIONS

1 Leon Campbell to Harlow Shapley, 15 October 1945, LC-Cor, 7, Shapley, H., AAVSO.

2 G. E. Ensor to Leon Campbell, 6 October 1940, LC-Cor, 3, Ensor, G. E., AAVSO.

3 Cyrus F. Fernald to Leon Campbell, 28 February 1942, LC-Cor, 3, Fernald, C. F., AAVSO.

4 Cyrus F. Fernald to Leon Campbell, 5 December 1942, LC-Cor, 3, Fernald, C. F., AAVSO; and Leon Campbell to Cyrus F. Fernald, 7 December 1942, LC-Cor, 3, Fernald, C. F., AAVSO.

5 Clinton B. Ford to Leon Campbell, 7 May 1942, LC-Cor, 3, Ford, C. B., AAVSO; and Leon Campbell to To Whom It May Concern, 7 May 1942, LC-Cor, 3, Ford, C. B., AAVSO.

6 Clinton B. Ford, *Some Stars, Some Music – The Memoirs of Clinton B. Ford*, Cambridge, Massachusetts: American Association of Variable Star Observers, 1986, 33–8.

7 Peggy Aldrich Kidwell, "Harvard astronomers in the second World War," *JHA* **21** (1990), 105–6. Other observatories were similarly impacted, see for example "Naval Award to McMath-Hulbert Staff," *S&T* **5** (1946), 6; Donald E. Osterbrock *et al.*, *Eye on the Sky: Lick Observatory's First Century*, Berkeley, California: University of California Press, 1988, pp. 226–30; David H. DeVorkin, "The maintenance of a scientific institution: Otto Struve, the Yerkes Observatory, and its Optical Bureau during the Second World War," *Minerva* **18** (1980), 595–623; and Donald E. Osterbrock, *Yerkes Observatory 1892–1950, The Birth, Near Death, and Resurrection of a Scientific Research Institution*, Chicago and London: The University of Chicago Press, 1997, pp. 245–65. Hoffleit describes her experiences with humor and provides a strong sense of the acceptance of professional women at the Aberdeen Proving Ground during the war, mentioning some of the other astronomers employed in the war effort, including Theodore Sterne and Edwin Hubble. E. Dorrit Hoffleit, *Misfortunes as Blessings in Disguise: The Story of My Life*, Cambridge, Massachusetts: American Association of Variable Star Observers, 2002, pp. 41–51.

8 Kidwell, *JHA*, **21**, (1990), 105–6; and Harlow Shapley, *Through Rugged Ways to the Stars*, New York: Charles Scribner's Sons, 1969, pp. 127–41. Dorrit Hoffleit's service during the war and for a short time thereafter was undertaken with mixed feelings. Although she agreed with the sentiments expressed by Shapley, Bok, and Payne-Gaposchkin, she also felt an obligation to help if the country needed her help. Her assignments included work similar to that of Martha Shapley at M.I.T. for a while before going to the Aberdeen Proving Ground, where Theodore Sterne from HCO was already working. Peggy Aldrich Kidwell, "Creating spaces in astronomy – Dorrit Hoffleit in Cambridge, Aberdeen, Nantucket and New Haven," in A. G. Davis Philip, *et al.* (Editors), *The Hoffleit Centennial: A Year of Celebration*, Schenectady, New York: L. Davis Press, 2006, pp. 3–10.

9 Leo Goldberg, and Lawrence H. Aller, "Donald Howard Menzel 1901–1976," *BMNAS* **47** (1991), 151, 161. The term "Ham" is still in use; some say it owes its origin to the fact that most early amateur radio enthusiasts equipped their radio shacks with Hammarlund radio transmitter/receiver sets.

10 Donald H. Menzel to Leon Campbell, 10 November 1943, LC-Cor, 6, Menzel, D. H., AAVSO.

11 Donald H. Menzel, *Elementary Manual of Radio Propagation*, New York: Prentice-Hall, Inc., 1948.

12 Leon Campbell to Donald H. Menzel, 15 October 1945, LC-Cor, 6, Menzel, D. H., AAVSO.

13 Amédée Dermul, "M. Eppe Loreta," *Gazette Astronomique, Bulletin de la Société d'Astronomie d'Anvers* **28** (1946), 5–6; "General notes," *PA* **54** (1946), 204; and "Eppe Loreta," *S&T* **5** (1946), 4.

14 Amédée Dermul, "Félix De Roy (1883–1942)," *Ciel et Terre* **58** (1942), 236; and Amédée Dermul, "M. Felix De Roy," *Gazette Astronomique, Bulletin de la Société d'Astronomie d'Anvers* **28** (1946), 3–4.

15 Domenico Benini to Leon Campbell, 20 November 1945, LC-Cor, 1, Benini, D., AAVSO; Giovanni B. Lacchini to Leon Campbell, 6 February 1946, LC-Cor, 5, Lacchini, G. B., AAVSO; and Giovanni B. Lacchini, and Luigi Jacchia, "Brief history of a grand old man of the AAVSO," 20 November 1945, LC-Cor, 1, Benini, D., AAVSO.

16 Cyrus F. Fernald to Leon Campbell, 1 October 1944, LC-Cor, 3, Fernald, C. F., AAVSO.

17 Charles W. Elmer to Leon Campbell, 17 May 1946, LC-Cor, 3, Elmer, Charles W., AAVSO. The affection Elmer held for Campbell, mentioned in Chapter 8, is visible in many ways in this letter.

18 Harry L. Bondy, "Desiderata," 6 October 1972, MWM-Org, 5, Papers and Talks, non-technical, historical (AAVSO),

AAVSO; and Neal J. Heines to Leon Campbell, 29 October 1939, LC-Cor, 4, Heines, N. J., AAVSO.

19 Leon Campbell to Neal J. Heines, 23 November 1939, LC-Cor, 4, Heines, N. J., AAVSO; Neal J. Heines to Leon Campbell, 8 May 1944, LC-Cor, 4, Heines, N. J., AAVSO; and Leon Campbell to Neal J. Heines, 16 May 1944, LC-Cor, 4, Heines, N. J., AAVSO.

20 Neal J. Heines, "A proposal for the establishment of a solar section in the American Association of Variable Star Observers," Paterson, New Jersey: August 1944, CBF, 6, Secretary File, 1927–1949, AAVSO; and Neal J. Heines, "[First AAVSO Solar Division report]," Paterson, New Jersey: 17 June 1945, CBF, 6, Secretary File, 1927–1949, AAVSO.

21 Neal J. Heines, "[First AAVSO Solar Division report]," Paterson, New Jersey: 17 June 1945, CBF, 6, Secretary File, 1927–1949, AAVSO.

22 Bart Fried, "H. B. Rumrill: An amateur's legacy," *S&T* **75** (1989), 86–7; William M. Kearons, "Photograph of the Sun," *PA* **36** (1928), plate 23; William M. Kearons, "How I observe the sun," *The Telescope* **Series II, vol. 7** (1940), 52–3; and Donald H. Menzel, *Our Sun*, Philadelphia: The Blakiston Company, 1949, pp. 26, 109–110.)

23 Neal J. Heines, "A proposal for the establishment of a solar section in the American Association of Variable Star Observers," Paterson, New Jersey: August 1944, CBF, 6, Secretary File, 1927–1949, AAVSO.

24 Neal J. Heines to Curvin H. Gingrich, 10 November 1944, LC-Org, 2, Solar Program, AAVSO.

25 Neal J. Heines, "[First AAVSO Solar Division report]," Paterson, New Jersey: 17 June 1945, CBF, 6, Secretary File, 1927–1949, AAVSO.

26 Alan H. Shapley, "American observations of relative sunspot-numbers in 1945 for application to ionospheric predictions," *PA* **54** (1946), 351–8.

27 Harry L. Bondy, "Desiderata," 6 October 1972, MWM-Org, 5, Papers and talks, non-technical, historical (AAVSO), AAVSO.

28 H.A.M., "Variable star observers," *AASB* **2** ([1945?]), 62.

29 Leon Campbell to John A. Fleming, 23 May 1945, LC Cor, 3, Fleming, J. A., AAVSO; and Leon Campbell to John A. Fleming, 25 January 1946, LC-Cor, 3, Fleming, J. A., AAVSO.

30 Neal J. Heines, Chairman, AAVSO Solar Division, *Instructions for Entering Sunspot-Data on Monthly Report-Forms*, Paterson, New Jersey: 1945; and Neal J. Heines, "[First AAVSO Solar Division report]," Paterson, New Jersey: 17 June 1945, CBF, 6, Secretary File, 1927–1949, AAVSO.

31 Herbert M. Harris, "Minutes of meeting of A.A.V.S.O. council," Cambridge, Massachusetts: 3 May 1946, CBF, 6, Secretary File, 1927–1949, AAVSO; and Neal J. Heines to Leon Campbell, 8 September 1946, LC-Cor, 4, Heines, N. J., AAVSO.

32 Leon Campbell to Harlow Shapley, 15 October 1945, LC-Cor, 7, Shapley, H., AAVSO.

33 Neal J. Heines, "Report of the Solar Division for the American Association of Variable Star Observers," Paterson, New Jersey: 1 May 1947, LC-Org, 2, Solar Program, AAVSO; and Neal J. Heines, "Report of the Solar Division for the American Association of Variable Star Observers," Paterson, New Jersey: 1 May 1948, LC-Org, 2, Solar Program, AAVSO.

34 Neal J. Heines to Leon Campbell, 30 June 1947, LC-Cor, 4, Heines, N. J., AAVSO.

35 David W. Rosebrugh, "The 27-year history of the Solar Division of the AAVSO," 15 September 1971, MWM-Org, 5, Papers and Talks, non-technical, historical (AAVSO), AAVSO.

36 Neal J. Heines, "Report of the Solar Division for the American Association of Variable Star Observers," 1 May 1948, LC-Org, 2, Solar Program, AAVSO.

37 J. Virginia Lincoln to Leon Campbell, 7 October 1948, LC-Org, 2, Solar Program, AAVSO.

38 Neal J. Heines, "Report of the Solar Division for the American Association of Variable Star Observers," Paterson, New Jersey: 13 October 1948, LC-Org, 2, Solar Program, AAVSO; and "Report of the Solar Division of A.A.V.S.O.," *PA* **54** (1946), 321–3.

39 Herbert M. Harris, "Minutes of meeting of A.A.V.S.O. council," Cambridge, Massachusetts: 11 October 1946, CBF, 6, Secretary File, 1927–1949, AAVSO.

40 Neal J. Heines to Leon Campbell, 31 October 1946, LC-Cor, 4, Heines, N. J., AAVSO; and Leon Campbell to Neal J. Heines, 1 November 1946, LC-Cor, 4, Heines, N. J., AAVSO.

41 Leon Campbell, "Tenth annual report of the Pickering Memorial Astronomer A. A. V. S. O. Recorder, 1940–41," *VC* **4** (1941), 11–12; and Thomas R. Williams, "Getting Organized: A history of amateur astronomy in the United States," unpublished Ph.D. dissertation, Rice University, 2000, pp. 247–52.

42 David W. Rosebrugh, "Minutes of the AAVSO council meeting, at Harvard College Observatory, October 6, 1944," LC-Org, 2, Council and meeting minutes, folder 2, AAVSO.

43 David W. Rosebrugh, "Minutes of the AAVSO council meeting, at Harvard College Observatory, October 6, 1944," LC-Org, 2, Council and meeting minutes, folder 2, AAVSO.
44 Leon Campbell to Roy A. Seely, 16 October 1944, LC-Cor, 7, Seely, R. A., AAVSO.
45 Leon Campbell to Herbert Harris, 22 February 1946, LC-Cor, 4, Harris, H. M., AAVSO.
46 David W. Rosebrugh, *VC* **4** (October 1946), 71–4.
47 David W. Rosebrugh, 'Minutes of the AAVSO council meeting, at Yale Observatory, June 17, 1945," LC-Org, 2, Council and meeting minutes, folder 2, AAVSO.
48 Leon Campbell to Herbert Harris, 7 May 1946, LC-Cor, 4, Harris, H. M., AAVSO.
49 James L. Russell to Leon Campbell, 29 October 1945, LC-Cor, 9, Q-R misc. letters, AAVSO.
50 Cyrus F. Fernald to James L. Russell, 4 November 1945, LC-Cor, 3, Fernald, C. F., AAVSO.
51 A few AAVSO visual observers have approached the level of 20 000 observations in one year, but it has never been exceeded. With modern robotic telescopes, photometric methods involving CCD photometers, and high-speed digital computers, individual observers will exceed Fernald's seemingly outrageous forecast.
52 Cyrus F. Fernald to Leon Campbell, 4 November 1945, LC-Cor, 3, Fernald, C. F., AAVSO; and Leon Campbell to Cyrus F. Fernald, 5 November 1945, LC-Cor, 3, Fernald, C. F., AAVSO.
53 Lawrence N. Upjohn to Leon Campbell, 11 May 1944, LC-Cor, 8, Upjohn, L. N., AAVSO; and Lawrence N. Upjohn to Leon Campbell, 4 May 1946, LC-Cor, 8, Upjohn, L. N., AAVSO. Upjohn was the nephew of the founder of Upjohn Pharmaceuticals Company, acted as its chief executive from 1932 to 1934, and since 1944 was Chairman of the Board.
54 Lawrence N. Upjohn to Leon Campbell, 4 May 1946, LC-Cor, 8, Upjohn, L. N., AAVSO; Leon Campbell to Lawrence N. Upjohn, 7 May 1946, LC-Cor, 8, U-V, AAVSO; and Leon Campbell to Lawrence N. Upjohn, 22 May 1946, LC-Cor, 8, U-V, AAVSO.
55 Leon Campbell to Lawrence N. Upjohn, 22 May 1947, LC-Cor, 8, U-V, AAVSO.
56 Lawrence N. Upjohn to Leon Campbell, 8 June 1947, LC-Cor, 8, U-V, AAVSO.
57 Neal J. Heines to Leon Campbell, and David W. Rosebrugh, 15 February 1946, LC-Cor, 4, Heines, N. J., AAVSO.
58 James Sayre Pickering published seven popular books on astronomy; *The Stars Are Yours*, *1001 Questions about Astronomy*, *Captives of the Sun*, *Asterisks*, *Windows to Space*, *Famous Astronomers*, and *Astronomy for Amateur Observers*. While working in the jewelry business as a credit manager, he sought and secured a part-time position as a lecturer at the American Museum-Hayden Planetarium, eventually becoming full-time as assistant astronomer and supervisor of customer relations. He also authored a number of television documentaries about astronomy. "James Pickering, Astronomer, dies," *New York Times*, New York, New York, 15 February 1969; and "Planetarium lecturer dies," *Sky and Telescope* **37** (April 1969), 221.
59 Neal J. Heines to Leon Campbell, 5 June 1946, LC-Cor, 4, Heines, N. J., AAVSO; James S. Pickering to Leon Campbell, 18 August 1946, LC-Cor, 6, Pickering, J. S., AAVSO; and Leon Campbell, "David Bedell Pickering 1873–1946," *PA* **54** (1946), 443–7.
60 David H. DeVorkin, "Serving the muse in Shapley's wake," A. G. Davis Philip *et al.*, Editors, *The Hoffleit Centennial: A Year of Celebration*, April 2006, Schenectady, New York: L. Davis Press, 2006, p. 12. According to DeVorkin, Shapley employed a fund-raising consultant in this time frame. From cover to cover, *Harvard College Observatory – The First Century* sounds a great deal like what such a consultant might suggest.
61 Anonymous, *Harvard College Observatory: The First Century*, Cambridge, Massachusetts: Harvard College Observatory, 1946, p. 20.
62 Anonymous, p. 18.
63 Anonymous, p. 18.
64 Owen Gingerich, "Through rugged ways to the galaxies," *Two Astronomical Anniversaries: HCO & SAO*, Cambridge, England: Science History Publications Ltd., 1990, p. 87.
65 Philip *et al.*, *In Shapley's Wake*, p. 12; and Morton Keller and Phyllis Keller, *Making Harvard Modern, The Rise of America's University*, Oxford, UK, and New York: Oxford University Press, 2007, pp. 239–40.
66 Examples of the severe expense management regime at HCO are too numerous to bear citation. Anyone interested in further researching this problem will find the necessary materials in Box 7, Shapley folder, of the AAVSO Correspondence files, and in Box 1, HCO Memos folder, of the Margaret W. Mayall Collection.
67 Lawrence N. Upjohn to Leon Campbell, 9 June 1949, LC-Cor, 8, Upjohn, L. N., AAVSO.
68 Harlow Shapley to Leon Campbell, 15 June 1949, LC-Cor, 8, Upjohn, L. N., AAVSO.
69 H. B. Allen to Leon Campbell, 27 June 1949, LC-Cor, 8, Upjohn, L. N., AAVSO.

70 Leon Campbell to Lawrence N. Upjohn, 30 June 1949, LC-Cor, 8, Upjohn, L. N., AAVSO.

71 Clinton B. Ford, "Minutes of the AAVSO Council meeting, at Harvard College Observatory, May 27, 1949," CBF, 6, Secretary File 1927–1949, AAVSO; and Clinton B. Ford, "Minutes of the AAVSO Council meeting, at Harvard College Observatory, October 14–16, 1949," CBF, 6, Secretary File 1927–1949, AAVSO.

72 The tables were prepared by Leon Campbell, although the report's authorship is "Margaret W. Mayall, Recorder" indicating a much later preparation date, after November 1949.

73 Cyrus F. Fernald to Leon Campbell, 12 December 1944, LC-Cor, 3, Fernald, C. F., AAVSO.

74 Cyrus F. Fernald to Leon Campbell, 13 October 1947, LC-Cor, 3, Fernald, C. F., AAVSO.

75 Leon Campbell to Cyrus F. Fernald, 3 April 1947, LC-Cor, 3, Fernald, C. F., AAVSO.

76 John L. Grinch to Leon Campbell, 14 March 1946, LC-Cor, 3, misc. G, AAVSO.

77 Leon Campbell to John L. Grinch, 15 March 1946, LC-Cor, 3, misc. G, AAVSO.

78 Another piece of evidence for the intractability of this problem is found in the Amateur Telescope Making books, recently republished by Willmann-Bell Publishing Company. With the three volumes reorganized and well-indexed, it is possible to see the amazing range of other devices that ATMs considered achievable and constructed, including polarizing monochrometers, spectometers, spherometers, and many other gadgets to satisfy the most avid amateur instrument builder's desire for challenge, but no wedge or other types of photometers. See Albert Graham Ingalls, *Amateur Telescope Making,* Richmond, Virginia: Willmann-Bell, Inc., 1996.

79 John J. Ruiz to Leon Campbell, 3 October 1947, HUG 4773.10, 22B, R Misc., HUA.

80 Gerald E. Kron, "The construction and use of a photomultiplier type of photoelectric photometer," *HarCi* No. 151 (1946), 10–29; Gerald E. Kron, "Popular photometry," *S&T* 6 (1947), 7–10; and David H. DeVorkin, "Electronics in astronomy: Early applications of the photoelectric cell and photomultiplier for studies of point-source celestial phenomena," *Proceedings of the IEEE* **73** (1985), 1205–20.

81 Leon Campbell to John J. Ruiz, 14 October 1947, HUG 4773.10, 22B, R Misc., HUA.

82 David W. Rosebrugh to Clinton B. Ford, 18 July 1949, CBF, 7, Secretary File, 1949–1952, AAVSO.

83 David W. Rosebrugh to Margaret W. Mayall, 15 October 1949, MWM-Cor, 11, Rosebrugh, D. W., folder 1, AAVSO.

84 Jesse L. Greenstein to Leon Campbell, 20 October 1947, LC-Cor, 3, misc. G, AAVSO; and Jesse L. Greenstein to Leon Campbell, 20 November 1947, LC-Cor, 3, misc. G, AAVSO.

85 Leon Campbell to Jesse L. Greenstein, 26 November 1947, LC-Cor, 3, misc. G, AAVSO; Leon Campbell to Jesse L. Greenstein, 13 January 1948, LC-Cor, 3, misc. G, AAVSO; and Jesse L. Greenstein to Leon Campbell, 19 January 1948, LC-Cor, 3, misc. G, AAVSO.

86 Paul W. Merrill, *The Nature of Variable Stars*, New York: The Macmillan Company, 1938; and Paul W. Merrill, *The Spectra of Long-Period Variable Stars*, Chicago: The University of Chicago Press, 1940.

87 Paul W. Merrill, "Some questions concerning long period variables," Leon Campbell *Studies of Long Period Variables,* Cambridge, Massachusetts: AAVSO, 1955, ix–xi.

88 Leon Campbell to Sergei Gaposchkin, 25 June 1945, LC-Cor, 3, Gaposchkin, S., AAVSO; and David W. Rosebrugh to Clinton B. Ford, 6 June 1951, CBF, 7, Secretary File, 1949–1952, AAVSO.

89 Margaret Harwood, "Fifty years at HCO," *S&T* 8 (1949), 191–3. The quote appears on p. 192.

90 Anna Sudaric Hillier, "Chronology of ATMoB," 1997; and James M. Gagan, "History of ATM as written in long hand by Jim Gagan on different sized note papers," in Anna Sudaric Hillier (Editor/Compiler/Historian of the Amateur Telescope Makers of Boston), *The Early Years: The History of the Amateur Telescope Makers of Boston,* 1979. During World War II, members of the ATMoB actually participated in the classified optical work of the HCO led by Dr. James G. Baker.

91 Leon Campbell to Owen Gingerich, 2 April 1947, LC-Cor, 3, Gingerich, O., AAVSO.

92 Owen Gingerich to Leon Campbell, 12 February 1948, LC-Cor, 3, Gingerich, O., AAVSO; and Owen Gingerich to Leon Campbell, 8 July 1948, LC-Cor, 3, Gingerich, O., AAVSO.

93 An intensity curve, as opposed to the usual visual light curve, is computed from the following relationship:

$$\log I = 0.4d$$

where I is the intensity of any point on the mean visual light curve, and d is the difference between the visual magnitude at that same point and the mean magnitude of the light curve at minimum. The minimum intensity is considered zero, and the maximum intensity is 100. Because the relationship is logarithmic, much of the detail in the light curve is obscured when the star is near minimum brightness, but the details in the light curve at maximum are enhanced or exaggerated. In effect, the intensity curve yields a better visual representation of the rate of energy release associated with the brightening of the star, but has no absolute relationship to the physical conditions in the star otherwise.

94 Leon Campbell to Willard P. Gerrish, 22 June 1948, LC-Cor, 3, Gerrish, W. P., AAVSO.

95 Leon Campbell to Leo Goldberg, 7 July 1949, LC-Cor,.3, misc. G, AAVSO; and Leon Campbell to Charlotte Greene, 27 June 1949, LC-Cor, 3, misc. G, AAVSO.

96 The personal letters to Campbell from his many friends in AAVSO and variable star astronomy were bound and, after a selection of them were read at the annual banquet held in his honor, the bound collection was presented to him as a gift he cherished.

97 Cyrus F. Fernald to Leon Campbell, 18 October 1949, LC-Cor, 3, Fernald, C. F., AAVSO.

98 "Leon Campbell," *S&T*, **10**, (1951), 210, 230; and Margaret W. Mayall, "Leon Campbell 1881–1951," *PA* **59** (1951), 356–8.

10 LEARNING ABOUT INDEPENDENCE

1 Margaret W. Mayall, "Annie J. Cannon & spectra of stars," March 1950, MWM-Org, 5, Papers and Talks, non-technical, historical, by MWM, AAVSO.

2 Leon Campbell to Harlow Shapley, 19 January 1949, LC-Cor, 7, Shapley, H., AAVSO.

3 Leon Campbell to Harlow Shapley, 4 March 1949, LC-Cor, 7, Shapley, H., AAVSO.

4 David W. Rosebrugh to Clinton B. Ford, 7 March 1949, CBF, 6, Secretary File, 1927–1949, AAVSO; and David W. Rosebrugh to Clinton B. Ford, 20 March 1949, CBF, 6, Secretary File, 1927–1949, AAVSO.

5 Clinton B. Ford to David W. Rosebrugh, 24 March 1949, CBF, 6, Secretary File, 1927–1949, AAVSO; and David W. Rosebrugh to Clinton B. Ford, 18 July 1949, CBF, 7, Secretary File, 1949–1952, AAVSO.

6 Clinton B. Ford to Charles H. Smiley, 15 June 1948, CBF, 6, Secretary File, 1927–1949, AAVSO; and Charles H. Smiley to Clinton B. Ford, 3 June 1948, CBF, 6, Secretary File, 1927–1949, AAVSO.

7 Clinton B. Ford to David W. Rosebrugh, 26 July 1949, CBF, 7, Secretary File, 1949–1952, AAVSO; and Leon Campbell to Harlow Shapley, 11 July 1949, LC-Cor, 7, Shapley, H., AAVSO. LaPelle joined the AAVSO in May 1948, perhaps in anticipation of Campbell's retirement. See also Thomas R. Williams, "Getting organized: U.S. amateur astronomy from 1860 to 1985," John Percy and Joseph B. Wilson (Editors), *Amateur-Professional Partnerships in Astronomy*, July 1999, University of Toronto, Toronto, Provo, Utah: Astronomical Society of the Pacific, 2000, pp. 322–6.

8 David W. Rosebrugh to Clinton B. Ford, 18 July 1949, CBF, 7, Secretary File, 1949–1952.

9 Harlow Shapley to Margaret W. Mayall, 12 August 1949, MM, 1, HCO-AAVSO correspondence and notes, AAVSO.

10 Margaret W. Mayall, "Interview conducted by Owen Gingerich," OHP, OHI 28323, CHP-NBL; *Iron Hill* (Sheriff, B. R.), *Maryland State Gazetteer and Baltimore City Business Directory for 1902 and 1903*, Baltimore, Maryland: R. L. Polk & Co., 1902; and Carie Ricoletta, "Maryland, Cecil, Elkton, Enumeration District 17," 10, Ancestry.com, http://search.ancestry.com/iexec?/?htx.

11 Margaret W. Mayall, "Margaret Mayall," *VV* (1966); E. Dorrit Hoffleit, "Margaret Walton Mayall, 1902–1995," *BullAAS* **28** (1996); Margaret W. Mayall, "MWM and the AAVSO," May 1978, MWM-Org, 5, Papers and Talks, non-technical, historical, by MWM, AAVSO; and Margaret W. Mayall, "Autobiographical essay and brief bibliography," undated [probably 1960], MWM-Org, 5, Papers and Talks, non-technical, historical, by MWM, AAVSO.

12 Leslie J. Comrie, and Margaret L. Walton, "Occultations during the lunar eclipse of 1924 August 14," *JBAA* **34** (1923–1924), 236–7; Charles H. Smiley, "Leslie John Comrie 1893–1950," *PA* **59** (1951), 115–17; and Margaret W. Mayall, "L. J. Comrie," *S&T* **10** (1951), 82.

13 Margaret W. Mayall, Owen Gingerich Interview. See Hermann A. Brück, "Dr. Annie J. Cannon," *Obs* **64** (1941), 113–15; and Pamela E. Mack, "Strategies and compromises: women in astronomy at Harvard College Observatory, 1870–1920," in Owen Gingerich and Michael Hoskin (Editors), *Two Astronomical Anniversaries: HCO & SAO*, Cambridge, England: Science History Publications Ltd., 1990, pp. 65–75.

14 Margaret W. Mayall, Owen Gingerich Interview.

15 Margaret W. Mayall, Owen Gingerich Interview.

16 Harlow Shapley, and Margaret L. Walton, "Investigations of Cepheid variables: I. The period-spectrum relation," *HarCi* No. 313 (1927), 1–7.

17 Margaret W. Mayall, "A new eclipsing star of unusual character," *AnHar* **105** (1937), 491–8; Harlow Shapley to Margaret W. Mayall, 23 July 1936, LC-Org, 4, M. W. Mayall Papers, (c), Notes from H. Shapley, AAVSO; and Harlow Shapley to Margaret W. Mayall, 7 November 1936, LC-Org, 4, M. W. Mayall Papers, (c), Notes from H. Shapley, AAVSO.

18 Margaret W. Mayall, and Mary H. Baker, "Spectral curves for thirty Cepheid variables," *HarCi* No. 436 (1940), 1–7.

19 Harlow Shapley, "Introduction," *AnHar* **113** (1943), 1; and E. Dorrit Hoffleit, "The Milton Bureau revisited," *JAAVSO* **28** (2000), 127–48.
20 Margaret W. Mayall, Owen Gingerich Interview.
21 See, for example, Margaret W. Mayall to Daniel M. Popper, 6 March 1946, LC-Org, 4, M. W. Mayall Papers, (c), General Corresp., AAVSO, and correspondence with other astronomers in this folder.
22 Otto Struve to Margaret W. Mayall, 3 December 1947, MM, 1, IAU Comission 29, AAVSO.
23 Margaret W. Mayall to Otto Struve, 17 December 1947, MM, 1, IAU Comission 29, AAVSO.
24 A. N. Vyssotsky to Margaret W. Mayall, 15 September 1948, MM, 1, Standard Stars for Spectral Type, AAVSO.
25 Otto Struve to To members of Commission 29 of the I.A.U., 13 January 1948, MM, 1, IAU Comission 29, AAVSO.
26 Otto Struve to Margaret W. Mayall, 31 May 1948, MM, 1, IAU Comission 29, AAVSO.
27 Jan H. Oort, *IAUT* **7** (1950), 67, 69, 298–9.
28 Oort, *IAUT*, **7** (1950), 528–43.
29 Adriaan Blaauw, *History of the IAU: The Birth and First Half-Century of the International Astronomical Union*, Dordrecht, The Netherlands: Kluwer Academic Publishers, 1994, p. 146.
30 Boris V. Kukarkin, "Report on the study of variable stars in the U.S.S.R. during the period 1948–1951," *IAUT* **8** (1954), 392–4; Margaret W. Mayall, "General catalogue of variable stars (book review)," *PA* **57** (1949), 47–8; and Harlow Shapley to Margaret W. Mayall, 13 October 1948, LC-Org, 4, M. W. Mayall Papers, (c), Book Reviews, AAVSO.
31 Margaret W. Mayall to H.C.O.C. [Harvard College Observatory Council], 24 May 1947, LC-Org, 4, M. W. Mayall Papers, (c), General Corresp., AAVSO.
32 Margaret W. Mayall 29 April 1949, LC-Org, 4, M. W. Mayall Papers, (c), General Corresp., AAVSO.
33 Harlow Shapley to Margaret W. Mayall, 29 November 1950, MM, 1, HCO-AAVSO correspondence and notes, AAVSO.
34 Margaret W. Mayall to Harlow Shapley, 13 August 1949, MM, 1, HCO-AAVSO correspondence and notes, AAVSO.
35 David W. Rosebrugh to Clinton B. Ford, 16 August 1949, CBF, 7, Secretary File, 1949–1952, AAVSO.
36 Clinton B. Ford to Margaret W. Mayall, 18 August 1949, CBF, 7, Secretary File, 1949–1952, AAVSO.
37 Margaret W. Mayall to Clinton B. Ford, 25 August 1949, CBF, 7, Secretary File, 1949–1952, AAVSO.
38 E. Dorrit Hoffleit, "Margaret Walton Mayall, 1902–1995," *BullAAS* **28** (1996), 1455–6.
39 David W. Rosebrugh to Clinton B. Ford, 18 July 1949, CBF, 7, Secretary File, 1919–1952, AAVSO.
40 Anonymous, "The Pickering astronomer," *HCOGossip* (1949), MM, 1, HCO Gossip Sheet, AAVSO.
41 Anonymous, "The little museum," *HCOGossip* (1949); Anonymous, "The new broom," *HCOGossip* (1949); and Anonymous, "Harvard College Observatory personnel lists," 1949–1953, MWM-Cor, 13, Stephansky, H. M., AAVSO.
42 Clinton B. Ford to Lewis J. Boss, 26 November 1949, CBF, 7, Secretary File, 1949–1952, AAVSO.
43 David W. Rosebrugh to Clinton B. Ford, 6 June 1951, CBF, 7, Secretary File, 1949–1952, AAVSO.
44 Paul W. Merrill, *The Nature of Variable Stars*, New York: The Macmillan Company, 1938.
45 Margaret W. Mayall, "Variable star spectroscopist," *S&T* **22** (1961), 127.
46 Paul W. Merrill to Margaret W. Mayall, 23 January 1950, MWM-Cor, 8, Merrill, P. W., AAVSO.
47 Margaret W. Mayall to Paul W. Merrill, 10 February 1950, MWM-Cor, 8, Merrill, P. W., AAVSO.
48 Margaret W. Mayall to Harlow Shapley, 9 January 1950, MWM-Cor, 8, Mayall, M. W., AAVSO.
49 Margaret W. Mayall, "Variable star notes from the American Association of Variable Star Observers," *PA* **58** (1950), 306–8.
50 Margaret W. Mayall to Leslie C. Peltier, 2 May 1950, MWM-Cor, 10, Peltier, L. C., AAVSO; Leslie C. Peltier to Margaret W. Mayall, March 1950–30 April 1950, MWM-Cor, 10, Peltier, L. C., AAVSO; and Margaret W. Mayall to Leslie C. Peltier, 02 May 1950, MWM-Cor, 10, Peltier, L. C., AAVSO.
51 Harlow Shapley to The Gaposchkins, 27 September 1950, MM, 1, HCO Memos, AAVSO.
52 Clinton B. Ford, Cambridge, Massachusetts: October 13–14, 1950, CBF, 7, Secretary File, 1949–1952, AAVSO.
53 Margaret W. Mayall to various, undated [1950 August], CBF, 7, Secretary File, 1949–1952, AAVSO; and Margaret W. Mayall, untitled note, 21 March 1951, MM, 1, HCO Memos, AAVSO.
54 Clinton B. Ford, AAVSO Council Minutes, October 13–14, 1950, CBF, 7, Secretary File, 1949–1952, AAVSO.
55 Clinton B. Ford, AAVSO Council Minutes, October 13–14, 1950, CBF, 7, Secretary File, 1949–1952, AAVSO.
56 Harlow Shapley, 1950, CBF, 7, Secretary File, 1949–1952, AAVSO.
57 Clinton B. Ford, Cambridge, Massachusetts: 51, CBF, 7, Secretary File, 1949–1952, AAVSO.

58 David W. Rosebrugh to Margaret W. Mayall, 4 April 1951, MWM-Cor, 11, Rosebrugh, D. W., 1951–1952, AAVSO.

59 Joseph Ashbrook, "Memorandum to Mrs. Mayall on proposed AAVSO observations of eclipsing variables," 10 October 1950, MWM-Cor, 1, Ashbrook, J., AAVSO.

60 Margaret W. Mayall to Cyrus F. Fernald, 27 February 1953, MWM-Cor, 4, Fernald, C. F., AAVSO.

61 Mayall, 27 February 1953; and Margaret W. Mayall to David W. Rosebrugh, 3 April 1951, MWM-Cor, 11, Rosebrugh, D. W., 1951–1952, AAVSO.

62 Harlow Shapley to Margaret W. Mayall, 18 May 1951, MWM-Cor, 8, Mayall, M. W., AAVSO; and Margaret W. Mayall, "Present state of Mr. Campbell's volume: H.A. 114," Cambridge, Massachusetts: 21 May 1951, MWM-Cor, 8, Mayall, M. W., AAVSO.

63 David W. Rosebrugh to Margaret W. Mayall, 29 May 1951, MWM-Cor, 11, Rosebrugh, D. W., 1951–1952, AAVSO; Margaret W. Mayall to David W. Rosebrugh, 20 June 1951, MWM-Cor, 11, Rosebrugh, D. W., 1951–1952, AAVSO; David W. Rosebrugh to Margaret W. Mayall, 21 October 1951, MWM-Cor, 11, Rosebrugh, D. W., 1951–1952, AAVSO; and David W. Rosebrugh to Margaret W. Mayall, and Helen Stephansky, 9 February 1952, MWM-Cor, 11, Rosebrugh, D. W., 1951–1952, AAVSO.

64 Frederick C. Leonard, "Curvin Henry Gingrich 1880–1951," *PA* **59** (1951), 343–7; and Margaret W. Mayall, "Leon Campbell 1881–1951," *PA* **59** (1951), 356–8. Leonard became a professional astronomer; he founded and chaired for many years the astronomy department at the University of California at Los Angeles.

65 Clinton B. Ford to Mrs. Curvin H. Gingrich, 16 November 1951, CBF, 7, Secretary File, 1949–1952, AAVSO.

66 David W. Rosebrugh, "History of the American Association of Variable Star Observers 1936 to 1951," October 1951, MWM-Org, 5, Papers and Talks, non-technical, historical (AAVSO), AAVSO.

67 R. Newton Mayall (Editor), "untitled [Abstracts of the Fall 1951 meeting of the AAVSO]," *AVSOA* (1951); and Margaret W. Beardsley, "40th Annual Meeting of the A.A.V.S.O. at Harvard College Observatory October 12–13, 1951," *VC* **5** (1951), 13–15.

68 John J. Ruiz, "The gremlins and my photometer," *S&T* **11** (1951), 43–5; See also William A. Baum, "Amateur Photometry," *The Griffith Observer* (1950), 11; John S. Hall, and John F. Jewett, "A simple DC photometer for photoelectric photometry," *S&T* **7** (1949), 169–71; and Gerald E. Kron, "Popular photometry," *S&T* **6** (1947), 7–10.

69 Frank Bradshaw Wood to John J. Ruiz, 7 December 1951, MWM-Cor, 12, Ruiz, J. J., AAVSO; Margaret W. Mayall to John J. Ruiz, 9 October 1951, MWM-Cor, 12, Ruiz, J. J., AAVSO; John J. Ruiz to Margaret W. Mayall, 10 December 1951, MWM-Cor, 12, Ruiz, J. J., AAVSO; and Margaret W. Mayall to John J. Ruiz, 3 January 1952, MWM-Cor, 12, Ruiz, J. J., AAVSO.

70 Laurence M. Gould, "Announcement," *PA* **59** (1951), i; and Leif J. Robinson, "Enterprise at Harvard College Observatory," *JHA* **21** (1990), 99.

71 Charles H. Smiley to Martha Stahr Carpenter, 31 December 1951, CBF, 7, Secretary File, 1949–1952, AAVSO.

72 Clinton B. Ford to Martha Stahr Carpenter, 10 January 1952, CBF, 7, Secretary File, 1949–1952, AAVSO.

73 David W. Rosebrugh to Helen M. Stephansky, 10 January 1952, CBF, 7, Secretary File, 1949–1952, AAVSO; and Clinton B. Ford to Hester W. Meech, 27 January 1952, CBF, 7, Secretary File, 1949–1952, AAVSO.

74 Clinton B. Ford to G. Howard Carragan, 13 January 1952, CBF, 7, Secretary File, 1949–1952, AAVSO.

75 Clinton B. Ford to Margaret W. Mayall, 13 January 1952, CBF, 7, Secretary File, 1949–1952, AAVSO.

76 E. Rose D'Arcy to Margaret W. Mayall, 21 April 1952, MM, 1, HCO-AAVSO correspondence and notes, AAVSO; and Margaret W. Mayall to David W. Rosebrugh, 31 March 1952, MWM-Cor, 11, Rosebrugh, D. W., 1951–1952, AAVSO. Ford and Mayall briefly considered publishing variable star notes in the "Amateur Astronomer" section of the *Scientific American* magazine. But *SA* was having financial difficulties of its own, and the idea was dropped. See Clinton B. Ford to Birt Darling, 8 February 1952, CBF, 7, Secretary File, 1949–1952, AAVSO; and Clinton B. Ford to Richard W. Hamilton, 16 February 1952, CBF, 7, Secretary File, 1949–1952, AAVSO.

77 G. Howard Carragan to Clinton B. Ford, 20 February 1952, CBF, 7, Secretary File, 1949–1952, AAVSO.

78 Clinton B. Ford to Margaret W. Mayall, 1 April 1952, CBF, 7, Secretary File, 1949–1952, AAVSO.

79 Clinton B. Ford, Cambridge, Massachusetts: 52, CBF, 7, Secretary File, 1949–1952, AAVSO.

80 Clinton B. Ford to David W. Rosebrugh, 8 June 1952, CBF, 7, Secretary File, 1949–1952, AAVSO; and R. Peter Broughton, *Looking Up: A History of the Royal Astronomical Society of Canada*, Toronto, Ontario: Dundurn Press, 1994, p. 99. The first "Variable Star Notes – American Association of Variable Star Observers" appeared in *JRASC* **46**, number 4 (July–August 1952), 151–4.

81 Margaret W. Mayall, "Variable Star Notes," *JRASC* **46** (1952a), 151–4; Margaret W. Mayall, "Variable Star Notes," *JRASC* **46** (1952b), 202–3; and John J. Ruiz, "Light curves of 12 Lacertae," *JRASC* **46** (1952), 203–5.

82 Joseph Ashbrook to Margaret W. Mayall, 16 January 1952, MWM-Cor, 1, Ashbrook, J., AAVSO; Joseph Ashbrook to Margaret W. Mayall, 16 August 1952, MWM-Cor, 1, Ashbrook, J., AAVSO; Margaret W. Mayall, "AE Aquarii month," *S&T* **11** (1952), 204; Margaret W. Mayall, "AE Aquarii month," 1952, MWM-Org, 5, Papers and Talks, technical, by MWM, AAVSO; and Leif J. Robinson, "Ashbrook, Joseph," in Thomas Hockey *et al.* (Editors), *Biographical Encyclopedia of Astronomers,* New York: Springer, 2007, pp. 65–6.

83 Sarah Lee Lippincott to Margaret W. Mayall, 24 November 1952, MWM-Cor, 8, Lippincott, S. L., AAVSO; Margaret W. Mayall to Sarah Lee Lippincott, 20 December 1952, MWM-Cor, 8, Lippincott, S. L., AAVSO; Sarah Lee Lippincott, "Red dwarf flare stars," *JRASC* **47** (1953), 24–7; and Mayall, 27 February 1953.

84 Margaret W. Mayall to David W. Rosebrugh, 16 July 1952, MWM-Cor, 11, Rosebrugh, D. W., 1951–1952, AAVSO.

85 A. Danjon, "Report of Commission 27 (Variable Stars)," *IAUT* **8** (1954), 390–2.

86 Boris V. Kukarkin, "Report on the study of variable stars," *IAUT* **8**, (1954), 392–4.

87 Paul W. Merrill, "Emission lines in the spectra of long-period variable stars," *JRASC* **46** (1952), 181–90.

88 Morton Keller and Phyllis Keller, *Making Harvard Modern: The Rise of America's University*, New York: Oxford University Press, 2007, pp. 78; 102–3; 108–9; 126; 129; 144–5); James B. Conant to Paul H. Buck, 29 July 1952, HU-PO-JBC, UAI 5.168 Official Correspondence 1933–1955, 450, Astronomy 1952–1953, HUA; and Paul H. Buck to Bart J. Bok, 24 March 1952, Papers of Harlow Shapley, 1906–1966, HUG 4773.10 Correspondence, Manuscripts, and Other Papers, ca. 1907–1965, Correspondence 1921–1965, Box 19B, Conant, James B., HUA.

89 Marion B. Folsom to Ralph S. Damon, 12 June 1952, HCO-DHM, UAV 630.37 Menzel correspondence 1953, Box 5, D, HUA; and "Observatory committee report," Cambridge, Massachusetts: 18 June 1952, HU-PO-JBC, UAI 5.168 Official Correspondence 1933–1955, 450, Astronomy 1952–1953, HUA.

90 J. Robert Oppenheimer, Observatory committee report, 18 June 1952, HU-PO-JBC, UAI 5.168 Official Correspondence 1933–1955, 450, Astronomy 1952–1953, HUA; and Harlow Shapley to Margaret W. Mayall, 18 June 1952, MM, 1, HCO Memos, AAVSO.

91 Oppenheimer, "Observatory committee report," 18 June 1952.

92 Osterbrock identified the third candidate as Otto Struve but he turned it down in favor of a position. Donald E. Osterbrock, "Young Don Menzel's amazing adventures at Lick Observatory," *JHA* **33** (2002), 110.

93 Keller and Keller, 108; and Harlow Shapley to Paul H. Buck, 1953, HCO-DHM, UAV 630.37 Menzel correspondence 1953, Box 8, H. Shapley, HUA.

94 Harlow Shapley to James B. Conant, 22 July 1952, HU-PO-JBC, UAI 5.168 Official Correspondence 1933–1955, 450, Astronomy 1952–1953, HUA. See recent historical work by astronomer/historian Donald E. Osterbrock; see Donald E. Osterbrock, "Herman Zanstra, Donald H. Menzel, and the Zanstra method of nebular astrophysics," *JHA* **32** (2001), 100, 103–6; and Donald E. Osterbrock, "Young Don Menzel's amazing adventures at Lick Observatory," *Journal for the History of Astronomy*, **33**, 100, 102.

95 Paul H. Buck to James B. Conant, 30 July 1952, HU-PO-JBC, UAI 5.168 Official Correspondence 1933–1955, 450, Astronomy 1952–1953, HUA.

96 James B. Conant to Paul H. Buck, 29 July 1952, HU-PO-JBC, UAI 5.168 Official Correspondence 1933–1955, 450, Astronomy 1952–1953, HUA.

97 Donald H. Menzel to James B. Conant, 8 August 1952, HU-PO-JBC, UAI 5.168 Official Correspondence 1933–1955, 450, Astronomy 1952–1953, HUA.

98 David W. Bailey to Donald H. Menzel, 12 August 1952, HCO-DHM, UAV 630.37 Menzel correspondence 1952, Box 1, A-G, HUA.

99 Helen M. Stephansky to Margaret W. Mayall, 18 August 1952, MWM-Cor, 13, Stephansky, H. M., AAVSO.

100 Margaret W. Mayall to Helen M. Stephansky, 1 September 1952, MWM-Cor, 13, Stephansky, H. M., AAVSO; and Margaret W. Mayall to Helen M. Stephansky, 12 September 1952, MWM-Cor, 13, Stephansky, H. M., AAVSO.

101 Sybil L. Chubb to Donald H. Menzel, 25 September 1952, HCO-DHM, UAV 630.37 Menzel correspondence 1952, Box 3, S, HUA.

102 Nathan M. Pusey to Donald H. Menzel, 21 August 1953, HCO-DHM, UAV 630.37 Menzel correspondence 1953, Box 7, P, HUA; and Donald H. Menzel to Charles C. Pyne, 16 November 1953, HCO-DHM, UAV 630.37 Menzel correspondence 1953, Box 7, P, HUA.

103 Donald H. Menzel, "Interim report on observatory problems," Cambridge, Massachusetts: 13 February 1953, HU-PO-JBC, UAI 5.168 Official Correspondence 1933–1955, 450, Astronomy 1952–1953, HUA.
104 Cecilia H. Payne-Gaposchkin to Donald H. Menzel, 16 October 1952, HCO-DHM, UAV 630.37 Menzel correspondence 1952, Box 1, A, HUA.
105 Margaret W. Mayall to AAVSO membership, 5 August 1953, MWM-Org, 7, AAVSO/HCO, group 1, folder a, AAVSO.
106 Clinton B. Ford, Cambridge, Massachusetts: 52, CBF, 8, Secretary File, 1952–1955, AAVSO.
107 Clinton B. Ford to Neal J. Heines, 11 November 1952, CBF, 8, Secretary File, 1952–1955, AAVSO; Clinton B. Ford to Richard W. Hamilton, 11 November 1952, CBF, 8, Secretary File, 1952–1955, AAVSO; Richard W. Hamilton to Clinton B. Ford, 16 November 1952, CBF, 8, Secretary File, 1952–1955, AAVSO; Ford, AAVSO Council Minutes, 17–18 October 1952, CBF, 8, Secretary File, 1952–1955, AAVSO; and Richard W. Hamilton to Clinton B. Ford, 16 November 1952, CBF, 8, Secretary File, 1952–1955.
108 Martha Stahr Carpenter to Margaret W. Mayall, August 1953, MWM-Org, 4, Fund Drive Endowment, folder 13, AAVSO.
109 Charles H. Smiley to Margaret W. Mayall, 17 November 1952, MWM-Org, 7, AAVSO/HCO, group 2, folder a, AAVSO.
110 Clinton B. Ford to Margaret W. Mayall, 23 October 1952, CBF, 8, Secretary File, 1952–1955, AAVSO.
111 Margaret W. Mayall to AAVSO membership, 5 August 1953, MWM-Org., 7, AAVSO/HCO, group 1, folder a, AAVSO.
112 Harlow Shapley, "Note to Mrs. Mayall," 18 June 1949, MM, 1, HCO Memos, AAVSO; and Donald H. Menzel, "Memorandum to all Observatory Personnel," 6 January 1953, MM, 1, HCO Memos, AAVSO.
113 John W. Streeter to Donald H. Menzel, 30 October 1952, HCO-DHM, UAV 630.37 Menzel correspondence 1952, Box 3, S, HUA.
114 Donald H. Menzel to John W. Streeter, 5 November 1952, HCO-DHM, UAV 630.37 Menzel correspondence 1952, Box 3, S, HUA.
115 Donald H. Menzel to HCO Council, 6 November 1952, HCO-DHM, UAV 630.37 Menzel correspondence 1953, Box 2, Harvard College Observatory Council, HUA.
116 Clinton B. Ford to Margaret W. Mayall, 3 November 1952, CBF, 8, Secretary File, 1952–1955, AAVSO.
117 Clinton B. Ford to Claude B. Carpenter, 20 November 1952, CBF, 8, Secretary File, 1952–1955, AAVSO.
118 James B. Conant, "Draft report on astronomy at Harvard," 11 November 1952, HU-PO-JBC, UAI 5.168 Official Correspondence 1933–1955, 450, Astronomy 1952–1953, HUA.
119 James B. Conant, "Tentative decisions concerning astronomy," 24 November 1952, HU-PO-JBC, UAI 5.168 Official Correspondence 1933–1955, 450, Astronomy 1952–1953, HUA.
120 Charles H. Smiley to Donald H. Menzel, 14 November 1952, HCO-DHM, UAV 630.37 Menzel correspondence 1952, Box 3, S, HUA.
121 Donald H. Menzel to Charles H. Smiley, 10 December 1952, HCO-DHM, UAV 630.37 Menzel correspondence 1952, Box 3, S, HUA.
122 Donald H. Menzel to James B. Conant, 11 December 1952, HCO-DHM, UAV 630.37 Menzel correspondence 1952, Box 1, C, HUA.
123 Clinton B. Ford to Richard W. Hamilton, 19 December 1952, CBF, 8, Secretary File, 1952–1955, AAVSO; and Clinton B. Ford to Margaret W. Mayall, 16 December 1952, CBF, 8, Secretary File, 1952–1955, AAVSO.
124 Margaret W. Mayall to various, 18 November 1952, MWM-Org, 7, AAVSO/HCO, group 1, folder c, AAVSO.
125 Margaret W. Mayall to Paul W. Merrill, 19 November 1952, MWM-Org, 7, AAVSO/HCO, group 1, folder c, AAVSO.
126 Margaret W. Mayall to Leo Goldberg, 19 November 1952, MWM-Org, 7, AAVSO/HCO, group 2, folder a, AAVSO.
127 Clinton B. Ford to Richard W. Hamilton, 21 January 1953, CBF, 8, Secretary File, 1952–1955, AAVSO.
128 Donald H. Menzel to Martha Stahr Carpenter, 1 December 1952, MM, 1, HCO-AAVSO correspondence and notes, AAVSO.
129 Martha Stahr Carpenter to Donald H. Menzel, 16 December 1952, HCO-DHM, UAV 630.37 Menzel correspondence 1952, Box 1, C, HUA.
130 Neal J. Heines to Clinton B. Ford, 2 December 1952, CBF, 8, Secretary File, 1952–1955, AAVSO.
131 Clinton B. Ford to Margaret W. Mayall, 3 December 1952, CBF, 8, Secretary File, 1952–1955, AAVSO.
132 Charles H. Smiley to James B. Conant, 8 December 1952, MWM-Org, 7, AAVSO/HCO, group 2, folder d, AAVSO.
133 Charles H. Smiley to Margaret W. Mayall, 19 December 1952, MWM-Org, 7, AAVSO/HCO, group 2, folder d, AAVSO.
134 Keller and Keller, 21.

135 Charles H. Smiley to Margaret W. Mayall, 19 December 1952, MSM-Org, 7, AAVSO/HCO, group 2, folder d, AAVSO.

136 Clinton B. Ford to Richard W. Hamilton, 19 December 1952, CBF, 8, Secretary File, 1952–1955, AAVSO.

137 Clinton B. Ford to Richard W. Hamilton, 19 December 1952, CBF, 8, Secretary File, 1952–1955, AAVSO.

138 Martha Stahr Carpenter to Clinton B. Ford, 15 January 1953, CBF, 8, Secretary File, 1952–1955, AAVSO; and Clinton B. Ford to Martha Stahr Carpenter, 9 January 1953, CBF, 8, Secretary File, 1952–1955, AAVSO.

139 Margaret W. Mayall to various astronomers, 18 November 1952, MWM-Org., 7, AAVSO/HCO, group1 folder c, AAVSO. Surprisingly, Mayall overlooked one potentially strong ally in Helen Sawyer Hogg, who for many years had strongly supported AAVSO. When asked by Cecilia Payne-Gaposchkin what she knew about what AAVSO was doing, Hogg was forced to respond, somewhat bitterly, that "I never hear from Margaret Mayall and hence know nothing about what she is thinking or feeling." In addition to probing what actions the AAVSO Council might be taking or contemplating for the HCO Council, Payne-Gaposchkin's letter also speculated that the External Review Committee was likely to question whether the HCO ". . . shall continue to be the home and sponsor of the AAVSO." The statement reveals little understanding on Payne-Gaposchkin's part that such details were well below the threshold of interest on the part of the External Review Committee. The AAVSO is never mentioned in the committee's reports or in President Conant's communication about the HCO with the Committee. Cecilia Payne-Gaposchkin to Helen Sawyer Hogg, 21 November 1952, HSH, B1994–0002, 6, 12, UTor; and Helen Sawyer Hogg to Cecilia H. Payne-Gaposchkin, 2 December 1952, HSH, B1994–0002, 6, 12, UTor.

140 Various letters to Margaret W. Mayall, January 1952, MWM-Org, 7, AAVSO/HCO, group 1, folder c, AAVSO.

141 Martin Schwarzschild to Margaret W. Mayall, 8 January 1953, MWM-Org, 7, AAVSO/HCO, group 1, folder c, AAVSO; also see Paul W. Merrill to Margaret W. Mayall, 24 November 1952, MWM-Org, 7, AAVSO/HCO, group 1, folder c, AAVSO; Jan H. Oort to Margaret W. Mayall, 5 January 1953, MWM-Org, 7, AAVSO/HCO, group 1, folder c, AAVSO; and D. J. O'Connell, J. S. to Margaret W. Mayall, 1 January 1953, MWM-Org, 7, AAVSO/HCO, group 1, folder c, AAVSO.

142 Donald H. Menzel to Harlow Shapley, 1 December 1952, MWM-Org, 7, AAVSO/HCO, group 1, folder a, AAVSO.

143 Harlow Shapley, "Statement about the A.A.V.S.O.," Cambridge, Massachusetts: 4 January 1953, MWM-Org, 7, AAVSO/HCO, group 2, folder e, AAVSO.

144 Shapley, "Statement about the A.A.V.S.O.," 4 January 1953. Several typed transcription copies of this statement exist in various files of the AAVSO archives; however, one of them appears to be a typed original copy that Shapley himself had sent to Mayall and on which he wrote in pencil: "Confidential sent to DHM." Each of the copies held by the AAVSO appear to be complete. However, a second copy of this statement found in the Harvard University archives, although containing the full number of pages and numbered sections, has about two-thirds of a blank page occurring in section seven, where Shapley had provided excerpts from professional astronomers' letters of support for the AAVSO (which Shapley copied from the AAVSO's evaluation report). Presumably the blank space in the Harvard archives copy was left either for selective entry of the quotes in support of AAVSO or the space was meant to be used for entry of letters of support that had been solicited by Menzel himself. Once the blank was filled in, it seems, Shapley's statement would then be suitable for forwarding to the Harvard hierarchy.

145 Margaret W. Mayall to ADW [Arville D. Walker], 10 January 1953, MWM-Org, 7, AAVSO/HCO, group 2, folder a, AAVSO.

146 Margaret W. Mayall to Charles H. Smiley, 12 January 1953, MWM-Org, 7, AAVSO/HCO, group 2, folder a, AAVSO.

147 Margaret W. Mayall, MWM-Org, 7, AAVSO/HCO, group 3, folder d, AAVSO; and Margaret W. Mayall to Harlow Shapley, 13 January 1953, MWM-Org, 7, AAVSO/HCO, group 2, folder d, AAVSO.

148 Charles H. Smiley to Margaret W. Mayall, 16 January 1953, MWM-Org, 7, AAVSO/HCO, group 2, folder a, AAVSO; Margaret W. Mayall to Charles H. Smiley, 12 January 1953, MWM-Org, 7, AAVSO/HCO, group 2, folder a, AAVSO; Margaret W. Mayall to Martha Stahr Carpenter, 15 January 1953, MWM-Org, 7, AAVSO/HCO, group 2, folder a, AAVSO; and Charles H. Smiley to Margaret W. Mayall, 15 January 1953, MWM-Org, 7, AAVSO/HCO, group 2, folder a, AAVSO.

149 Charles H. Smiley to Harlow Shapley, 13 January 1953, Papers of Harlow Shapley, 1906–1966, HUG 4773.10 Correspondence, Manuscripts, and Other Papers, ca. 1907–1965, Correspondence 1921–1965, Box 22D, Smiley, Charles H. (Dr.), HUA.

150 Cyrus F. Fernald to Charles H. Smiley, 20 January 1953, CFF, 1, AAVSO Reorganization, AAVSO.

151 Charles H. Smiley, "Report on the value of past and future activities of the American Association of Variable Star Observers," Cambridge, Massachusetts: 15 January 1953, CBF, 8, Secretary File, 1952–1955, AAVSO.
152 Margaret W. Mayall to Cyrus F. Fernald, 21 January 1953, CFF, 1, AAVSO Reorganization, AAVSO.
153 Margaret W. Mayall to Clinton B. Ford, 21 January 1953, CBF, 8, Secretary File, 1952–1955, AAVSO.
154 Clinton B. Ford to Richard W. Hamilton, 21 January 1953, CBF, 8, Secretary File, 1952–1955.

11 EVICTION FROM HARVARD COLLEGE OBSERVATORY

1 Margaret W. Mayall to Cyrus F. Fernald, 21 April 1953, MWM-Cor, 4, Fernald, C. F., AAVSO.
2 Anonymous, "An impromptu meeting of the A. A. V. S. O.," *The MAS Bulletin* **2** (1935), 8.
3 Margaret W. Mayall, "Interview conducted by Owen Gingerich," OHP, OHI 28323, CHP-NBL, part 1, p. 15.
4 Harlow Shapley to Margaret W. Mayall, 12 August 1953, MWM-Org, 4, Fund Drive Endowment, folder 13, AAVSO.
5 Leon Campbell to Harlow Shapley, 19 January 1949, LC-Cor, 7, Shapley, H., AAVSO.
6 Donald H. Menzel to [HCO] Observatory Personnel, 20 January 1953, MWM-Org, 7, AAVSO/HCO, group 1, folder f, AAVSO.
7 Donald H. Menzel to [HCO] Observatory Personnel, 2 February 1953, MWM-Org, 7, AAVSO/HCO, group 1, folder f, AAVSO.
8 Donald H. Menzel to Walter Baade, 10 February 1953, HCO-DHM, UAV 630.37 Menzel correspondence 1953, Box 4, B, HUA.
9 Sybil L. Chubb to Donald H. Menzel, 11 February 1953, HCO-DHM, UAV 630.37 Menzel correspondence 1953, Box 5, C, HUA.
10 Donald H. Menzel, "Interim report on observatory problems," Cambridge, Massachusetts: 13 February 1953, HU-PO-JBC, UA I 5.168 Official Correspondence 1933–1955, 450, Astronomy 1952–1953, HUA.
11 Menzel, "Interim report," 8.
12 E. Dorrit Hoffleit, *Misfortunes as Blessings in Disguise: The Story of My Life*, Cambridge, Massachusetts: AAVSO, 2002), pp. 53–54; 64; 69 and 70. Suffering other such indignities – being forced to move from a comfortable office to a much smaller one; losing her long-held volunteer columnist position at *Sky and Telescope* (the loss of which she felt Menzel had something to do with) – by 1956, Hoffleit left Harvard for Yale. "Realizing that if I stayed associated with the current Observatory administration," she wrote, "I might actually grow insane."
13 Hoffleit, *Misfortunes*, 65. Menzel also favored keeping the Observatory's "Open Nights," when the public was allowed to attend lectures and look through the HCO telescopes.
14 Menzel, "Interim report," 4–5.
15 Menzel, "Interim report," 4–5.
16 Clinton B. Ford to Martha Stahr Carpenter, 27 January 1953, CBF, 8, Secretary File, 1952–1955, AAVSO.
17 Clinton B. Ford to Birt Darling, 9 February 1953, CBF, 8, Secretary File, 1952–1955, AAVSO.
18 Richard W. Hamilton to Clinton B. Ford, 15 February 1953, CBF, 8, Secretary File, 1952–1955, AAVSO.
19 Margaret W. Mayall to Warren Weaver, 11 February 1953, MWM-Org, 7, AAVSO/HCO, group 2, folder 3, AAVSO.
20 Bart J. Bok to Harvard Observatory Council, and Harlow Shapley, 20 February 1953, HCO-DHM, UAV 630.37 Menzel correspondence 1953, Box 4, Bok, Bart J., HUA.
21 Margaret W. Mayall to Donald H. Menzel, 11 February 1953, MWM-Org, 7, AAVSO/HCO, group 2, folder a, AAVSO.
22 Margaret W. Mayall to Mary Van Meter, 12 February 1953, MWM-Cor, 8, Mayall, M. W., AAVSO.
23 Anonymous, "From 'Endowment funds of Harvard University June 30 1947 page 372'," 30 June 1947, MWM-Org, 7A, AAVSO/HCO, group 3, folder g, AAVSO.
24 Margaret W. Mayall to Martha Stahr Carpenter, 19 February 1953, MWM-Org, 7, AAVSO/HCO, group 2, folder f, AAVSO.
25 Mayall, 19 February 1953.
26 Mayall, 19 February 1953.
27 Donald H. Menzel to Cecilia Payne-Gaposchkin, 2 March 1953, HCO-DHM, UAV 630.37 Menzel correspondence 1953, Box 5, G, HUA.
28 Otto Struve to Donald H. Menzel, 17 February 1953, HCO-DHM, UAV 630.37 Menzel correspondence 1953, Box 4, Amer. Assoc., HUA.
29 Donald H. Menzel to Henry Norris Russell, 4 March 1953, HCO-DHM, UAV 630.37 Menzel correspondence 1953, Box 7, R, HUA.
30 Henry Norris Russell to Donald H. Menzel, 10 March 1953, HCO-DHM, UAV 630.37 Menzel correspondence 1953, Box 8, Visiting Committee, HUA.

31 Harlow Shapley to Donald H. Menzel, 20 February 1953, HCO-DHM, UAV 630.37 Menzel correspondence 1953, Box 8, H. Shapley, HUA.
32 Donald H. Menzel to Harlow Shapley, 25 February 1953, HCO-DHM, UAV 630.37 Menzel correspondence 1953, Box 8, H. Shapley, HUA.
33 Harlow Shapley to Warren Weaver, 20 February 1953, Papers of Harlow Shapley, 1906–1966, HUG 4773.10 Correspondence, Manuscripts, and Other Papers, ca. 1907–1965, Correspondence 1921–1965, Box 23A, Weaver, Warren (Dr.), HUA.
34 Margaret W. Mayall to Robert M. Greenley, 4 February 1953, MWM-Cor, 5, Greenley, R. M., AAVSO.
35 Clinton B. Ford to Martha Stahr Carpenter, 15 February 1953, CBF, 8, Secretary File, 1952–1955, AAVSO.
36 David W. Rosebrugh to Clinton B. Ford, 22 February 1953, CBF, 8, Secretary File, 1952–1955, AAVSO; and David W. Rosebrugh to Margaret W. Mayall, 22 February 1953, MWM-Cor, 11, Rosebrugh, D. W., 1953–1954, AAVSO.
37 Margaret W. Mayall to David W. Rosebrugh, 27 February 1953, MWM-Cor, 11, Rosebrugh, D. W., 1953–1954, AAVSO.
38 Margaret W. Mayall to Cyrus F. Fernald, 27 February 1953, MWM-Cor, 4, Fernald, C. F., AAVSO.
39 Margaret W. Mayall to various, undated [1953 March], MWM-Org, 7, AAVSO/HCO, group 1, folder a, AAVSO.
40 Charles H. Smiley to Lawrence N. Upjohn, 6 February 1953, MWM-Org, 7, AAVSO/HCO, group 2, folder e, AAVSO.
41 Lawrence N. Upjohn to Charles H. Smiley, 26 February 1953, MWM-Org, 7, AAVSO/HCO, group 2, folder e, AAVSO.
42 Margaret W. Mayall to Charles H. Smiley, 6 March 1953, MWM-Org, 7, AAVSO/HCO, group 2, folder e, AAVSO.
43 Charles H. Smiley to Margaret W. Mayall, 2 March 1953, MWM-Org, 7, AAVSO/HCO, group 2, folder e, AAVSO.
44 Harlow Shapley to Margaret W. Mayall, 4 March 1953, MWM-Org, 7, AAVSO/HCO, group 2, folder e, AAVSO.
45 Margaret W. Mayall to Percy W. Witherell, 2 March 1952, MWM-Cor, 11, Rosebrugh, D. W., 1951–1952, AAVSO.
46 Martha Stahr Carpenter to Margaret W. Mayall, 3 March 1953, MWM-Cor, 2, Carpenter, M. S., AAVSO.
47 Clinton B. Ford, Notes taken at the informal gathering, 20 March 1953, CBF, 8, Secretary File, 1952–1955, AAVSO.
48 Ford, Notes, 20 March 1953.
49 Clinton B. Ford to Richard W. Hamilton, 21 January 1953, CBF, 8, Secretary File, 1952–1955, AAVSO.
50 Ford, Notes, 20 March 1953.
51 Ford, Notes, 20 March 1953.
52 Helen M. Stephansky to Percy W. Witherell, 10 March 1953, MWM-Cor, 15, Witherell, P. W., 1952–1958, AAVSO.
53 Richard W. Hamilton to Helen M. Stephansky, 15 March 1953, MWM-Cor, 5, Hamilton, R. W., folder 1, AAVSO.
54 Margaret W. Mayall to Jason J. Nassau, 12 March 1953, MWM-Cor, 9, Nassau, J. J., AAVSO.
55 Clinton B. Ford to Margaret W. Mayall, 15 March 1953, CBF, 8, Secretary File, 1952–1955, AAVSO.
56 Clinton B. Ford to Margaret W. Mayall, 28 March 1953, CBF, 8, Secretary File, 1952–1955, AAVSO.
57 Ford, 15 February 1953.
58 Helen M. Stephansky to Richard W. Hamilton, 27 April 1953, MWM-Cor, 5, Hamilton, R. W., folder 1, AAVSO.
59 Harvard Observatory Council, "untitled: [report on the future status of the AAVSO at Harvard University]," Cambridge, Massachusetts: 16 February 1953, MWM-Org, 7, AAVSO/HCO, group 1, folder a, AAVSO.
60 E. Dorrit Hoffleit to Margaret W. Mayall, August 1953, MWM-Org, 4, Fund Drive Endowment, folder 13, AAVSO.
61 Paul H. Buck to Donald H. Menzel, 18 February 1953, MWM-Org, 7, AAVSO/HCO, group 1, folder a, AAVSO; and Paul H. Buck to Donald H. Menzel, 11 March 1953, MWM-Org, 7, AAVSO/HCO, group 1, folder a, AAVSO.
62 Mary Van Meter to Donald H. Menzel, 12 March 1953, HCO-DHM, UAV 630.37 Menzel correspondence 1953, Box 8, V, HUA; and Mary Van Meter to Donald H. Menzel, 19 March 1953, HCO-DHM, UAV 630.37 Menzel correspondence 1953, Box 8, V, HUA.
63 Donald H. Menzel to Martha Stahr Carpenter, 1 April 1953, HCO-DHM, UAV 630.37 Menzel correspondence 1953, Box 5, C, HUA.
64 Margaret W. Mayall to AAVSO membership, 5 August 1953, MWM-Org, 7, AAVSO/HCO, group 1, folder a, AAVSO; and Donald H. Menzel to unspecified [AAVSO], 8 April 1953, CBF, 8, Secretary File, 1952–1955, AAVSO.
65 Harvard Observatory Council, "[report on the future status of the AAVSO at Harvard University]."
66 Harvard Observatory Council, "[report on the future status of the AAVSO at Harvard University]," 2.
67 Harvard Observatory Council, "[report on the future status of the AAVSO at Harvard University]," 2.
68 Harvard Observatory Council, "[report on the future status of the AAVSO at Harvard University]."
69 Helen M. Stephansky *et al.*, "Miscellaneous Notes of the General Meeting in Library Concerning the AAVSO April 6, 1953, 3:30 P.M.," April 1953, MWM-Org, 7A, AAVSO/HCO, group 3, folder g, AAVSO.

70 Margaret W. Mayall to Robert R. McMath, April 1953, MWM-Org, 7, AAVSO/HCO, group 2, folder f, AAVSO.
71 Margaret W. Mayall to Watson Davis, 7 April 1953, MWM-Org, 7, AAVSO/HCO, group 2, folder f, AAVSO.
72 Margaret W. Mayall, "News Release," 1953, MWM-Org, 7, AAVSO/HCO, group 1, folder a, AAVSO.
73 Margaret W. Mayall, "News Release."
74 Margaret W. Mayall, "News Release," *JRASC* **47** (1953), 119.
75 Donald H. Menzel to Margaret W. Mayall, 14 April 1953, MWM-Org, 7, AAVSO/HCO, group 2, folder e, AAVSO.
76 Robert C. Cowen, "Amateurs build telescopes by hand," *The Christian Science Monitor,* 7 May 1953, CBF, 8, Secretary File, 1952–1955, AAVSO.
77 Robert R. McMath to Margaret W. Mayall, 18 April 1953, HCO-DHM, UAV 630.37 Menzel correspondence 1953, Box 7, Mc, HUA.
78 Robert R. McMath to Donald H. Menzel, 18 April 1953, HCO-DHM, UAV 630.37 Menzel correspondence 1953, Box 7, Mc, HUA. McMath's sympathetic response to Menzel was natural. As solar astronomers, McMath and Menzel had, by this point in time, met frequently. Their first contact may have come as early as the solar eclipse that both observed successfully at Fryeburg, Maine, on August 31, 1932. At that eclipse, as an amateur astronomer, McMath successfully captured a motion picture of the eclipse's progress over time. He attracted the attention of professional astronomers with his elaborate drive mechanism for that purpose. Both McMath and Menzel were involved in not only solar eclipse chasing and other aspects of solar astronomy during the 1930s, but also served in Washington, DC, during World War II. At that time McMath's responsibilities included work on radio propagation with the Joint Chief's of Staff and for the Navy on developing a low-altitude bomb sight. Menzel was deeply involved in radio propagation and frequency allocation problems for the Navy. Robert Raynolds McMath, "Autobiographical Material for Robert (R)aynolds McMath (As requested by the Home Secretary, National Academy of Sciences)," 25 September 1958, NAS; Robert Raynolds McMath to William Hammond Wright, 28 September 1932, Correspondence files, MLS-LO; and Leo Goldberg, and Lawrence H. Aller, "Donald Howard Menzel 1901–1976," *BMNAS* **47** (1991), 156, 160–161.
79 Margaret W. Mayall to Jason J. Nassau, 20 April 1953, MWM-Cor, 9, Nassau, J. J., AAVSO.
80 Mayall, 21 April 1953.
81 Mayall, 21 April 1953.
82 Cyrus F. Fernald to Margaret W. Mayall, 22 April 1953, MWM-Cor, 4, Fernald, C. F., AAVSO.
83 Margaret W. Mayall to Charles H. Smiley, 22 April 1953, MWM-Cor, 13, Smiley, C. H., AAVSO.
84 Stephansky, 27 April 1953. Stephansky's name for Menzel – "his nibs" – means "the boss."
85 Harlow Shapley to Margaret W. Mayall, 23 April 1953, MWM-Org, 7, AAVSO/HCO, group 3, folder d, AAVSO.
86 Margaret W. Mayall to Martha Stahr Carpenter, 25 April 1953, MWM-Org, 7, AAVSO/HCO, group 2, folder d, AAVSO.
87 Anonymous, "Report of committee: draft of proposal for the support of astronomy by the National Science Foundation," *AJ* **56** (1951), 147–8; Jesse L. Greenstein, "The NSF and Astronomy," *S&T* **12** (1953), 282, 285; and Frank K. Edmondson, *AURA and Its US National Observatories*, Cambridge, UK and New York: Cambridge University Press, 1997, pp. 7–8.
88 Margaret W. Mayall to Martha Stahr Carpenter, 28 April 1953, MWM-Org, 7, AAVSO/HCO, group 2, folder d, AAVSO.

12 ACTIONS AND REACTIONS

1 Richard W. Hamilton to Helen M. Stephansky, 20 September 1953, MWM-Cor, 5, Hamilton, R. W., folder 1, AAVSO.
2 E. Dorrit Hoffleit to AAVSO membership, 16 April 1953, CBF, 8, Secretary File, 1952–1955, AAVSO.
3 Helen M. Stephansky to Richard W. Hamilton, 11 May 1953, MWM-Cor, 5, Hamilton, R. W., folder 1, AAVSO.
4 Anonymous to Percy W. Witherell, 11 May 1953, MWM-Cor, 15, Witherell, P. W., 1952–1958, AAVSO.
5 Richard W. Hamilton to Helen M. Stephansky, 13 May 1953, MWM-Cor, 5, Hamilton, R. W., folder 1, AAVSO.
6 David W. Rosebrugh to Clinton B. Ford, 10 May 1953, CBF, 8, Secretary File, 1952–1955, AAVSO.
7 Hamilton, 13 May 1953; Morton Keller and Phyllis Keller, *Making Harvard Modern: The Rise of America's University*, New York: Oxford University Press, 2007, pp. 102–3.
8 Rosebrugh, 10 May 1953.
9 Clinton B. Ford to David W. Rosebrugh, 17 May 1953, CBF, 8, Secretary File, 1952–1955, AAVSO.
10 AAVSO, undated [1953], CBF, 8, Secretary File, 1952–1955, AAVSO.
11 Clinton B. Ford, "Minutes of the AAVSO Council meetings held May 22, 1953," 22 May 1953, CBF, 8, Secretary File, 1952–1955, AAVSO.

12 Clinton B. Ford, "Minutes of the AAVSO general meeting held May 23, 1953," 23 May 1953, CBF, 8, Secretary File, 1952–1955, AAVSO.

13 R. Newton Mayall, "Papers presented at the spring meeting May 23, 1953," *AVSOA* (1953).

14 Clinton B. Ford to Margaret W. Mayall, 1 June 1953, CBF, 8, Secretary File, 1952–1955, AAVSO.)

15 Harlow Shapley to Margaret W. Mayall, 13 November 1951, MWM-Org, 19, Office of Naval Research Proposals and Contracts (1), AAVSO; and Margaret W. Mayall, "Variable star notes," *JRASC* **47** (1953), 166–70.

16 Helen M. Stephansky to Clinton B. Ford, 16 June 1953, CBF, 8, Secretary File, 1952–1955, AAVSO; Clinton B. Ford to Helen M. Stephansky, 18 June 1953, CBF, 8, Secretary File, 1952–1955, AAVSO; Margaret W. Mayall, "Variable star notes," *JRASC*, **47**, 166–70; and Margaret W. Mayall to Watson Davis, 5 June 1953, MWM-Cor, 9, National Science Foundation (NSF), folder 1, AAVSO.

17 Mayall, 5 June 1953, and other letters in this folder.

18 Margaret W. Mayall to Alan T. Waterman, 4 August 1953, CBF, 8, Secretary File, 1952–1955, AAVSO.

19 Margaret W. Mayall to Clinton B. Ford, 28 July 1953, CBF, 8, Secretary File, 1952–1955, AAVSO.

20 Clinton B. Ford to Margaret W. Mayall, 1 August 1953, CBF, 8, Secretary File, 1952–1955, AAVSO.

21 Margaret W. Mayall to AAVSO membership, 5 August 1953, MWM-Org, 7, AAVSO/HCO, group 1, folder a, AAVSO.

22 David W. Bailey to Margaret W. Mayall, 14 August 1953, MWM-Org, 7A, AAVSO/HCO, group 3, folder i, AAVSO; and Margaret W. Mayall to various, August 1953, MWM-Org, 7A, AAVSO/HCO, group 3, folder i, AAVSO.

23 Margaret W. Mayall to Helen Sawyer Hogg, 15 August 1953, MWM-Adm, 6, JRASC Notes (1), AAVSO.

24 Helen M. Stephansky to Richard W. Hamilton, 31 August 1953, MWM-Cor, 5, Hamilton, R. W., folder 1, AAVSO.

25 Harlow Shapley to Margaret W. Mayall, 12 August 1953, MWM-Org, 4, Fund Drive Endowment, folder 13, AAVSO.

26 Frank J. Kelly to Margaret W. Mayall, 20 August 1953, MWM-Org, 7A, AAVSO/HCO, group 3, folder i, AAVSO.

27 Walter P. Reeves to Margaret W. Mayall, 11 September 1953, MWM Org, 7A, AAVSO/HCO, group 3, folder i, AAVSO.

28 Lewis J. Boss to Margaret W. Mayall, 30 August 1953, MWM-Cor, 2, Boss, L. J., AAVSO.

29 Charles P. Olivier to Margaret W. Mayall, 24 August 1953, MWM-Org, 7A, AAVSO/HCO, group 3, folder i, AAVSO.

30 Charles P. Olivier to David W. Bailey, August 1953, MWM-Org, 7A, AAVSO/HCO, group 3, folder i, AAVSO.

31 Walter Scott Houston to Margaret W. Mayall, 8 September 1953, MWM-Org, 7A, AAVSO/HCO, group 3, folder i, AAVSO.

32 Walter Scott Houston to David W. Bailey, 8 September 1953, MWM-Org, 7A, AAVSO/HCO, group 3, folder i, AAVSO.

33 Theodore R. Treadwell to David W. Bailey, 25 August 1953, MWM-Org, 7A, AAVSO/HCO, group 3, folder i, AAVSO.

34 Margaret W. Mayall to Paul W. Merrill, 1 September 1953, MWM-Cor, 8, Merrill, P. W., AAVSO.

35 Alice H. Farnsworth to Margaret W. Mayall, 14 September 1953, MWM-Org, 7A, AAVSO/HCO, group 3, folder i, AAVSO.

36 Donald H. Menzel to Margaret W. Mayall, 14 September 1953, MWM-Org, 7A, AAVSO/HCO, group 3, folder g, AAVSO.

37 Charles H. LeRoy to Margaret W. Mayall, 27 June 1953, MWM-Org, 7, AAVSO/HCO, group 1, folder b, AAVSO; Clinton B. Ford to Margaret W. Mayall, 24 September 1953, CBF, 8, Secretary File, 1952–1955, AAVSO; Anonymous, "Summary report of A.A.V.S.O. activities October 1952-October 1953," Proceedings, Astronomical League Convention, September 4–7, 1953, Washington, DC (1954), 94–5; and Grace G. Scholz to Clinton B. Ford, 5 August 1953, CBF, 8, Secretary File, 1952–1955, AAVSO.

38 Claude B. Carpenter to Clinton B. Ford, 26 September 1953, CBF, 8, Secretary File, 1952–1955, AAVSO. LaPelle knew better; in 1949 he attempted to persuade the AAVSO to write a portion of a training manual for a fledgling AL observer training program. He was unsuccessful in persuading Campbell and others to do so, but with help, likely again from Rosebrugh, he succeeded in convincing a few AAVSO members to cooperate in the preparation of a variable star chapter for the draft manual. The manual was never published after it was completely written and edited. Thomas R. Williams, "Getting Organized: A history of amateur astronomy in the United States," unpublished Ph.D. dissertation, Rice University, 2000, 314–16.

39 Sergei Gaposchkin, "How to estimate the brightness of a star," *S&T* **12** (1953), 283–85.

40 Hamilton, 20 September 1953.

41 Velma Adams to Donald H. Menzel, 18 August 1953, HCO-DHM, UAV 630.37 Menzel correspondence 1953, Box 7, Menzel and Adams, HUA.

42 Donald H. Menzel, September 1953, HCO-DHM, UAV 630.37 Menzel correspondence 1953, Box 4, Amer. Assoc., HUA.

43 Velma Adams to Donald H. Menzel, 21 August 1953, HCO-DHM, UAV 630.37 Menzel correspondence 1953, Box 4, Amer. Assoc., HUA.
44 Nathan M. Pusey to Donald H. Menzel, 21 August 1953, HCO-DHM, UAV 630.37 Menzel correspondence 1953, Box 7, P, HUA.
45 David W. Bailey to Donald H. Menzel, 24 August 1953, HCO-DHM, UAV 630.37 Menzel correspondence 1953, Box 4, Bailey, David, HUA. One of the letters in response to – but not in favor of – Mayall's circular letter was from John W. Streeter, whom the HCO Council had enlisted 10 months earlier to find an alternative home for the AAVSO. (John W. Streeter to David W. Bailey, 27 August 1953, HCO-DHM, UAV 630.37 Menzel correspondence 1953, Box 4, Bailey, David, HUA.) Streeter's negative assertions regarding Campbell's stewardship of the AAVSO bear more than a little truth. However, they should have more appropriately been addressed, well before this time, to Mayall herself in encouraging her to broaden AAVSO's observing program. As we have already seen, Mayall had set out on that path, but her progress was rudely interrupted by the HCO Council's decision to evict AAVSO from the observatory.
46 Donald H. Menzel to David W. Bailey, 3 September 1953, HCO-DHM, UAV 630.37 Menzel correspondence 1953, Box 4, Amer. Assoc., HUA.
47 David W. Bailey to Donald H. Menzel, 10 September 1953, HCO-DHM, UAV 630.37 Menzel correspondence 1953, Box 4, Bailey, David, HUA.
48 Bailey, 10 September 1953.
49 Donald H. Menzel to Margaret W. Mayall, 15 September 1953, HCO-DHM, UAV 630.37 Menzel correspondence 1953, Box 4, Amer. Assoc., HUA.
50 Bailey, 10 September 1953.
51 David W. Bailey to Donald H. Menzel, 18 September 1953, HCO-DHM, UAV 630.37 Menzel correspondence 1953, Box 4, Bailey, David, HUA.
52 Velma Adams to Donald H. Menzel, 3 September 1953, HCO-DHM, UAV 630.37 Menzel correspondence 1953, Box 7, Menzel and Adams, HUA.
53 Velma Adams to Donald H. Menzel, 9 September 1953, HCO-DHM, UAV 630.37 Menzel correspondence 1953, Box 7, Menzel and Adams, HUA.
54 Donald H. Menzel to Edward Reynolds, 9 September 1953, HCO-DHM, UAV 630.37 Menzel correspondence 1953, Box 7, R, HUA.
55 Menzel, 14 September 1953.
56 Edward Reynolds to Donald H. Menzel, 28 September 1953, HCO-DHM, UAV 630.37 Menzel correspondence 1953, Box 7, R, HUA.
57 Harlow Shapley to McGeorge Bundy, 19 September 1953, Papers of Harlow Shapley, 1906–1966, HUG 4773.10 Correspondence, Manuscripts, and Other Papers, ca. 1907–1965, Correspondence 1921–1965, Box 19A, Bundy, McGeorge, HUA.
58 O. M. Shaw to McGeorge Bundy, 21 September 1953, HCO-DHM, UAV 630.37 Menzel correspondence 1953, Box 4, Bundy, McGeorge, HUA. Note that Shaw's letter is directed to Bailey and the corporation. In referring to "they," Shaw clearly means the HCO Council.
59 Donald H. Menzel to Lawrence Terry, 23 September 1953, HCO-DHM, UAV 630.37 Menzel correspondence 1953, Box 8, T, HUA.
60 Donald H. Menzel to McGeorge Bundy, 25 September 1953, HCO-DHM, UAV 630.37 Menzel correspondence 1953, Box 4, Bundy, McGeorge, HUA.
61 Donald H. Menzel to McGeorge Bundy, 28 September 1953a, HCO-DHM, UAV 630.37 Menzel correspondence 1953, Box 4, Bundy, McGeorge, HUA. Since the AAVSO was, and had always been, an organization of both amateur and professional astronomers, this use of "amateur" in a pejorative sense reflects Menzel's own outlook on the AAVSO more than the actual status of the organization. What Rosebrugh obviously meant was "independent," but Rosebrugh's own use of such language, although surprising, is perhaps consistent with his behavior throughout this sorry episode in personal and organizational behavior.
62 Donald H. Menzel to David W. Bailey, 29 September 1953, HCO-DHM, UAV 630.37 Menzel correspondence 1953, Box 4, Bailey, David, HUA.
63 Margaret W. Mayall to McGeorge Bundy, 28 September 1953, MWM-Org, 7, AAVSO/HCO, group 1, folder b, AAVSO.
64 Margaret W. Mayall to Margaret W. Beardsley, 28 September 1953, MWM-Org, 7A, AAVSO/HCO, group 3, folder g, AAVSO.
65 Margaret W. Mayall, "Interview conducted by Owen Gingerich," OHP, OHI 28323, CHP-NBL, part 2, page 17; Domingo Taboada to Margaret W. Mayall, 6 October 1951, MWM-Cor, 13, Taboada, D., AAVSO; Bart J. Bok to Domingo Taboada, 6 October 1953, MWM-Cor, 13, Taboada, D., AAVSO; Marco A. Moreno Corral and Maria G. Lopez Molina, *Domingo Taboada Roldan, Estudioso del Espacio y del Tiempo*, Puebla, Mexico: Fundacion Protectora

de Puebla "Aurora Marin de Taboada," 1992; and Donald H. Menzel, "Sojourn in Mexico," *S&T* 1 (1942), 3–5.

66 Mayall, 28 September 1953.

67 C. M. Huffer to Margaret W. Mayall, 29 September 1953, MWM-Org, 7, AAVSO/HCO, group 1, folder b, AAVSO.

68 Margaret W. Mayall to Clinton B. Ford, 30 September 1953, CBF, 8, Secretary File, 1952–1955, AAVSO.

69 Margaret W. Mayall to Martha Stahr Carpenter, 30 September 1953, MWM-Org, 7A, AAVSO/HCO, group 3, folder g, AAVSO.

70 Margaret W. Mayall to Peter M. Millman, 29 September 1953, MWM-Cor, 9, Millman, P. M., AAVSO. According to Owen Gingerich, the revolving table was not destroyed – as the paragraph suggests – but was sent to Alan Shapley in Colorado, who planned to return it to HCO at some later date. The whereabouts of the table is unknown today.

71 Mayall, 30 September 1953.

72 Donald H. Menzel to McGeorge Bundy, 28 September 1953b, HCO-DHM, UAV 630.37 Menzel correspondence 1953, Box 4, Bundy, McGeorge, HUA.

73 Harlow Shapley to Margaret W. Mayall, undated [1953 October], MWM-Org, 7A, AAVSO/HCO, group 3, folder g, AAVSO; "One hundred and seventh annual report of the Director of the astronomical observatory of Harvard College for the year ending September 30, 1952," Cambridge, Massachusetts: 1953; and Anonymous [probably Margaret W. Mayall], "Presidents of AAVSO," undated [1953], MWM-Org, 7A, AAVSO/HCO, group 3, folder g, AAVSO.

74 President and Fellows of Harvard College, MWM-Org, 7A, AAVSO/HCO, group 3, folder i, AAVSO.

75 Domingo Taboada to Margaret W. Mayall, 1 October 1953, MWM-Org, 7A, AAVSO/HCO, group 3, folder g, AAVSO; and Margaret W. Mayall, Owen Gingerich Interview, part 2, page 17.

76 Helen M. Stephansky, transcriber, "AAVSO council meeting – October 9, 1953," MWM-Org, 7A, AAVSO/HCO, group 3, folder f, AAVSO.

77 Taboada, 1 October 1953; and Margaret W. Mayall, Owen Gingerich Interview, part 2, page 17.

78 Margaret W. Mayall to Robert M. Greenley, 26 August 1953, MWM-Cor, 5, Greenley, R. M., AAVSO.

13 IN SEARCH OF A HOME

1 Paul W. Merrill, "Questions concerning long-period variable stars," October 1953, MWM-Adm, 3, Campbell Volume, AAVSO.

2 Jocelyn Gill, "Harvard conclave, October 9–10, 1953," *VC* 5 No. 11 (1954), 1–4.

3 Margaret W. Mayall, "Variable star notes," *JRASC* 47 (1953), 249–52.

4 Gill, "Harvard conclave"; Margaret W. Mayall, *Special Announcement*, Oct 1953; and R. Newton Mayall, "The Campbell memorial volume," *AVSOA* (Oct 1953), 2.

5 Margaret W. Mayall, "Twenty-second annual report of the Recorder 1952–1953," *VC* 5 (1954), 2–4.

6 Clinton B. Ford, "Minutes of the general meetings," 13 December 1953, CBF, 8, Secretary File, 1952–1955, AAVSO.

7 Ford, "Minutes," 13 December 1953.

8 E. Dorrit Hoffleit to Paul W. Merrill, 15 August 1953, MWM-Cor, 8, Merrill, P. W., AAVSO.

9 Paul W. Merrill to Margaret W. Mayall, 3 September 1953, MWM-Cor, 8, Merrill, P. W., AAVSO.

10 Merrill, "Questions"; and Paul W. Merrill, "Some questions concerning long period variable stars," in Campbell, Leon, *Studies of Long Period Variables*, Cambridge, Massachusetts: AAVSO, 1955, ix–xi.

11 E. Dorrit Hoffleit to Paul W. Merrill, 12 October 1953, MWM-Adm, 3, Campbell Volume, AAVSO.

12 David W. Rosebrugh to Harlow Shapley, 21 July 1953, MWM-Cor, 11, Rosebrugh, D. W., 1953–1954, AAVSO; and David W. Rosebrugh to Clinton B. Ford, 21 July 1953, CBF, 8, Secretary File, 1952–1955, AAVSO.

13 Hoffleit, 15 August 1953.

14 E. Dorrit Hoffleit, "On errors of mean light-curves," *AVSOA* (Oct 1953), 4–5.

15 Ford, "Minutes," 13 December 1953.

16 Margaret W. Mayall to Duncan E. Macdonald, 16 October 1953, MWM-Org, 7, AAVSO/HCO, group 3, folder a, AAVSO. The primary authorship of this draft appeal is uncertain: it appears to have been composed on Margaret Mayall's personal typewriter, yet the document as a whole is not up to her usual standard of careful composition and word choice; the overblown language and awkward style of many passages seems to be that of her husband, Newton. At the end of the document, noted in pencil in Margaret Mayall's hand, is a list of enclosures: "Paper by Merrill / AAVSO Membership Feb 1953 / Var. Comments Vol 5, No 1 [reporting on the May 1950 meeting at State College, Pennsylvania] / Max and Min L.P.Var. 1953 / Const and by-laws / Quarterly Rep No 12 / AAVSO Abstracts Oct 18, 1952 / 22 Annual Report / Report re: Boyden Sta." Judging from a penciled note on one corner of the draft, the appeal

was to be delivered to Duncan Macdonald at BU on Friday, October 16.

17 Margaret W. Mayall to C. M. Huffer, 20 October 1953, MWM-Org, 7, AAVSO/HCO, group 1, folder b, AAVSO.

18 Peter M. Millman to Harvard Corporation, 21 October 1953, MWM-Org, 7A, AAVSO/HCO, group 3, folder i, AAVSO.

19 McGeorge Bundy to Donald H. Menzel, 13 October 1953, Records of the Harvard College Observatory, UAV 630.37 Menzel correspondence 1953, Box 4, Bundy, McGeorge, HA.

20 Donald H. Menzel to McGeorge Bundy, 22 October 1953, Records of the Harvard College Observatory, UAV 630.37 Menzel correspondence 1953, Box 4, Bundy, McGeorge, HA. There is no evidence that Harvard followed up on Menzel's suggestion.

21 Edward Boit to Margaret W. Mayall, 26 October 1953, MWM-Org, 7A, AAVSO/HCO, group 3, folder i, AAVSO; O. M. Shaw to Edward Boit, 9 October 1953, MWM-Org, 7A, AAVSO/HCO, group 3, folder i, AAVSO; and President and Fellows of Harvard College, MWM-Org, 7A, AAVSO/HCO, group 3, folder i, AAVSO.

22 Neal J. Heines, "Letter," *AAVSO Solar Division Bulletin* (1952), 209–210.

23 Harlow Shapley to Margaret W. Mayall, undated [1953 June], MWM-Org, 3, Solar Program Letters, folder 1, AAVSO.

24 Neal J. Heines to Margaret W. Mayall, July 1953, MWM-Cor, 6, Heines, N. J., AAVSO.

25 Alan H. Shapley to Neal J. Heines, 9 July 1953, MWM-Org, 3, Solar Program Letters, folder 1, AAVSO.

26 Neal J. Heines, "Special bulletin," *AAVSO Solar Division Bulletin* (1953).

27 David W. Rosebrugh to Margaret W. Mayall, 30 August 1953, MWM-Org, 7A, AAVSO/HCO, group 3, folder i, AAVSO.

28 Alan H. Shapley to Martha Stahr Carpenter c/o M. W. Mayall, 2 October 1953, MWM-Org, 7A, AAVSO/HCO, group 3, folder g, AAVSO.

29 Margaret W. Mayall to Martin Schwarzschild, 10 October 1953, MWM-Cor, 6, IAU, AAVSO.

30 Margaret W. Mayall to David W. Rosebrugh, 1 December 1953, MWM-Org, 3, Solar Program Letters, folder 1, AAVSO.

31 Margaret W. Mayall to Solar Observers, 7 December 1953, MWM-Org, 3, Solar Program Letters, folder 1, AAVSO.

32 Margaret W. Mayall, 11 December 1953, MWM-Org, 3, Solar Program Letters, folder 1, AAVSO.

33 Margaret W. Mayall, "untitled," December 1953, MWM-Org, 3, Solar Program Letters, folder 1, AAVSO.

34 Margaret W. Mayall to Ralph N. Buckstaff, 23 December 1953, MWM-Org, 3, Solar Program Letters, folder 1, AAVSO; and other letters in this folder.

35 Clinton B. Ford, "Minutes of the council meetings," 10 December 1953, CBF, 8, Secretary File, 1952–1955, AAVSO.

36 Clinton B. Ford to Martha Stahr Carpenter, 15 October 1953, CBF, 8, Secretary File, 1952–1955, AAVSO.

37 Clinton B. Ford to Margaret W. Mayall, 6 November 1953, CBF, 8, Secretary File, 1952–1955, AAVSO; Clinton B. Ford to Martha Stahr Carpenter, 9 November 1953, CBF, 8, Secretary File, 1952–1955, AAVSO; and Clinton B. Ford to Margaret W. Mayall, 22 November 1953, CBF, 8, Secretary File, 1952–1955, AAVSO.

38 Clinton B. Ford to Paul W. Merrill, 6 November 1953, CBF, 8, Secretary File, 1952–1955, AAVSO.

39 Paul W. Merrill to Clinton B. Ford, 13 November 1953, CBF, 8, Secretary File, 1952–1955, AAVSO.

40 Clinton B. Ford, "A proposed letter to foundations appealing for funds for the continuance and expansion of the work of the A.A.V.S.O. (first revision of the original letter written in March 1953)," 22 November 1953, CBF, 8, Secretary File, 1952–1955, AAVSO.

41 Margaret W. Mayall to Clinton B. Ford, 25 November 1953, CBF, 8, Secretary File, 1952–1955, AAVSO; Charles H. Smiley to Clinton B. Ford, 14 December 1953, CBF, 8, Secretary File, 1952–1955, AAVSO; Martha Stahr Carpenter to Clinton B. Ford, 20 December 1953, CBF, 8, Secretary File, 1952–1955, AAVSO; and Martha Stahr Carpenter to AAVSO membership, undated [1953 November], MWM-Org, 7, AAVSO/HCO, group 3, folder b, AAVSO.

42 Margaret W. Mayall to Watson Davis, 25 November 1953, MWM-Cor, 3, Davis, W., AAVSO.

43 Margaret W. Mayall to Martin Schwarzschild, 25 November 1953, MWM-Cor, 12, Schwarzschild, M., AAVSO.

44 Mayall, 25 November 1953.

45 Mayall, 25 November 1953.

46 Clinton B. Ford to Claude B. Carpenter, 3 December 1953, CBF, 8, Secretary File, 1952–1955, AAVSO.

47 Alan R. Earls, "BU and the original eye in the sky," *Bostonia* (2003). In later years, the PRL was responsible for photographing the Soviet Union's Sputnik satellite in 1957, and for the design of a giant camera used in the U2 spy plane.

48 Ford, 3 December 1953.

49 Boston University, *Harold C. Case 1951–1967*, 2005, http://www.bu.edu/inauguration/presidents/index.html.

50 Donald H. Menzel to Margaret W. Mayall, 8 December 1953, MWM-Org, 7, AAVSO/HCO, group 1, folder a, AAVSO.

51 Margaret W. Mayall to Clinton B. Ford, 9 December 1953, CBF, 8, Secretary File, 1952–1955, AAVSO.

52 Bundy, 13 October 1953.

53 Margaret W. Mayall to Duncan E. Macdonald, 14 December 1953, MM, 1, HCO-AAVSO correspondence and notes, AAVSO.

54 Margaret W. Mayall to Cyrus F. Fernald, 31 December 1953, MWM-Cor, 4, Fernald, C. F., AAVSO.

55 Mayall, 31 December 1953.

56 Mayall, 31 December 1953.

57 Mayall, 31 December 1953.

14 SURVIVAL ON BRATTLE STREET

1 Margaret W. Mayall to Cyrus F. Fernald, 16 March 1954, MWM-Cor, 4, Fernald, C. F., AAVSO.

2 Anonymous [probably Margaret W. Mayall], "Friends of the AAVSO!!!," 1953, MWM-Org, 7, AAVSO/HCO, group1, folder f, AAVSO.

3 Helen M. Stephansky to Richard W. Hamilton, 15 January 1954, MWM-Cor, 5, Hamilton, R. W., folder 1, AAVSO; and Margaret W. Mayall to Charles H. Smiley, 26 June 1959, MWM-Cor, 13, Smiley, C. H., AAVSO.

4 Clinton B. Ford to Martha Stahr Carpenter, 25 January 1954, CBF, 8, AAVSO Secretary File, 1952–1955, AAVSO. Collector and AAVSO member Robert Ariail eventually purchased the Post refractor.

5 AAVSO, "Log of the C. A. Post memorial telescope," LC-Org, 5, HCO Logbooks, Post Telescope, AAVSO.

6 Margaret W. Mayall, "Variable star notes," *JRASC* 48 (1954), 70–3.

7 Stephansky, 15 January 1954.

8 Stephansky, 15 January 1954.

9 Helen M. Stephansky to Richard W. Hamilton, 24 February 1954, MWM-Cor, 5, Hamilton, R. W., folder 1, AAVSO.

10 McGeorge Bundy to Margaret W. Mayall, 4 January 1954, MWM-Org, 7, AAVSO/HCO, group 1, folder b, AAVSO.

11 Duncan E. Macdonald to Margaret W. Mayall, 6 January 1954, MWM-Org, 7, AAVSO/HCO, group 1, folder b, AAVSO.

12 Stephansky, 15 January 1954.

13 Margaret W. Mayall to Martha Stahr Carpenter, 15 January 1954, MWM-Cor, 2, Carpenter, M. S., AAVSO.

14 Helen M. Stephansky to Richard W. Hamilton, 27 January 1954, MWM-Cor, 5, Hamilton, R. W., folder 1, AAVSO; and Margaret W. Mayall to Peter van de Kamp, 29 January 1954, MWM-Org, 15, Witherell, P. W., folder 1, AAVSO.

15 Margaret W. Mayall to Harry L. Bondy, 5 February 1954, MWM-Org, 3, Solar Program Letters, folder 2, AAVSO.

16 Clinton B. Ford to Percy W. Witherell, 8 February 1954, CBF, 8, Secretary File, 1952–1955, AAVSO. $500 000 would be more than $3.9 million in today's dollars.

17 Clinton B. Ford, 15 February 1954, CBF, 8, Secretary File, 1952–1955, AAVSO; Clinton B. Ford to Margaret W. Mayall, 3 March 1954, CBF, 8, Secretary File, 1952–1955, AAVSO; and Margaret W. Mayall to Clinton B. Ford, 9 March 1954, CBF, 8, Secretary File, 1952–1955, AAVSO; AAVSO Endowment Fund Committee to AAVSO members and friends, undated [February 1954], CBF, 8, Secretary File, 1952–1955, AAVSO.

18 Watson Davis to Clinton B. Ford, 15 February 1954, CBF, 8, Secretary File, 1952–1955, AAVSO; and AAVSO Endowment Fund Committee, endowment fund brochure, 1954, MWM-Org, 4, Fund Drive Endowment, folder 6, AAVSO. One of the sponsors, Watson Davis, head of the Science Service news bureau, sent Ford a list of action items closing with, "We want to cooperate to the fullest extent possible."

19 Clinton B. Ford, "Proposed letter to foundations appealing for funds to continue and expand the work of the A.A.V.S.O. (third revision)," February 1954, CBF, 8, Secretary File, 1952–1955, AAVSO; and Clinton B. Ford, "Minutes of the interim meeting of the A.A.V.S.O. council held at 4 Brattle Street, Cambridge, Mass., February 27, 1954," 27 February 1954, CBF, 8, Secretary File, 1952–1955, AAVSO.

20 AAVSO Endowment Fund Committee 3 March 1954, CBF, 8, Secretary File, 1952–1955, AAVSO.

21 Margaret W. Mayall to Clinton B. Ford *et al.*, 2 March 1954, CBF, 8, Secretary File, 1952–1955, AAVSO.

22 Cyrus F. Fernald to Margaret W. Mayall, 11 March 1954, MWM-Cor, 4, Fernald, C. F., AAVSO.

23 Mayall, 16 March 1954.

24 Cyrus F. Fernald to Margaret W. Mayall, 18 March 1954, CFF, 1, AAVSO Reorganization, AAVSO.

25 Margaret W. Mayall, "Variable star notes," *JRASC* 48 (1954), 70–3.

26 Anonymous, "Science and public policy," *Physics Today* 7 (1954), 6–7.

27 Helen M. Stephansky to Percy W. Witherell, 2 April 1954, MWM-Cor, 15, Witherell, P. W., folder 1, AAVSO.

28 G. E. Jones to Clinton B. Ford, 14 April 1954, CBF, 8, Secretary File, 1952–1955, AAVSO; and Clinton B. Ford to Percy W. Witherell, 22 April 1954, CBF, 8, Secretary File, 1952–1955, AAVSO.

29 Clinton B. Ford, "Minutes AAVSO council meetings," 7 May 1954, CBF, 8, Secretary File, 1952–1955, AAVSO.
30 Roy A. Seely, "The AAVSO Spring Meeting at Columbia University, New York," *VC* 5 No. 13 (1954), 1–4.
31 Margaret W. Mayall to Clinton B. Ford, 31 August 1954, CBF, 8, Secretary File, 1952–1955, AAVSO.
32 Seely, "The AAVSO Spring Meeting"; R. Newton Mayall (ed.), "Papers presented at the Spring Meeting, May 7–8, 1954," *AVSOA* **Spring** (1954); and Margaret W. Mayall to Philip J. Warner, 12 May 1954, MWM-Org, 7, AAVSO/HCO, group 3, folder c, AAVSO.
33 C. E. Kenneth Mees to Margaret W. Mayall, 7 May 1954, CBF, 8, Secretary File, 1952–1955, AAVSO.
34 Margaret W. Mayall to C. E. Kenneth Mees, 12 May 1954, MWM-Org, 7, AAVSO/HCO, group 3, folder c, AAVSO.
35 Clinton B. Ford to various, 20 August 1954, MWM-Org, 4, Fund Drive Endowment, folder 1, AAVSO.
36 Clinton B. Ford to Francis Reynolds, 23 September 1954, CBF, 8, Secretary File, 1952–1955, AAVSO.
37 Mayall, 31 August 1954.
38 Clinton B. Ford to Margaret W. Mayall, 8 September 1954, CBF, 8, Secretary File, 1952–1955, AAVSO.
39 Margaret W. Mayall to Clinton B. Ford, 24 September 1954, CBF, 8, Secretary File, 1952–1955, AAVSO.
40 Mayall, 24 September 1954; Anonymous [Helen M. Stephansky], untitled ["Thanks to Robert Dunn . . . "], *AVSOB* (1954); and Clinton B. Ford to Margaret W. Mayall, 28 September 1954, CBF, 8, Secretary File, 1952–1955, AAVSO.
41 Helen M. Stephansky to Percy W. Witherell, 15 December 1954, MWM-Cor, 15, Witherell, P. W., folder 1, AAVSO; Donald H. Menzel to Margaret W. Mayall, 6 December 1954, MWM-Org, 7, AAVSO/HCO, group 1, folder f, AAVSO; and Donald H. Menzel to Margaret W. Mayall, 19 November 1954, MWM-Org, 7, AAVSO/HCO, group 1, folder f, AAVSO.
42 Stephansky, 15 December 1954.
43 Cyrus F. Fernald to Roy A. Seely, 16 December 1954, CFF, 1, AAVSO Committee Appointments, AAVSO.
44 Charles H. Smiley to Cyrus F. Fernald, 31 December 1954, MWM-Cor, 4, Fernald, C. F., AAVSO.
45 Cyrus F. Fernald to Charles H. Smiley, 4 January 1955, MWM-Cor, 4, Fernald, C. F., AAVSO. The articles appeared in Vol. 48, Nrs. 5 and 6, respectively.
46 Margaret W. Mayall to Charles H. Schauer, 14 January 1955, CBF, 8, Secretary File, 1952–1955, AAVSO.
47 Clinton B. Ford to Lewis J. Boss, 13 February 1955, CBF, 8, Secretary File, 1952–1955, AAVSO.
48 Richard W. Hamilton to Cyrus F. Fernald, 15 May 1955, CFF, 1, AAVSO Meetings, AAVSO.
49 Clinton B. Ford to C. E. Kenneth Mees, 23 March 1955, MWM-Org, 4, Fund Drive Endowment, folder 11, AAVSO.
50 Lawrence N. Upjohn to Clinton B. Ford, 27 March 1955, CBF, 8, Secretary File, 1952–1955, AAVSO.
51 Clinton B. Ford to Lawrence N. Upjohn, 31 March 1955, MWM-Org, 4, Fund Drive Endowment, folder 5, AAVSO.
52 Margaret W. Mayall to Cyrus F. Fernald, 2 April 1955, MWM-Cor, 4, Fernald, C. F., AAVSO.
53 Mayall, 2 April 1955.
54 Helen M. Stephansky to Richard W. Hamilton, 12 April 1955, MWM-Cor, 5, Hamilton, R. W., folder 1, AAVSO.
55 Carl Schauer to Margaret W. Mayall, 18 April 1955, MWM-Org, 4, Fund Drive Endowment, folder 5, AAVSO.
56 Cyrus F. Fernald to Margaret W. Mayall, 4 May 1955, MWM-Cor, 4, Fernald, C. F., AAVSO.
57 Cyrus F. Fernald to Percy W. Witherell, 9 June 1955, CFF, 1, AAVSO Meetings, AAVSO; Cyrus F. Fernald to Richard W. Hamilton, 5 June 1955, CFF, 1, AAVSO Meetings, AAVSO; Cyrus F. Fernald to Margaret Harwood, 4 June 1955, CFF, 1, AAVSO Meetings, AAVSO; and Margaret Harwood to Cyrus F. Fernald, 14 May 1955, CFF, 1, AAVSO Meetings, AAVSO.
58 Clinton B. Ford, "Minutes of the council meetings," May 1955, CBF, 8, Secretary File, 1952–1955, AAVSO.
59 Clinton B. Ford, "Minutes of the general meetings," May 1955, CBF, 8, Secretary File, 1952–1955, AAVSO.
60 Charles LeRoy to Clinton B. Ford, 1 June 1955, CBF, 8, Secretary File, 1952–1955, AAVSO.
61 Clinton B. Ford, "Minutes of the Finance Committee meeting," May 1955, CBF, 8, Secretary File, 1952–1955, AAVSO.
62 Fernald, 9 June 1955.
63 Clinton B. Ford to Margaret W. Mayall, 19 June 1955, CBF, 8, Secretary File, 1952–1955, AAVSO; and Cecilia Payne-Gaposchkin, *Variable Stars and Galactic Structure*, London: Athlone Press, 1954.
64 Helen M. Stephansky to Richard W. Hamilton, 18 June 1954, MWM-Cor, 5, Hamilton, R. W., folder 1, AAVSO.
65 Margaret W. Mayall to Margaret Harwood, 28 June 1955, MWM-Adm, 3, Campbell Volume, AAVSO.
66 Margaret W. Mayall to Clinton B. Ford, 11 July 1955, CBF, 9, Secretary File, 1955–1957, AAVSO.
67 Cyrus F. Fernald to Clinton B. Ford, 21 July 1955, CBF, 9, Secretary File, 1955–1957, AAVSO.
68 Helen M. Stephansky to Richard W. Hamilton, 24 August 1955, MWM-Cor, 5, Hamilton, R. W., folder 1, AAVSO.

69 Cyrus F. Fernald to Margaret Harwood, 24 October 1955, CFF, 1, AAVSO Committee Appointments, AAVSO; and Clinton B. Ford, "Minutes of the council meetings," October 1955, CBF, 9, Secretary File, 1955–1957, AAVSO. Reeve's $300 contribution was sufficient to cover Helen Stephansky's wages for the period in question.

70 Cyrus F. Fernald to David W. Rosebrugh, 4 November 1955, CFF, 1, AAVSO Committee Appointments, AAVSO.

71 Margaret W. Mayall to Alan T. Waterman, 25 November 1955, MWM-Org, 19, National Science Foundation (NSF) proposals and contracts, 1, AAVSO.

72 Margaret W. Mayall to Office of Naval Research, 1 December 1955, MWM-Org, 19, Office of Naval Research (ONR) proposals and contracts, 1, AAVSO.

73 Margaret W. Mayall 1956, MWM-Org, 19, National Science Foundation (NSF) proposals and contracts, 1, AAVSO.

74 Margaret W. Mayall to Percy W. Witherell, 8 March 1956, MWM-Org, 19, National Science Foundation (NSF) proposals and contracts, 1, AAVSO.

75 Helen M. Stephansky to Cyrus F. Fernald, 9 March 1955, CFF, 1, AAVSO Budget and Funding, AAVSO.

76 Margaret W. Mayall to Cyrus F. Fernald, 10 March 1956, MWM-Org, 19, National Science Foundation (NSF) proposals and contracts, 1, AAVSO.

77 Cyrus F. Fernald to Richard W. Hamilton *et al.*, 12 March 1956, CBF, 9, Secretary File, 1955–1957, AAVSO.

78 Clinton B. Ford, "Minutes of the council meetings," May 1956, CBF, 9, Secretary File, 1955–1957, AAVSO; and Margaret W. Mayall to Harlow Shapley, 5 October 1956, MWM-Cor, 12, Shapley, H., AAVSO.

79 Margaret W. Mayall to Office of Naval Research, 10 December 1956, MWM-Org, 19, Office of Naval Research (ONR) proposals and contracts, 1, AAVSO; Margaret W. Mayall to Bache Fund, 13 December 1956, MWM-Org, 19, Office of Naval Research (ONR) proposals and contracts, 1, AAVSO; Edwin B. Wilson 10 July 1956, MWM-Org, 19, Office of Naval Research (ONR) proposals and contracts, 1, AAVSO; and Margaret W. Mayall 14 December 1956, MWM-Org, 19, Office of Naval Research (ONR) proposals and contracts, 1, AAVSO.

80 Margaret W. Mayall to Percy W. Witherell, 7 January 1957, MWM-Org, 19, National Science Foundation (NSF) proposals and contracts, 1, AAVSO.

81 Margaret W. Mayall to Richard W. Hamilton, 12 March 1957, MWM-Org, 19, Office of Naval Research (ONR) proposals and contracts, 1, AAVSO.

82 Clinton B. Ford to Francis Reynolds, 9 April 1957, CBF, 9, Secretary File, 1955–1957, AAVSO.

83 Clinton B. Ford, "Minutes of the council meetings," May 1957, CBF, 9, Secretary File, 1955–1957, AAVSO.

84 Clinton B. Ford, "Minutes of the council meetings," October 1958, CBF, 9, Secretary File, 1955–1957, AAVSO.

85 Dwight E. Gray to Margaret W. Mayall, 4 December 1958, MWM-Org, 19, National Science Foundation (NSF) proposals and contracts, 2, AAVSO.

86 Margaret W. Mayall to Dwight E. Gray, 11 December 1958, MWM-Org, 19, National Science Foundation (NSF) proposals and contracts, 2, AAVSO.

87 Clinton B. Ford, "Minutes of the council meeting," October 1960, CBF, 9, Secretary File, 1955–1957, AAVSO.

88 Clinton B. Ford, "Minutes of the general meeting," October 1960, CBF, 9, Secretary File, 1955–1957, AAVSO; and Helen M. Stephansky to Margaret W. Mayall, 16 November 1960, MWM-Cor, 13, Stephansky, H. M., AAVSO.

89 Margaret W. Mayall to Alan T. Waterman, 18 October 1960, MWM-Org, 19, Office of Naval Research (ONR) proposals and contracts, 3, AAVSO.

90 Clinton B. Ford, "Minutes of the council meetings," October 1962, CBF, 9, Secretary File, 1955–1957, AAVSO.

91 Clinton B. Ford, "Minutes of the council meetings," April 1963, CBF, 9, Secretary File, 1955–1957, AAVSO.

92 Clinton B. Ford, "Minutes of the council meetings," October 1963, CBF, 9, Secretary File, 1955–1957, AAVSO.

15 AAVSO ACHIEVEMENTS

1 Cyrus F. Fernald to Margaret W. Mayall, 11 March 1954, MWM-Cor, 4, Fernald, C. F., AAVSO.

2 Margaret W. Mayall, "Variable star notes," *JRASC* 48 (1954), 21–3.

3 Fernald, 11 March 1954.

4 Margaret W. Mayall, "23rd annual report of the Recorder for year ending September 30, 1954," *VC* 5, No. 14 (1954), 1–4; and Boris V. Kukarkin 26 August 1954, MWM-Cor, 7, Kukarkin, B. V., AAVSO. The founding director of the RASNZ-VSS, Frank N. Bateson, was eventually elected an Honorary Member of AAVSO.

5 Mayall, "23rd annual report."

6 Margaret W. Mayall to Clinton B. Ford, 9 March 1954, CBF, 8, Secretary File, 1952–1955, AAVSO.

7 Clinton B. Ford, March 1954, CBF, 8, Secretary File, 1952–1955, AAVSO; and Jerome Ellison, "The mysteries of the night," *The Saturday Evening Post* (March 6, 1954), 36–9.

8 Helen M. Stephansky, "Spare time space men," *Mechanix Illustrated* (April 1954), 180–1, 222.

9 Anonymous [C. L. Stong], "Two astronomical matters," *SciAm* **190** (1954), 100–6.

10 Anonymous [Helen M. Stephansky], "untitled ["It's an ill wind that blows no good . . .]" *AVSOB* (1954).

11 Anonymous, "Struck by the stars," *Life* (25 July 1955), 87–91.

12 Boris V. Kukarkin, "Report of Commission 27 (Variable Stars)," *IAUT* **9** (1957), 368–85. In the same report (p. 376) Kukarkin states, "The Variable Star Section of the British Astronomical Association was not active at all. Attempts by the President of Commission 27 to start systematic correspondence with any of its leaders were unsuccessful."

13 Clinton B. Ford to Claude B. Carpenter, 29 November 1955, CBF, 9, Secretary File, 1955–1957, AAVSO.

14 Margaret Harwood to Cyrus F. Fernald, 10 October 1955, CFF, 1, AAVSO Committee Appointments, AAVSO.

15 Margaret W. Mayall, "Report on the I.A.U.," *AVSOA* (1955).

16 Mayall, "Report on the I.A.U."; Margaret W. Mayall, "Variable star notes," *JRASC* **49** (1955), 248–50; and Margret W. Mayall, "24th annual report of the recorder for the year ending September 30, 1955," October 1955, CBF, 9, Secretary File, 1955–1957, AAVSO.

17 Anonymous, "Famous woman astronomer back from Dublin," *Cambridge Chronicle*, Cambridge, Massachusetts, 13 October 1955, News about AAVSO people file, 1, AAVSO.

18 Clinton B. Ford, "Minutes of the general meetings of the A.A.V.S.O. held at the Springfield Museum of Natural History, October 21–22, 1955," 22 October 1955, CBF, 9, Secretary File, 1955–1957, AAVSO; Mayall, "24th annual report"; and Ford, 29 November 1955 (also see numerous letters by HMS during this period, in CBF 9).

19 Cyrus F. Fernald to P. O. Parker, 4 November 1955, CFF, 1, AAVSO Committee Appointments, AAVSO.

20 H. B. Allen to Leon Campbell, 27 June 1949, LC-Cor, 8, Upjohn, L. N., AAVSO.

21 Margaret W. Mayall to Clinton B. Ford, 30 December 1955, CBF, 9, Secretary File, 1955–1957, AAVSO; various, material pertaining to the publication of *Studies of Long Period Variables*, October 1955, LC, 3, Campbell Volume, AAVSO; Paul W. Merrill, "Questions concerning long-period variable stars," October 1953, MWM-Adm, 3, Campbell Volume, AAVSO; Paul W. Merrill, "Some questions concerning long period variable stars," in Campbell, Leon, *Studies of Long Period Variables*, Cambridge, Massachusetts: AAVSO (1955), ix–xi; Margaret W. Mayall, "Preface," in Campbell, Leon, *Studies of Long Period Variables*, Cambridge, Massachusetts: AAVSO, 1955, v; and Leon Campbell, *Studies of Long Period Variables*, Cambridge, Massachusetts: AAVSO, 1955.

22 Clinton B. Ford to Cyrus F. Fernald, 10 December 1955, CBF, 9, Secretary File, 1955–1957, AAVSO.

23 R. Newton Mayall to Cyrus F. Fernald, 26 March 1956, MWM-Org, 2, Satellite Program, Committee, AAVSO; and Cyrus F. Fernald to R. Newton Mayall, 27 March 1956, CFF, 1, AAVSO Committee Appointments, AAVSO.

24 Margaret W. Mayall to Cyrus F. Fernald, 9 March 1956, MWM-Org, 19, National Science Foundation (NSF) proposals and contracts, 1, AAVSO.

25 Fred L. Whipple to Margaret W. Mayall, 5 April 1956, MWM-Org, 2, Satellite Program, Committee, AAVSO, and other correspondence in this folder; and Patrick McCray, *Keep Watching the Skies! The Story of Operation Moonwatch and the Dawn of the Space Age*, Princeton, New Jersey and Oxford, England: Princeton University Press, 2008, 87.

26 Max Good, "Name of Manhattan established by efforts of Moonwatch team," *The Manhattan Mercury*, Manhattan, Kansas, 15 December 1957, p. 15. WSH, 11, Moonwatch Program, AAVSO, and other letters, articles, etc. pertaining to Houston's Manhattan, Kansas Moonwatch program.

27 Smithsonian Astrophysical Observatory, *Bulletin for Visual Observers of Satellites* **1** (July 1956), and following issues.

28 Margaret W. Mayall 10 January 1956, CBF, 9, Secretary File, 1955–1957, AAVSO.

29 Mayall, 9 March 1956.

30 R. Newton Mayall, "What goes on!," MWM-Org, 5, Papers and talks, non-technical, historical (AAVSO), AAVSO.

31 Helen M. Stephansky to Clinton B. Ford, 23 May 1956, CBF, 9, Secretary File, 1955–1957, AAVSO.

32 Clinton B. Ford, "Minutes of the A.A.V.S.O. Council meetings held at the Museum of Natural History, Springfield, Massachusetts, October 19–20, 1956," October 1956, CBF, 9, Secretary File, 1955–1957, AAVSO.

33 Margaret W. Mayall 21 October 1956, CBF, 9, Secretary File, 1955–1957, AAVSO.

34 Frank M. Bateson, *Paradise Beckons*, Waikanae, New Zealand: Heritage Press, 1989, 155.

35 Margaret W. Mayall, and Frank M. Bateson, "Summary of meeting between the directors of the A.A.V.S.O. and the Variable Star Section of the Royal Astronomical Society of New Zealand," MWM-Cor, 1, Bateson, F. M., AAVSO; Bateson, *Paradise beckons*; and Frank M. Bateson to Margaret W Mayall, 1957, MWM-Cor, 1, Bateson, F. M., AAVSO.

36 I. L. Thomsen to Frank M. Bateson, 27 November 1952, MWM-Cor, 1, Bateson, F. M., AAVSO; Bart J. Bok to

Frank M. Bateson, 31 March 1958, MWM-Cor, 1, Bateson, F. M., AAVSO; Bart J. Bok to Frank M. Bateson, 9 July 1959, MWM-Cor, 1, Bateson, F. M., AAVSO; Frank M. Bateson to Margaret W. Mayall, 15 May 1956, MWM-Cor, 1, Bateson, F. M., AAVSO; Frank M. Bateson to Bart J. Bok, 30 July 1956, MWM-Cor, 1, Bateson, F. M., AAVSO; Frank M. Bateson, "Variable star observing in New Zealand," *JRASC* **52** (1958), 241–9; Frank M. Bateson, "Variable star observing in New Zealand," *JRASC* **53** (1959), 1–6; Alan Plummer, and Mati Morel, "The VSS RASNZ variable star charts: a story of co-evolution," *JAAVSO* **38** (2010), 123–31; and Frank M. Bateson to Margaret W. Mayall, 23 November 1957, MWM-Cor, 1, Bateson, F. M., AAVSO. Bateson sought the help of Bart Bok, then Director of the Commonwealth Observatory, Mount Stromlo, Australia, in determining magnitude standards for bright long-period southern variables. Bok wanted to help initially, but by 1958 – perhaps overwhelmed with his new duties as Director of Mount Stromlo Observatory – he tried to discourage Bateson from enlisting his help, suggesting instead that he obtain what he needed from HCO's plate collection, saying "it seems in a way very silly that we at Mount Stromlo, with very limited photographic equipment, should take photographs all over again to take the place of fine existing photographic plates now safely stored away in Harvard Observatory plate collection. However, this is your problem, not mine." Bok also suggested that it would be best in the long run if the New Zealand amateurs, with professional help, could set up their own observatory. By contrast, the Yerkes and Lick observatories provided Bateson with the sequence data he needed, and, he wrote, "where plates are not available they are taking them for us." Furthermore, when Bateson requested prints from HCO, Director Menzel [at Mayall's suggestion], "agreed to supply the material without charge." In 1959, only after Bok received urging from Margaret Mayall, who, according to Bok, "was instrumental in getting us started on standard sequences for your use," did Bok manage to complete work on two sequences for Bateson.

37 Margaret W. Mayall, "Variable star notes," *JRASC* **51** (1957), 109–12.

38 John J. Ruiz, "A Photoelectric Light Curve of mu Herculis," *PASP* **69** (1957), 261.

39 Clinton B. Ford, "Minutes of the A.A.V.S.O. Council meetings held at Amherst College, Amherst, Mass., October 4–5, 1957," October 1957, CBF, 9, Secretary File, 1955–1957, AAVSO.

40 Anonymous, "Blair Medal awarded at Western convention," *S&T* **17** (1957), 77.

41 Donald H. Menzel to Margaret W. Mayall, 30 December 1957, MWM-Cor, 8, Menzel, D. H., AAVSO.

42 Margaret W. Mayall to Arthur Grad, 28 May 1958, MWM-Org, 19, Office of Naval Research (ONR) proposals and contracts, 1, AAVSO.

43 Clinton B. Ford, "Minutes of the general meeting of the A.A.V.S.O. held at Maria Mitchell Observatory, Nantucket, Mass., Saturday, June 14, 1958," June 1958, CBF, 9, Secretary File, 1955–1957, AAVSO.

44 Boris V. Kukarkin, "Report of Commission 27 (Variable Stars)," *IAUT* **10** (1958), 398–431. The Mayalls probably paid for this trip. Inveterate travelers, Margaret and Newton Mayall made the voyage in their usual roundabout fashion: sailing to Holland, then on to Moscow, from there to Norway, then back to Holland, and probably visiting other points on their return to the United States. Their entire trip lasted about 3 months. They made their 1952 IAU excursion to Rome, and their 1955 trip to Ireland in similar fashion. See, for example, Margaret W. Mayall to Helen M. Stephansky, 2 September 1958, MWM-Cor, 13, Stephansky, H. M., AAVSO and other letters in this folder. Much later, in October 1973, the council approved a motion "that travel funds be paid to the Director for trips to IAU meetings, since such trips are an implied duty of the Director." Clinton B. Ford, "Minutes of the A.A.V.S.O. Council meetings held at Blantyre Castle, Lenox, Mass., October 19–20, 1973," October 1973, CBF, 10, Minutes 1964–1990, AAVSO.

45 Clinton B. Ford, "Minutes of the meeting of the A. A. V. S. O. Council held at the Adler Planetarium, Chicago, Illinois, May 29, 1959," May 1959, CBF, 9, Secretary File, 1955–1957, AAVSO.

46 Margaret W. Mayall, "Miscellany," *AVSOA* (1959), 9. Note that at this time, Robinson was still a member of the Los Angeles Astronomical Society and had not yet started work at *Sky & Telescope*.

47 Margaret W. Mayall to Clinton B. Ford, 22 November 1960, MWM-Cor, 4, Ford, C. B., AAVSO.

48 Clinton B. Ford, "Minutes of the general meetings of the A. A. V. S. O. held at Harvard College Observatory, Cambridge, Mass., October 12–14, 1961," October 1961, CBF, 9, Secretary File, 1955–1957, AAVSO; R. Newton Mayall, "Papers presented at the 50th anniversary meeting, October 12–15, 1961," *AVSOA* (1961), 1–25; and Margaret W. Mayall, "Variable star notes," *JRASC* **56** (1962), 37–39.

49 Margaret W. Mayall to Clinton B. Ford, 24 October 1961, MWM-Cor, 4, Ford, C. B., AAVSO.

50 R. Newton Mayall, "The Story of the AAVSO," *RPA* **55** (1961), 4–9. *RPA* was an outgrowth of Zahner's earlier

enterprise, *The Monthly Evening Sky Map*, which he purchased and revived in 1959. At that time, and at the suggestion of *Sky & Telescope* publisher Charles Federer, Zahner's intent was to orient *Sky Map*, and later *RPA*, as a magazine dedicated much more to observational amateur astronomy, in contrast to the strong flavor of amateur telescope making that had come to dominate *Sky & Telescope*, especially after the retirement of *Scientific American* editor Albert G. Ingalls in 1955.

51 Donald D. Zahner to Margaret W. Mayall, 7 September 1959, MWM-Cor, 15, Zahner, D. D., AAVSO; and Margaret W. Mayall to Donald D. Zahner, 24 September 1959, MWM-Cor, 15, Zahner, D. D., AAVSO.

52 Donald D. Zahner, "Book Review: The Sky Observer's Guide," *MESM* **54** (1960), 17; Mayall, "The Story of the AAVSO"; Margaret W. Mayall, "On three overlooked variables in Cancer," *RPA* **56** (1961), 28–30; Donald D. Zahner to Margaret W. Mayall, 17 September 1961, MWM-Cor, 15, Zahner, D. D., AAVSO; and Margaret W. Mayall to Donald D. Zahner, 23 September 1963, MWM-Cor, 15, Zahner, D. D., AAVSO.

53 Clinton B. Ford to Margaret W. Mayall, 19 November 1962, MWM-Cor, 4, Ford, C. B., AAVSO; and Clinton B. Ford to Margaret W. Mayall, 5 February 1964, MWM-Cor, 4, Ford, C. B., AAVSO.

54 Margaret W. Mayall, "Variable star programs for amateur observers," *Kleine Veroffentlichungen der Remeis-Sternwarte Bamberg* (1962), 131–33.

55 Martin Schwarzschild to Margaret W. Mayall, 31 December 1963, MWM-Cor, 12, Schwarzschild, M., AAVSO. Historian David DeVorkin provides a bit more detail on Schwarzschild's Stratoscope experiments in David H. DeVorkin, *Race to the Stratosphere, Manned Scientific Ballooning in America*, New York, Berlin, Heidelberg, London, Paris, Tokyo: Springer-Verlag, 1989, p. 303.

56 John S. Hall to Margaret W. Mayall, 17 July 1953, MWM-Org, 2A, Photoelectric Photometry Program, AAVSO.

57 Lewis J. Boss to Margaret W. Mayall, 21 January 1954, MWM-Cor, 2, Boss, L. J., AAVSO.

58 Clinton B. Ford, "Minutes of the A.A.V.S.O. Council meetings held Brown University, Providence, Rhode Island, October 8–9, 1954," October 1954, CBF, 8, Secretary File, 1952–1955, AAVSO.

59 Cyrus F. Fernald to Lewis J. Boss, 4 November 1954, CFF, 1, AAVSO Committee Appointments, AAVSO.

60 Lewis J. Boss to Cyrus F. Fernald, 31 January 1955, CFF, 1, AAVSO Committee Appointments, AAVSO.

61 Boss, 31 January 1955; Lewis J. Boss to Claude B. Carpenter, 31 January 1955, CFF, 1, AAVSO Committee Appointments, AAVSO; and Lewis J. Boss to John J. Ruiz, 31 January 1955, CFF, 1, AAVSO Committee Appointments, AAVSO.

62 Clinton B. Ford to Lewis J. Boss, 13 February 1955, CBF, 8, Secretary File, 1952–1955, AAVSO.

63 See, for example John J. Ruiz to Margaret W. Mayall, 29 September 1953, MWM-Org, 2A, Photoelectric Photometry Program, AAVSO and other letters in this folder.

64 Margaret W. Mayall to Domingo Taboada, 3 June 1954, MWM-Cor, 13, Taboada, D., AAVSO; and John J. Ruiz to Clinton B. Ford, 11 May 1951, CBF, 7, Secretary File, 1949–1952, AAVSO.

65 Lewis J. Boss, "Second annual report of the photo-electric photometer committee," 11 October 1956, CBF, 9, Secretary File, 1955–1957, AAVSO; and AAVSO Photoelectric Photometer Committee, *The Photoelectric Photometer*, 1956, MSM-Org, 2A, Photoelectric Photometry Program, AAVSO PEP Handbook.

66 Lewis J. Boss to Clinton B. Ford, 16 October 1956, CBF, 9, Secretary File, 1955–1957, AAVSO; Boss, Lewis J., "Second annual report of the photo-electric photometer committee," 11 October 1956, CBF, 9, Secretary File, 1955–1957; Lewis J. Boss, "Supplement to second annual report of the photo-electric photometer committee," 16 October 1956, CBF, 9, Secretary File, 1955–1957, AAVSO; and John J. Ruiz 25 October 1956, CBF, 9, Secretary File, 1955–1957, AAVSO.

67 Lewis J. Boss, "Semi-annual report of the photo-electric photometer committee," 15 May 1959, CBF, 9, Secretary File, 1955–1957, AAVSO; AAVSO Photoelectric Photometer Committee, *The Photoelectric Photometer*, January 1958.

68 Lewis J. Boss to Margaret W. Mayall, 15 June 1959, MWM-Cor, 2, Boss, L. J., AAVSO; Donald Engelkemeir to Margaret W. Mayall, 23 September 1959, MWM-Org, 2A, Photoelectric Photometry Program, AAVSO PEP Handbook, AAVSO; and Margaret W. Mayall to Donald Engelkemeir, 16 August 1960, MWM-Org, 2A, Photoelectric Photometry Program, AAVSO PEP Handbook, AAVSO.

69 Lewis J. Boss to Margaret W. Mayall, 18 March 1960, MWM-Cor, 2, Boss, L. J., AAVSO.

70 Clinton B. Ford, "Minutes of the A. A. V. S. O. Council meetings held at the Hayden Planearium, American Museum of Natural History, New York, N. H., May 27–28, 1960," May 1960, CBF, 9, Secretary File, 1955–1957, AAVSO; and Lewis J. Boss to Margaret W. Mayall, 21 April 1967, MWM-Cor, 2, Boss, L. J., AAVSO.

71 Harry L. Bondy, "Editorial," *AVSOSolB* (January 1954), 1–2.

72 Margaret W. Mayall to Clinton B. Ford, 31 August 1954, CBF, 8, Secretary File, 1952–1955, AAVSO.

73 Ford, AAVSO Council Minutes, October 1954.

74 Ford, AAVSO Council Minutes, October 1954.

75 Margaret W. Mayall to Cyrus F. Fernald, 22 October 1954, MWM-Cor, 4, Fernald, C. F., AAVSO.

76 Thomas A. Cragg to Cyrus F. Fernald, 25 January 1955, CFF, 1, AAVSO Committee Appointments, AAVSO.

77 Cyrus F. Fernald to Harry L. Bondy, 10 March 1955, CFF, 1, AAVSO Committee Appointments, AAVSO; and Harry L. Bondy to Cyrus F. Fernald, 14 March 1955, CFF, 1, AAVSO Committee Appointments, AAVSO.

78 Clinton B. Ford, "Minutes of the A. A. V. S. O. Council meetings, West Boylston, Mass., October 11–12, 1963," October 1963, CBF, 9, Secretary File, 1955–1957, AAVSO.

79 Helen M. Stephansky to Margaret W. Mayall, 29 August 1955, MWM-Cor, 13, Stephansky, H. M., AAVSO.

80 Harry L. Bondy, "Neal J. Heines June 21, 1892 August 21, 1955," *AVSOSolB* (August–September 1955), 2–3.

81 David W. Rosebrugh, "The 27-year history of the Solar Division of the AAVSO," 15 September 1971, MWM-Org, 5, Papers and Talks, non-technical, historical (AAVSO), AAVSO.

16 BREATHING ROOM ON CONCORD AVENUE

1 Clinton B. Ford, "Our directors – past and present," *JAAVSO* **2** (1973), 50.

2 R. Newton Mayall, *The Story of the AAVSO, Part Three*, Cambridge, Massachusetts: AAVSO, 1961; Curtis E. Anderson, *The Story of the AAVSO, Part Four*, Cambridge, Massachusetts: AAVSO, 1964; and Clinton B. Ford, "Minutes of the general meeting of the A. A. V. S. O. held at the McDonnell Planetarium, St. Louis, Missouri, April 18, 1964," April 1964, CBF, 10, Minutes 1964–1990, AAVSO.

3 Margaret W. Mayall to Ralph K. Dakin, 4 August 1964, MWM-Cor, 8, Mayall, M. W., AAVSO.

4 Clinton B. Ford, "Minutes of the general meetings of the A. A. V. S. O. held in Woods Hole, Massachusetts, October 9–10, 1964," October 1964, CBF, 10, Minutes 1964–1990, AAVSO.

5 Clinton B. Ford, "Minutes of the A. A. V. S. O. Council meeting held at the Headquarters of the Royal Astronomical Society of Canada, Toronto, Ontario, Canada, May 21, 1965," May 1965, CBF, 10, Minutes 1964–1990, AAVSO.

6 Clinton B. Ford, "Minutes of the meeting of the A. A. V. S. O. Finance Committee held at the Toronto Education Centre Building, Toronto, Ontario, Canada, May 22, 1965," May 1965, CBF, 10, Minutes 1964–1990, AAVSO.

7 R. Newton Mayall, "Proposed AAVSO Headquarters (preliminary plans)," October 1965, CBF, 4A, AAVSO HQ Plans, AAVSO; R. Newton Mayall, "Proposed AAVSO Headquarters (alternate plans)," May 1966, CBF, 4A, AAVSO HQ Plans, AAVSO; Earl E. Lhamon, "The Schoonover public observatory," *S&T* **28** (1964), 356–7; Thomas A. Cragg, "Ford observatory in California," *S&T* **31** (1966), 138–41; and Clinton B. Ford to Margaret W. Mayall, 13 June 1965, MWM-Cor, 4, Ford, C. B., AAVSO.

8 Margaret W. Mayall to Clinton B. Ford, 1 July 1965, MWM-Cor, 4, Ford, C. B., AAVSO.

9 Mayall, 1 July 1965.

10 Leon Campbell, "Weekly Reports 370," Arequipa, Peru: 25 April 1915, HCO-BS, UAV 630.100, 6, 7, HUA; and Edward C. Pickering to Leon. Campbell, 5 June 1915, HCO-BS, UAV 630.100, Box 3, HUA.

11 Henry Norris Russell *et al.*, *Astronomy, A Revision of Young's Manual of Astronomy: Volume II – Astrophysics and Stellar Astronomy*, Boston: Ginn and Company, 1927, 777–98; Arthur B. Wyse, "Algol the demon-star," *ASP Leaflets* **2** (1936), 138–9; and Alan H. Batten, *Binary and Multiple Systems of Stars*, Oxford, UK: Pergamon Press, 1973.

12 Alexander William Roberts, "Density of Close Double Stars," *ApJ* **10** (1899), 308–14; Henry Norris Russell, "Densities of the variable stars of the Algol type," *ApJ* **10** (1899), 315–18.

13 Henry Norris Russell, and Harlow Shapley, "On darkening at the limb in eclipsing variables, I," *ApJ* **36** (1912), 239–54; Harlow Shapley, "The orbits of eighty-seven eclipsing binary stars," *ApJ* **38** (1913), 158–74; and Josef Kallrath and Eugene Milone, *Eclipsing Binary Stars: Modeling and Analysis*, New York: Springer, 2009, pp. 15–21. We are grateful to Professors George Wallerstein and Albert Linnell for their assistance in understanding this history.

14 Alan H. Batten, "Why observe stellar eclipses?," *S&T* **19** (1960), 464–66; and anonymous, "The eclipsing variable star RZ Cassiopeiae," *S&T* **20** (1960), 310.

15 David B. Williams, "Memorandum of discussion with David B. Williams," 1 July 2007, author Williams files.

16 Williams, "Memorandum."

17 Williams, "Memorandum"; and Thom Gandet, "Re: Tom Kandedt: a Pittsburgh Amateur Astronomer???," to Thomas R. Williams, 9 December 1931. Gandet became a member of the AAVSO in 2000 and remained actively involved in 2010.

18 Williams, "Memorandum."

19 Clinton B. Ford, "Minutes of the general meeting of the A. A. V. S. O. held at Harvard College Observatory, Cambridge, Massachusetts, October 9, 1965," October 1965, CBF, 10, Minutes 1964–1990, AAVSO.

20 Leif J. Robinson, "Minima of eclipsing variables," *IBVS* (1965a); Leif J. Robinson, "Minima of eclipsing variables," *IBVS* (1965b); Leif J. Robinson, "Minima of eclipsing variables," *IBVS* (1965c); Leif J. Robinson, "Current elements of RZ Cassiopeiae," *IBVS* (1965); and Leif J. Robinson, "Minima of RZ Cassiopeiae," *IBVS* (1965).

21 Margaret W. Mayall to David B. Williams, 15 April 1969, MWM-Cor, 15, Williams, D. B., AAVSO.

22 "Eclipsing Binary Committee," Cambridge, Massachusetts: May 1969, MWM-Org, 1, Committee Reports, AAVSO.

23 Williams, "Memorandum."

24 Williams, "Memorandun."

25 Margaret W. Mayall, "Variable star notes," *JRASC* **48** (1954), 21–3; and Boris V. Kukarkin 26 August 1954, MWM-Cor, 7, Kukarkin, B. V., AAVSO.

26 Margaret W. Mayall, "Variable star notes: a century of observations of U Geminorum," *JRASC* **51** (1957), 165–9; and George S. III Mumford, "The Dwarf Novae – 1," *S&T* **23** (1962), 71–4.

27 Robert P. Kraft, "Exploding Stars," *SciAm* (1962), 54–63; and Robert P. Kraft, "Binary stars among cataclysmic variables. I. U Geminorum stars," *ApJ* **135** (1962), 408–23.

28 Cragg, 138–141. Dedicated on August 23, 1965, the Ford Observatory is located on Mount Peltier at an elevation of 7474 feet above sea level near the eastern end of Table Mountain, San Gabriel Mountains. The Ford observing group called the location "Mt. Peltier," but the name is not an official geographical designation. The land is part of the Angeles National Forest, near Wrightwood, California (near the northern edge of metropolitan Los Angeles).

29 W. Krzeminski, "The eclipsing binary U Geminorum," *ApJ* **142** (1965), 1051–67.

30 Thomas A. Cragg to Margaret W. Mayall, 6 March 1965, MWM-Org, 20, U Gem stars, AAVSO.

31 Thomas A. Cragg to Margaret W. Mayall, 5 April 1965, MWM-Org, 20, U Gem stars, AAVSO.

32 Cragg, 5 April 1965.

33 Larry C. Bornhurst, "Dwarf novae as binary systems," *JRASC* **60** (1966), 32–3.

34 Merle F. Walker, "Nova DQ Herculis (1934): an eclipsing binary with very short period," *PASP* **66** (1954), 230–2; Merle F. Walker, "Nova T Aurigae 1891: a new short-period eclipsing binary," *ApJ* **138** (1963), 313–19; Alfred H. Joy, "Radial-velocity measures of SS Cygni at minimum light," *ApJ* **124** (1956), 317–32; and Robert P. Kraft, "Binary stars among cataclysmic variables. III. Ten old novae," *ApJ* **139** (1964), 457–75.

35 Margaret W. Mayall, "Visual observations of U Geminorum stars," *Kleine Veroffentlichungen der Remeis-Sternwarte Bamberg* **4** (1965), 241–44.

36 "Observable eclipse minima predicted for 074922 U Geminorum," Suffield, Connecticut: 24 April 1965, CBF, 3, U Gem Eclipses (2), AAVSO.

37 Ford, "Minutes general meeting," October 1965.

38 Clinton B. Ford, "Minutes of the A. A. V. S. O. Council meetings held at Maria Mitchell Observatory, Nantucket, Massachusetts, October 14–15, 1966," October 1966, CBF, 10, Minutes 1964–1990, AAVSO. See also Planning & Research Associates to AAVSO, 1 May 1965, MWM-Org, 1A, Chart Program, folder 6, AAVSO; and Kerriann H. Malatesta, and Charles E. Scovil, "The history of AAVSO charts, Part I: The 1880s through the 1950s', *JAAVSO* **34** (2005), 81–101.

39 Clinton B. Ford to Margaret W. Mayall, 3 April 1960, MWM-Cor, 4, Ford, C. B., AAVSO.

40 Charles P. Olivier, "On the epochs and periods of 24 variable stars," *AN* **285** (1959), 148; and Charles P. Olivier, "Epochs and periods of certain variable stars," *AN* **284** (1958), 185–6.

41 Charles P. Olivier to Margaret W. Mayall, 26 January 1961, MWM-Cor, 4, Ford, C. B., AAVSO.

42 Clinton B. Ford, "Preparation of new AAVSO charts for faint variables from Flower Observatory sequence data," *AVSOA* (1966), and subsequent abstracts; and Kerriann H. Malatesta *et al.*, "The history of AAVSO charts, part II: the 1960s through 2006," *JAAVSO* **35** (2007), 377–89.

43 Clinton B. Ford, "Minutes of the general meetings of the A. A. V. S. O. held at the Adler Planetarium, Chicago, Illinois on May 28–29, 1966," May 1966, CBF, 10, Minutes 1964–1990, AAVSO.

44 Ford, "Council minutes," October 1966.

45 This statement oversimplifies the information recorded on each card, which of course would have to be the star's identification number, a date and time of the observation, and an identification of the observer as well as the observation itself.

46 Ford, "Council minutes," October 1966; Margaret W. Mayall, "Interview conducted by Owen Gingerich," OHP, OHI 28323, CHP-NBL; and Margaret W. Mayall, "The invasion of the grey giant," *AVSOA* (1967).

47 Clinton B. Ford, "Minutes of the general meetings of the A. A. V. S. O. held at the Springfield Museum of Science,

Springfield, Massachusetts, October 20–21, 1967," October 1967, CBF, 10, Minutes 1964–1990, AAVSO; and Clinton B. Ford, "Minutes of the A.A.V.S.O. Council meetings held at the Springfield Museum of Science, Springfield, Massachusetts, October 20–21, 1967," October 1967, CBF, 10, Minutes 1964–1990, AAVSO.

48 Clinton B. Ford, "Minutes of the A. A. V. S. O. Council meeting held in Lima, Ohio, May 31, 1968," May 1968, CBF, 10, Minutes 1964–1990, AAVSO.

49 Margaret W. Mayall, "AAVSO Report 28," Cambridge, Massachusetts: December 1970; and Barbara L. Welther, "Computer plots of AAVSO observations," *AVSOA* (1970).

50 "Minutes of the A. A. V. S. O. general meetings held in Lima, Ohio, May 31–June 1, 1968," June 1968, CBF, 10, Minutes 1964–1990, AAVSO.

51 Margaret W. Mayall, "Eclipses of U Geminorum," *CoKon* **65** (1969), 377–80.

52 Margaret W. Mayall to David B. Williams, 13 November 1969, MWM-Cor, 15, Williams, D. B., AAVSO; and Margaret W. Mayall, "Greetings," *JAAVSO* **2** (1973), 49. The last issue of *RPA* was the August 1961 issue, Volume 63, number 558. These volume and issue numbers may seem out of line in view of the short publishing life of *RPA*, but Zahner had continued both series of numbers from those of the original publication he acquired, *The Monthly Evening Sky Map*.

53 Thomas R. Cave, Jr., "Memorandum of discussion with Thomas R. Williams," 9 November 1998, author Williams file; and Donald D. Zahner, "Memorandum of discussion with Thomas R. Williams," May 1999, author Williams file. Tom Cave related this story in brief and put us in touch with Zahner who confirmed it in great detail.

54 Clinton B. Ford to Margaret W. Mayall, 26 November 1950, MWM-Cor, 4, Ford, C. B., folder 1, AAVSO.

55 Margaret W. Mayall to Clinton B. Ford, 1 May 1964, MWM-Cor, 4, Ford, C. B., folder 3, AAVSO.

56 Margaret W. Mayall to Carolyn J. Hurless, 4 January 1968, MWM-Cor, 6, Hurless, C. J., AAVSO, and other letters in this folder.

57 Clinton B. Ford, "Annual programming for observations of faint variable stars," *AVSOA* (1965); and John E. Bortle, "AAVSO Centennial History," to Thomas R. Williams, 28 December 2009.

58 Charles E. Scovil to Michael Saladyga, 23 October 2009.

59 John E. Bortle to Michael Saladyga, 6 November 2009.

60 H. B. Webb, *Webb's Atlas of the Stars (2nd edition)*, Long Island, N. Y.: H. B. Webb, 1945.

61 Bortle, "AAVSO Centennial History."

62 Charles E. Scovil, "The history of the AAVSO Variable Star Atlas Project," 8 April 1974, JAM-Org, 26, Star atlas contracts, AAVSO; and *Star Atlas of Reference Stars and Non-stellar Objects*, Smithsonian Astrophysical Observatory, Cambridge, Massachusetts: MIT Press, 1969.

63 Margaret W. Mayall to Charles E. Scovil, 1 May 1972, MWM-Cor, 12, Scovil, S. E., AAVSO.

64 Katherine L. Haramundanis to Charles E. Scovil, 18 April 1972, AAVSO, AAVSO Corr-JAM, Box 26, folder, Star atlas contracts, AAVSO Corr-JAM. An interesting sidelight connection to AAVSO here is that Haramundanis was the daughter of Cecilia Payne and Serge Gaposchkin.

65 Margaret W. Mayall to David G. Black, Jr., 9 October 1972, JAM-Org, 4, Research Corporation, AAVSO; and Charles E. Scovil, "History," 1–2.

66 Charles E. Scovil to Margaret W. Mayall, 23 April 1972, JAM-Org, 26, Star atlas contracts, AAVSO; and Clinton B. Ford, "Minutes of the council meetings," October 1972, CBF, 10, Minutes 1964–1990, AAVSO, 2.

67 Charles E. Scovil, "History."

68 Bortle, 6 November 2009.

69 Margaret W. Mayall to Charles E. Scovil, 7 June 1972, MWM-Cor, 12, Scovil, S. E., AAVSO; and Scovil, 23 October 2009, personal correspondence. According to Charles Scovil, "We [i.e., himself, Ford, Bortle, Lowder, and Glenn] were all members and we had proposed the publications [*Circular* and *Journal*], so we were quite happy with the way MWM accepted the ideas. As far as I know, no one was unhappy. If there were any [disagreements], they never made it public."

70 Clinton B. Ford, "Minutes of the A. A. V. S. O. Council meeting held in Tucson, Arizona, May 26, 1972," May 1972, MWM-Org, 1, Council and Meeting Minutes, AAVSO; and Margaret W. Mayall, "From the Director of the AAVSO," *JAAVSO* **1** (1972), 1.

71 Clinton B. Ford, "Minutes of the general meetings of the A.A.V.S.O., Nantucket, Massachusetts, October 14, 1972," October 1972, CBF, 10, Minutes 1964–1990, AAVSO.

72 Clinton B. Ford, "Minutes of the A.A.V.S.O. Council meetings, Lenox, Massachusetts, October 19–20, 1973, May 1973, CBF, 10, Minutes 1964–1990, AAVSO.

73 Ford, "Council minutes," October 1973.

74 Ford, "Council minutes," October 1973.

75 Clinton B. Ford, "Minutes of the AAVSO Council meetings, Woods Hole, Massachusetts, October 16–17, 1970," October 1970, CBF, 10, Minutes 1964–1990, AAVSO.

76 State of New York Surrogate's Court 1 March 1971, MWM-Org, 4, Fund Drive Endowment, folder 12, AAVSO. The importance of this bequest may be seen in the fact that in 2009 dollars, this amount is equivalent to $2.6 million.
77 Margaret W. Mayall 9 March 1971, MWM-Org, 4, Fund Drive Endowment, folder 12, AAVSO.
78 Anonymous, "$10-million left by math teacher," *The New York Times,* New York, New York, 10 March 1971, p. 42. MWM-Org, 4, Fund Drive Endowment, folder 12, AAVSO. This could not have been a surprise to Clinton Ford even though it may have been to the other members of the AAVSO council. Although there is no indication in anything that Clint wrote that hinted that the AAVSO simply needed to hang on until his father died, it seems likely that he knew this was coming and it formed a basis for his optimism over the years of the AAVSO's struggle. Ford really was a pillar of support for those years before, as much as after, this bequest.
79 Clinton B. Ford, "Minutes of the A. A. V. S. O. Council meetings, Strasenburgh Planetarium, Rochester, New York, Friday April 30, 1971," April 1971, CBF, 10, Minutes 1964–1990, AAVSO.
80 Alan H. Shapley 30 March 1971, MWM-Org, 4, Fund Drive Endowment, folder 12, AAVSO; Ida Barney 14 March 1971, MWM-Org, 4, Fund Drive Endowment, folder 12, AAVSO; John J. Ruiz March 1971, MWM-Org, 4, Fund Drive Endowment, folder 12, AAVSO; and other letters in this folder.
81 Margaret W. Mayall to Bart J. Bok, 17 August 1971, MWM-Cor, 2, Bok, B. J., AAVSO.
82 Clinton B. Ford, "Minutes of the general meeting of the A.A.V.S.O., Steward Observatory, Tucson, Arizona, May 27, 1972," May 1972, CBF, 10, Minutes 1964–1990, AAVSO.
83 Clinton B. Ford, "Minutes of general meeting of the A.A.V.S.O., University of Virginia, Charlottesville, Virginia, May 19, 1973," May 1973, CBF, 10, AAVSO Minutes 1964–1990, AAVSO.
84 Clinton B. Ford, "Minutes of the general meetings of the A.A.V.S.O., Harvard College Observatory, Cambridge, Massachusetts, October 16, 1971," October 1971, CBF, 10, Minutes 1964–1990, AAVSO.
85 Margaret W. Mayall to Bart J. Bok, 28 January 1972, MWM-Cor, 2, Bok, B. J., AAVSO.
86 Ford, "Council minutes," May 1972; Margaret W. Mayall, "Annual report of the Director 30 September 1972," *JAAVSO* 1 (1972), 80–6; and Margaret W. Mayall, "Annual report of the Director 30 September 1972," *JAAVSO* 2 (1973), 92–8.

17 THE GROWTH OF A DIRECTOR

1 Margaret W. Mayall to Clinton B. Ford, 19 September 1972, MWM-Cor, 4, Ford, C. B., folder 3, AAVSO.
2 Margaret W. Mayall to Bart J. Bok, 28 January 1972, MWM-Cor, 2, Bok, B. J., AAVSO.
3 The liberalization of the Turkish state after World War I created opportunities for women that had not previously existed. The Treaty of Lausanne of July 24, 1923, led to international recognition of the sovereignty of the newly formed Republic of Turkey. That nation formed as one of the successor states of the Ottoman Empire that collapsed as a result of its participation in World War I on the side of Germany and the Habsburg Empire. After World War II, demands by the Soviet Union for military bases in the Turkish Straits prompted US President Harry S. Truman to declare what became known as the Truman Doctrine in 1947. This doctrine spelled out American intentions to guarantee the security of Turkey and Greece; it resulted in massive economic and military support to both countries. Since that time, there has been a constant flow of people and resources between the United States and Turkey.
4 Yusef Akyüz, "Memorandum of discussion with Thomas R. Williams, 3 Nov 09," 4 November 1909; and Yusef Akyüz, "Your mother's name?" to Thomas R. Williams, 9 November 2009.
5 E. Dorrit Hoffleit, "Two Turkish lady astronomers," *JAAVSO* 33 (2005), 127.
6 Elizabeth O. Waagen, "Janet Akyüz Mattei and the AAVSO," *JAAVSO* 33 (2005), 106–7; and Hoffleit, "Two Turkish lady astronomers," 127.
7 Janet Akyüz, "RR Lyrae type variables with two periods," *AAVSO Abstracts* (1969).
8 Hoffleit, "Two Turkish lady astronomers," 127; and John R. Percy, "Janet Akyüz Mattei (1943–2004)," *JRASC* 98 (2004), 215–16.
9 Margaret W. Mayall, "Greetings," *JAAVSO* 2 (1973), 49.
10 Margaret W. Mayall to Clinton B. Ford, 19 September 1972, MWM-Cor, 4, Ford, C. B., folder 3, AAVSO.
11 Clinton B. Ford to AAVSO Council, 26 May 1972, MWM-Org, 1, Committee Reports, Search Committee, AAVSO. The Director Search Committee comprised: Clinton B. Ford (Chairman), Dr. Owen Gingerich, Dr. Dorrit Hoffleit, Dr. Helen S. Hogg, and Thomas A. Cragg.

12 Clinton B. Ford to AAVSO Council, 26 May 1972, MWM-Org, 1, Committee reports, search committee, AAVSO.
13 Clinton B. Ford, "Minutes of the Council meeting," May 1972, MWM-Org, 1, Council and Meeting Minutes, AAVSO; S. W. McCuskey to Clinton B. Ford, 16 May 1973, MWM-Org, 1, Committee Reports, Search Committee, AAVSO; Clinton B. Ford to Owen Gingerich, 26 July 1973, MWM-Org, 1, Committee Reports, Search Committee, AAVSO; and H. M. Gurin to H. M. Gurin, 4 August 1972, MWM-Org, 1, Committee Reports, Search Committee, AAVSO.
14 Clinton B. Ford to Margaret W. Mayall, 24 January 1973, MWM-Cor, 4, Ford, C. B., folder 3, AAVSO.
15 Clinton B. Ford, "Minutes of the Council meeting," May 1973, MWM-Org, 1, Council and Meeting Minutes, AAVSO; and Clinton B. Ford to various, 13 June 1973, MWM-Org, 1, Committee Reports, Search Committee, AAVSO.
16 Clinton B. Ford, "Minutes of the Council meetings," October 1973, MWM-Org, 1, Council and Meeting Minutes, AAVSO.
17 Clinton B. Ford, "Minutes of the general meeting," October 1973, MWM-Org, 1, Council and Meeting Minutes, AAVSO.
18 AAVSO members and friends, *Mayall Retirement Album*, Cambridge, Massachusetts: AAVSO, 1973; and Janet Akyüz Mattei to various, August 1974–31 August–September 1974, JAM-Org, 7, letters of appreciation, AAVSO.
19 Margaret W. Mayall, "Greetings," *JAAVSO* **2** (1973), 49.
20 Philip C. Steffy to Janet Akyüz Mattei, 6 November 1973, JAM-Cor, 16A, Steffey, P. C., AAVSO.
21 Clinton B. Ford, "Minutes of the AAVSO Council Meetings held at The University of Manitoba, Winnepeg, Manitoba, Canada, on June 28–30, 1974," 28–30 June 1974, CBF, 10, Minutes 1974–1981, AAVSO.
22 Jean-Louis Tassoul and Monique Tassoul, *A Concise History of Solar and Stellar Physics*, Princeton and Oxford: Princeton University Press, 2004, pp. 146–7 and 228–9. See also W. Strohmeier, *Variable Stars*, Oxford, UK: Pergamon Press, 1972, 84–7.
23 Philip C. Steffey to Janet Akyüz Mattei, 22 March 1980, JAM-Cor, 16A, Steffey, P. C., AAVSO; Philip C. Steffey to Leif J. Robinson, November 1983, JAM-Cor, 16A, Steffey, P. C., AAVSO; and Janet Akyüz Mattei to Philip C. Steffey, 6 Dec 1983, JAM-Cor, 16A, Steffey, P. C., AAVSO.
24 Elizabeth O. Waagen, to Thomas R. Williams, personal communication, 2010. One of the college graduates on the AAVSO payroll at that time, Karen Meech, returned to school to earn a Ph.D. in astronomy but remained interested in variable stars throughout her career, hosting the AAVSO's Pan-Pacific meeting in 2004.
25 Janet Akyüz Mattei, undated (1974?), AAVSO, JAM-Cor, 26, Star atlas contracts, AAVSO. This page, in Mattei's handwriting, reflects considerable skepticism on her part about the project, with notes such as "Who is it going to serve, beginner, binocular obs., Professional" and "PEP mag & visual mag . . . [Brian] Marsden, [Luigi] Jacchia, [Martha and Bill] Lillers, [?] McCorsky all objected unless two systems reduced to each scale." The main objection, from her perspective, seemingly reflecting those of others, was Scovil's proposed plan to enter star magnitudes from the USNO *Photoelectric Visual Magnitudes Catalogue* (Blanco *et al.*) that had recently been published, mixing PEP and visual magnitudes in a manner unacceptable to Mattei.
26 Charles E. Scovil, "The history of the AAVSO Variable Star Atlas Project," 8 April 1974, JAM-Cor, 26, Star atlas contracts, AAVSO.
27 Clinton B. Ford to Richard H. Davis, 23 April 1972, JAM-Cor, 26, Star atlas contracts, AAVSO.
28 Richard H. Davis to Clinton B. Ford, 29 March 1974, JAM-Cor, 26, Star atlas contracts, AAVSO. In framing his initial objections in this manner, Davis revealed his deeper concern, one that emerged more strongly soon thereafter, for whether or not the funding arrangement for the variable star atlas could withstand detailed scrutiny under the provisions of the federal tax code. The concept that a donee should receive "full and fair value" from any donation claimed as a deduction from taxable income by the donor existed as a well-tested concept, indeed a bed-rock tax fundamental, in the federal tax law well before this project was initiated.
29 Richard H. Davis to Charles E. Scovil, 11 May 1974, JAM-Cor, 26, Star atlas contracts, AAVSO; and Richard H. Davis to Clinton B. Ford, 11 April 1974, JAM-Cor, 26, Star atlas contracts, AAVSO.
30 Richard H. Davis to Charles E. Scovil, 10 April 1974, JAM-Cor, 26, Star atlas contracts, AAVSO.
31 Richard H. Davis to Clinton B. Ford, 5 April 1974, JAM-Cor, 26, Star atlas contracts, AAVSO.
32 Clinton B. Ford to Richard H. Davis, 8 April 1974, JAM-Cor, 26, Star atlas contracts, AAVSO.
33 Clinton B. Ford to Janet Akyüz Mattei, 14 April 1974, JAM-Cor, 26, Star atlas contracts, AAVSO.
34 Richard H. Davis, "Draft AAVSO variable star atlas project contract No. 1," Cambridge, Massacusetts: 1 May1974,

JAM-Cor, 26, Star atlas contracts, AAVSO; and Richard H. Davis to Janet Akyüz Mattei, 16 April 1974, JAM-Cor, 26, Star atlas contracts, AAVSO.

35 Charles E. Scovil to Richard H. Davis, 1 May 1974, JAM-Cor, 26, Star atlas contracts, AAVSO.

36 Richard H. Davis to Charles E. Scovil, 11 May 1974, JAM-Cor, 26, Star atlas contracts, AAVSO.

37 Clinton B. Ford to Janet Akyüz Mattei, 15 May 1974, JAM-Cor, 26, Star atlas contracts, AAVSO.

38 Charles E. Scovil to Richard H. Davis, 28 May 1974, JAM-Cor, 26, Star atlas contracts, AAVSO.

39 Richard H. Davis to Clinton B. Ford, 1 June 1974, JAM-Cor, 26, Star atlas contracts, AAVSO. Davis' resignation and the appointment of a temporary replacement were announced by Scovil and Ford at the Council meeting in Winnipeg, Saskatchewan, Canada a few weeks later.

40 Charles E. Scovil to Janet Akyüz Mattei, 7 June 1974, JAM-Cor, 26, Star atlas contracts, AAVSO.

41 Charles E. Scovil to AAVSO Council Members, 7 June 1974, JAM-Cor, 26, Star atlas contracts, AAVSO.

42 Clinton B. Ford, "Minutes of the AAVSO Council Meetings held at The University of Manitoba, Winnepeg, Manitoba, Canada, on June 28–30, 1974," 28–30 June 1974, CBF, 10, Minutes 1974–1981, AAVSO.

43 R. Newton Mayall to Charles E. Scovil, 30 July 1975, JAM-Cor, 26, Star atlas contracts, AAVSO. As an experienced architect who owned his own business, Newton Mayall was, of course, very skilled as a project manager and schooled in the preparation of such schedules.

44 Charles E. Scovil to R. Newton Mayall, 26 August 1975, JAM-Cor, 26, Star atlas contracts, AAVSO.

45 Charles E. Scovil to R. Newton Mayall, 8 January 1976, JAM-Cor, 26, Star atlas contracts, AAVSO.

46 It appears that the original question to Mayall was premised on the idea that the combination of Antonīn Bečvár's *Atlas Eclipticalis*, *Atlas Borealis*, or *Atlas Australis*, coupled with the SAO *Star Catalog*, would be all that an observer might need, if the SAO atlas itself was not adequate.

47 Charles E. Scovil to R. Newton Mayall, 8 January 1976, JAM-Cor, 26, Star atlas contracts, AAVSO.

48 Clinton B. Ford, "Committee Reports: Variable Star Atlas," *JAAVSO* 4 (1976), 112.

49 Clinton B. Ford, "Committee Reports: Variable Star Atlas," *JAAVSO* 7 (1978) 55–40.

50 Clinton B. Ford, "Committee Reports: Variable Star Atlas," *JAAVSO* 8 (1979), 91; and Leif J. Robinson to Janet Akyüz Mattei, 28 March 1979, JAM-Org, 26,, Chart Program, *S&T* Corresp., AAVSO.

51 Janet Akyüz Mattei, "Annual report of the Director for fiscal year 1979–1980," *JAAVSO* 9 (1980), 103–12.

52 Clinton B. Ford, "Minutes of the AAVSO Finance Committee and Council Meetings held at Brookline and Cambridge, Mass. October 28 and 30, 1982," 28 October 1982, CBF, 10, Minutes 1982–1990, AAVSO.

53 The time commitment involved in preparing *JAAVSO* constitutes only one obvious conflict of the many that must surely have entered into Scovil's failure to meet the original schedule for *The AAVSO Variable Star Atlas*. As will be seen, it was not the only AAVSO work that conflicted with his need to stay focused on the project.

54 Charles E. Scovil, "Memorandum of Discussion with Thomas R. Williams," 31 October 2004.

55 Clinton B. Ford, "Minutes of the AAVSO Council Meetings held at the Treadway Williams Inn, Williamstown, Massachusetts," 18 October 1974, CBF, 10, Minutes 1974–1981, AAVSO.

56 Charles A. Whitney, "The radii of delta Cephei and eta Aquilae. II," *Astrophysical Journal* **122** (1955), 385–9; Charles A. Whitney, "Note on the discontinuous velocity-curves of Population II Cepheids," *Annales d'Astrophysique* **18** (1955), 376–7; Charles A. Whitney, "Stellar pulsation. I. Momentum transfer by compression waves of finite amplitude," *Annales d'Astrophysique* **19** (1956), 34–43; Charles A. Whitney, "Stellar pulsation. II. A kinematic model for the atmosphere of W Virginis," *Annales d'Astrophysique* **19** (1956), 142–64; and John P. Cox, and Charles A. Whitney, "Stellar pulsation. IV. A semitheoretical period-luminosity relation for classical Cepheids," *Astrophysical Journal* **127** (1958), 561–72.

57 Clinton B. Ford, "Minutes of the AAVSO Council Meetings held at the Rodeway Inn, Decatur, Georgia, May 23, 1975," 23 May 1975, CBF, 10, Minutes 1974–1981, AAVSO.

58 Scovil and Bob Leitner continued to perform related preparation work at Stamford such as optical character recognition of submitted typed material, initial preparation of figures, and related tasks as a service to AAVSO Headquarters as late as 1991.

59 Clinton B. Ford, "Minutes of the AAVSO Council Meetings held at the Rodeway Inn, Decatur, Georgia, May 23, 1975," 23 May 1975, CBF, 10, Minutes 1974–1981, AAVSO.

60 Paul Sventek to Thomas R. Williams, personal communication.

61 The first crude mimeographed issue of the *AAVSO Circular* included observations of only stars in the cataclysmic variables or R CrB variable categories. Soon the *Circular* was being printed on pre-printed letterhead paper and presented a far more professional appearance.

62 Richard H. Davis to John E. Bortle, 8 January 1974, JAM-Org, 27, AAVSO Circular Misc. Corr., AAVSO.

63 John E. Bortle to Janet Akyüz Mattei, 5 September 1974, JAM-Org, 27, AAVSO Circular Misc. Corr., AAVSO.

64 Janet Akyüz Mattei to John E. Bortle, 10 September 1974, JAM-Org, 27, AAVSO Circular Misc. Corr., AAVSO.

65 Janet Akyüz Mattei to John E. Bortle, 14 February 1974, JAM-Org, 27, AAVSO Circular Misc. Corr., AAVSO.

66 David R. Florkowski to John E. Bortle, 9 July 1974, JAM-Org, 27, AAVSO Circular Misc. Corr., AAVSO.

67 John E. Bortle to David R. Florkowski, 11 July 1974, JAM-Org, 27, AAVSO Circular Misc. Corr., AAVSO.

68 Janet Akyüz Mattei to John E. Bortle, 1974, JAM-Org, 27, AAVSO Circular Misc. Corr., AAVSO.

69 Marvin E. Baldwin, "Minima of Eclipsing Binary Stars I," *JAAVSO* 3 (1974), 60–9.

70 Marvin E. Baldwin, "Minima of Eclipsing Binary Stars, II," *JAAVSO* 4 (1975), 86–91; and Marvin E. Baldwin, "Minima of Eclipsing Binary Stars, III," *JAAVSO* 4 (1976), 29–40.

71 Charles A. Whitney, *JAAVSO* 4 (Spring–Summer 1975), 1.

72 Marvin E. Baldwin, "Minima of Eclipsing Binary Stars, II," *JAAVSO* 4 (1975), 60.

73 Marvin E. Baldwin, "Eclipsing Binary (committee report)," *JAAVSO* 7 (1979), 95.

74 Personal communication. Thomas R. Williams had the privilege of seeing this all set up and demonstrated by Samolyk during a visit to the MAS Observatory site.

75 Clinton B. Ford, "Minutes of the AAVSO Council Meetings held at the Treadway Williams Inn, Williamstown, Massachusetts," 18 October 1974, CBF, 10, Minutes 1974–1981, AAVSO.

76 Clinton B. Ford, "Minutes of the AAVSO Council Meetings held at the Bel Air West Motor Hotel, St. Louis, Missouri May 28, 1976," 26 May 1976, CBF, 10, Minutes 1974–1981, AAVSO.

77 George L. Fortier, "Committee Reports: Occultations," *JAAVSO* 5 (1977), 107.

78 Clinton B. Ford, "Minutes of the AAVSO Council Meetings held at the Rodeway Inn, Decatur, Georgia, May 23, 1975," 23 May 1975, CBF, 10, Minutes 1974–1981, AAVSO; and Clinton B. Ford, "Minutes of the AAVSO Council Meetings held at the Bel Air West Motor Hotel, St. Louis, Missouri May 28, 1976," 26 May 1976, CBF, 10, Minutes 1974–1981, AAVSO. Formalization of the IOTA organization triggered Council recognition that the AAVSO Occultation Committee no longer served a useful purpose and should be disbanded. Richard Nugent (Editor), *Chasing the Shadow: The IOTA Occultation Observer's Manual,* http://www.poyntsource.com/IOTAmanual/index.htm.

79 Clinton B. Ford, "Minutes of the AAVSO Council Meetings held at the Rodeway Inn, Decatur, Georgia, May 23, 1975," 23 May 1975, CBF, 10, Minutes 1974–1981, AAVSO. Landis had been at work for some time building his telescope and photometer and mastering the techniques involved in not only finding the variable star, but also making a reliable PEP measurements. Art Stokes, with help and encouragement from PEP observer Larry Lovell, did a nice job of bringing Landis along to this point.

80 David Skillman to Janet Akyüz Mattei, 29 January 1979, JAM-Org, 1, Photoelectric Photometry Program, Data Correspondence, AAVSO.

81 David Skillman to Janet Akyüz Mattei, 1979, JAM-Org., 1, Photoelectric Photometry Program, Data Reports, folder 1, AAVSO. The Canadian professional astronomer referred to by Skillman was probably John Percy of the University of Toronto. The meeting was likely Percy's first; he spoke as president of the Royal Astronomical Society of Canada. Percy would later play a significant role in AAVSO.

82 Clinton B. Ford, "Minutes of the AAVSO Council Meetings held at the Bel Air West Motor Hotel, St. Louis, Missouri May 28, 1976," 26 May 1976, CBF, 10, Minutes 1974–1981, AAVSO.

83 Janet Akyüz Mattei to David Skillman, 12 November 1979, JAM-Org., 1, Photoelectric Photometry Program, Data Reports, folder 1, AAVSO.

84 David Skillman to Douglas S. Hall, 27 April 1980, JAM-Org, 1, Photoelectric Photometry Program, Data Correspondence, AAVSO.

85 Douglas S. Hall to David Skillman, 28 March 1980, JAM-Org, 1, Photoelectric Photometry Program, Data Correspondence, AAVSO.

86 David Skillman to Janet Akyüz Mattei, 23 March 1981, JAM-Org, 1, Photoelectric Photometry Program, Data Correspondence, AAVSO.

87 It is unclear from the record why Whitney took it upon himself to get involved at this point, likely because the tensions between two groups he admired seemed to be rising to the detriment of all involved, as well as to astronomy. Mattei never responded in writing to his letter, nor to Genet, but Whitney's initiative had at least called the attention of everyone involved to the risks of allowing the situation to evolve in a negative direction and invited active communication. Charles A. Whitney to Janet Akyüz Mattei *et al.*, JAM-Cor, 9A, IAPPP, AAVSO; and Russell M. Genet to Janet Akyüz Mattei, JAM-Cor, 9A, IAPPP, AAVSO.

88 Braulio Iriate *et al.*, "Five color photometry of bright stars," *Sky and Telescope* **30** (1965), 21; and Thomas G. McFaul, "Photoelectric photometry with a commercially available solid-state photometer," *JAAVSO* **8** (1979), 64–6.

89 Thomas G. McFaul to Marvin E. Baldwin, 23 January 1982, JAM-Org, 1, Photoelectric Photometry Program, Data Correspondence, AAVSO.

90 Harold J. Stelzer to Janet Akyüz Mattei, 20 July 1980, JAM-Cor, 16A, Steltzer, H. J., AAVSO; Harold J. Stelzer to Janet Akyüz Mattei, 15 July 1981, JAM-Org, 1, Photoelectric Photometry Program, Data Correspondence, AAVSO; Harold J. Stelzer to Janet Akyüz Mattei, 23 June 1982, JAM-Cor, 16A, Steltzer, H. J., AAVSO; Janet Akyüz Mattei to Harold J. Stelzer, 9 June 1982, JAM-Cor, 16A, Steltzer, H. J., AAVSO; and Harold J. Stelzer to Janet Akyüz Mattei, 5 January 1982, JAM-Cor, 16A, Steltzer, H. J., AAVSO.

91 Salary administrators at several astronomical institutions identified four comparable jobs in terms of the scope of the scientific work, the number of staff supervised, and the magnitude of the financial expenditures controlled by the position; for example, the supervisor of the HCO radio observatory at Fort Davis, Texas.

92 Thomas R. Williams, "An unfinished but closed chapter in AAVSO history," *JAAVSO* **33** (2005), 118–26.

18 LEARNING THE ROPES THE HARD WAY

1 Riccardo Giacconi, "Remarks at the dedication of the new AAVSO Headquarters," *JAAVSO* **15** (1986), 296.

2 Thomas R. Williams to Carl A. Anderson, 20 December 1979, JAM-Cor, 18, Williams, Thomas R. (1), AAVSO.

3 Clinton B. Ford to Thomas R. Williams, 27 January 1980, JAM-Cor, 18, Williams, Thomas R. (1), AAVSO.

4 Clinton B. Ford, Secretary, AAVSO, "Minutes of the AAVSO Couincil meeting held at the Hilton Airport Inn, Houston, Texas on May 2, 1980," 2 May 1980, JAM-Org, 1, Council and Meeting Minutes, AAVSO; and Clinton B. Ford to AAVSO Council members, 22 August 1980, CBF, 10, Minutes 1974–1981, AAVSO.

5 Thomas R. Williams to Arthur J. Stokes, 29 November 1981, JAM-ORG, 3, Futures Study, AAVSO.

6 Arthur J. Stokes to Thomas R. Williams, 7 December 1981, JAM-Org, 3, Futures Study, AAVSO; and Thomas R. Williams to Walter Scott Houston *et al.*, April 1982, JAM-Org, 3, Futures Study, AAVSO.

7 Janet Akyüz Mattei, "The present status of the AAVSO," undated, JAM-Org, 3, Futures Study, AAVSO.

8 The term X-ray applies to a broad range of the electromagnetic spectrum, ranging from wavelengths just shorter than extreme ultraviolet to much shorter wavelengths. X-rays at the shortest end of this portion of the spectrum are termed *hard* X-rays and generally possess energy levels not quite equivalent to gamma radiation. In contrast, *soft* X-rays are at the longer wavelength end of the X-ray spectrum and exhibit energy levels just above that of extreme ultraviolet radiation.

9 France C. Córdova, "Highlights from the HEAO-1 survey of cataclysmic variable stars," *JAAVSO* **7** (1978), 57–63. See also Frances C. Córdova, "The path to discovery," *Astronomy* **7** (1979), 24–8.

10 Martine Mouchet, "CV Observations," to Janet Akyüz Mattei, 8 January 1982, JAM-Org, 3, Special Programs, IUE-CV (8) Mouchet, AAVSO; and Janet Akyüz Mattei, "CV Observations," to Derek Jones and Czerny, 8 January 1982, JAM-Org, 3, Special Programs, IUE-CV (8) Mouchet, AAVSO.

11 J. E. Pringle to Janet Akyüz Mattei, 8 December 1981, JAM-Org, 3, Special Programs, IUE-CV (14) Pringle, AAVSO; and J. E. Pringle to Janet Akyüz Mattei, 5 October 1981, JAM-Org, 3, Special Programs, IUE-CV (14) Pringle, AAVSO.

12 Janet Akyüz Mattei, "Annual report of the director for fiscal year 1979–1980," *JAAVSO* **9** (1980), 106.

13 Janet Akyüz Mattei, "Notes from the IUE Symposium," 30 March–1 April 1982, JAM-Org, 2A, Papers and talks, technical, IUE Symposium, AAVSO.

14 E. Dorrit Hoffleit, "Two Turkish lady astronomers," *JAAVSO* **33** (2005), 129.

15 Jean-Louis Tassoul and Monique Tassoul, *A Concise History of Solar and Stellar Physics*, Princeton and Oxford: Princeton University Press, 2004, pp. 172–83; and John R. Percy, "Highlights of variable star astronomy 1900–1986," *JAAVSO* **15** (1986), 126–32.

16 Janet Akyüz Mattei, "Annual Report of the director for fiscal year 1981–1982," *JAAVSO* **12** (1983), 83–104.

17 Robert S. Hill, "AAVSO data processing: Ten years of computerization," *JAAVSO* **6** (1977), 12–14.

18 Elizabeth O. Waagen, "Data management at AAVSO Headquarters," *JAAVSO* **13** (1984), 23–6.

19 Clinton B. Ford, "Minutes of the AAVSO Finance Committee and Council Meetings held at Brookline and Cambridge, Mass. October 28 and 30, 1982," 28 October 1982, CBF, 10, Minutes 1982–1990, AAVSO.

20 Evans' supernova discoveries came at such a pace that the plaque had already been modified with several additions

before the plaque for his first discovery could be presented to him at the Nantucket meeting. Recognizing that Evans was likely to continue with his successes, two hooks were installed at the bottom of the plaque from which additional awards could be hung.

21 Clinton B. Ford, Secretary, AAVSO, "Minutes of the AAVSO Couincil meetings held at the Holiday Inn, Gateway Center, Iowa State University, Ames, Iowa, May 24, 1984," 24 May 1984, JAM-Org, 1, Council and Meeting Minutes, AAVSO)

22 Clinton B. Ford to Ernst H. Mayer, 7 January 1985, JAM-Cor, 7, Ford, C. B. folder 7, AAVSO; and Clinton B. Ford to Janet Akyüz Mattei, 29 December 1984, JAM-Cor, 7, Ford, C. B. folder 6, AAVSO.

23 Clinton B. Ford, Secretary, AAVSO, "Minutes of the AAVSO Council meetings held at the Lemieux Library of Seattle University, Seattle Washington, 21 June 1985," JAM-Org, 1, Council and Meeting Minutes, AAVSO.

24 According to her husband Mike Mattei, in private conversations with Ford, he had on several occasions urged Janet Mattei to "Go find another building," but she was reluctant to take Ford at his word. When Ford repeated this instruction a third time, Janet Mattei altered the guidance to Keith Danskin.

25 Clinton B. Ford, Secretary, AAVSO, "Minutes of the AAVSO Council meetings held at Mt. Holyoke College, South Hadley, Massachusetts, October 31–November 2, 1985," 1 October 1985, JAM-Org, 1, Council and Meeting Minutes, AAVSO. Although Mayer participated in the building dedication, he stopped observing and soon resigned from the Association. This outstanding observer's decision to leave variable star astronomy represented a tragic loss for the Association.

26 Clinton B. Ford, Secretary, AAVSO, "Minutes of the AAVSO Council meetings held at Mt. Holyoke College, South Hadley, Massachusetts, October 31–November 2, 1985," 1 October 1985, JAM-Org, 1, Council and Meeting Minutes, AAVSO.

27 Giacconi, "Highlights of the 75th Anniversary," 296.

19 MANAGING WITH RENEWED CONFIDENCE

1 Janet Akyuz Mattei, "Annual report for the Director for Fiscal Year 1986–1987," *JAAVSO* **16** (1987), 328.

2 (Storm Dunlop and M. Gerbaldi (Editors), *Stargazers – The contributions of amateurs to astronomy – Proceedings of the International Astronomical Union Colloquium 98*, 20 June 1987, Paris, France (Berlin: Springer-Verlag, 1988).

3 Janet Akyüz Mattei, "Annual report for the Director for Fiscal Year 1986–1987," *JAAVSO* **16** (1987), 156.

4 Private discussions with George Fortier, and Michael Mattei, October and November 2004, notes in author Williams' files.

5 Janet Akyüz Mattei, "Carolyn J. Hurless, 1934–1987: AAVSO's enthusiastic ambassador," *JAAVSO* **16** (1987), 35–6.

6 Janet Akyüz Mattei, "Annual report for the Director for Fiscal Year 1986–1987," *JAAVSO* **16** (1987), 156.

7 Janet Akyüz Mattei, "Report on meetings with HIPPARCOS input catalogue team on AAVSO support of HIPPARCOS observations of long period variable stars," *JAAVSO* **18** (1989), 70–1.

8 Janet Akyüz Mattei, "Annual report of the Director for the fiscal year 1989–1990," *JAAVSO* **19** (1990), 175–6.

9 Adriaan Blaauw, "International cooperation and coordination in astronomical research," in John R. Percy, *et al.* (Editors), *Variable Star Research: An International Perspective,* Cambridge, UK: Cambridge University Press, 1992, pp. 3–7.

10 John E. Isles, "The analysis of observations of Mira stars," in John R. Percy, *et al.* (Editors), *Variable Star Research: An International Perspective,* Cambridge, UK: Cambridge University Press, 1992, pp. 231–41.

11 Marie-Odile Mennessier, "Predicting the behavior of variable stars," in John R. Percy, *et al.* (Editors), *Variable Star Research: An International Perspective,* Cambridge, UK: Cambridge University Press, 1992, pp. 247–54.

12 Margarita Karovska, "International cooperation for coordinated studies of Mira variables," in John R. Percy, *et al.* (Editors), *Variable Star Research: An International Perspective,* Cambridge, UK: Cambridge University Press, 1992, pp. 255–8.

13 John Percy, *et al.* (Editors), *Variable Star Research: An International Perspective*, July 1990, The Free University, Brussels, Belgium (Cambridge, UK: Cambridge University Press, 1992).

14 Charles A. Whitney, *JAAVSO* **19** (1990a), number 1.

15 Charles A. Whitney, *JAAVSO* **19** (1990b), number 2.

16 John R. Percy, to Janet Akyüz Mattei, 14 November1990, JAM-Cor, 14, Percy, J. R., AAVSO.

17 Janet Akyüz Mattei, "Annual Report of the Director for the Fiscal Year 1990–1991," *JAAVSO* **20** (1991), 263; Janet Akyüz Mattei, "Annual Report of the Director for the Fiscal Year 1991–1992," *JAAVSO* **21** (1992), 168; and Janet Akyüz Mattei, "Annual Report of the Director for the Fiscal Year 1993–1994," *JAAVSO* **22** (1994), 174.

18 Janet Akyüz Mattei, "Annual Report of the Director for the Fiscal Year 1994–1995," *JAAVSO* **23** (1995), 155–6; and Janet Akyüz Mattei, "Annual Report of the Director for the Fiscal Year 1995–1996," *JAAVSO* **24** (1996), 150. The finished project drew wide praise from the teachers who applied the materials in classrooms. *AAVSO Hands-On Astrophysics: Variable Stars in Science, Math, and Computer Education*, Cambridge, Massachusetts: AAVSO, 1997. A recently revised edition of HOA, accessible through the AAVSO Web site, is now called *AAVSO Variable Star Astronomy*.

19 Priscilla J. Benson, and Irene R. Little-Marenin, "Comparison of water maser emission and the visual light curve," *JAAVSO* **17** (1988), 111–14; Kristin A. Blais *et al.*, "Comparison of light variability and water maser emission in V Canum Venaticorum," *JAAVSO* **17** (1988), 115–17; Rebecca P. Webb *et al.*, "Comparison of light variability and water maser emission of R Ceti," *JAAVSO* **17** (1988), 118–19; and Tara C. Woods *et al.*, "Comparison of light variability and water maser emission in T Ursae Majoris," *JAAVSO* **17** (1988), 120–1.

20 Richard Berry *et al.*, *The CCD Camera Cookbook*, Richmond, Virginia: Willmann-Bell, Inc., 1994. The 1994 book was not Berry's first effort to interest amateur astronomers in this technology. He had published two previous books on CCD cameras, but the emphasis had originally been on imaging and image processing as opposed to photometry. See also Richard Berry, *Introduction to Astronomical Image Processing*, Richmond, Virginia: Willmann-Bell, Inc., 1991; and Richard Berry, *Introduction to Astronomical Image Processing*, Richmond, Virginia: Willmann-Bell, Inc., 1992.

21 Thomas R. Williams, "Clinton Banker Ford, 1913–1992," *Bulletin of the American Astronomical Society* **26** (1994), 1602–3.

22 Dorrit Hoffleit, "Clinton Banker Ford, 1913–1992," *JAAVSO* **21** (1992), 144–6.

23 Dorrit Hoffleit, "Margaret Walton Mayall, 1902–1995," *Bulletin of the American Astronomical Society* **28** (1996), 1455–6.

24 It should be noted that the Ford estate involved assets in three states, which greatly complicated the process of handling the disbursements. In addition to that complexity, the distribution laid out in Ford's will was challenged in Court by one of the smaller beneficiaries who felt it was due a greater share.

25 See, for example, Martha Hazen, AAVSO Secretary, "Minutes of the Meeting of the AAVSO Council held at the Headquarters Office, Cambridge, Mass. April 19, 1997," 19 April 1997, Council Minutes, 1997, AAVSO.

26 A copy of the certificate, dated January 22, 1993, and related correspondence is contained in the AAVSO Archives, AAVSO Cor-JAM, Box 12 folder Mattei, J. A.

27 Martha Hazen, AAVSO Secretary, "Minutes of the Meeting of the AAVSO Council held at the Headquarters Office, Cambridge, Mass. April 19, 1997," 19 April 1997, Council Minutes, 1997, AAVSO.

28 Lee Anne Willson, "What makes a star act like a Mira? – and other conclusions from dynamical atmosphere models," Charles A. Whitney and Elizabeth O. Waagen (Editors), *Variable Stars: New Frontiers*, 1997, *Sion-St. Luc, Switzerland*, Cambridge, Massachusetts: AAVSO, 2006, 62–71.

29 Brian Warner, "Cataclysmic Variables: From radio to gamma-rays," in Whitney and Waagen, *New Frontiers*, 98–106; E. M. Sion, "New insights on cataclysmic variables from HST spectroscopy," in Whitney and Waagen, *New Frontiers*, 107–22; and Irina Voloshina and Tatiana Khruzina, "Parameters of the dwarf nova SS Cygni obtained from *UBV* photoelectric light curve analysis," in Whitney and Waagen, *New Frontiers*, 141–3.

30 Hans Martin Schmid, "Variability and orbital parameters for symbiotic stars," in Whitney and Waagen, *New Frontiers*, 159–62.

31 Olga A. Tsiopa, "The variability of stars – Supernova precursors," in Whitney and Waagen, *New Frontiers*, 153–7.

32 G. Meynet *et al.*, "The François-Xavier Bagnoud Observatory," in Whitney and Waagen, *New Frontiers*, 214–21.

33 Wolfgang Quester, "Support of the Bundsdeutsche Arbeitsgeminschaft für Veränderliche Sterne (BAV)," in Whitney and Waagen, *New Frontiers*, 209; Miroslav Zejda, "The Brno Regional Network of Observers (B.R.N.O.)-Variable Star section of the Czech Astronomical Society-and its activities," in Whitney and Waagen, *New Frontiers*, 211–13; Arunas Kucinskas, "Astronomical activities in Lithuania," in Whitney and Waagen, *New Frontiers*, 223–5; Jaime R. Garcia, "Variable star observing in Latin America; Past and future," in Whitney and Waagen, *New Frontiers*, 208; M. Danie Overbeek, "Variables down under," in Whitney and Waagen, *New Frontiers*, 210; and Keiichi Saijo and Seiichi Sakuma, "Ten years of VSOLJ," in Whitney and Waagen, *New Frontiers*, 210.

34 Albert V. Holm, "Issues and strategies for the Future," in Whitney and Waagen, *New Frontiers*, 237–44.

20 EXPANDING THE SCIENTIFIC CHARTER

1 Quoted in "1: Internet Presence: the AAVSO website," part of Elizabeth O. Waagen, "Annual report of the Director for the fiscal year 2002–2003," *JAAVSO* **32** (2004), 145.

2 Beck represents a special case in that she routinely worked full time at the AAVSO for 7 months each year, spending the other 5 months at sea as a licensed mate on sailing ships.

3 Arne Henden *et al.*, "Discovery of the optical afterglow of GRB030323," *JAAVSO* **32** (2004), 19–22.

4 Janet Akyüz Mattei, *AAVSO Newsletter*, Cambridge, Massachusetts, October 2003, 2.

5 See, for example, Arto Oksanen *et al.*, "Discovery and observations of the optical afterglow of GRB 071010B," *JAAVSO* **36** (2008), 53–9.

6 At this juncture, we might also point out that a subtle but important change is reflected in the GRB activity and its subsequent morphing into the High-Energy Network. This is essentially a task force run by the AAVSO Headquarters rather than having a Council-appointed committee chairman and recruited committee members.

7 Caspar H. Hossfield, "A simple, easy-to-build VLF receiver (Abstract)," *JAAVSO* **31** (2002), 175. Unfortunately, Hossfield died a short time after this awareness of his prior contributions arose. See Carl E. Feehrer, "Casper H. Hossfield (1918–2002)," *AAVSO Solar Bulletin*, Cambridge, Massachusetts, November 2002; and in *JAAVSO* **31** (2003), 171–2.

8 Eugene Milone, and Andrew T. Young, "Infrared passbands for precise photometry of variable stars by amateur and professional astronomers," *JAAVSO* **36** (2008), 110–26.

9 Mary Ann Kadooka, and James Bedient, "TOPS telescope projects on variable stars and other objects," *JAAVSO* **31** (2002), 39–47.

21 BRIDGING THE GAP

1 Elizabeth O. Waagen, "Janet Akyüz Mattei and the AAVSO," *JAAVSO* **33** (2005), 106–7.

2 Established in 1831 and known as America's first garden cemetery, the Mount Auburn Cemetery served as the final resting place for such distinguished local residents as academic leaders and presidential cabinet member McGeorge Bundy; nurse and hospital reformer Dorthea Dix; religious leader Mary Baker Eddy; presidents of Harvard, including Charles William Eliot and Abbot Lawrence Lowell; poets, including Amy Lowell; architects like Buckminster Fuller; many jurists, including Supreme Court Justice Felix Frankfurter; scientists, including Nobel prize winning physicist Julian Schwinger and inventor Edwin Land; artists like Winslow Homer; and musicians and composers like Randall Thompson. Waagen also listed many distinguished astronomers who are also buried in the Mount Auburn Cemetery.

3 Kerriann H. Malatesta *et al.*, "The AAVSO data valadation project," *JAAVSO* **34** (2006), 238–50.

4 The meeting, advertised by the Astronomical League in their accustomed format as AstroCon 2004, received strong support from the Eastbay Astronomical Society, the Astronomical Association of Northern California, and the San Jose Astronomical Society.

5 Josef Smak, and Elizabeth O. Waagen, "The 1985 Superoutburst of U Geminorum. Detection of Superhumps," *AcA* **54** (2004), 931–7; and Allen W. Shafter *et al.*, "A recurrence time versus orbital period relation for the Z Camelopardalis stars," *PASP* **117** (2005), 433–42.

6 Papers delivered at the conference in honor of Bateson were all published in a special number of *Southern Stars, The Journal of the Royal Astronomical Society of New Zealand*, 44, 1, March 2005.

22 ACCELERATING THE SCIENCE – THE HENDEN ERA BEGINS

1 Arne Henden, "Annual report of the Director," *AAVSO Annual Report 2006–2007*, (2007), 37.

2 Arne A. Henden and Ronald H. Kaitchuck, *Astronomical Photometry*, New York: Van Nostrand Reinhold Company, 1982; and Arne A. Henden and Ronald H. Kaitchuck, *Astronomical Photometry: A Text and Handbook for the Advanced Amateur and Professional Astronomer*, Richmond, Virginia: Willmann-Bell Inc., 1995–2008. Note that the first edition of this well-recognized and widely used volume was completed while Henden was a graduate student in Bloomington, Indiana.

3 The actual closing dates for the transactions were as follows: for the purchase of 49 Bay State Road, December 27, 2006; for the sale of 25 Birch Street, March 29, 2007.

4 See also E. Dorrit Hoffleit, *Misfortunes as Blessings in Disguise: The Story of My Life*, Cambridge, Massachusetts: American Associations of Variable Star Observers, 2002; E. Dorrit Hoffleit, "Two Turkish lady astronomers," *JAAVSO* **33** (2005), 127–9; and Kristine Larsen, "An interview with Dorrit Hoffleit," *JAAVSO* **37** (2009), 52–69.

5 Kerriann H. Malatesta, and Charles E. Scovil, "The history of AAVSO charts, Part I: The 1880s through the 1950s," *JAAVSO* **34** (2005), 81–101; and Kerriann H. Malatesta *et al.*, "The history of AAVSO charts, part II: the 1960s through 2006," *JAAVSO* 35 (2007), 377–89.

6 John Toone, "British Variable Star Associations 1848–1908," *Journal of the British Astronomical Association* **120** (2010), 135–51.

7 We are grateful to Christopher Watson for his detailed description of this whole process in an e-mail message: Christopher Watson, "VSX: The beginning," to Thomas R. Williams, August 2010.

8 Watson provided this description of VSX when a beta version was first made available to the variable star observing community in 2006. Christopher Watson, "VSX," to AAVSO Observers, 15 March 2006.

9 AAVSO observer Jeff Hopkins of Phoenix, Arizona, had been one of the few observers of the previous eclipse. Hopkins completed an exhaustive set of photoelectric photometric observations of the star from 1982 to 1984.

10 The Citizen Sky project has allowed the AAVSO to expand into new areas and technologies. For example, the Citizen Sky website required new tools to support intensive collaboration between participants. As a result, it was programmed using a new open-source web content management system called Drupal (http://drupal.org). This involved a substantial learning curve for staff; however, the end result was so positive that a new AAVSO website was produced using the same system in 2010.

11 That the gender ratio and age skew toward an older, male audience is typical for the amateur astronomy community. However, Citizen Sky participants are still somewhat younger and include more females than is typical of subscribers to *Sky & Telescope* magazine, the premier magazine in the market. *Sky & Telescope* reports that 95% of their subscriber readership is male, with a mean age of 51 years. New Track Media, 2009.

12 The use of the term "section" rather than Committee reflects the use of such terminology in other similar organizations such as the British Astronomical Association (BAA) and the Association of Lunar and Planetary Observers (ALPO).

Bibliography

NOTES ON REFERENCES

In writing this book, extensive use has been made of archival resources, for which documentation in each citation should include the name of the collection, a call for the specific storage box or file, and the name and location of the repository. To include all this information in each citation would, of necessity, involve an excessive addition to the length of each individual citation and therefore of the book. Accordingly, we have taken the following approach to this problem to conserve space. First, for each archival item, we list in the notes for each chapter the complete citation information in abbreviated form. Below this explanation, we provide the details that those abbreviations represent. Second, for letters, newspaper articles, manuscript material, and other archival material, the bibliography will not repeat that information; it appears only in the chapter notes. This bibliography will catalog only the published material to which reference is made in the citations, not the archival materials.

We take that step not only to conserve space in the book, but also to allow for the inclusion of certain important published materials in the bibliography that were helpful in writing the book but not cited in individual chapter citations.

The following summarizes the abbreviations used for archival materials in individual citations found within the chapter notes:

AAVSO The American Association of Variable Star Observers Archive, 49 Bay State Road, Cambridge, Massachusetts. AAVSO Collections are: **ECP**, materials relevant to the era before the founding of the AAVSO; **WTO**, materials relevant to the William Tyler Olcott era; **LC**, materials relevant to the Leon Campbell era, **MWM**, materials relevant to the Margaret Walton Mayall era; **JAM**, materials relevant to the Janet Akyüz Mattei era; and **AAH**, materials relevant to the Arne Henden era.

Categories within each of the above AAVSO collections are: **Cor**, Correspondence; **Org**, Organization, including such things as meeting minutes, treasurer files, committee reports, and so on; **Adm**, Administration, including meeting notices, papers relating to publications and mailings, business items, and so on.

AAVSO Special Collections cited include: **AA**, AAVSO observing reports; **CBF**, Clinton B. Ford; **CFF**, Cyrus F. Fernald; **EFS**, Edwin F. Sawyer; and **MM**, Margaret W. Mayall; and **WSH**, Walter Scott Houston.

CHC Cambridge Historical Commission Archives, 831 Massachusetts Avenue, #2, Cambridge, Massachusetts.

CHP-NBL Center for the History of Physics, Niels Bohr Library and Archives, One Physics Ellipse, College Park, Maryland 20740–3843. NBL special collections include the **OHP**, Oral History Project.

CHS Chicago Historical Society Museum and Archives, 1601 North Clark Street, Chicago, Illinois.

HUA Harvard University Archives, Pusey Library, Harvard Yard, Cambridge, Massachusetts; including HUA collections cited as:

HU-PO, Harvard University, Presidents Office, Records of the President of Harvard University, including **-ALL**, Abbott Lawrence Lowell; **-JBC**, James Bryant Conant; and **-NMP**, Nathan Marsh Pusey.

HCO, The official records of the Harvard College Observatory; within HCO are sub-collections representing the eras of each observatory director as well as other official records: **-BS**, Records of the HCO-Boyden Station; **-ECP**, Edward C. Pickering era, **-HS**, Harlow Shapley era, **-BJB**, Bart J. Bok (associate director under Shapley and Menzel), and **–DHM**, Donald H. Menzel. The archive calls for these collections are UAV 630.16 and UAV 630.17.X. Observatory correspondence not related to actual administration of the observatory but to astronomy and other relevant topics is organized in separate collections, generally identified by calls as UAV 630.17.5 Series Two.

HUG, the personal papers of individuals largely associated with Harvard University, but not their official records as part of the university. Thus HUG includes the personal correspondence, diaries, scrapbooks, and other memorabilia of observatory directors, astronomers, administrators, and others, each having a separate call number.

LAC Library and Archives of Canada, Ottawa, Ontario, Canada, K1A 0N4, including **JSP**, J. S. Plaskett Papers.

MAS Archives of the Milwaukee Astronomical Society, Milwaukee, Wisconsin.

MLS-LO Mary Lea Shane Archives, Lick Observatory, Santa Cruz, California.

NAS Archives of the National Academy of Science, Washington, DC.

OLA Otis Library Research and Genealogy Collections, 263 Main Street, Norwich, Connecticut.

TCA Trinity College Archives, The Watkinson Library, Trinity College, 300 Summit Street, Hartford, Connecticut, including the **AFC**, Alumni Files Collection, **SC**, Special Collections.

USNO Library and archives of the US Naval Observatory, Washington, DC.

UTor University of Toronto Archives, 120 St. George Sreet, Toronto, ON M5S 1A5, Canada.

Yerkes Archives of the Yerkes Observatory, University of Chicago Special Collections Research Center, 1100 East 57th Street, Chicago, IL 60637, including Yerkes collections cited as **OFD**, Office of the Director.

File and box call numbers should be explored with the archives staff at each of the above repositories.

JOURNAL TITLE ABBREVIATIONS

For the same reasons of brevity in the endnote citations, we have abbreviated the titles of most of the journals we have cited. To the maximum extent possible, these abbreviations match those applied in the SAO/NASA Astrophysics Data System (ADS) on line archives of the literature of astronomy:

Formal journal title	Abbreviation
AAVSO Abstracts	AVSOA
AAVSO Bulletin	AVSOB
AAVSO Solar Bulletins	AVSOSolB
Acta Astronomica	AcA
American Association for the Advancement of Science Bulletin	AASB
Annales d' Astrophysique	AnAp
Annals of the New York Academy of Science	NYASA
Annals of the Harvard College Observatory	AnHar
Astronomical Journal	AJ
Astrophysical Journal	ApJ
Astronomy and Astro-Physics	A&A
Biographical Memoirs of the National Academy of Science	BMNAS
Communications of the Konkoly Observatory, Hungary	CoKon
Harvard College Observatory Annual Reports	HarAR
Harvard College Observatory Circulars	HarCi
Information Bulletin on Variable Stars	IBVS
The Irish Astronomical Journal	IrAJ
Journal of the American Association of Variable Star Observers	JAAVSO
Journal of the British Astronomical Association	JBAA
Journal for the History of Astronomy	JHA
Journal and Proceedings of the Royal Society of New South Wales	RSNSW
Journal of the Royal Astronomical Society of Canada	JRASC
Memoirs of the British Astronomical Association	MmBAA
Memoirs of the BAA, Reports of the Observing Sections	MmBAAROS
Memoirs of the Royal Astronomical Society	MmRAS
Monthly Evening Sky Map	MESM
Monthly Notices of the Royal Astronomical Society	MNRAS
Monthly Register of the Society for Practical Astronomy	MRSPA
The Observatory	Obs
Physics Today	PhT
Popular Astronomy	PA

Formal journal title	Abbreviation
Publications of the Astronomical Society of the Pacific	PASP
Proceedings of the Royal Irish Academy of Science	PRIAA
Review of Popular Astronomy	RPA
Science	Sci
The Sidereal Messenger	SidM
Sky and Telescope; Sky & Telescope	S&T
Transactions of the International Astronomical Union	IAUT
Transactions of the Royal Irish Academy	RIAT
Variable Comments	VC
Variable Views	VV

SELECTED READING

We have deliberately concentrated our references in the endnotes for each chapter, where correct bibliographic information about our sources will be found. Originally, we hoped to use the bibliography to not only summarize those references, but also to add a useful selection of other books and reference materials that we believed were relevant if not directly cited. Space and time limitations prevent us from doing that. We do present key published books that were essential to our understanding of various elements of this history.

The following is a listing of selected books and book chapters that are cited in this volume. It is not a complete list of all the books, just the more relevant for purposes of this history:

Bailey, Solon I., *The History and Work of Harvard Observatory*, New York: McGraw-Hill, 1931.

Batten, Alan H., *Binary and Multiple Systems of Stars*, Oxford, UK: Pergamon Press, 1973.

Berry, Richard, *Introduction to Astronomical Image Processing*, Richmond, Virginia: Willmann-Bell, Inc., 1992.

Berry, Richard, Veikko Kanto, and John Munger, *The CCD Camera Cookbook*, Richmond, Virginia: Willmann-Bell, Inc., 1994.

Blaauw, Adriaan, *History of the IAU: The Birth and First Half-Century of the International Astronomical Union*, Dordrecht, The Netherlands: Kluwer Academic Publishers, 1994.

Broughton, R. Peter, *Looking Up: A History of the Royal Astronomical Society of Canada*, Toronto, Ontario: Dundurn Press, 1994.

Campbell, Leon, *Studies of Long Period Variables*, Cambridge, Massachusetts: AAVSO, 1955.

Campbell, Leon, and Jacchia, Luigi, *The Story of Variable Stars*, Philadelphia: The Blakiston Company, 1941.

Carter, Bill, and Carter, Merri Sue, *Latitude: How American Astronomers Solved the Mystery of Variation*, Annapolis, Maryland: Naval Institute Press, 2002.

DeVorkin, David H. (Editor), *The American Astronomical Society's First Century*, Washington, DC: The American Astronomical Society, 1999.

———, *Henry Norris Russell: Dean of American Astronomers*, Princeton, New Jersey: Princeton University Press, 2000.

Evans, David S., *Under Capricorn: A History of Southern Hemisphere Astronomy*, Bristol, UK and Philadelphia: Adam Hilger, 1988.

Fahy, Thomas P., *Richard Scott Perkin and the Perkin–Elmer Corporation*, unknown: privately published at Perkin–Elmer Corporation, 1987.

Furness, Caroline E., *An Introduction to the Study of Variable Stars*, Boston: Houghton Mifflin Company, 1915.

Gingerich, Owen, "Through Rugged Ways to the Galaxies," in Gingerich, Owen, and Hoskin, Michael (Editors), *Two Astronomical Anniversaries: HCO & SAO*, Cambridge, England: Science History Publications Ltd., 1990, pp. 76–88.

Gingerich, Owen, and Kenneth R. Lang, *A Source Book in Astronomy and Astrophysics, 1900–1975*, Cambridge, Massachusetts, and London, England: Harvard University Press, 1979.

Hagen, Rev. Johann Georg, S. J., *Die veränderlichen sterne (vol. 1, part 1)*, Freiburg, Germany: Herdersche Verlagshandlung, 1913.

Hearnshaw, J. B., *The Measurement of Starlight: Two Centuries of Astronomical Photometry*, Cambridge, UK: Cambridge University Press, 1996.

Henden, Arne A., and Ronald H. Kaitchuck, *Astronomical Photometry*, New York: Van Nostrand Reinhold Company, 1982.

———, *Astronomical Photometry: A Text and Handbook for the Advanced Amateur and Professional Astronomer*, Richmond, Virginia: Willmann-Bell Inc., 1995–2008.

Herschel, John Frederick William, *A Treatise on Astronomy*, London: Longman, Rees, Orme, Brown, Green &

Longman, Paternoster Row; and John Taylor, Upper Gower Street, 1833.

Hoffleit, E. Dorrit, *Misfortunes as Blessings in Disguise: The Story of My Life*, Cambridge, Massachusetts: American Associations of Variable Star Observers, 2002.

Hoskin, Michael A., *Stellar Astronomy*, Chalfont St. Giles, Bucks, England: Science History Publications, 1982.

Isles, John E., *Variable Stars*, Hillside, New Jersey and Aldershot, Hants, UK: Enslow Publishers, Inc., 1990.

Jones, Bessie Zaban, and Lyle Gifford Boyd, *The Harvard College Observatory, The First Four Directorships, 1839–1919*, Cambridge, Massachusetts: Harvard University Press, 1971.

Kallrath, Josef, and Eugene Milone, *Eclipsing Binary Stars: Modeling and Analysis*, New York: Springer, 2009.

Keller, Morton, and Phyllis Keller, *Making Harvard Modern: The Rise of America's University*, Oxford, UK and New York: Oxford University Press, 2007.

Kidwell, Peggy Aldrich, "Harvard Astronomers and World War II – Disruption and Opportunity," in Elliott, Clark A., and Rossiter, Margaret W. (Editors), *Science at Harvard University: Historical Perspectives*, Bethlehem, Pennsylvania: Lehigh University Press, 1992, pp. 285–302.

Lankford, John, *American Astronomy – Community, Careers and Power, 1859–1940*, Chicago and London: The University of Chicago Press, 1997.

Mack, Pamela E., "Strategies and Compromises: Women in Astronomy at Harvard College Observatory, 1870–1920," in Gingerich, Owen, and Hoskin, Michael (Editors), *Two Astronomical Anniversaries: HCO & SAO*, Cambridge, England: Science History Publications Ltd., 1990, p. 65–75.

Marché, Jordan D. II, *Theaters of Time and Space*, New Brunswick, New Jersey, and London: Rutgers University Press, 2005.

Mennessier, Marie-Odile, "Predicting the Behavior of Variable Stars," in Percy, John R., Mattei, Janet Akyüz., and Sterken, Christiaan (Editors), *Variable Star Research: An International Perspective*, Cambridge, UK: Cambridge University Press, 1992, pp. 247–54.

Menzel, Donald H., *Our Sun*, Philadelphia: The Blakiston Company, 1949.

Merrill, Paul W., *The Nature of Variable Stars*, New York: The Macmillan Company, 1938.

Morison, Samuel Eliot (Editor), *The Development of Harvard University since the Inauguration of President Eliot, 1869–1929*, Cambridge, Massachusetts: Harvard University Press, 1930.

———, *Three Centuries of Harvard, 1636–1936*, Cambridge, Massachusetts: Harvard University Press, 1936.

Morris, Richard B. *Encyclopedia of American History*, New York: Harper & Row, Publishers, 1965, pp. 538–40.

Olcott, William Tyler, *The Romance of the Southern Cross*, Miami, Florida: Privately published, 1935.

Osterbrock, Donald E., *Yerkes Observatory 1892–1950, The Birth, Near Death, and Resurrection of a Scientific Research Institution*, Chicago and London: The University of Chicago Press, 1997.

Osterbrock, Donald E., John R. Gustafson, and W. J. Shiloh Unruh, *Eye on the Sky, Lick Observatory's First Century*, Berkeley, California: University of California Press, 1988.

Payne, Cecilia H., *The Stars of High Luminosity*, New York and London: McGraw-Hill Book Company, Inc., 1930.

———, *Stellar Atmospheres; A Contribution to the Observational Study of High Temperature in the Reversing Layers of Stars*, Cambridge, Massachusetts: Harvard University Press, 1925.

Payne-Gaposchkin, Cecilia, *Variable Stars & Galactic Structure*, London: Athlone Press, 1954.

Payne-Gaposchkin, Cecilia, and Gaposchkin, Sergei, *Variable Stars*, Cambridge, Massachusetts: The Observatory, 1938.

Percy, John R., *Understanding Variable Stars*, Cambridge, UK: Cambridge University Press, 2007.

Rosseland, Svein, *The Pulsation Theory of Variable Stars*, Oxford, UK: The Clarendon Press, 1949.

———, *Theoretical Astrophysics*, Oxford, UK: The Clarendon Press, 1936.

Shapley, Harlow, *Through Rugged Ways to the Stars*, New York: Charles Scribner's Sons, 1969.

Smith, Robert W., *The Expanding Universe: Astronomy's 'Great Debate' 1900–1930*, Cambridge, UK: Cambridge University Press, 1982.

Strohmeier, W., *Variable Stars*, Oxford, UK: Pergamon Press, 1972.

Sutton, S. B., *Cambridge Reconsidered: $3\frac{1}{3}$ Centuries on the Charles*, Cambridge, Massachusetts and London, England: The MIT Press, 1976.

Tassoul, Jean-Louis, and Monique Tassoul, *A Concise History of Solar and Stellar Physics*, Princeton and Oxford: Princeton University Press, 2004.

Webb, H. B., *Webb's Atlas of the Stars (2nd edition)*, Long Island, NY: H. B. Webb, 1945.

Index

Note: Page numbers shown in **bold** typeface refer to figures, while those shown as *italics* refer to tables of information.

Printed in the United States
by Baker & Taylor Publisher Services